THE WEATHER OBSERVER'S HANDBOOK
Second Edition

This handbook provides a comprehensive, practical and independent guide to all aspects of making weather observations. The second edition has been fully updated throughout with new material, new instruments and technologies, and the latest reference and research materials. Traditional and modern weather instruments are covered, including how best to choose and to site a weather station, how to get the best out of your equipment, how to store and analyse your records and how to share observations. The book's emphasis is on modern electronic instruments and automatic weather stations. It provides advice on replacing 'traditional' mercury-based thermometers and barometers with modern digital sensors, following implementation of the UN Minamata Convention outlawing mercury in the environment. *The Weather Observer's Handbook* will again prove to be an invaluable resource for both amateur observers choosing their first weather instruments and professional observers looking for a comprehensive and up-to-date guide.

Dr Stephen Burt commenced his professional career in meteorology within the UK Met Office in 1977, followed by a series of private-sector marketing roles within the computer industry before returning to academia, eventually retiring from the Department of Meteorology at the University of Reading in 2018 (where he remains a Visiting Fellow). His meteorological research interests span instruments, observation techniques and methods, climatological data rescue, case studies of notable weather events and long weather records and their sites; in all, he has authored or jointly authored over 50 peer-reviewed journal papers to date, and four other books. Stephen is a Fellow of the UK's Royal Meteorological Society and a member of the American Meteorological Society, the Irish Meteorological Society and the Scientific Instruments Society. He is also an accomplished photographer with several hundred published photographs to his name.

From reviews of the first edition:

' … a goldmine of information.'

Geoff Jenkins, Weather

'I would highly recommend this comprehensive weather-observing guide to hobbyists, professionals, teachers and college instructors.'

Sytske Kimball, Bulletin of the American Meteorological Society

' … a handy reference for any of its intended users (from backyard enthusiasts to professionals involved in research).'

T. Colleen Farrell, Bulletin of the Canadian Meteorological and Oceanographic Society

Endorsement quotes for the second edition:

'I like this book very much. I am a lifelong weather observer and user of weather data both as an amateur and a career professional. This second edition answers so many of the questions I have had over the decades and the many questions I continue to field. At last, I have a single book that I can point people to, comprehensive, scientifically rigorous, yet very readable.'

Nolan Doesken, Colorado State Climatologist, former President of the
American Association of State Climatologists, and founder of the
Community Collaborative Rain, Hail and Snow network

'This is a very informative book for both amateur and professional meteorological observers.'

Steven Colwell, Chair of the Royal Meteorological Society's Special
Interest Group on Meteorological Observing Systems

'As climate change has an ever-increasing impact on our lives, this book is an essential tool to ensure accuracy and consistency of our weather records. It is difficult to imagine a more useful publication in the world of weather observation. This updated edition of The Weather Observer's Handbook is concisely and expertly written, with clear explanations of often complex matters. In Dr Burt's hands all the weather parameters are explored in detail, but the management of the collected data is particularly well covered. The updated version of The Weather Observer's Handbook will meet the demands of amateurs and professionals across the world for the next decade.'

Roger Bailey, Chair, British Standards Technical Committee for Precipitation

'Stephen is the foremost authority on all aspects of observing the weather. Building upon the successes of its first edition, this newly updated edition of The Weather Observer's Handbook is the essential resource for anyone interested in ground-based observational meteorology, whether using the latest equipment or traditional instrumentation, amateur or professional. The handbook provides a fully comprehensive, detailed and practical guide to all aspects of observing the weather.'

Eddy Graham, Atmospheric Scientist and Editor-in-Chief of Weather

'If you still have some doubts about the robustness and reliability of the data on which we base our knowledge of meteorology and climate, this book will definitively and clearly explain how instruments work properly and measurements are accurate. Stephen Burt brings us on a journey across meteorological instrumentation, starting from a very easy approach, let's say from an "amateur perspective", up to professional measurement procedures and high-level systems, passing through a variety of users' needs … In a world where distributed networks, low-cost sensors and citizens' science data will increase their role in weather and climate analysis, disseminating good practice in meteorological observation, with a constant taste for "metrological rigour", is surely a valuable mission, and is fully achieved by this book.'

Andrea Merlone, Research Director at the Italian Institute
for Research in Metrology

The Weather Observer's Handbook

Second Edition

Stephen Burt

Shaftesbury Road, Cambridge CB2 8EA, United Kingdom

One Liberty Plaza, 20th Floor, New York, NY 10006, USA

477 Williamstown Road, Port Melbourne, VIC 3207, Australia

314–321, 3rd Floor, Plot 3, Splendor Forum, Jasola District Centre, New Delhi – 110025, India

103 Penang Road, #05–06/07, Visioncrest Commercial, Singapore 238467

Cambridge University Press is part of Cambridge University Press & Assessment, a department of the University of Cambridge.

We share the University's mission to contribute to society through the pursuit of education, learning and research at the highest international levels of excellence.

www.cambridge.org
Information on this title: www.cambridge.org/9781009260541

DOI: 10.1017/9781009260534

First published 2012
Second edition 2024

A catalogue record for this publication is available from the British Library

A Cataloging-in-Publication data record for this book is available from the Library of Congress

ISBN 978-1-009-26054-1 Hardback
ISBN 978-1-009-26058-9 Paperback

Additional resources for this publication at www.cambridge.org/burt2.

For Helen, Fiona and Jennifer

Mr Hook[e] produced a part of his new weather Clock which he had been preparing which was to keep an Account of all the Changes of weather which should happen, namely the Quarters and points in which the wind should blow 2ly the strength of the Wind in that Quarter. 3ly The heat and cold of the Air. 4ly The Gravity and Levity of the Air. 5ly the Dryness and moisture of the Air. 6ly The Quantity of Rain that should fall. 7ly The Quantity of Snow or Hail that shall fall in the winter. 8ly the times of the shining of the Sun. This he was desired to proceed with all to finish he hoped to doe within a month or six weeks.

From Royal Society Journal Book (JBO/6), dated 5 December 1678
Reproduced by kind permission of the Royal Society Archives

Contents

Foreword

I spent the latter half of my career at the UK's National Physical Laboratory (NPL), and one part of my duties there involved making ultra-precision measurements of temperature. Working from the first principles of physics, my colleagues and I created an apparatus the size of a small room that could detect the tiny errors in the temperature measurements made by all the other practical thermometers on Earth. A single temperature measurement might take perhaps three days. The bulk of the work involved making a list of ways in which our apparatus might be in error, and then thinking of ways to evaluate how large that error might be. Few people are fortunate enough to work on such a fascinating and arcane project.

Since historical weather observations play a key role in our understanding of climate change, another strand of my work at NPL involved investigating the uncertainties around *meteorological* temperature measurements. In retrospect, I now understand that my work on ultra-precision measurements may have led me to approach this area from an unhelpfully narrow perspective. Applying the same 'first principles' approach of '*making a list of ways in which meteorological temperature measurements might be in error*' results in a list so long that the endeavour might seem doomed to failure. And yet, somehow weather observers have been busy making measurements day in and day out for hundreds of years. What I had failed to understand is the great value of meteorological measurements *despite* their difficulties and uncertainties.

Thus in 2012, I devoured the first edition of *The Weather Observer's Handbook* with relish. The book uniquely combined Stephen Burt's decades of experience running a weather observatory, his practical first-hand knowledge of modern meteorological techniques, and an outstanding technical and historical perspective. Reading the book, I learned what I should have already known: that to make reliable weather observations one must combine scientific knowledge of the measuring process with an appreciation of engineering reality. It is this multidisciplinary approach which has been at the heart of making reliable and meaningful measurements in the challenging environments to which meteorological instruments are exposed.

For anyone who seeks to make weather observations, either in support of an auxiliary activity or out of a regard for the measurements as an end in themselves, there is no better guide than this *Handbook*. The first edition helped me both in my research activities and also at home, where I maintain a weather station in my garden. In the last few years of my career at NPL I was fortunate enough to work with Stephen Burt, and together we found that, even after hundreds of years, there were still insights to be gained about the relatively simple task of making temperature measurements in air. I am looking forward to the second edition of *The Handbook* in the certainty that there will be new insights and developments in the long story of humanity's observations of the weather.

Michael de Podesta MBE, FInstP.
Teddington, January 2024

Acknowledgements

There have been very many significant developments in meteorological monitoring since the first edition of this book was published in 2012, including once-in-a-lifetime changes in instrument type and function (the statutory withdrawal of mercury-based devices, specifically thermometers and barometers), alongside the continuing rise in capability and deployment of automated systems. These major changes, together with many others of similar import, necessitated a line-by-line revision and update of the existing text, and large sections of this book have been completely rewritten as a result. In doing so, I have been fortunate in being able to draw upon on the willing help and assistance of many individuals and organisations around the world, and it is a great pleasure to be able to acknowledge your contributions – whether helping answer my questions, providing a photograph or reference to published work, or reading and commenting upon draft chapters. I hope and trust you will recognise your input in the pages following: I am most grateful to you all.

There are a few individuals who I would particularly like to thank by name for their contributions and support. Firstly, my Editor Matt Lloyd, of Cambridge University Press in New York: Matt was my Editor for the first edition of this book, and his enthusiasm and commitment provided the motivation to bring this second edition to fruition. Thanks also to tireless copyeditor Bret Workman (in Columbus, Ohio) and project manager Geethanjali Rangaraj (at Integra in Pondicherry, India) together with Jenny van der Meijden (senior content manager, Cambridge University Press in the UK) for their good-natured determination in surmounting the occasional obstacle during the publishing process. This book would have been incomplete without up-to-date information kindly provided from numerous national weather services around the globe, specifically NOAA in the United States, the UK Met Office (particular thanks to Melyssa Wright, at that time Observations Network manager, who went 'the extra mile' to answer my questions and to ensure network information and maps were fully up-to-date), Met Éireann in Dublin, Deutsche Wetterdienst, the Hong Kong Observatory, and the Australian Bureau of Meteorology in Melbourne. Thanks are also due to the very many instrument suppliers and manufacturers – too many to list by name here, but full details are given in Appendix 3 – who kindly answered my questions and provided photographs and specifications of their sensors and instruments. Few people know more about Davis Instruments than John Dann from UK reseller Prodata Weather Systems (weatherstations.co.uk), and John's help was particularly valuable in preparing the all-new Chapter 13. Steve Colwell from the British Antarctic Survey in Cambridge, UK, and Jo Cole and Eloïse Chambers from the British Antarctic Survey base at Rothera, together with Tom Matthews from King's College, London, provided priceless background information and photographs regarding the difficulties of running weather stations in the most extreme of environments

(Antarctica and Mount Everest, respectively): at the opposite extreme Christopher C. Burt (US weather historian, Oakland, California – no relation) again provided invaluable background regarding extreme heatwaves and record rainfalls in the United States. Roger Bailey (chairman, British Standards Hydrology working group) and Richard Griffith (private observer, Caithness, Scotland) kindly read through complete early drafts of specialist chapters, providing detailed and constructive feedback – as did others who requested anonymity, who have my gratitude nonetheless. My colleague Giles Harrison within the University of Reading provided many useful discussions and comments on various aspects related to improving the accuracy of air temperature measurements. And to round off, Nolan Doesken (ex-Colorado State University, and co-founder of CoCoRaHS), together with Henry Reges and Steve Hilberg from CoCoRaHS, kept me up-to-date on their organisation, which achieves an enviable and efficient synergy between professional and amateur spheres in meteorology within the United States.

Finally, I owe an enormous debt to my family — my wife Helen and my two daughters, Fiona and Jennifer. Without your tireless love, support, and understanding, I could not have started, far less completed, this latest book project. Thank you.

Stephen Burt
Berkshire, England
www.measuringtheweather.net
February 2024

Photograph credits – all photographs are by the author, unless stated otherwise. I am grateful to all the individuals who contributed photographs, and to the various organisations who permitted usage of their material within this book. Unless otherwise stated, Copyright of images used remains with the owner. Every attempt has been made to trace the Copyright owners of all images used within this book. If any Copyright clearances have been overlooked, please write to the Publishers with details, and any omissions will be corrected in future editions.

Author's note

Throughout this book, suggestions and the occasional recommendation are completely **independent of manufacturer or supplier influence**. No sponsorship, paid 'product placement' or other incentives were requested or offered by any of the companies whose products are referred to in this book. Although it is not possible to be fully conversant with every instrument or system described, wherever possible usage details are from first-hand experience. System specifications and performances have been taken from published manufacturer literature or websites, except where specifically stated otherwise. As product specifications inevitably change over time, potential purchasers are advised to check manufacturer literature or websites for the latest information.

If you use this book to help choose an automatic weather station, or the components of one, please mention this to your reseller or dealer when you make your purchase.

For the latest product information, updated equipment reviews, useful references and downloadable material related to this book, please visit the author's website www.measuringtheweather.net.

The information in this book is given in good faith. No liability can be accepted for any loss, damage or injury occasioned as a result of using this book or any of the information contained within, howsoever caused.

Abbreviations, footnotes and references

Abbreviations are defined within the text when first used; they are listed below only when used in more than one chapter.

Footnotes (indicated by superscripted symbols [*] [†] and so on) are given at the foot of the page to which they refer.

References and further reading are indicated within the text by bracketed numerals thus: [9]. They suggest sources of material or further reading for those who seek more detail on the topic. Numbered references are listed after the Appendices.

AMSL	Above Mean Sea Level
ASOS	Automated Surface Observing System
AWS	Automatic weather station
CIMO	Conference on Instruments and Methods of Observation, the WMO committee responsible for defining and publishing global instrumental and observation standard guidelines ('the CIMO guide')
CoCoRaHS	Community Collaborative Rain, Hail and Snow network
COOP	Cooperative Observer Program (US)
DST	Daylight Savings Time
DWD	*Deutscher Wetterdienst* – the German state weather service
GMT	Greenwich Mean Time, or Z time (equivalent to UTC for all practical purposes)
KNMI	*Koninklijk Nederlands Meteorologisch Instituut* – the Dutch state weather service
LAT	Local Apparent Time
MMTS	Maximum-Minimum Temperature System (sometimes known as NIMBUS)
MSL	Mean Sea Level (*see also* AMSL)
NOAA	National Oceanic and Atmospheric Administration (US)
NWS	National Weather Service (US)
PRT	Platinum Resistance Thermometer
RTD	Resistance Temperature Device
SPICE	Solid Precipitation Intercomparison Experiment (WMO)
SRG	Standard Rain Gauge (US)
TBR	Tipping-Bucket Raingauge
USB	Universal Serial Bus (a communications port on computers)
USCRN	US Climate Reference Network
USRCRN	US Regional Climate Reference Network
UTC	Coordinated Universal Time (equivalent to GMT for all practical purposes)
WMO	World Meteorological Organization

PART I

THE BASICS

1 Why measure the weather?

Of all the physical sciences, meteorology depends more than any other on frequent, accurate and worldwide measurements. Every day, millions of weather measurements are made by people and automated sensors across the globe, on land, over the oceans, in the upper reaches of the atmosphere and from space, providing the raw data essential to supercomputer-based weather forecasting models that are vital to modern economies. Meteorology (and its statistical cousin, climatology) is one of the few sciences where both amateurs and professionals make significant contributions.

Society 'measures the weather' for many different reasons: as well as input to weather and climate forecasts, it is a vital part of aviation safety, critical in detecting and quantifying climate change, keeping tabs on typhoons and hurricanes, monitoring the ebb and flow of pollutants and arctic ice, and hundreds of other applications of enormous benefit to the community. Accurate surface measurements provide essential 'ground truth' to the broad-brush data sweeps of remote sensing capabilities such as satellite and radar. Weather records are made in every country and region in the world – from the hottest deserts to the coldest ice sheets in Antarctica, from densely populated city centres to the most remote mountaintops. The latter category includes a network of six automatic weather stations (AWS) on Mt Everest (**Figure 1.1**), from Phortse at 3810 m (about 640 hPa) and Base Camp at 5315 m (about 530 hPa), to the highest weather station on Earth at Everest's Bishop Rock, 8810 m above mean sea level (about 340 hPa), just below the summit at 8849 m [1,2,3].

For many, professionals and amateurs alike, measuring the weather is also an absorbing long-term interest, guaranteed to deliver something different every day of every year. Records kept by individuals and organisations alike assist in the scientific analysis of all types of weather phenomena, and can become a permanent part of a nation's weather record.

About this book

This book has been written with four main audiences in mind:

- Professional users, including local or state authorities and other statutory bodies
- Weather enthusiasts and amateur meteorologists
- Schools, colleges and universities
- Weather-dependent outdoor activity professions and organisations.

Figure 1.1 Installing one of the world's highest weather stations, the 'Balcony Station' at 8430 m above sea level on the south col of Mt Everest, just below the summit at 8849 m. (Photograph courtesy of the Ev-K2-CNR Committee archive)

The aim of this book is to provide **useful, independent and practical guidance** on most aspects of weather observing, with emphasis on instrumental observations. Throughout, the information and standards set out in this volume are based upon the authority and guidance of the Geneva-based World Meteorological Organization's (WMO) latest *Guide to Meteorological Instruments and Methods of Observation* publication, otherwise known as the WMO CIMO guide (CIMO being the Commission for Instruments and Methods of Observation) [4]. By necessity in its role as a global reference and being written for a professional, specialised audience, the CIMO guide is very detailed and often highly technical. In contrast, this book is intended as the 'everyday CIMO guide', with content sufficiently detailed for most purposes, but where individual topics can be referenced quickly and easily back to WMO standards and guidelines as necessary. The focus of this new edition is primarily on the selection and use of modern electronic instruments and automatic weather stations (AWS), although sections on the history and use of 'traditional' instruments within the twenty-first-century context have been retained where appropriate.

Professional users

There are many 'professional' users who need reliable and accurate weather information, whether at a single site or from multiple locations, whose needs can be served by one or more properly sited AWS. Typical users include local or state authorities managing road maintenance (including winter gritting or snow clearance), landfill management, airport or airfield weather systems, environmental monitoring as part of civil engineering projects, outdoor field study centres and many more. For professional users requiring environmental records, perhaps as part of new statutory requirements, this handbook provides independent guidance on choice of systems, siting of sensors and suggestions on data collection and handling processes. The information gathered needs to be both manageable and relevant, while meeting both the appropriate standards of measurement and exposure and budget restrictions. It also includes advice on how to document the site and instruments in use (and any changes over time), to minimise possible future downstream technical or legal challenges relating to the records made.

Weather enthusiasts and amateur meteorologists

For individuals who are new to the fascinating science of measuring the weather, this book is intended to help guide your choice of what may be your first item or items of weather-measuring equipment. It explains the important things to look out for, what can be measured within particular budgets, how best to site your instruments, and how to start collecting and sharing data. Whether your site and equipment is basic and sheltered, or extensive and well exposed, this book provides help to improve the quality and comparability of your observations to attain, or even surpass, the standards set out by the World Meteorological Organization.

For those who already have experience of weather observing and who already own a basic AWS and are looking to add additional sensors or upgrade to a more capable system, or those who are considering upgrading an AWS to complement or replace existing 'traditional' instrumentation (particularly 'legacy' mercury thermometers and barometers), this book provides assistance and suggestions on choosing and siting appropriate equipment. It is also a practical day-to-day observing reference handbook to help get the most out of your instruments and your interest.

Schools, colleges and universities

The installation of automated weather-monitoring equipment offers the chance to include weather observations at all levels within the educational curriculum, from early schooldays to post-doctoral levels. Weather measurements are often made more relevant and interesting to the student by virtue of the readings being made at the school or college site, particularly where both real-time and archived records can be made easily available. From elementary school to university, the observations can be used immediately (especially so in an interesting spell of weather, such as a heatwave or flood event) or in a variety of curricular activities such as numeracy, IT, telecommunications, climate change as considered within the school curriculum, severe weather awareness training and alarms, office software packages (particularly spreadsheets), statistics and website design, in addition to conventional science, geography, and mathematics courses. There are dozens of websites giving examples of the installation and use of school weather stations: this book provides independent assistance on choosing and siting suitable systems and making best use of the data collected.

Many of those who have gone on to study meteorology further, and became professional meteorologists, picked up the 'weather observing' bug at school (including the author). The importance of encouraging curious young minds to observe and take an active and enquiring interest in their physical environment on a day-to-day basis, and the 'bigger picture' of global climate change with its risks and consequences, cannot be underestimated.

Weather-dependent outdoor activity professions and organisations

Many organisations or clubs need site-specific weather information; for example, field study centres, yacht clubs, gliding or parachuting clubs and private aviation airfields, as well as windsurfers and microlight pilots. In some cases, particularly microlight and gliding clubs, members may live a considerable distance from the main club operations and value the opportunity to be able to view live weather conditions at the site on a club website before making a decision on whether to travel to the club that day. Farmers and other professions largely at the mercy of the weather also need accurate and timely weather information, perhaps from more than one site within a local area. Many such organisations or businesses may not have previously considered their own weather station or monitoring network as being cost-effective. Today, respectable quality weather data in real time is available from inexpensive, easily maintained and robust systems. Modern electronic weather stations connected to the Internet can provide local or distant-reading output facilities quickly, cheaply and reliably; this book outlines what is available and where to site the instruments for best results.

Topics covered

Although the main focus of this book is on electronic instruments, current 'traditional' or legacy weather instruments – largely non-digital – and their development are also covered in this handbook. Many of these are being replaced, or indeed have already been, by their digital equivalents. Some still have an important part to play, not least in providing continuity with existing and historical records, but in many countries automated systems now supply the majority of meteorological measurements. This trend has been greatly accelerated by statutory restrictions on the manufacture and sale of mercury-based instruments, such as traditional thermometers and barometers, under UN Environment Programme regulations following the Minamata Convention, signed by 128 countries in 2013 and which came into force in 2020 [5]. Mercury is a powerful toxin, and while (understandably) both legal and WMO advice is to eliminate completely the use of mercury-based instruments, a very few exceptions may be sanctioned at long-period sites where a period of a few years of overlap records are beneficial to assess the consistency and homogeneity of the 'new' instrumentation alongside traditional methods. With such 'replacement' situations becoming more the norm, this book provides suggestions on how best to minimise the discontinuity of record that may happen when such changes are introduced – although in all cases the network administrator (such as NOAA in the United States, the Bureau of Meteorology in Australia or the Met Office in the United Kingdom) should also be consulted at the earliest opportunity.

This book covers a wide range of weather station systems, sensors and associated technology, from $100 (US) to upwards of $2,000 (nominal price guidelines correct at the time of going to press). It does not cover homemade instruments or remote-reading sensors lacking any means of logging (such as wireless temperature and humidity displays), nor does it cover in detail professional systems costing thousands of dollars upwards (for which more specific presales advice should be sought from the manufacturer). It covers land- and surface-based systems only. Sensors and logging equipment for aircraft, drones or buoys have very different characteristics and are outside the scope of this book.

Geographical coverage

As previously stated, this book covers equipment, standards and measurement methods for surface meteorological observations as set out by the World Meteorological Organization (WMO), based in Geneva, Switzerland [4]. The details of some measurements and methods differ slightly from country to country, and where applicable this book provides specifics relevant particularly within the United States, the United Kingdom and the Republic of Ireland. The majority of the book is also relevant outside these geographies, but readers in other regions should check the availability of products and the detail of country-specific equipment, specifications and siting recommendations with their national meteorological service prior to implementation. Details of, and links to, the world's national weather services are available on the WMO website at https://community.wmo.int/members.

Abbreviations, references, footnotes and further reading

Abbreviations and technical terms are kept to a minimum: where used, they are defined at first use and indexed. The most frequently used abbreviations are listed at the front of the book for easy reference. References and suggestions for further reading are included towards the end of the book for readers who may wish to delve further into these topics. Specific references are indicated within the text by a number within square brackets (thus [9]) and are listed in numerical order at the end of the book. Footnotes are indicated by symbols (thus * †) with the appropriate text appearing at the foot of the page on which the footnote appears.

A number of sample and template spreadsheets referred to within the book are available online at www.measuringtheweather.net. These are referenced to the appropriate point in the text and are available for free download. They can then be customised to your specific requirements.

Units

Meteorology is necessarily an international science, and consistent units are required for information exchange and understanding. In this book WMO recommendations for units are used in preference, with bracketed alternatives where necessary; for instance, wind speeds can be expressed in metres per second (m/s), knots (kn), miles per hour (mph) or kilometres per hour (kph), depending upon the country and requirement. Conversions between different units are given in **Appendix 6**.

Automatic weather stations

In this book an automatic weather station (AWS) is defined as any system which creates and archives a digital (computer-readable) record of one or more weather

Figure 1.2 The US National Weather Service Automated Surface Observing System (ASOS) sensor package located at Pocatello, Idaho (42.917°N, 112.567°W, 1356 m above MSL, WMO station no. 72578). From left to right, the instruments shown are – heated tipping-bucket raingauge (TBR) within wind shield: aspirated temperature and humidity sensors: present weather sensor: 10 m wind mast with heated ultrasonic wind sensor: data collection panel (big box): laser ceilometers: freezing precipitation sensor (little tilted box), and finally the visibility sensor. See also www.weather.gov/asos. (Photograph by Gary Wicklund)

'elements', such as air temperature, precipitation, sunshine, wind speed or other parameters.

In its simplest form, an AWS can be a single sensor integrated with a small, inexpensive electronic data recorder (a 'datalogger' or simply 'logger'), which may simply be a memory chip within a circuit board, device or console. Loggers that can record only one input signal, or 'channel', are therefore 'single-channel loggers'; those that can handle two or more are 'multi-channel'. Most such systems can be left exposed as a complete package, including the logger within a suitable enclosure, perhaps for several months in a remote location, before the recorded data are retrieved. The most advanced AWSs (**Figure 1.2**) are completely automated remote multi-element, single-site observing systems built around a sophisticated (user-)programmable datalogger, requiring only the minimum of human attention and maintenance, self-powered by a solar cell array or wind turbine and communicating observations at regular intervals over a telecommunications system to a collecting centre. Telecommunications may be via satellite in remote areas.

Most of the world's meteorological services are increasingly adopting such systems to replace costly human observers. But even with today's most sophisticated technology and sensors, human observers are still required for many weather observing tasks; AWSs are still very poor at telling the difference between rain and wet snow, nor can they report shallow fog just starting to form across the low-lying parts of an airfield or see distant lightning flashes on the horizon warning of an approaching thunderstorm. Human weather observers will continue to be required for a long while to come!

The makers of the observations

Fascination with the changes in day-to-day weather is nothing new, although weather records were by necessity purely descriptive until the invention of meteorological instruments in the mid-seventeenth century [6]. Probably the oldest known weather diary in the Western world is that of Englishman William Merle, who kept notes on the weather in Oxford and in north Lincolnshire from 1337 to 1344 [7, 8]. In North America, the earliest surviving systematic weather records are those made by Rev. John Campanius Holm, a Lutheran minister originally from Sweden, who made observations at Fort Christina in New Sweden (near present-day Wilmington, Delaware) in 1644–45 [9]. (Today, the prestigious National Oceanic and Atmospheric Administration NOAA John Campanius Holm Award is given for outstanding accomplishments in the field of cooperative meteorological observations.)

During the Renaissance, the invention of instruments to measure the temperature and pressure of the atmosphere, and later other elements, made it possible to track the changes in weather conditions more accurately, and more consistently, between different observers and locations. Galileo invented the air thermoscope around 1600; Santorio added a scale to make it a thermometer in 1612. The first liquid-in-glass thermometer (in a form we would recognise today) was invented by Ferdinand II, the Grand Duke of Tuscany, in 1646, while Evangelista Torricelli invented the mercury barometer in 1644.

Surprisingly perhaps, what we would now call multi-element automated weather stations began to appear very early in the history of meteorological instruments. Sir Christopher Wren (1632–1723) is best known today as the architect of London's St Paul's Cathedral, but in his early career he was a noted astronomer and mathematician [10,11,12], a founding member of the Royal Society in London in 1660 [13], and a prolific instrument designer. Together with Robert Hooke (1635–1703) he designed and built many weather instruments, including Hooke's sophisticated mechanical 'weather clock' in the 1670s [14] (see Box, *Wren and Hooke: the first automated weather station, 1678*). The earliest surviving rainfall records in the British Isles were made by Richard Towneley at Towneley Hall near Burnley, Lancashire, in northern England from January 1677 [15]; the raingauge used was based upon Wren's 1662/63 design of the tipping-bucket type. Modern varieties of Wren's instrument are used to measure rainfall at hundreds of thousands of locations across the globe today (see Chapter 6, *Measuring precipitation*).

As materials and methods evolved, meteorological instruments became more practical, robust, reliable and cheaper, and thus were used more widely, carried to the New World on the ships of the European superpowers of the day. The once ubiquitous Six's maximum-minimum thermometer was invented by James Six in 1782 [19], and although these instruments ceased to be used for accurate climate recording over 150 years ago, they can still be found today in many a gardener's greenhouse (**Figure 1.3**). In the early nineteenth century one of the first amateur meteorologists, apothecary Luke Howard of London, popularly known as 'the inventor of clouds' [20], owned a magnificent – and very expensive – 'clock-barometer', or mercury barograph [21]. Records from this instrument survive today; a very similar instrument, made for Great Britain's King George III in 1763–65 by Alexander Cumming (c. 1732–1814), remains in the Royal Collection [22].

Wren and Hooke: the first automated weather station, 1678

Christopher Wren's long friendship and professional collaboration with Robert Hooke spawned many designs for instruments to 'measure the weather'. Wren is acknowledged as the inventor – around 1662/3 [16] – of the tipping-bucket mechanism for measuring rainfall, the principle of which is still used in today's instruments. Hooke was a polymath with a superb ability for translating ideas into practical working devices [14, 17, 18], and he built many weather instruments, including what can only have been the first automated weather station, as the following extract from the Royal Society Journal Book (JBO/6), dated 5 December 1678, shows:

> Mr Hook[e] produced a part of his new weather Clock which he had been preparing which was to keep an Account of all the Changes of weather which should happen, namely the Quarters and points in which the wind should blow 2ly the strength of the Wind in that Quarter. 3ly The heat and cold of the Air. 4ly The Gravity and Levity of the Air. 5ly the Dryness and moisture of the Air. 6ly The Quantity of Rain that should fall. 7ly The Quantity of Snow or Hail that shall fall in the winter. 8ly the times of the shining of the Sun. This he was desired to proceed with all to finish he hoped to doe within a month or six weeks.

Reproduced by kind permission of the Royal Society Archives

Figure 1.3 A Six's maximum-minimum thermometer; this formerly in the Temperate House at Kew Gardens in west London. (Photograph by the author)

Weather instruments benefited from the enormous technological and manufacturing advances made between the late eighteenth and late nineteenth centuries. Many of today's instruments date from this period (see timescale in **Figure 1.4**) [23] including the Stevenson screen (see Box, *The lighthouse Stevensons*). As late as the first decade of the current century, a meteorological observer from the late nineteenth century would have found little difficulty in making a standard observation in almost any weather station in the world, but as traditional instruments are rapidly superseded by newer electronic equipment, our time-traveller would today find it increasingly difficult even to recognise many instruments.

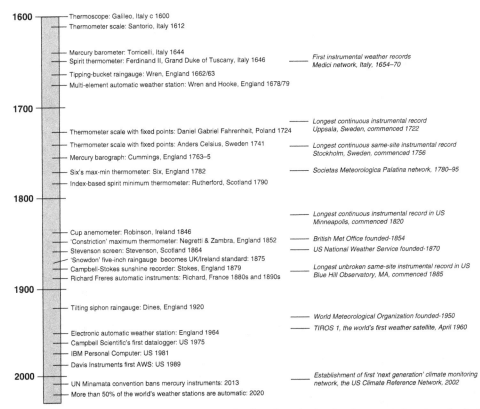

Figure 1.4 Four-hundred-year timeline showing key dates in the development of meteorological instruments and weather recording. For sources, see references in the text.

Recording meteorological instruments continued to be developed and improved during the eighteenth and nineteenth centuries, but while many ingenious designs were invented, almost all relied upon mechanical components and thus were, to a greater or lesser degree, subject to friction. They were also often extremely expensive (many were made to order or in very small numbers), and difficult to maintain in good working order when exposed to the elements. For these reasons few were made – and even fewer have survived, even in museums.

The lighthouse Stevensons

Thomas Stevenson (1818–87) was a marine engineer; a member of the famous Stevenson engineering dynasty which built most of Scotland's lighthouses [24, 25], and the father of Robert Louis Stevenson (1850–94), author of *Treasure Island* (1881), *Kidnapped* (1886) and *The Strange Case of Dr Jekyll and Mr Hyde* (1886). In a brief note in the *Journal of the Scottish Meteorological Society* in 1864 [26] he described the form of thermometer shelter which still bears his name – a white-painted double-louvred box which protected the thermometers inside from rain-fall, sunshine and infrared radiation from Earth and sky. At the time there were dozens of proposed designs for thermometer screens, some of which had been in use for a decade or more [23]. It was only a series of painstaking trials undertaken by the Reverend Charles Griffith at Strathfield Turgiss rectory in Hampshire, England, in the late 1860s and early 1870s [27], comparing air temperatures measured in Stevenson's screen with other models (**Figure 1.5a**), that eventually led to its adoption (with only minor amendments to the original design) as the preferred method for taking air temperature measurements by the Royal Meteorological Society in 1884 [28]. The *de facto* British standard spread rapidly to the rest of Britain's empire and then to the rest of the world (**Figure 1.5**) as the enclosure was simple, easy to make locally, robust and gave good protection from the tropics to the poles. The basic principles – a white, louvred, roofed enclosure – remain common to many thousands of thermometer screens in daily use through-out the world today (**Figure 1.6**). Today, the original wooden construction has increasingly been superseded by low-maintenance plastic variants, as described in more detail in Chapter 5, *Measuring the temperature of the air*.

(a)

(b)

Figure 1.5 Early models of thermometer screen.
(a) This photograph was taken at Berkhamsted, Hertfordshire, England, on 29 July 1896, and shows two Stevenson screens (centre of picture) adjacent to a much larger modified Glaisher stand (an earlier and more open pattern of thermometer screen). The observer is Edward Mawley, President of the Royal Meteorological Society 1896–98. (Courtesy Royal Meteorological Society Archive). (b) Cotton Region Shelter (see Chapter 5) and young observer at US Weather Bureau observing site at Granger, Utah, c. 1930. (Courtesy NOAA/Department of Commerce National Weather Service Collection wea00903)

Figure 1.6 Stevenson-type thermometer screens in use on three continents and in the south polar regions. (a) US 'cotton region shelter' thermometer screen at Blue Hill Observatory, Massachusetts (Courtesy of Blue Hill Observatory). (b) Stevenson-type thermometer screen on the seafront at Cannes in the south of France (photograph by the author). (c) Stevenson screen (Bilham variant) in the Botanic Gardens in Christchurch, New Zealand (photograph by the author). (d) Metspec aluminium-and-plastic Stevenson screen at Rothera research station, Antarctica (photograph by Eloïse Chambers, British Antarctic Survey).

Figure 1.7 A thermograph from the London makers Short & Mason, 1913.

The end of the nineteenth century saw the advent of relatively inexpensive, mass-produced, single-element mechanical recording instruments using clock-driven paper charts, such as the barograph and thermograph (**Figure 1.7**), and, later, various forms of automated rainfall recorders. These instruments revolutionised automated weather recording [29]. As a result, for over a century 'automatic' weather records were obtained by tedious and time-consuming manual analysis of paper-based records from these single-element instruments. Very few remain in regular use today, superseded by electronic systems much closer to Wren and Hooke's original concept of the 'weather clock' – namely, a single mechanism to record multiple sensors – one which was 350 years ahead of its time.

The modern AWS, consisting essentially of an electronic logging system connected to a variety of electrical sensors, began to take form in the early 1960s (**Figure 1.8** [30]). It was the advent of cheap computing power, advances in data storage and telecommunications and the reduction in size and power requirements of electronic components resulting from the personal computer revolution during the 1980s that enabled the economies of scale needed for prices to fall sharply. One of today's mid-range 'personal' AWS would have cost tens of thousands of dollars in the mid-1980s. Today, automatic high-quality, multi-element, remote wireless data display and logging are available for less than the price of a new clock-driven thermograph, with enormous additional benefits in data reliability, accuracy and precision.

The longest-running weather observations in the world

The earliest versions of many 'traditional' meteorological instruments were invented in Europe in the seventeenth or eighteenth century – the mercury barometer, mercury thermometer and various instruments to measure rainfall all made their first appearance in recognisably modern form around this time, and people began to keep instrumental weather records. At first these were individual efforts, experimental and sporadic, rarely lasting for more than a few years in any one place, and of course with widely differing standards of exposure, accuracy and (not least) units. The earliest surviving instrumental weather records are from the Medici network, based in Florence, Italy, covering the period 1654 to 1670. Thanks to tremendous historical detective work, these early records have since been recovered and made available for modern analysis [31]. We also have daily observations of a mercury barometer in Pisa, Italy, in 1657–58 [32], and for 1694 there are sufficient surviving barometric pressure records across Europe for outline daily synoptic weather maps to be prepared. An almost complete daily pressure record has been assembled for locations in Paris back to 1670, and for London since 1692 [33, 34] with a few gaps back to Robert Hooke's diaries starting in 1672.

In the early eighteenth century, regular and systematic weather records commenced in various places in Europe, often as part of the observational routine at astronomical observatories. Many of these observatories are still in existence, and at a few locations continuous weather observations have been made in much the same location for 200 years or more. Although most long-period records have been made in towns or cities, and as a result individual temperature records have been affected by urban growth and urban warming to a greater or lesser extent, the observations are still the most detailed and comprehensive records we have for ongoing evidence of climate change since the seventeenth century.

Figure 1.8 An early prototype automatic weather station at the then Institute of Hydrology in Wallingford, England, in the late 1960s. (Photograph by Ian Strangeways)

The longest temperature record in the world: 1659 to date

In 1953, the British climatologist Gordon Manley (1902–80) published his first paper on what became known as the Central England Temperature (CET) series in the Royal Meteorological Society's *Quarterly Journal* [35]. Manley's extensive and painstaking research assembled scattered early instrumental temperature records and descriptive weather diaries to produce a chronology of mean monthly temperatures covering the period 1698 to 1952 representative of a roughly triangular area of England enclosed by Lancashire, London and Bristol. A second, longer paper in 1974 [36] extended the series back to 1659 – about the time the earliest thermometers appeared in England – and brought it up-to-date. Other records that had come to light in the intervening 20 years also allowed for corrections or improved estimates to the existing series. Since Manley's death the series has been kept up to date by the Hadley Centre, part of the UK Met Office, and today the series forms the longest instrumental record of temperature in the world. A similar monthly rainfall record, the 'England and Wales Precipitation' series, extends back to 1766 [37].

Early English temperature records are sufficiently numerous that an expanded *daily* temperature series back to 1778 has been assembled [38]. Daily 'maximum' and 'minimum' CET are also available back to 1878; since 1974 the data have been adjusted slightly to allow for urban warming.

Centennial sites

Manley's work pieced together many disparate records to produce a figure representative of a region rather than a single location. There are other long composite series of temperature, rainfall and/or pressure records representative of other cities or regions in Europe extending back to the eighteenth century, but there are a perhaps surprising number of locations where instrumental weather records are still made today in the *same* location, or very nearly so, where continuous observations commenced at least 100 years ago. The World Meteorological Organization's Centennial Observing Stations programme [39] was established in 2017 to identify and recognise these important observing locations. To qualify, the national meteorological service of the country concerned nominates site(s) meeting a number of criteria, chief amongst which are a period of record of not less than 100 years (with a minimum 90 per cent record availability), and which continues to date. Moves or relocations of the weather station over the period considered must have been minor and without significant impact on the homogeneity of the data series considered (which could be temperature, sunshine, rainfall and so on, or a combination of other elements), and in addition the data series themselves, together with information relating to the site and instruments, must be made available for scientific research. At the time of going to press, 291 stations have been recognised in this manner, although it remains a pity that the station records themselves, or links thereto, are not yet available through this useful WMO initiative.

The following paragraphs identify and briefly discuss some of the best-documented long records in Europe, North America and in Asia and the Pacific region. Details of the site and records therefrom have been taken from published work and/or the WMO Centennial Stations listings.

Europe

Locations of the sites listed in this section are shown in **Figure 1.9**.

Uppsala, Sweden – 1722 to date

59.847°N, 17.635°E, 25 m above sea level

The oldest mostly continuous same-site meteorological records in the world are those from Uppsala, Sweden, about 65 km (40 miles) north of Stockholm, where records commenced in 1722 [40]. The earliest meteorological observations in the Nordic countries were initiated around 1720 by the Society of Science in Uppsala, when Professor Erik Burman started the observations there, assisted by the young Anders Celsius (1701–44), who took over responsibility for the records in 1729.

The oldest surviving observatory journal dates from the year 1722, although there are some gaps in the Uppsala record until 1773. Before 1751 a variety of thermometer scales were used, amongst them of course Celsius' own thermometer with the first Celsius scale, and this very fact of the record enabled researchers to check the calibration and consistency of the early records. Celsius' own observatory, the Celsiushuset, still stands in central Uppsala. In September 1853, the observation site was moved to a then newly built astronomical observatory situated in open fields outside the town, about 1 km north-west of the original site (**Figure 1.10a**). Further changes of site took

Figure 1.9 The locations of some of the longest-running meteorological records in Europe – see text for sources.

place in October 1865, June 1952 and August 1959, all within a few hundred metres of each other, as the original rural site had become increasingly urbanised as Uppsala expanded over the years (its population today is around 150,000). In August 1959 the observing site moved to the Department of Meteorology at Uppsala University, and in January 1998 a further move took place 1.4 km further south to 'Geocentrum' at the university's Department of Earth Sciences. Today the observations are made hourly using a Campbell Scientific AWS (**Figure 1.10b**). In September 2022, an international symposium was held at Uppsala University to celebrate the completion of 300 years of records [41]. In this long record, the extremes of temperature to date have been –39.5 °C in January 1875, and 37.4 °C in July 1933.

Padova (Padua), Italy – 1725 to date

45.402°N, 11.869°E, 20 m above sea level

Meteorological observations commenced as part of the astronomical observational routine in Padova (Padua) in northern Italy in 1725 [42]. Until 1767, observations were made by various individual observers in their own dwellings within the town, but from 1768 to 1962 the records were kept at the *Specola* complex in the centre of Padua (**Figure 1.11**). Today, the records are maintained at the nearby Botanical Gardens.

The temperature record has been carefully reconstructed and homogenised, taking account of multiple instruments, calibrations, observers, observing sites and practices, making this the oldest record in southern Europe.

Figure 1.10 Weather records have been made in Uppsala, Sweden, since 1722. (a) The instruments in Observatory Park, where measurements were made 1853–1952 (Uppsala University). (b) The current site at Uppsala University Geocentrum. (Photograph by the author)

Figure 1.11 The Specola complex in Padua, Italy, where weather observations commenced in 1768. (Photograph courtesy of Dario Camuffo, Institute of Atmospheric Sciences and Climate, Padua)

Stockholm, Sweden – 1756 to date

59.342°N, 18.055°E, 38 m above sea level

Weather observations began at the old astronomical observatory in Stockholm in 1754. Complete daily mean series of air temperature and barometric pressure have been reconstructed from the original observational data for the period 1756 to date [39, 43]. In 2006 the observatory completed 250 years of records, the longest unbroken same-site observation series in the world.

The first observer was the secretary of the Royal Swedish Academy of Sciences, the astronomer and statistician Pehr Wilhelm Wargentin (1717–83). He lived on the second floor in the then newly built observatory and placed his thermometer outside one of his windows. Wargentin is a well-known figure in Swedish scientific history as the father of Swedish population statistics, and also for his studies of Jupiter's moons. When the observatory was renovated and extended in 1875, the thermometer was moved to a metal cage outside a window on the first floor. The current observation site, dating from 1960, lies only a few metres from the building (**Figure 1.12**).

Since 1901, the highest temperature observed has been 35.4 °C (on 6 August 1975), the lowest –28.2 °C (on 25 January 1942). Less rigorous early records show 36 °C on 3 July 1811 and –32 °C on 20 January 1814.

Figure 1.12 The current weather station at Stockholm Observatory. (Photograph by the author)

Kremsmünster, Austria – 1762 to date

48.05°N, 14.13°E, 389 m above sea level

Kremsmünster Abbey is a Benedictine monastery founded in 777. Much of the present vast complex, one of the largest in Austria, was rebuilt from the middle of the seventeenth century, including an astronomical observatory completed in 1750. A near-continuous series of daily meteorological observations exists at the site from 1762; this was one of the first sites to be awarded WMO Centennial Station status in 2017. The earliest records were made in an unheated tower open to the air, but today's observations are made by an AWS in standard fashion within the abbey's grounds. For an eight-year period, temperature records were restarted in the original exposure in order to provide comparisons between old and new records [44]. This reference also includes details of many other long-period sites within the Alpine region of central Europe, including details of multi-site records in Basel and Geneva (Switzerland) from 1760, and from Turin (Italy) also from 1760.

Milan, Italy – 1763 to date

45.471°N, 9.189°E, 121 m above sea level

The Astronomical Observatory of Brera (OAB) in Milan was founded in 1762, and daily meteorological observations have been made here since 1763. It is the oldest scientific institution in Milan, and remains one of the top astronomical research institutes in the world.

Although observations have always been made at the observatory, many changes of instruments, station location and observation methods over the years render the original observations series far from consistent. Fortunately, detailed metadata (records of the instruments and their exposure) were kept. A meticulous research programme conducted at the University of Milan [45, 46] re-examined all of the original records to produce a complete and homogeneous daily series of maximum, mean, and minimum temperatures, and daily mean barometric pressures, covering the period from 1763 to date. The site became a WMO Centennial Station in June 2021.

Prague, Czechia – 1775 to date

50.086°N, 14.416°E, 191 m above sea level

Regular meteorological observations commenced in the vast Baroque complex of the former Jesuit College in Prague's Old Town, the Clementinum (or Klementinum), in 1752, although there are breaks in the record until 1775; the daily records are unbroken since January 1784 [47, 48]. Observations continue at the same site (now the Czech National Library) today, in much the same surroundings as they were more than 250 years ago, with observations made at the 'Mannheim hours' of 0700, 1400 and 2100 local time (see below).

Two thermometer screens are in use, similar to the original eighteenth-century models rather than today's standards – a louvred screen located on the first floor of the north side of the south annex and another on the flat roof of the east annex. Rainfall amounts and sunshine duration are also measured here (**Figure 1.13**). Although measurements at his site were and still are influenced by a number of factors (such as the location of the measuring instruments within the complex, and its position in the centre of Prague), for modern science, they represent a unique and extremely valuable source of information on central European weather and climate conditions during modern history.

Since 1775, the temperature extremes at the site have been 37.8 °C (on 27 July 1983) and −27.6 °C (on 1 March 1785).

Figure 1.13 Meteorological instruments in the Prague Clementinum, where records have been kept since 1752. (Photographs by kind courtesy of the Czech Hydrometeorological Institute)

Hohenpeissenberg, Germany – 1781 to date

47.801°N, 11.010°E, 977 m above sea level

Hohenpeissenberg is the oldest mountain observatory in the world, and possesses one of the longest reliable observational records of any location [49, 50]. It is located about 80 km south-west of Munich at an altitude of just under 1000 m (**Figure 1.14**) and has an interesting and varied history. It was one of the first sites accepted into the WMO Centennial Stations initiative in 2017.

Meteorological observations were first made here in 1758/59, but regular and uninterrupted records started on 1 January 1781 as one of the stations in the Societas Meteorologica Palatina observation network, established by the Meteorological Society of the Palatinate with the support and funding of Karl Theodor, Elector of the Palatinate. This was the world's second international climate observation network (Florence's Medici Network in 1654–70 was the first): it consisted of 39 stations extending from eastern America to the Ural Mountains, and from Greenland to the Mediterranean [51]. The Societas Meteorologica Palatina established standardised instruments, observing procedures and observation times (the so-called Mannheim hours of 0700, 1400 and 2100) for the first time. (Observations made at the Mannheim hours are still used for today's climatological records at Hohenpeissenberg.) Augustinian monks from the nearby Rottenbuch monastery made the observations; Hoher Peissenberg was a place of pilgrimage and a subsidiary convent.

The Palatinate came under occupation in 1792, during the Austrian-French war. This brought an end to the Societas Meteorologica Palatina, although fortunately the Augustinian Canons decided to continue the meteorological observations. Following secularisation in 1803, the Bavarian Academy of Sciences assumed responsibility for the station and appointed the parish priest of Hohenpeissenberg as the responsible observer. In 1838, the observatory came under the responsibility of the Royal Observatory of Munich, and in 1878 part of the Bavarian State Weather

Figure 1.14 The current meteorological instrument site at Hohenpeissenberg Observatory in southern Germany: meteorological observations have been made here without significant interruption since January 1781. (Photograph courtesy of Stefan Gilge, Deutscher Wetterdienst)

Service. In 1934, the Meteorological Service of the Third Reich assumed responsibility for the site, which was expanded into a main weather observation location in late 1937, commencing synoptic observations. In 1940, the station was relocated a short distance from the existing monastery buildings into newly built premises on the western side of the mountain. In the closing days of the Second World War, southern Bavaria came under attack from the Allied armies. Observations at Hohenpeissenberg were interrupted by artillery fire on 28 April 1945, and had to cease altogether on 2 May because of the danger to the observers, but were restarted on 14 May. On 1 April 1946, the station was incorporated into the network of the newly founded West German state weather service, the Deutscher Wetterdienst (DWD), and in March 1950 the site was formally upgraded to that of a meteorological observatory.

The range of instrumentation and observing routines has expanded considerably since. Records of atmospheric ozone commenced in 1967, a weather radar was installed in 1968, later upgraded to Doppler capabilities, and observations of trace atmospheric gases commenced. In 1994, the observatory became a part of WMO's Global Atmosphere Watch (GAW) programme.

Armagh Observatory, Northern Ireland – 1794 to date

54.353°N, 6.650°W, 62 m above sea level

The astronomical observatory at Armagh, built in 1790, is the oldest scientific institution in Northern Ireland [52]. Intermittent observations of the weather have been made on this site since 1784, prior to the building of the observatory: more

Figure 1.15 Armagh Observatory in Northern Ireland; meteorological observations began here in December 1794. (Photograph by the author)

systematic daily observations of temperature and barometric pressure commenced in December 1794. Although there are some gaps in the early years, and numerous changes of instrument and site around the observatory, the records are largely complete from 1833 to the present day. They represent the longest series of continuous weather records in Ireland. All of the records, including both current observations and scanned copies of the original manuscript records, are available on the observatory website www.armagh.space/weather and have been extensively documented [53, 54, 55]. Armagh Observatory was declared a WMO Centennial Station in 2018, **Figure 1.15** [39].

The site lies approximately 1 km north-east of the centre of the ancient city of Armagh, at the top of a drumlin (a small hill moulded by glacial action) in an estate of natural woodland and parkland of some 7 ha. The observatory is still largely surrounded by countryside similar to that which has existed since its foundation. The population of Armagh has increased relatively little in 200 years (2022: 14,590) and so the observatory suffers from little or no urban micro-climatic effects.

The third director of the observatory, Thomas Romney Robinson, appointed in 1832, made many experiments in other fields of science. One of his most enduring interests was the study of meteorology and in particular the measurement of wind speed. He invented the cup anemometer (see Chapter 9, *Measuring wind speed and direction*), a device that is still widely used throughout the world. An original Robinson anemometer can still be seen in operation on the roof of the Observatory.

The Radcliffe Meteorological Station, Oxford, England – 1813 to date

51.761°N, 1.264°W, 63 m above sea level

The University of Oxford is the oldest university in the English-speaking world, founded in 1249: the Radcliffe Observatory was established as part of the university in 1772. John Radcliffe (1652–1714) was a British physician who bequeathed

Figure 1.16 The Radcliffe Meteorolocical Station, Oxford, England. Meteorological observations have been made here since 1772; an unbroken daily temperature series exists from November 1813, and rainfall from January 1828. (Photograph by the author)

property to various charitable causes, including St Bartholomew's Hospital in London and University College, Oxford [56]. Several landmark buildings in Oxford, including the Radcliffe Camera, the Radcliffe Infirmary (now the John Radcliffe Hospital) and the Radcliffe Observatory, were named after him. The observatory site, situated in the walled garden of Green Templeton College in Woodstock Road adjacent to the old observatory building, possesses the longest same-site series of temperature and rainfall records in the British Isles (**Figure 1.16**). Irregular observations of air temperatures, barometric pressure and rainfall amounts exist from 1760, before the founding of the Observatory itself; a daily air temperature record is unbroken from November 1813, and daily rainfall from January 1828. The records have recently been extensively documented in book form [57]. Amongst other records, a unique near-200 year detailed record of thunderstorm occurrence continues today [58].

Dr Thomas Hornsby, then Savilian Professor of Astronomy at the university, approached the Radcliffe Trustees with a request for funds for the erection and equipping of an astronomical observatory in 1768. Building began in 1772, although it was not completed until 1794. The central feature of the building is the octagonal tower, 33 metres high, an adaptation of the Tower of the Winds at Athens. It is widely acknowledged as one of the most beautiful buildings in Oxford.

Initially, observations of air temperature (from thermometers mounted on the north wall of the observatory) were made three times daily to determine atmospheric refraction effects on star positions. From 1849, meteorological observations were made in their own right; thermometers were exposed in a thermometer screen at ground level on the north lawn within the observatory's walled garden.

Air temperatures, rainfall and sunshine records are measured today at the same place, and in very much the same way, as they have been since 1879. Since July 1935 the station has been known as 'The Radcliffe Meteorological Station,

Oxford'. In 1978 the site became part of what is today Green Templeton College, with the university's School of Geography and the Environment responsible for the daily observations. A university decree guarantees the continuation of the observations 'as long as they are deemed to be of scientific value' [57, 59]. The Radcliffe site became a WMO Centennial site in 2020.

Since daily maximum and minimum temperatures were first recorded in 1815, the extremes at the Radcliffe Meteorological Station, Oxford, have been 38.1 °C on 19 July 2022 and −17.8 °C on 24 December 1860. The full daily series of temperature, precipitation and sunshine records are available on the Oxford University School of Geography website at www.geog.ox.ac.uk/research/climate/rms/.

Durham University Observatory, England – 1843 to date

54.768°N, 1.584°W, 102 m above sea level

Meteorological records began at the Durham University Observatory in 1841, although the earliest surviving logbook dates from July 1843. As at Oxford, the observations were originally made as part of the astronomical routine of the observatory; astronomical records ceased at the outset of World War II in 1939. The long series of records were fortunately rediscovered by Gordon Manley [60], the first professor of Geography at Durham in 1928, and developed into the longest series of temperature and pressure observations in north-east England. Manual observations were discontinued in October 1999, since when observations have been made by a Met Office AWS (**Figure 1.17**). The full history of the meteorological records at Durham have been published [61] and the entire twice-daily observation series is available on the Durham University website at https://durhamweather.webspace .durham.ac.uk. Durham University Observatory became a WMO Centennial site in 2023.

Figure 1.17 The Durham University Observatory AWS. Meteorological observations have been made at this site since 1843. (Photograph by the author)

Since daily maximum and minimum temperatures were first recorded in 1843, the extremes at the Durham University Observatory have been 36.9 °C on 19 July 2022 and −18.0 °C on 8 February 1895.

The oldest weather records in North America

The earliest known instrumental weather records within today's United States began in Cambridge, Massachusetts, in 1715 and in Boston, Massachusetts, a decade later [9]. Morning and afternoon temperature observations survive from Philadelphia, Pennsylvania, from 1731 to 1732. Dr John Lining kept daily records of temperature, barometric pressure and rainfall in Charleston, South Carolina, between 1737 and 1753, to investigate relationships between weather and illness; these records were continued until 1759 by another doctor, Lionel Chambers. Early records include those of two presidents – George Washington (1732–99, first president 1789–97) kept a detailed weather diary at Mount Vernon, Virginia, in the late 1700s, while Thomas Jefferson (1743–1826, third president 1801–09) kept an almost continuous daily record of weather conditions from 1776–1816, mostly at Williamsburg and Monticello in Virginia. Washington's entries were diary-style, whereas Jefferson usually noted temperatures more systematically at numerous times during the day [9].

The earliest record on the National Centers for Environmental Information (NCEI) database is that for Nottingham in Prince George's County, Maryland, with records from August 1753 to December 1757. Morrisville in Pennsylvania has records from January 1790 to December 1859.

It was not until after the Surgeon-General of the United States issued an order in 1814 for each Army post surgeon to 'keep a diary of the weather … noting everything of importance relating to the medical topography of his station' that any form of systematic observations began to be made in America. The Smithsonian Institution ran its own network of weather reporting sites between 1849 and 1874. Many additional weather observing sites were established at or shortly after the foundation of the US Weather Bureau in 1870, but it was not until the adoption of a uniform plan of observations in 1895 and the printing of monthly climate reports from 1896 that standardisation across the different networks was finally secured.

Probably the longest continuous current records for any US city are those for Chapel Hill in North Carolina and for Minneapolis, Minnesota, both of which extend back to 1820. In Minneapolis, records commenced at Fort Snelling in 1820 [62] and continue today at Minneapolis-St Paul International Airport (since 1938). There are many hundreds of records extending back to 1872 or earlier (although many have moved from one site to another within the town or city within the period of record), including those for San Francisco (1850 for precipitation and 1870 for temperature) and Des Moines, Iowa (1865). The longest single-site records in the United States are those from Central Park, New York (1869), and from Blue Hill Observatory in Massachusetts (1885), which are covered in more detail below.

Many less-continuous and multi-site early records are known for Canada, sufficient to prepare a near-unbroken daily temperature series for the St Lawrence valley in Quebec as far back as 1798 [63, 64]. The oldest continuous record still in existence from the Environment Canada historical data website https://climate.weather.gc.ca/historical_data/search_historic_data_e.html is for

Cornwall, Ontario, where the record is almost unbroken from 1887; the longest Canadian record on the WMO Centennial Stations listing is from Ottawa, dating from 1889 [39].

Central Park, New York – 1869 to date

40.779°N, 73.969°W, 40 m above sea level

The longest records made in the United States on (almost) the same site are those for Central Park in New York, where Dr Daniel Draper began keeping records in December 1868. Originally observations were made at the Arsenal Building on 5th Avenue (between 63rd and 64th Streets), but since January 1920 they have been made at Belvedere Castle, Transverse Road (near 79th and 81st Streets) [65]. The distance between the two sites is just less than 1500 metres (0.9 miles), although the record is generally regarded as being fairly homogeneous. The equipment has been automated since the late 1980s (**Figure 1.18**).

Figure 1.18 The Automated Surface Observing Site (ASOS) weather station in New York's Central Park (photograph by Simon Lee, Columbia University).

Blue Hill Meteorological Observatory, Massachusetts – 1885 to date

42.212°N, 71.114°W, 193 m above sea level

The Blue Hill Meteorological Observatory is located at the top of a scenic range of hills 15 km (10 miles) south-south-west of Boston, Massachusetts (**Figure 1.19**), and describes itself as the 'home of the longest climate record in the nation' [66]. The observatory was founded in 1885 by Abbott Lawrence Rotch as a private scientific centre for the study and measurement of the atmosphere and was the site of many pioneering weather experiments and discoveries: the earliest kite soundings of the atmosphere in North America (1890s) and early development of the radiosonde (1930s) both took place here. The first weather observations were made here on 1 February 1885. They have remained

Figure 1.19 The Blue Hill Meteorological Observatory, Massachusetts. Observations commenced here on 1 February 1885 and continue to this day. (Photograph by Mike Iacono, Blue Hill Observatory)

unbroken, and on the same plot, ever since, the most homogeneous climate record in North America.

Construction of the observatory was started by Rotch in 1884 using his own private funds. The original structure consisted of a two-storey circular tower and an adjoining housing unit; extensions were added in 1889 and 1902. Native stone, gathered from the summit of the Great Blue Hill, was used for the buildings, while copper sheathing was used for roofing. The original stone tower was demolished in 1908 and a new reinforced three-storey late Gothic Revival concrete tower, 6 m wide and 10 m high and with a crenelated top, was constructed in its place. The site was declared a National Historic Landmark in 1989.

The observatory retains barometers and other instrumentation dating from the late nineteenth century – these instruments remain in use alongside modern instrumentation, preserving the accuracy and integrity of the long record period. Blue Hill is also unique in North America in possessing a long sunshine record made with the Campbell–Stokes sunshine recorder (see Chapter 11). The National Oceanic and Atmospheric Administration (NOAA) also operates automated remote-reading climate monitoring equipment at the site. The observatory was declared a WMO Centennial Observing Station in its first global listing in May 2017 [39].

Since records commenced in February 1885, the highest recorded temperature has been 38.3 °C (101 °F) on 10 August 1949, equalled on 2 August 1975, and the lowest −29.4 °C (−21 °F) on 9 February 1934. During the Great New England Hurricane of 1938, the observatory recorded one of the highest wind gusts yet recorded by surface instruments anywhere in the world – 186 mph (83 m/s, 162 knots).

South America

The oldest observation series known for South America is that for Quinta Normal in Chile, where records commenced in 1857, followed by two sites in Argentina,

Santigo Del Estero Aero (1873) and San Luis Aero (1874) [39]. Further details of all other sites in South America with records commencing earlier than 1920 are available on the WMO Centennial Stations website [39].

Africa

The oldest continuous meteorological record in Africa is from Cape Agulhas in South Africa, where records commenced in 1855. Records began at two stations in Mauritius (Pamplemousses and Labourdonnais) in 1862, with another two sites (Beau Vallon Cour and Constance) dating from 1865. Observations on Tenerife, at the Santa Cruz de Tenerife site, also began in 1865. As with the sites in South America, further details of all other sites in Africa with records commencing earlier than 1920 are available on the WMO Centennial Stations website [39].

Asia

China has a wealth of documentary climate data, although instrumental records are scattered and short in length until the latter part of the nineteenth century [67, 68]. The earliest known measurements were made by Cunningham at Xiamen (Amoy) from October 1698 to January 1699, and a continuous record at Zhousan, Zhejiang Province between 1700 and 1702. A composite record from Beijing exists from 1724, including fragmentary records made there in 1743, followed by a longer series of air temperature, pressure, and rainfall made there by missionary Joseph Amiot between 1757 and 1762 [69]. The Russian observatory network also made measurements at their Consulate in Beijing from 1841. In India, records from Nungambakkam date from 1792; several other sites in India possess records dating back 150 years or more, including Mumbai (1841) and Thiruvananthapuram (1853). Two sites in Kazakhstan, Kazakhstan and Fort-Shevchenko, possess continuous records back to 1848. As with the sites in South America and Africa, further details of all other sites in Asia with records commencing earlier than 1920 are available on the WMO Centennial Stations website [39].

Hong Kong Observatory, China – 1884 to date

22.302°N, 114.174°E, 32 m above sea level

One of Asia's oldest and most complete meteorological records is that from the Hong Kong Observatory in Kowloon, where observations began in 1884 and continue on the same site today (**Figure 1.20**). The Hong Kong Observatory was amongst the first sites to be awarded WMO Centennial Station status in May 2017, proudly recorded on a plaque on the Observatory's front entrance. Today, in addition to maintaining the site's long meteorological record, the original historic colonial-era buildings house a modern forecasting centre for the South China Sea providing, amongst many other services, commercial aviation forecasts and a public forecasting service for the 7.6 million residents of Hong Kong and the Pearl River Delta (including typhoon warnings and research programmes).

Figure 1.20 Hong Kong Observatory. The thatched screen has been in use since 1884; conventional Stevenson screen temperature measurements are made at the King's Park site, not far from the main Observatory. (Photograph by the author)

Figure 1.21 Meteorological instruments at the historic Sydney Observatory site. (Photograph by the author)

Australia and New Zealand

Australia's first weather records were made at Dawes Point in Sydney Harbour, where observations of temperature, pressure, wind and cloud cover were made from September 1788 to December 1791, the first three years of British settlement in Australia. Other records survive for various sites in Sydney from the 1820s, prior to the start of observations in July 1858 at the historic Observatory Hill site (**Figure 1.21**) overlooking Sydney Harbour, where they continue today [70].

The oldest Australian records on the WMO Centennial Stations listing [39] are those for Mt Boninyong in western Victoria (745 m altitude), where records began in 1856, and Wooltana Station (600 km north of Adelaide in South Australia) and Yamba, a coastal town in northern New South Wales: both sites' records began in 1877. A single, homogenous daily temperature series for Adelaide (South Australia) has been published covering the period from 1859 to the present [71], while a composite series exists for Perth in Western Australia from 1830 [72].

New Zealand's oldest weather records are from Hokitika, South Island, where records began in 1865. A site at Lincoln Broadfield, a few kilometres south-west of Christchurch in South Island, possesses records from 1881.

Times of change ...

The way we measure weather has changed rapidly over the last two or three decades. As with any change, both opportunities and threats present themselves. Most meteorological observers brought up on mercury-in-glass thermometers and clock-driven autographic instruments with inky paper charts on clockwork drums have already retired. New sensors and measurement methods have evolved and are still evolving, some completely novel, and all offering improved ease of use, accuracy and cost-efficiency – although not always longevity. It should be remembered that many 'traditional' instruments, such as mercury thermometers or barometers, were brought to a high level of precision, reliability and ease of use over the course of almost 400 years, while the first modern prototype AWSs appeared only a few decades ago. As a result, 'sensor and instrument churn' is likely to be faster in the coming decades than it was over the last half-century. The retirement of mercury-based instruments mandated by the Minamata Convention [5] is already complete in many regions of the world: it is to be hoped that careful overlap periods using both 'old and new' methods of measurement were undertaken during the changeover to minimise risks to the homogeneity of the few genuinely long-period weather records we possess. Reliable and consistent long-period records are essential to provide an accurate assessment of past and present climate change, and to check the reliability of both past and future climate models. A carefully maintained weather record of 25 years is useful; one of 250 years many times more so. Only consistent long records can help answer questions such as 'How is our climate changing?' and 'Are extremes of climate becoming more frequent?'

Why are instrumental and observing standards necessary?

Standards are needed in order to be able to compare observations between sites. Only by minimising or eliminating measurement differences owing to varying exposures, instrumentation or observing methods can your own observations be directly comparable with those from the next village, or the next continent, or the last century.

So-called traditional methods of measuring weather elements evolved in Europe and North America during the latter nineteenth century and throughout the twentieth. Careful intercomparisons between instruments, exposures and observation methods determined which provided the best combinations of ease of use, cost, accuracy and suitability for the climatic regime. Such intercomparisons have continued with the new generation of digital meteorological sensors whose output comes via a datalogger, such that we can now be sure that solid-state pressure sensors, for example, can provide data to the same or better levels of accuracy than mercury barometers more safely, more cheaply, more easily and without requiring a human observer (see Chapter 7, *Measuring atmospheric pressure*). Not all comparative overlap records have been so clear-cut, however: the replacement of other sensors has been more problematic, particularly sunshine (Chapter 11, *Measuring sunshine and solar radiation*), where new electronic sensors sometimes give very different results from traditional instruments. Ill-thought-out replacement schemes have irreparably damaged the climatological continuity of many long sunshine records.

As a result of the need to maintain some form of practical standardisation for purposes of comparison with other observations, each of Chapters 5–11 in this book, describing how to measure a particular element of the weather, starts with a short description of the currently accepted standard method or methods of making observations of that element, based upon World Meteorological Organization standards and guidelines as set out in the WMO CIMO guide [4]. The instruments, siting and observing practices involved in doing so are described, followed by methods of siting AWS sensors and adopting measurement and observing methods to obtain as nearly as possible 'standard' results. For most elements, 'perfect' (and unchanging) sites are hard to come by, and compromises in exposure or instrumentation may be required. Some compromises are permissible where effects on the readings obtained will be relatively small, and where these are known allowances can sometimes be made. Other compromises may render the records made largely meaningless, and they should therefore be avoided. Each chapter attempts to outline the permissible and the impossible in this regard.

This approach will benefit both first-time purchasers of such systems, who may have little or no prior knowledge of where, why or how to site their instruments to provide observations that can become genuinely useful beyond purely local record-keeping, as well as those looking to expand existing weather measurements. Those looking to establish a new professional weather monitoring site from scratch will also find useful guidance on current best practices.

The future

It is very likely that many, if not most, of today's standard methods of measuring weather will continue to evolve over the coming decades as improved technology and lower-cost measurement methods continue to be introduced. Some countries have established 'fresh start' weather measuring sites, equipped from the outset with modern electronic instruments, on good sites that offer both excellent exposures and a reasonable expectation of record continuity for decades or even centuries to come [73, 74, 75]. Careful consideration today in choice and exposure of sensors, loggers and data archiving – all subjects covered in more detail in this book – will go a long way towards both 'future-proofing' and maximising the practical benefits of a well-sited AWS.

The first step is to list your specific requirements for weather measurements, then decide the best balance of budget and equipment to meet those requirements. How best to do this is set out in the following chapters.

Further Reading

Knowles Middleton's three excellent and very readable books on the history of meteorological instruments cover the entire spectrum of invention from the wildly impractical to the brilliantly simple. W.E. Knowles Middleton: *The history of the barometer* (1964), *A history of the thermometer and its uses in meteorology* (1966) and *Invention of the meteorological instruments* (1969) were all published by Johns Hopkins, Baltimore.

Invention has been out of print for two decades or more, but good second-hand copies surface occasionally in online second-hand booksellers such as Abebooks.com. *Thermometer* and *Barometer* have a new lease of life in print-on-demand editions, available at much lower prices than the original editions; AbeBooks is also a good place to start.

Ian Strangeways' *Measuring the natural environment* (Cambridge University Press, Second Edition, 2003) provides an excellent and readable account of most meteorological instruments, both 'traditional' and digital types, and very usefully picks up roughly where Middleton leaves off. For those seeking a greater theoretical understanding of meteorological instruments, Giles Harrison's *Meteorological measurements and instrumentation* (Wiley, 2014) provides a comprehensive and up-to-date introduction at undergraduate or masters module level.

2 Choosing a weather station

There are many different varieties of automatic weather stations (AWSs) available, and a huge range of different applications for them. To ensure any specific system satisfies any particular requirement, it is vital to consider carefully, in advance of purchase, what it needs to accomplish, and from there prioritise the features and benefits of suitable systems to choose the best solution from those available. The choices can be complex, and a number of important factors may not be immediately obvious to the first-time purchaser. Deciding a few months down the line that the unit purchased is unsuitable and difficult to use (or simply does not do what you want it to) is likely to prove an expensive mistake, as very few entry-level and budget systems can be upgraded or expanded.

This chapter suggests a structured way to do this:

- Decide what the primary use of the system will be;
- Review relevant decision-making factors as outlined in this chapter, and prioritise accordingly;
- Balance requirements against budget, identify potential suppliers and models; and only then –
- Purchase the most appropriate system.

Armed with a list of key requirements from this chapter, Chapter 3, *Buying a weather station*, provides a short guide to AWS products and services available in both North America and Europe.

Throughout this and the next chapter, the following loose definitions of AWS systems by budget level are suggested (see **Table 2.1**). Most systems fit comfortably within one of these price/performance bands: note that prices quoted are indicative only (correct at the time of going to press) and exclude local sales taxes, value added tax, delivery costs and optional sensors or fittings. Other models may be rebadged equivalents from the manufacturer or supplier shown. Particularly at the budget end, brands and products come and go quite frequently and the main players are likely to change during the lifetime of this book.

Note, however, that an AWS doesn't have to be the first rung on the weather measurement ladder. Short of funds? Not sure whether you'll keep the records going and don't want to spend a lot until you have tried it out for a few months? Not sure where to start? See Box, *Limited budget? Sheltered site?* in this chapter.

Important note

Throughout this book, suggestions and recommendations are completely independent of manufacturer or supplier influence. Details of the systems/sensors described in this book are almost entirely based upon the author's professional and personal experience of those items: no sponsorship, incentives, paid-for 'product placement' or special treatment were requested or offered by any of the companies whose products are referred to in this book.

If you use this book to help choose an automatic weather station or other weather-related equipment, please mention this to your reseller or dealer when you make your purchase.

Table 2.1 *Categories and main brands of automatic weather stations. Prices (in US dollars) are broadly approximate, correct at the time of going to press and exclude local sales taxes and shipping costs. Not all products will be available or supported in every geographical region.*

	Entry-level single-element	Entry-level AWS	Budget AWS	Mid-range AWS	Advanced and professional systems
Typical price range	$100 or less	$250 or less	$250 to about $750	$750 to about $2,500	$2,500 upwards
	← *Built to a price point* →			← *Built to a specification* →	
Typical brands	USB dataloggers such as Elitech TechnoLine La Crosse + *Numerous Chinese brands and products*	TechnoLine Ecowitt (Fine Offset) NetAtmo Bresser Ventus + *Numerous Chinese brands and products*	Ambient Weather AcuRite Davis Instruments NetAtmo Ventus ***Specialist products*** Nielsen-Kellerman /Kestrel, Gemini/ Tinytag	Davis Instruments Ambient Weather	Campbell Scientific Lambrecht meteo Environmental Measurements Ltd (EML) Met One Instruments Vaisala

Step 1: What will the system be used for?

All AWSs maintain a digital record of one or more weather elements (air temperature, rainfall, sunshine, wind speed and so on). Being clear from the outset what the essential system requirements are, and whether they are likely to change over time, will quickly help narrow the search for suitable products. This applies equally to both business and consumer purchases.

Many first-time purchasers rush into buying the first system that appears to satisfy their immediate requirements (and the available budget) without adequately considering future needs, only to regret the decision some weeks or months later when limited functionality, expandability or build quality results in frustration. It is better, of course, to be sure of what is needed – and what is not – at the outset to avoid subsequent disappointment. Writing off the initial system after only a short time to buy a more capable system will cost more (in both financial terms and duplicated installation effort) than if the desired system characteristics had been clearly thought through beforehand.

It is also important to regard money spent on the chosen system as a medium- to long-term investment. With careful consideration given to the robustness and longevity of system components and supplier reputation, with appropriate maintenance (and occasional inevitable sensor replacements) a lifetime of 10 or even 20 years for mid-range or professional systems is certainly achievable. Take this into account when making your decisions.

Typical uses for AWSs

Automatic weather stations are used for many different purposes; some of the main reasons for using them are given in the following list. These are not mutually exclusive – requirements for any particular application may include several of the following points – and neither is it an exhaustive list.

- Home/hobby weather interest – either starting from scratch, or expanding an existing set of manual instruments
- Augmentation or automation of existing weather monitoring equipment – at airfields or gliding clubs, or field centres, for example
- Replacement of paper-based autographic recording instruments
- Remote weather monitoring with distant-reading or website display facilities
- Absence and backup cover – unattended and/or more frequent observational records (weekend and holiday cover for schools, for example)
- Significant weather event logging
- Statutory requirement to maintain records of certain weather elements
- Replacement of human observer(s) to reduce costs
- Long-term climatological monitoring to 'official standards' (if the intention is to establish an official-standard site providing data to a regional or state weather service network, the views of the relevant network authority should be sought prior to purchasing equipment or establishing the site).

Consider which of these are most relevant to your requirements, and then review the decision criteria in Step 2 below. The relative importance of each factor will differ for each requirement, and of course you may wish to add others of your own.

Advantages of AWSs

Most modern AWSs will measure and log a number of weather elements reasonably well with minimum user intervention: even low-budget entry-level systems can provide worthwhile results, provided care and attention is paid to siting the instruments. All such systems provide a number of clear advantages, as follows:

Cost-effective deployment

The huge decrease in cost and improvements in accuracy in AWSs in the last three decades means that such systems can be relied upon to provide both more frequent observations and better spatial coverage.

These advantages are combined when AWSs are located in remote or inaccessible locations (such as hilltops or mountain sites) where regular human observations are impractical or impossible owing to distance from settlements, difficulty of access or frequent severe weather conditions. Similar advantages also apply to suburban gardens or backyards: the reduced cost and reasonable accuracy of most modern systems enables more people, of all age groups, to observe and measure the weather. These individuals and systems may group into formal or informal networks to enable the study of (for example) urban climates or severe storms in unprecedented detail and, increasingly, in real time. There is more on this aspect in Chapter 19, *Sharing your observations.*

Lower resource costs

The lower costs of AWSs, and their ability to run for long periods with little or no human intervention, also means that a valuable observation record can be obtained from sites where the cost of deploying human observers might otherwise rule this out.

There are many civil and military airfields, for example, where automated weather observation systems, run by the state weather service or the airport itself, provide 24×7 instrument-based weather observations, even though the site may not be staffed during weekends and on public holidays; indeed, these are probably now in the majority. Although automated systems cannot (yet!) provide the full range of observations possible with a human observer, the presence of such systems does enable many of these sites to offer a complete (365 days per year) climatological record covering at least the major elements such as temperature and precipitation. The same is true for amateur meteorologists and for school sites, where the deployment of an AWS can eliminate gaps in coverage caused by vacation periods during the year, removing what was often the largest single obstacle to maintaining an unbroken school observational record.

Improved sensors

The development of modern systems has led to a vast expansion in the range, accuracy and sensitivity of sensors. Many of these are completely novel, being based upon very different physical measurement characteristics than those of the instruments they replace. Of particular importance here are new instruments that have largely replaced mercury-based instruments such as thermometers and barometers, following the UN's Minamata Convention [5] and national and international legislation outlawing the manufacture and sale of mercury products. Many such replacement products are also better suited to modern digital acquisition systems.

Objective digital data

Output of objective, accurate and computer-ready digital data is a major advantage of AWSs, and of course observations are available as frequently as required. Manual observations can suffer from conscious or unconscious bias; some people subconsciously avoid certain decimal places when reading glass thermometers, for example, while some instruments show considerable 'interpretation' variations between observers.

A notorious example of the latter is the measurement and tabulation of sunshine cards from the Campbell–Stokes sunshine recorder, where the variation between human analysts can amount to 10 or 15 per cent, particularly on days of broken sunshine. With such variability, it can be difficult to be sure whether observed differences in sunshine duration between sites or over varying time periods are due to genuine climatic effects or simply observer/analysis variations. The introduction of new electronic sunshine sensors, while providing a clearly different determination of sunshine from that given by 'traditional' instruments, offers objective, consistent and repeatable measurements between sites (more in Chapter 11, *Measuring sunshine and solar radiation*).

The sheer torrent of data that can result from AWSs can present problems of its own, of course: Chapter 17, *Collecting and storing data*, offers suggestions and methods to optimise, analyse and archive the 'data avalanche' to ensure maximum benefit from the information generated. The investment of a little forethought in a data management strategy enables useful long-term climatological databases to be built quickly with little or no additional effort.

The installation of an AWS should not be thought of as providing the only source of weather observations: indeed, the author's experience is that an AWS is an important addition to, rather than a replacement for, existing visual observations of clouds, precipitation type, snow depth and the like. Chapter 14, *Non-instrumental weather observing*, goes into more detail on this important point.

'As good or better' record quality

All the preceding would be for nothing if the observational output from such systems were inferior to historical measurements made using 'conventional' or 'traditional' instruments and methods. Here the quality of record varies significantly by weather element. Some modern sensors, such as chip-based barometric pressure sensors, offer both health and safety advantages and improved digital output standards over conventional mercury barometers. Very few observing standards are still tied to traditional instrument benchmarks and methods, and increasingly digital sensors represent 'best practice' precision and accuracy. However, careful long-term planning is needed to minimise the impact on existing long-period records where major changes of instruments take place, such as replacing a manual observations programme with automated data capture.

Disadvantages of AWSs

Automatic weather stations clearly offer many advantages, but current or future users should bear in mind two major disadvantages, both of which involve the potential for significant and irretrievable data loss.

DATA LOSS OWING TO SYSTEM FAILURES In the case of conventional instruments, damaging or breaking one instrument would not have led to loss of the entire record; for instance, accidentally breaking a traditional mercury thermometer may have meant the loss of a couple of weeks of record until a replacement could be obtained, but it would not have affected the readings of other thermometers, or the raingauge, for example. With AWSs, however, a fault in a critical component, particularly the datalogger function, may lead to total data loss – not just for the period of the fault, or a single element, but possibly the entire record stored in the system memory. Sometimes the very occasions when the records are most useful – during an exceptional cold spell, or in a severe windstorm, for example – are when the systems and the sensors are operating at or beyond their design specifications, and are more likely to fail.

Such problems can also arise from the most mundane of causes. The author's experience to date has included the unannounced failure of internal batteries, electrical shorting caused by rabbits nibbling through cable insulation, voltage spikes from close lightning strikes, and numerous other similarly unpredictable events. For remote or unattended sites, even a 'back garden' weather station during a short period away from home, this can be a serious drawback, as often the loss may not be obvious until the next data download – by which time it is too late, of course. A frequent (at least daily and ideally hourly) automated download interval, wherever possible, should quickly highlight any actual or imminent problems and thereby minimise any loss of record.

DATA LOSS OWING TO SENSOR FAILURE Electronic systems involve physical components that still require checking and maintenance. Sensor failures on a well-specified system that has been carefully installed and regularly maintained should be infrequent, but obviously any permanent sensor failures will result in data loss. Regular sensor replacement appears to be a fact of life on some low-build-quality entry-level and budget systems. Particular attention needs to be paid to exterior connections on cabled systems to avoid ongoing data loss problems, which can be very difficult to trace, particularly where damaged or erratic connections may affect more than one sensor (see Box, *High-quality connections* in this chapter). A sensible precaution is to keep a small stock of key spare parts and sensors, particularly batteries, especially if they are less easy-to-find components or are non-standard. Doing so will reduce or eliminate record loss while waiting for replacement parts to arrive, for example, and will also extend the system life should a critical component fail after the manufacturer has discontinued the model (or the supplier has ceased trading), resulting in that component becoming unavailable. It may also be prudent to keep older equipment in place for a time where it is feasible or practical to do so – perhaps manual instruments that the AWS has replaced – both for 'parallel running' intercomparison checks and as a backup in case of system failure.

More typical and insidious temporary failures include blockage of the tipping-bucket raingauge funnel, often from bird droppings, or jamming of the tipping-bucket mechanism by insect action. Either can cause inaccurate or missing rainfall data for days or weeks, depending upon download and maintenance intervals. Best practice is to fit 'bird spikes' around the raingauge rim, and also to set up and log *two* tipping-bucket raingauges alongside a standard manual raingauge (more in Chapter 6, *Measuring precipitation*). The chance of both gauges being blocked simultaneously is very slight – other than during snowfalls – and the value closest to the manual raingauge total is more likely to be correct.

Step 2: Decision factors for AWSs

The extent to which any specific requirement is met will depend on a number of factors, the most common of which are specified in the following list. Each of these factors is briefly outlined within the following sections. As each system will have its own requirements, they are not arranged in any particular priority order. Which are most important to you?

- How good is the exposure where the instruments will be located?
- How many weather elements do you want to measure?
- Will all the sensors be exposed in one place, or will they be sited separately?
- Is there a requirement for backup system(s) and conventional instruments?
- Does the system need to be capable of being expanded over time?
- What sensors are required – 'standard' (built-in) or specialist sensors?
- Will it be cabled or wireless?
- Will it be computer-based or use only a dedicated logger or Internet connection?
- What degree of automation is sought?
- What degree of accuracy and precision is sought?
- How often should the information be updated?
- How robust does the system need to be? What is its desired or expected lifetime?
- Is the system 'mission-critical'?
- Is climatological continuity/compatibility/parallel running to World Meteorological Organization CIMO standards [4] a requirement?
- How important is ease of setup?
- What computing facilities and expertise are available?
- Will the system be mains (utility) powered or from renewables (solar, wind)?

Finally, of course, there is the question of budget. Although this may well be *the* deciding factor, to avoid the risk of potentially frustrating and expensive under-specification it is better to consider each of the decision factors in turn before making the final budget decision. More detail on the capabilities that can be expected from different budget ranges is given in the following chapter.

How good is the exposure where the AWS will be located?

Careful consideration should also be given to the suitability of the site where the measurements will be made. It is pointless spending hundreds or even thousands of dollars on a sophisticated and flexible AWS if the location where it will be used is poorly exposed to the weather it is attempting to measure.

In general terms, a budget AWS exposed in a good location will give more representative results than a poorly exposed top-of-the-range system. Worthwhile observations *can* be made with budget instruments in limited exposures, but a very sheltered site may not justify a significant investment in precision instruments, as the site characteristics may increase uncertainty and/or limit how representative the records are. More information on siting instruments is given in subsequent chapters.

How many weather elements do you want to measure?

Entry-level systems will typically include sensors for air temperature, and sometimes humidity too, barometric pressure, rainfall, and occasionally wind speed and direction. More advanced systems tend to offer both higher-quality sensors (improved accuracy, build quality and robustness) and an expanded set of sensors. Typically these might include additional temperature sensors for ground and earth temperatures, a solar radiation or sunshine sensor or a higher-quality radiation shield for measuring air temperatures (see Chapter 5, *Measuring the temperature of the air*), but at a higher price. Advanced systems permit a fully bespoke system to be built

Limited budget? Sheltered site?

Measuring the weather does not mandate a minimum spend of thousands of dollars on instruments, and neither does it require a plot the size of a small airfield to expose them properly. Observations made with robust, accurate and well-exposed weather instruments are of course an ideal to aim for, but if funds are short and a perfect spot in which to deploy them is simply not available, the records obtained can still be useful and interesting to make.

Most amateur meteorologists started out with one or two simple instruments (usually measuring air temperature and/or rainfall) and added more over time as budget and space allowed. Many started making their own weather records at school, sometimes influenced by a school 'weather club' or a memorable weather event – a heavy thunderstorm, a gale or a severe winter, perhaps. (Well before the advent of electronic instruments, the author began making his own observations at school, with a homemade raingauge, a Six's max-min thermometer and a bespoke design of thermometer screen made in school woodwork classes.) Moving from an apartment with a balcony to a house with a garden, perhaps eventually to one with a larger or less sheltered garden, often permits an improved exposure over time. The records made may not be fully comparable throughout, but they will be your own, and they quickly build up as the years roll by.

There are numerous instruments available at lower price points than entry-level AWSs. These days, wireless electronic temperature sensors, widely available for just a few dollars, will give passably accurate results when shielded from sunshine and precipitation, and making a raingauge is no more difficult than it was when I was at school. USB-based sensors to measure temperature and humidity with built-in dataloggers are readily and cheaply available from companies such as Elitech, and provided that careful thought is given to how and where they are exposed, the results can be surprisingly good. Nielsen-Kellerman's range of handheld Kestrel weather meters offer a range of measurement options at prices starting well under $100 at the time of going to press, from simple digital-display wind sensors to fairly sophisticated portable AWSs little larger than a mobile phone (see Chapter 3, *Buying a weather station*, for more details).

An observing strategy does not need to include measuring and logging every conceivable meteorological element at day one: it can and should expand over time as experience, resources and budget allow. It costs nothing to observe and record clouds, for example. The important thing is to give measuring the weather a try, and start keeping your own records.

with a wide range of additional sensors, to measure almost anything from cloud base height to snow depth, but at a price to match.

A comprehensive weather monitoring system will include sensors for air, ground or grass and earth temperatures, barometric pressure, rainfall, relative humidity (from which dew point can be derived), wind speed and direction, sunshine duration and solar radiation intensity. For any particular budget, a choice has to be made between the number of sensors and their quality, accuracy and robustness. Depending upon requirements and budget, it may be advisable to specify a few, high-quality sensors covering the key elements, at least to start with, rather than a wider range of cheaper sensors that may be of limited accuracy or poor build quality.

Will all the sensors be exposed in one place, or will they be sited separately?

For best results, the various sensors need to be sited separately – anemometers and sunshine recorders are best exposed well above ground level, while most national guidelines specify the raingauge should be installed at or close to ground level, for example. Take into account whether the various sensors can be separately deployed in this way when choosing your system. Most basic entry-level and budget systems include all or most of the sensors in one integrated 'all-in-one' instrument package, inevitably forcing compromises on instrument exposure. There are situations where a single integrated unit may be preferable or easier to install, of course, and where the sensor accuracy is 'good enough', in which case this design of system may be perfectly suited to the requirement.

Is there a requirement for backup system(s) or conventional instruments?

Although this book is primarily concerned with automated electronic weather-logging systems, there are two circumstances where conventional or 'traditional' instruments are likely to remain in use for some time to come at some sites. These are so-called 'standard measurements' and others that could be termed 'backup measurements'.

For some weather elements, today's current 'standard measurement' may still be defined in terms of 'conventional instruments' (more details are given in the following chapters, by element). If it is a requirement to provide truly standards-based or standards-traceable measurements, then for most elements the conventional instrument(s) must be deployed alongside the AWS system to provide these, even if the conventional instruments are read only occasionally (for example, at a weekly 'maintenance visit' observation) and used only to provide record continuity and backup, calibration checking or to identify any gross errors. This is mainly the case for rainfall measurements; tipping-bucket raingauges (TBRs) tend to be less reliable at providing consistently reliable rainfall totals, for reasons explained in Chapter 6, *Measuring precipitation.* A co-located standard raingauge is normally a requirement as a check on totals from the TBR.

Is there a need for a backup system?

The risk of total data loss from AWSs arising from the failure of a vital component or interruption of power supply is small, but when it does happen, the risk of losing the lot is considerable. Where high data availability is required, some form of 'backup' measurement system should be deployed alongside the AWS. During their lifetime, existing conventional instruments may usefully combine this role alongside that of providing 'standard system' calibration and error checks. A standard raingauge should *always* be deployed as a 'checkgauge' alongside an AWS, even at remote sites where the gauge may be read only occasionally. Period accumulations should be compared with the total from the same period derived from the TBR and any significant (> 5 per cent) discrepancies identified. Where the period rainfall accumulation is known, a total failure of the TBR will not result in a break in the climatological record, although obviously daily or sub-daily records may not be available for some or all of the period of defective record.

How reliable is 'reliable'?

Although modern systems are highly reliable when correctly installed and operated, no system, datalogger or sensor is ever 100 per cent reliable. An availability of 99 per cent sounds impressive, but in reality this corresponds to around 88 hours per year (or nearly 4 days) 'missing or defective record'. If the periods of missing record were randomly distributed, this would be enough to spoil rather than ruin a year's records: but if the breaks were not randomly distributed, for example records were lost every time heavy rain fell or when the temperature fell below a certain level, then this would introduce a statistical bias into the remaining records, invalidating any climatological analyses based on that station's data.

A realistic availability target is 99.9 per cent or better, or 9 hours or less 'missing or defective record' in any 12 month period. To attain or exceed this, a backup system, perhaps a smaller or lesser-specified system or one based on conventional instruments, should be considered to 'shadow' the main system and to provide records in any periods when the main system is out of order or undergoing maintenance or calibration. The second system should be completely independent from the main system and ideally should not share sensors, logger, cable runs or power supplies, so that the failure of one component will not degrade or bring down both systems. Any periods of missing or defective record from the main system can then be backfilled using the backup record. Periods of substitution should be indicated in the station metadata (see Chapter 16, *Metadata: what is it, and why is it important?*).

Battery backup An essential requirement for all systems is the ability to operate all vital components (specifically the datalogger memory function and most vital sensors, not necessarily the data display) on battery power in the event of interruption to the primary power supply – whether from mains (utility) power or from renewable sources such as a solar panel. Battery backup must provide for at least 24 hours operation without needing to be recharged or replaced, and for remote or unattended sites a week or more is advisable – it may not be feasible to reach a hilltop or remote upland site for several days after a heavy snowstorm, for example. Ironically, even professional networks are occasionally caught out by widespread power outages resulting from severe weather conditions. The good news is that the power consumption of modern electronic systems is so small that many entry-level and budget AWSs can be kept running for up to 24 hours with a small 9 v battery. Ensure, therefore, that battery backup is included in the manufacturer's specifications – and replace the backup battery/batteries at least every 12 months even if they have not been used in that period.

More sophisticated systems are likely to be battery-powered in any case, with utility or renewable power sources used to keep batteries topped-up via a suitable recharging system. It is important, however, to monitor battery condition regularly (the datalogger itself can usually do this) and replace batteries before they fail, because failures can be sudden and unpredictable. Sealed lead–acid batteries have a typical lifetime of 2–3 years with a daily recharge cycle, for example, but should be replaced at the first sign of reduced charge capacity, as this may be an indicator of imminent failure. Most systems also include a small button-cell lithium battery to retain memory for short periods when the main battery is disconnected; they also need to be checked regularly and replaced as soon as they begin to run down. Keep a spare handy.

Does the system need to be capable of being expanded over time?

To get the best out of any AWS investment, think carefully before purchase about how the system may need to change or grow over the next 5–10 years. The initial requirement may be simply to log air temperature and rainfall once per hour, for example: most budget systems will provide basic capabilities of this nature with ease, and low-frequency monitoring of this type will meet many needs perfectly satisfactorily. It is important to be aware, however, that most budget AWSs come with a fixed range of sensors which cannot be added to, nor in most cases can the sensors themselves be replaced with alternative (better) instruments. If, some time after installation, the initial specification expands to add more elements to those measured, or the replacement of an existing sensor with an improved one is required (for instance, a more accurate or faster response temperature sensor), it may be impossible to upgrade the system. Under such circumstances, the only option is probably complete system replacement. Not only is this expensive, but the de-installation of the original system and reinstallation of the higher-specification system may require considerable additional investment in time and resources.

For first-time AWS purchasers or those with very sheltered sites, a basic system to help decide what to measure, or to assess the site's suitability for weather measurements, may represent a sound investment, and future expansion may not be a prime consideration.

What sensors are required – 'standard' (built-in) or specialist sensors?

For most entry-level, budget and mid-range AWSs, the choice and selection of sensors is dictated by the manufacturer and little or no choice in specific instruments is possible.

To use more sophisticated, accurate or robust sensors, monitor more elements than are offered in a pre-configured 'all-in-one' package, or perhaps integrate existing instruments into a new AWS setup, the best advice is to consider first specifying a suitable datalogger, one that will handle the required number and type of sensor inputs (more on datalogger types and considerations in Chapter 13, *AWS data flows, display and storage*). Sensors appropriate to each required element can be identified, checking of course whether the logger and logger software will support them, and the configuration then built up item by item. Pre-sales support available from the manufacturers or resellers of more advanced systems can often provide guidelines or recommendations on supported configurations.

A 'datalogger + sensors' approach is also likely to be preferable where there is a requirement for the sensors to be located in two or more locations (air temperature and rainfall at ground level, say, with wind and sunshine sensors on a rooftop or mast some distance away), to avoid siting compromises necessary with entry-level and budget 'all-in-one' instrument packages.

Upgrade the raingauge!

Upgrading sensors to higher accuracy or specification is impossible on most 'packaged' systems, with one exception – the tipping-bucket raingauge (TBR) unit. Many entry- and budget-level systems include as standard a 1.0 mm/0.04 inch capacity 'tipping spoon' or tipping-bucket raingauge, which is much too coarse for accurate weather monitoring (see Chapter 6, *Measuring precipitation*).

Almost all TBRs generate a simple 'pulse' output, requiring only a straightforward two-wire connection. It is therefore usually very easy to connect in a higher-spec unit – 0.2 mm or 0.01 inch capacity is ideal – to replace the supplied model. Don't forget to adjust the calibration setting in the logger software to reflect the higher resolution sensor.

Will it be cabled or wireless?

Wireless systems are without doubt quicker and easier to set up, avoiding the need for trailing cables, wiring connectors and the like (**Figure 2.1**). For most systems of this type, exterior setup is merely a matter of siting the sensors appropriately and establishing communication with the 'base station', usually located indoors. The network link may be Wi-Fi, cell phone or low-power, high-frequency radio, depending upon system needs and distances involved. As ever, you get what you pay for: some budget systems are limited to 25–30 m line-of-sight reception, some manufacturers claim 300 m or more. Wireless repeaters are available for some AWS models, and these can extend the range to a kilometre or more.

Wireless: US versus European specifications

It is important to note that different wireless frequencies are used within US and European markets. North American specification wireless AWS products transmit at 915 MHz, whereas most European models use 433 MHz (some, such as the Davis Instruments Vantage Vue, use 868 MHz). In Europe, 915 MHz is reserved for mobile telephony, for various defence and national security applications and for the emergency services. Be warned – importing a US model AWS into Europe and thereby generating unauthorised transmissions or interference on this frequency may not be treated very sympathetically by the relevant authorities!

Quoted reception capabilities are usually those available under ideal conditions and without intervening obstructions between transmitter and receiver. Where conditions are less than ideal (multiple line-of-sight obstructions, very thick exterior walls or interference from other wireless systems on the same frequency, or sometimes the weather itself), the signal may become scrambled or drop out altogether, resulting in erroneous or missing data. Data from wireless systems can become unreliable in certain conditions, particularly where the transmitter and receiver are operating close to their maximum operating distance. For some systems this can be as little as 25 metres at best: a minimum Wi-Fi or wireless specification of 100 m or so is advisable, even in a domestic or suburban setting. Reductions in range, or more frequent data dropouts, can also occur if the transmitter battery is running down, and if not changed quickly total data loss may occur for an extended period. This will become a problem on a domestic system if it occurs during a two-week vacation away from home, for example, or during cold weather when battery life may be reduced significantly. (The best solution is always to check thoroughly all connections and batteries well before an expected absence of more than a few days; however, don't do this the day before departure, just in case the new battery fails very soon after it is brought into use.)

Some AWSs are available in either cabled or wireless configurations. Some wireless models update sensor readings less frequently than the equivalent cabled system, to preserve battery life; one wireless system on the market updates only every 90 seconds or so compared with about 10 times that frequency for its cabled equivalent. A high data rate is essential for some elements (particularly the accurate recording of wind gusts) but less so for others: check manufacturer's specifications closely, and see also the relevant chapter covering each parameter in turn for more on this point.

Cabled systems are a little more complicated to set up, in that the cable run needs to be securely and safely laid out, and robust weatherproof connections made. The number of connections should be kept to the absolute minimum to reduce the potential for wiring problems, but where the sensors are some distance from the logger a length of extension cable will normally be required over and above the length of cable supplied with the sensor or sensor package. The maximum cabling distance may be as little as 30 m for some systems, although more normally up to 100 m is supported. Check supported configurations with your supplier carefully before purchase if the distance between sensors and logger is more than a few tens of metres. Remember when measuring this to factor in the actual length of the cable run, not the line-of-sight distance.

Establishing a cabled system is likely to involve most, if not all, of the following activities:

- Setting up reliable and weatherproof cable connections, preferably using a suitably weatherproofed exterior junction box
- Preparing a weatherproofed exit gland for the cable itself from the nearest building
- Securing external cable runs against wind and weather (particularly instruments in exposed locations, such as anemometers and sunshine sensors)
- Preparing trenches for burying cables where they run across grass or soil (**Figure 2.1**) or arranging suitable conduit to prevent trip hazards or accidental damage from strimmers, lawnmowers, children and so on.

In some cases, it may be necessary to enclose all external cable in tough plastic or metal conduit to avoid risk from insect or vermin attack (squirrels and moles appear rather partial to cable insulation), from vandalism, ground settlement or vehicular access, or to satisfy health and safety requirements, particularly in schools or public-access areas.

Installing a cabled AWS certainly involves more setup work than wireless units, but has two advantages. The first is that, when done carefully, the installation work should not need to be repeated. (If possible, use wiring conduits to preserve access to the cable and any connections in the event of maintenance or replacement being

Figure 2.1 Installing a cabled AWS can involve considerable preparation work in digging trenches for cable runs. ... (Photograph by the author)

required: this will also greatly simplify the installation of additional cables should the system be expanded subsequently.) The second is that a cabled system with sound connections is both weather-independent (not liable to potential signal disruption in severe conditions) and powered directly from the computer or logger, thus avoiding the risk of data loss associated with battery failure on wireless transmitters. In addition, data sampling rates on cabled systems can be substantially higher than on wireless equivalents. Finally, some elements cannot be measured using wireless sensors (earth temperatures, for example) and some form of cabled connection will be required in these circumstances. Note also that some manufacturers offer both cabled and wireless versions of the same system; the cabled versions are usually slightly less expensive.

In all cases, *screened* cable should be used, particularly for long runs or where electrical or radio-frequency interference may be a problem to the milliamp or even microamp currents involved. Screened cable, as the name suggests, screens or shields the current-carrying cables with an outer sheath of braided wire mesh or similar, which is then earthed at both termination points. Electrical interference will manifest itself as 'noise' on the signal – a temperature sensor may change by a few tenths of a degree Celsius every few seconds, for example, or a raingauge may show spurious tip counts. Screened cable will normally eliminate such problems. The source of the electrical interference may be difficult to determine – close proximity to electrical mains wiring, domestic heating or air-conditioning installations and Wi-Fi 'mesh' computer network-ing can be troublesome in domestic situations – and may also be weather-dependent. Long runs of above-ground unscreened cable are particularly vulnerable to induced transient spikes caused by lightning, and these can play merry havoc with observations during electrical storms, sometimes introducing entirely spurious signals and throwing the reality of sections of the logged data into doubt. Screened cable is more expensive than standard cable, but it is essential for most installations – and certainly cheaper and easier than repeating an existing installation with screened cable at a later date.

The wire mesh jacket of screened cable also provides some degree of armour for the cable, and deterrence to insects and animals, although where there is danger from lawnmowers, strimmers and the like it is advisable to enclose the cable in tough plastic or aluminium conduit to provide protection.

High-quality connections are essential

The cause of many, if not most, AWS sensor dropouts or wildly incorrect readings can be traced to poor connections, whether this be poor Wi-Fi coverage or a flaky cable junction. In the latter case, pre-built or packaged systems with little or no cabling connections are less at risk, but as soon as a cable needs extending or a sensor has to be replaced, there is an increased risk of poor connections affecting output. On simpler systems, with only a few wiring strands to join, cables can be extended using terminal blocks or similar. Davis Instruments sell small 'crimp connectors' from 3M (Davis part no. 7960) which can be used for permanent and waterproof connection of fine wires or data cables; they are simple and easy to fit, although a ferrule connection is better for narrow-gauge wiring. These connectors are also available direct from 3M and their distribution partners as Scotchlok™ UY2 IDC connectors.

It is advisable to use a minimum size of cable strand in cabled systems. The cores in some multicore cable systems comprise cables consisting of just a few strands of very fine wire, little more than a human hair in diameter. These are very difficult to work with, to take solder and to guarantee secure connections. Thicker cables are more expensive but much easier to work with and ensure much more reliable connections. A sensible minimum is a cable strand diameter of 0.8 mm and cross-sectional area of 0.5 mm^2 or more, corresponding to American Wire Gauge (AWG) of 20 or less.

More complex wiring is best joined using a DIN rail and suitable connectors (**Figure 2.2a**), housed within a weatherproof external junction box as necessary, with cable entry and exits via sealed cable glands matched to the external cable diameter. Fine cables should be terminated within the DIN connector itself using a bootstrap ferrule or crimp connector (**Figure 2.2b**), available in various sizes to suit, providing a simple, low-cost and much more secure fixing method. *Avoid underground junction boxes at all costs* – no matter how well sealed, water will eventually seep in and cause short-circuits, possibly damaging components or electronics modules. Retrieving and repairing problems caused by waterlogged junction boxes costs much more in time and resources than the initial expense of a longer cable run of the requisite length. Enclosing underground cable runs in suitable protection against accidental damage from groundwork or moles is also cost-effective; lengths of 30 mm plastic pipe from a plumber's merchant are cheap and effective. Make a sketch map of where the cables run to assist in finding them again if any remedial work becomes necessary at a future date.

In all cases, consider future expansion. If the system will be expanded over time, with additional sensors for example, much future effort can be saved by building in duplicated, terminated wiring, redundant cable glands and spare DIN rail connections at the outset. And remember – the more cable junctions there are, the more points of failure exist and the more that need to be checked in event of flaky connections. Wherever possible, use a single cable run.

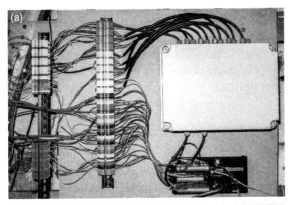

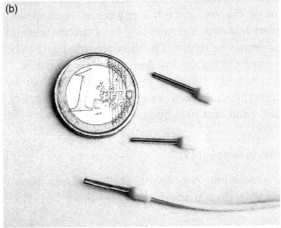

Figure 2.2 (a) DIN rail connectors and (b) ferrule or crimp connectors, useful for terminating fine cables and making the resulting connection much more robust. For scale, the 1 € coin is 23 mm in diameter. (Photograph by the author)

Will it be computer-based or use a dedicated logger?

Most weather monitoring requires availability 24 hours per day, 365 days per year. A dedicated datalogger provides standalone logging capability, a computer connection being required only occasionally for downloading data. With suitable battery backup, a datalogger provides independence both from mains power and from reliance on a dedicated computer connection. A dedicated datalogger, capable of being connected to and swiftly downloaded from a desktop or laptop computer, or sometimes from a handheld device, is much the easiest and most reliable method to ensure unbroken records from an AWS.

Most entry-level and budget AWSs do *not* include a dedicated battery-powered or battery-backed datalogger as part of the system, and may require a permanently connected host computer (desktop, laptop or equivalent device) to 'record' the data flowing from the sensors. Because the computer has to remain powered up at all times (and cannot be allowed to drop into a 'sleep' state at any time, including disks or USB ports, to avoid missing downloads), this can entail considerable power consumption. For true 24×7 capability, additional investment in backup power supplies (usually a battery-backed 'uninterruptable power supply' will suffice) to cover mains power failures is advisable, for of course failure of the Wi-Fi connection and/or PC will result in immediate cessation of record. Small, low-power-consumption PCs or iPad-type devices are cheap enough to consider their use as a 'dedicated' AWS PC, although their processing power may be insufficient for more advanced graphics or wireless Internet updates.

How much automation is required?

All but the most basic systems are capable of being left to run and record without attention for at least a few days, and for most 'domestic' installations where the logger is connected to, and downloaded regularly by, a directly connected computer (cabled or wireless connection) this degree of automation is adequate. For more remote locations such as isolated mountains or deserts, or even a city-centre rooftop site, where the site is not visited at least every few days, a greater degree of automation and a remote telecommunications capability may be required to allow recorded data to be downloaded and transmitted automatically at regular intervals to another location.

Transmission by telephone landline, via a mobile telephone network or by satellite are all options depending upon the equipment being used and your supplier, site access and availability of services – and of course budgets.

What degree of accuracy and precision is sought?

The terms 'accuracy' and 'precision' (sometimes 'resolution' in place of 'precision') are often used interchangeably, yet they do not mean the same thing (see Box, Precision versus accuracy).

The treatment of calibration and instrumental uncertainty is considered later (Chapter 15, *Calibration*), but for now it is sufficient to consider how accurate and how precise the AWS observations need to be. Many users will be content with being able to measure air temperatures within 2 degrees Celsius, or rainfall within 20 per cent of the 'standard' measurement methods, for example. This may be perfectly adequate for many purposes, particularly for new users, or if the exposure of the instruments is far from ideal and cannot be improved upon.

Where the requirement is to provide measurements conforming to standard practices such as those set out by the World Meteorological Organization or national meteorological or hydrological networks, thereby enabling accurate comparisons to be made with other sites or historical records, then tighter tolerances are called for. For such applications the correct exposure and siting of the instruments become as important as the absolute accuracy of the sensors themselves, and typical standards will be ±0.2 degC for air temperature sensors and ±2–5 per cent for rainfall, for example.

The choice of system involves making decisions on the level of accuracy required. As might be expected, higher accuracy generally comes at a price. This is not to say that entry-level systems cannot produce consistent results to a high level of accuracy, particularly if elementary calibration tests can be carried out at installation and at least annually thereafter (see Chapter 15, *Calibration*), but where high accuracy is a mandatory requirement, it is more likely to be achievable from an AWS within the medium or higher price ranges.

Precision versus accuracy

It is a very frequent mistake in our digital world to assume that just because a number is specified very precisely, the value quoted is accurate. This applies to all measurements, of course, but is particularly important in weather measurements.

As an example: let us say I have two digital clocks in front of me. One says the time is 10:23:46, the other says it is 10:19. The first is very precise, but is it accurate? Both clocks show different times, and clearly one must be wrong.

If I were to check the time using a third source of known accuracy, perhaps from an atomic clock via my smartphone, and find that at the time I observed the clocks, the exact time was 10:18:46, then it is apparent that although the first measurement is precise, it is not accurate. The second measurement is less precise, but it is more accurate.

Every measurement has an associated uncertainty. An AWS may display or log an air temperature of (say) 16.34 °C (a precision or resolution of two decimal places), but if the sensor is accurate only to ± 2 degC, then we can only say the temperature is somewhere between 14 and 18 °C. In this case, quoting the value even to a single decimal place is clearly unjustified. If the sensor was a more accurate one, with an uncertainty at that temperature of ± 0.2 degC (required for professional-quality sensors), then we can say with greater confidence that the temperature lies between 16.1 and 16.5 °C.

With careful sensor calibration, it is possible to reduce uncertainty further. Observations of air temperature are most often quoted to 0.1 degC resolution, although this precision is not always justified by the accuracy of the sensor itself (or indeed by the recency of its calibration).

Similar arguments apply to raingauges. The accuracy of tipping-bucket rain-gauges can vary enormously in heavy rainfall, yet one popular brand of AWS specifies the maximum rate of rainfall to a precision of 0.1 millimetres per hour in its display (an implied precision of better than 0.1 per cent above 100 mm/h): this for a system that probably cannot deliver better than 20–30 per cent accuracy in such circumstances.

When comparing uncertainty between two sensors, unless one or both are accurately calibrated, observed differences between the two may be entirely spurious. Consider two temperature sensors similarly exposed in different parts of a nursery garden, for example: one regularly indicates temperatures 2 degrees higher than the other. Does that mean the sensor that reads higher is located in a warmer part of the garden? Maybe – or perhaps the sensor simply reads 2 degrees too high. Perhaps both of the sensors are accurate only to ± 1 degree, in which case any difference up to 2 degrees may simply be due to a combination of instrumental uncertainty.

Much the same applies to all weather measurements, although in professional work uncertainty is assumed to be small (normally because the instruments have been regularly calibrated; see Chapter 15, *Calibration*), and consequently errors or uncertainty ranges are not usually quoted. Just because they are not quoted does not mean uncertainty does not exist, however, and quoting any particular uncorrected reading to a higher precision than its calibrated accuracy is unjustified.

How often will the information be updated?

For many weather elements, an update interval of a minute or even two is sufficient on most occasions (see *What is the difference between 'sampling interval' and 'logging interval'?*). For some elements, such as earth temperatures, once per hour (perhaps even once per day) is normally sufficient to record significant changes. For elements that change rapidly, particularly wind speed, a fast sampling interval is essential for comprehensive monitoring. Wind gusts, for instance, are most often considered as the 'highest 3 second running mean wind speed' [4]. It is clear that any system that updates on a time scale considerably longer than this – and some systems update only every 2–3 minutes – will not pick up the fine detail of the wind structure, and as a consequence reported gusts will be much lower than those from faster-response systems. If monitoring of changes in wind speed and direction in particular is an important consideration, as it will be in an airfield setting for example, then the sampling time of the system must be no more than a few seconds.

What is the difference between 'sampling interval' and 'logging interval'?

The *sampling interval* (sometimes called the update interval) is the length of time between readings of that particular sensor. It can vary from element to element – wind speed needs to be sampled much more frequently than earth temperatures, for example – but typically is between 1 second and 1 minute on most AWSs.

The *logging interval* is the frequency with which means or extremes of the sampled values are archived by the logging device – not every individual sample will be archived. The logging interval can be the same as the sampling interval, or a large multiple of it, but never less. For example, an AWS logging hourly means of wind speed might sample the anemometer every second: the hourly mean would therefore be the average of the 3,600 1 second samples. The same AWS might sample an earth thermometer just once per hour, and store that single sample. In this case the logging interval is the same but the sampling intervals are very different.

Entry-level AWSs may have a single, fixed logging interval: budget and mid-range systems will normally permit selection from a range from minutes to hours, depending upon the requirement, with 5–15 minutes being typical. More advanced dataloggers can log at more than one logging interval – perhaps storing data for a few rapidly changing elements every minute (such as air temperature, precipitation, wind speed and direction for example), while storing others that change less frequently every hour (such as earth temperatures, or hourly rainfall totals), in addition to providing a once-daily summary of means and extremes. Doing so greatly reduces the volume of memory and data storage required – there is very little point in frequently logging elements that change only slowly.

The chosen sampling and logging intervals should also, where possible, take into account the *response time* of individual sensors. The response time is the interval required for any given sensor to take up either 63 per cent or 95 per cent of a step change in input signal. There is more on the importance of sensor response time in **Appendix 1**, but for now it is sufficient to appreciate that sampling a sensor every second when its 63 per cent response time is (say) 60 s will inevitably mean that some samples will reflect only a partial adjustment to the change in signal, rather than the final value of that change, until the response time interval has been attained.

How robust does the system need to be? What is its desired or expected lifetime?

It is obvious that the weather instruments themselves, the means by which they are exposed (such as an anemometer mounting), and all the cable junctions and connections along the way need to be robust enough to stand up to (literally) the worst the weather can throw at them. There are few things in observational meteorology more frustrating than losing records of what may be a once-in-a-lifetime gale because an anemometer bracket has blown down, for example.

Choose the monitoring system with both the intended usage and expected lifetime in mind, together with the expected climatic conditions at the site. An AWS monitoring conditions in the south of Florida will clearly be exposed to very different conditions to one on a clifftop in the west of Ireland, or in northern Manitoba: less obviously an anemometer mounted at roof level even in a suburban environment will receive a much greater degree of 'weather stress' than a similar instrument mounted lower down in a sheltered garden setting.

The robustness of the system in the expected climatic conditions should be carefully considered prior to purchase. Comparing notes with existing users on Internet newsgroups is often a good way to do this – see Chapter 19, *Sharing your observations* for sources – as very few manufacturers give any useful measure of reliability (such as a 'mean time between failure'). The most exposed sensors, usually the anemometer and wind vane, are the most likely to fail early; even more expensive/professional sensors require replacement every 10 years or so – more frequently in windier locations. Other parts of the instrumentation and mechanics are also vulnerable. Plastic mountings and cable ties can become brittle and snap easily after just a few months exposure to the ultraviolet radiation in sunshine, bearings can freeze up, seize or jam, connectors can admit water and short out – the list is almost endless. One thing is certain – the weather *will* eventually win the corrosion battle!

A failure in one sensor over time may not be catastrophic in modular systems, where a replacement can be plugged in quite easily, but if replacements become required regularly the costs – and lost data – will soon mount up. It is also a fact of life that replacement parts on entry-level systems often become difficult or impossible to obtain after only a few years, as new systems are introduced regularly and older models (and their spares) are withdrawn from sale. Careful siting helps – avoid mounting instruments on a fence if that is the most likely structure to blow down in a gale, for example – but the more robust the sensor and its mountings, the longer the expected lifetime and the fewer corrosion-related incidents to be expected. Regular inspections and proactive maintenance will keep weathering-related problems to a minimum and will often provide early warning of potential failure points.

Over the longer term, experience to date strongly suggests, sadly, that the useful working lifetime of electronic AWSs is less than that of most traditional 'manual' instruments such as thermometers and raingauges. Liquid-in-glass thermometers, when used with care and their calibration checked from time to time, can provide reliable measurements for 50 years or more, although of course mercury legislation has now hastened their replacement: standard copper or steel raingauges will last at least as long with a little care and maintenance. Even wooden Stevenson screens should give 20–30 years' service, provided the woodwork is kept in good order. In contrast, the expected lifetime of current AWSs varies between no more than 12 months for some entry-level units and 15 or 20 years, perhaps more, for well-built mid-range and professional units, although some sensor replacements should be expected within this timeframe.

Not surprisingly, there is a relationship between build quality, longevity and price. Trying a budget package for a couple of years before deciding whether to move on to a more advanced system is a sensible approach: robustness may not then be the most important element in such a decision, but purchasing an inexpensive AWS and expecting it to last for many years, particularly in an exposed location or one subject to wide climatic extremes, is wishful thinking. It should also be borne in mind that, particularly at the budget end of the market, products are introduced and withdrawn very frequently: spares or support for such systems may suddenly become unavailable during the product's lifetime, or the manufacturer or supplier may themselves go out of business.

Is the system 'mission-critical'?

Many AWSs provide weather inputs to other systems – catchment flash-flood warning systems perhaps, or to provide continuous monitoring and display of wind direction and speed at an airfield as part of the air traffic control system. For mission-critical applications such as these, particularly where lives may be at risk, the availability of the weather-monitoring system itself is paramount. Robust measures to ensure availability may be mandated as part of the system specification, particularly in severe weather. In these situations a remote, duplicate and independently powered failsafe/backup capability may be necessary.

Is climatological continuity/compatibility/parallel running to 'official standards' a requirement?

Where compliance to current WMO or state weather service standards is mandatory, then the choice of both instruments and exposure must be made with this in mind. More details on how this is achieved are given in the following chapters.

Where an AWS is to be used to augment, automate or replace traditional measurement methods, it is essential carefully to overlap and compare both sets of instrumentation, for at least a full seasonal cycle (12 months) and preferably longer, to identify any systematic instrumentation differences which could otherwise damage or destroy the continuity of the record.

Where differences identified as a result of the overlap are climatologically small/insignificant (expert advice should be sought on this), a 12 month overlap period is likely to be sufficient. Where differences are large or highly variable, the overlap period for these elements should be extended. For long-period sites, those with a record of say, 50 years or more, an overlap period of 10 per cent of the record length is more appropriate. Where a site move is being considered in addition to an instrumentation change, the overlap should ideally cover both 'new' and 'old' instruments at the 'old' site together with the 'new' instruments at the 'new' site. The overlap period should again be at least 10 per cent of the existing record, and a minimum of 12 months.

How important is ease of setup?

Ease of setup and deployment is a powerful deciding factor for many purchasers, and here the pre-packaged hardware and software components of budget-level and mid-range systems have many advantages. One of the most obvious considerations is whether to specify 'wireless' or 'cabled' systems: both have their own advantages and disadvantages, as reviewed earlier.

Most software accompanying basic systems is easy to use and icon- or menu-driven. More advanced or customisable systems may require some programming ability, although most suppliers can provide customised/built-to-order packages covering installation, setup and programming. The extra costs can easily double the basic system price, however. More details are given in Chapter 13, *AWS software*.

What computing facilities and expertise are available?

Most systems can operate entirely standalone, but all have a finite memory capacity: once the memory is full, the earliest data stored will be overwritten in turn by later data. To make maximum use of the collected data, it normally has to be downloaded into a permanent storage medium before the memory becomes full. This can be undertaken with a direct computer connection (cabled or wireless), with a portable data collection unit (laptop, iPad/tablet computer, even some smartphones) or via a telecom connection (dedicated landline, mobile telephone transmitter or even direct to satellite in remote areas).

Most systems are easy to install and configure, but the more advanced systems may require some knowledge of datalogger programming, telecommunications protocols and so on. The suppliers of such systems normally offer an optional installation service covering sensors, logger and software installation, but the extra costs can be substantial, and the learning curve for self-teaching is often very steep. In schools or colleges and universities, in-house student or computing resources may be available, and may indeed make an attractive student project.

To display the output of any system in real-time (or near real-time) on the Internet, a dedicated weather station web page can be set up, or an auto-download of the logged data can be sent to a site which accepts inputs from multiple AWSs (see Chapter 19, *Sharing your observations*). For the non-technical user without a background in website programming, these operations are greatly simplified if a largely 'pre-configured' system is selected, because most of the required software is included in the package and little else is required beyond installing the sensors, hooking up to a suitable computer and running through a menu-driven configuration utility.

Will the system be mains-powered or solar-powered?

Most 'domestic' AWS systems are mains/utilitypowered, usually a low-voltage supply via a transformer, with battery backup sufficient to allow sensor input and logging to continue for at least 24 hours – the longer the better, to allow for possible extended power outages in severe weather conditions. Snowstorms, windstorms and major electrical storms can result in power spikes or lengthy power outages, even in urban or suburban areas, and must be allowed for even if they are very infrequent.

More sophisticated systems will normally be entirely battery-powered, usually from rechargeable high-capacity, long-life, sealed lead acid batteries recharged either by mains power, where this is available, or from solar cells or a small wind turbine. Sites remote from mains power need to be completely self-powered, usually using a combination of solar cells and wind power combined with lead–acid batteries, the battery capacity being sufficient to keep the system working for long periods in the event of prolonged periods of low solar radiation (or when snow covers the solar cells). Where power requirements are substantial, for telecommunications or sensor heating/de-icing for example, the manufacturer's advice on suitable power supplies should be sought prior to installation.

Step 3: Balance requirements against budget

The last item to consider is possibly the most important – namely, the available budget.

There are a few excellent basic systems for $100 or less which will measure, display and log just one or two elements to tolerable accuracy. One of these may be perfectly adequate for a first-time purchaser, or for a present for a friend or relation to 'dip a toe in the water' of measuring the weather. Other users require more sophisticated, capable, expandable and robust systems, which, depending on requirements, may cost ten or a hundred times that of an entry-level system. Not surprisingly, the more accurate, expandable, robust and flexible systems (with good post-sales support, should it be needed) tend to be more expensive.

For both private individuals and businesses, the money spent on a system is best viewed as a multi-year investment. Provided care is taken in exposure and siting, and with occasional maintenance, a mid-range system should last 10 or 20 years or more with little further outlay required beyond the initial purchase price. For professional users, budgeting for a capable and robust system which should give many years of trouble-free service will reduce future servicing and maintenance costs and minimise downtime. A WMO publication [76] provides a useful generic tender specification for professional AWSs.

One-minute summary – *Choosing a weather station*

- There are many different varieties of automatic weather stations (AWSs) available, and a huge range of different applications for them. To ensure any specific system satisfies any particular requirement, consider carefully, in advance of purchase, what are the main purposes for which it will be used, then consider and prioritise the features and benefits of suitable systems to choose the best solution from those available.
- The choices can be complex, and a number of important factors may not be immediately obvious to the first-time purchaser. Deciding a few months down the line that the unit purchased is unsuitable and difficult to use (or simply does not do what you want it to) is likely to prove expensive, as very few entry-level and budget systems can be upgraded or expanded.
- Decide firstly what the AWS will mainly be used for: some potential uses may not be immediately obvious. Once that is clear, review the relevant decision-making factors as outlined in this chapter, then prioritise them against your requirements.
- An AWS does *not* have to be the first rung on the weather measurement ladder. Short of funds? Not sure whether you'll keep the records going and don't want to spend a lot until you have given it a few months? Not sure where to start? Different options are covered in this and subsequent chapters.

- Consider firstly whether the site where the instruments will be used is suitable. There is little value in spending hundreds or thousands of dollars on a sophisticated and flexible AWS if the location where it will be used is poorly exposed to the weather it seeks to measure. In general, a budget AWS exposed in a good location will give more representative results than a poorly exposed top-of-the-range system. Worthwhile observations *can* be made with budget instruments in limited exposures, but a very sheltered site may not justify a significant investment in precision instruments, as the site characteristics may limit the accuracy and representativeness of the readings obtained.

- Carefully consider the key decision areas. Should the system be cabled, or wireless? Is it easy to set up and use? How many sensors are offered, and how accurate and reliable will they be? Are all the sensors mounted in one 'integrated' system, or can they be positioned separately for the optimum exposure in each case? Do the records obtained need to conform to WMO CIMO or national meteorological authority standards? Examples and suggestions are given in this chapter.

- Finally – and this should be the last step – match the available budget against the requirements and specifications outlined in previous steps. Consider that a reasonable mid-range or advanced system, when used with care and maintained, should last for 10 or even 20 years, and budget accordingly. There are many 'cheap and cheerful' systems available, but will they outlast their warranty period?

3 Buying a weather station

Up to this point, this book has largely treated AWSs as a single category. Of course, that is not the case and there are enormous differences in functionality and capability between basic and advanced models. The general rule that 'you get what you pay for' holds true for AWSs as well as for most other products, but in any price category some systems *are* better than others and it pays to check available products (and product reviews) carefully against the requirements outlined in the previous chapter to ensure the best fit for your needs.

The number, range and rate at which new models are introduced make it impossible for any printed work to provide up-to-date details or reviews of every AWS currently available on the market. This chapter outlines typical system specifications within broad capability and budget boundaries. When used with the prioritised assessments of functionality from the previous chapter, it should provide pointers to the main brands, products and suppliers.

What products are available?

The six broad product and budget categories shown in **Table 3.1** were introduced in the previous chapter. The cut-off feature for inclusion in the table is a means of *logging* data from one or more sensors to a personal computer (generally Windows or Mac-based, although some Linux versions exist), or via the computer to an Internet or cloud-based database, and the facility thereafter to *export* logged data to another application (such as a spreadsheet or database). Display-only systems are not included, nor build-your-own hobbyist electronic systems, such as those built around a Raspberry Pi or Arduino microprocessor, as the vast range of suppliers and components (and their limited shelf life) make generalisations and recommendations almost impossible.

For many first-time purchasers, it can be difficult aligning expectations of sensor and AWS performance and specifications with the products available within a particular budget, and thereby gaining a clearer understanding of 'value for money' at any price point. With this in mind, **Table 3.2** suggests realistic specifications for AWS systems within four very loose 'user profiles'. These are intended only as a pragmatic starting point to what is practical and affordable within various budget and site restraints rather than being overly prescriptive – for instance, with a limited budget it is probably better to concentrate on air temperature and rainfall observations, as wind speed and wind direction instruments are more complex and expensive to site and operate. Entry-level systems are particularly suited to those

Table 3.1 *Broad categories and typical brands of automatic weather stations. Prices (in US dollars) are approximate, correct at the time of going to press and exclude local sales taxes and shipping costs. Not all products will be available or supported in every geographical region. UK/European product prices for US brands are often marked up 20–50 per cent from the US dollar ticket price. (This table is identical to Table 2.1.) See also product positioning in Figure 3.1.*

	Entry-level single-element	Entry-level AWS	Budget AWS	Mid-range AWS	Advanced and professional systems
Typical price range	$100 or less	$250 or less	$250 to about $750	$750 to about $2,500	$2,500 upwards
Typical brands	←————————Built to a price point————————→			← Built to a specification →	
	USB dataloggers such as Elitech TechnoLine La Crosse *+ Numerous Chinese brands and products*	TechnoLine NetAtmo Bresser Ventus *+ Numerous Chinese brands and products*	Ambient Weather AcuRite Davis Instruments NetAtmo Ventus ***Specialist products*** Nielsen Kellerman /Kestrel, Gemini/ Tinytag	Davis Instruments Ambient Weather	Campbell Scientific Lambrecht meteo Environmental Measurements Ltd (EML) Met One Instruments Vaisala

Table 3.2 *Suggested realistic minimum specification levels for weather measuring equipment within broad 'user profiles'; see text for detail. Be aware of meaningless percentage specifications, such as 'Temperature ±2%'.*

	Starter	Hobbyist	Amateur	Professional
Suggested AWS category	Single element or Entry-level	Budget	Mid-range	Advanced
Air temperature	± 2 degC	± 1 degC	± 0.5 degC	± 0.2 degC
Rainfall	± 20%	± 10%	± 5%	± 2%
Wind speed	Estimate using Beaufort scale	Estimate using Beaufort scale, or use handheld instrument	± 10%	± 5%
Wind direction	Estimate using wind vane	Estimate using wind vane	± 10 degrees	± 5 degrees
Air pressure	± 2 hPa	± 1 hPa	± 0.5 hPa	± 0.2 hPa
Sunshine	–	–	± 10%	± 5%

looking to make a start in weather measurement, with the option of 'trading up' to more capable and accurate systems over time. (See also Box, *Weather instruments as gifts*.) In all cases, non-instrumental weather observing is a useful supplement to all categories and budget level (see Chapter 14, *Non-instrumental weather observing*), and does not depend upon budget.

'*Starter*' – suitable for budget-conscious buyers and first-time buyers: those who seek to measure only a single parameter (such as temperature or rainfall), for

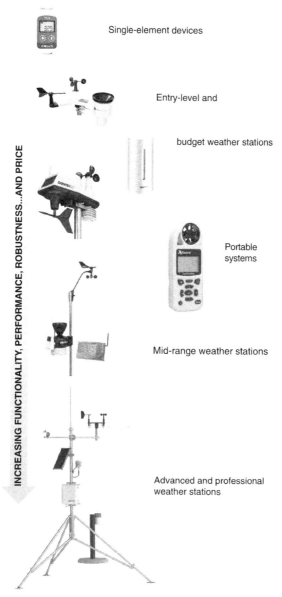

Single-element devices

Entry-level and

budget weather stations

Portable
systems

Mid-range weather stations

Advanced and professional
weather stations

Figure 3.1 Suggested visual product positioning hierarchy

those with a limited, non-standard or very sheltered site, or those for whom
accuracy and comparability with other observations is not the highest priority
'*Hobbyist*' – recommendations for those with a slightly higher budget, or with
sheltered sites
'*Amateur*' – mid-range systems, suitable for sites that range from slightly sheltered
to well-exposed, and for those looking to make medium- or long-term records
which will be comparable with other sites

'*Professional*' – systems suitable for serious amateur or professional long-term weather monitoring applications, where the proposed location of the instruments is reasonably well-exposed (more on site and exposure aspects in the following chapter) and where compliance to formal standards may be mandatory.

Product reviews

A number of the products covered in this book have been reviewed, by this author and others, and reviews and links can be found in **Appendix 3** and on www.measuringtheweather.net. Beware of online weather station reviews of consumer-level products on 'gadget' sites, many of which appear to have evaluated products over a few days (at best), rather than over at least a few months, possess little or no understanding of how to expose meteorological instruments, and rarely have access to reference instruments against which to assess accuracy and reliability.

Weather instruments as gifts

Weather instruments make ideal gifts for those who are interested in making their own observations – whether for a teenage child or grandchild, for a special anniversary, or as a leaving or retirement gift for a work colleague: there are many different possibilities. Traditionally such gifts might have included an aneroid barometer, or a display barograph: modern weather stations offer a much wider range of options and interest for a similar outlay. There are very few gifts that offer something different every day and provide sustained interest – perhaps even stimulate a career in meteorology or start a new retirement hobby. But what is the best choice when it comes to deciding what to buy?

When it comes to gifts, budget is probably the single most important criterion, followed by intended lifetime. A gift to stimulate or encourage interest in observations in a grandchild may serve its purpose (or not!) in a matter of months; a retirement gift should ideally last for many years.

For gifts up to about $150, a single-element system would be most appropriate – perhaps a small wireless raingauge or a logging wireless temperature sensor with display (the latter will also require some form of shading from the Sun, and to protect delicate electronics from the elements). Where the budget available is more substantial, a multi-element AWS would be an ideal choice. Generally, the larger the budget, the higher the quality, accuracy and lifetime of the system. This chapter provides more details on the capability of products within each budget category, advice on what to look for when choosing a system, and outlines some of the most popular products available.

Table 3.3 sets out specifications of *typical* products representative of each category as at the time of writing, but these can be expected to change during the lifetime of this book, particularly in the entry-level and budget categories. For this reason **Table 3.3** is best considered only as a snapshot overview of capability and specification within each category, and products referred to should not necessarily be taken as recommendations or even a complete list. Very often, some specifications will differ between US and European markets – for instance, wireless frequency, tipping-bucket raingauge capacities of 0.01 inch and 0.2 mm, respectively, date formats and so on. Note, though, that most permit easy selection of desired units, for instance °C or °F for temperature, inches of mercury or millibars (hPa) for barometric pressure and so on.

Where quoted, prices quoted are indicative only (correct at the time of going to press) and exclude local sales taxes/value added tax, delivery or installation costs and optional sensors or fittings except where shown.

Which product/s best suit my needs? Or 'Which is the best weather station?'

A Google search for 'best weather stations' will bring up a variety of listings for the most popular or best-reviewed weather station products within any country or region, but it is important to bear in mind that such listings and reviews are not necessarily independent (unlike this book), and that the range of products considered rarely extends outside consumer-level or budget price points. It is very difficult to ascertain reliable information on the relative popularity or sales of AWS hardware and software components because manufacturers are understandably reluctant to divulge such commercially sensitive information, but accordingly it is near-impossible to state confidently that such-and-such a brand is the market leader in any particular product category or geography. Over time, manufacturers gain or lose market share, and occasionally go out of business. Most products in these categories from American manufacturers are available in western European countries, often through country distributors rather than a local subsidiary office: sometimes, prices may reflect a considerable premium on the US dollar list price, and spares and support may be more difficult to obtain from minor players.

Best prices? Or best support?

Shopping around will often lead to better prices, but the lowest price may not represent the best value. For relatively complex and long-lasting products such as these, pre-sales and particularly post-sales support should be an important part of the purchasing decision. Is the reseller a registered dealer for that manufacturer, or are they selling 'grey' stock imported cheaply from fire-damaged stock in Asia? How long have they sold products from this manufacturer? Do they have access to the manufacturer's dealer support and technical helpdesks? Can they help with setup, installation and software questions? Do they stock spare parts and accessories? Most importantly, will they still be around if there's a problem under warranty?

In most cases, it is better to pay a little more for a product from a legitimate dealer who offers post-sales support, spares and warranty cover than to risk becoming stuck with a dead system and no support.

Table 3.3 *AWS specification table and typical products in each category. Prices are typical reseller levels correct at the time of going to press (excluding sales tax, delivery and optional extras) and will vary by country or region. Check latest specifications and prices for yourself before making purchase decisions.*

		Basic	Entry-level	Budget	Budget	Budget	Portable	Mid-range	Mid-range	Advanced
	Brand	Elitech	(Various)	Ambient Weather	Ambient Weather	Netatmo	Kestrel	Davis Instruments	Davis Instruments	Campbell Scientific
	Model	RC–5 RC–5+	Typical generic	WS–2902c	WS–5000	Smart Home Weather Station	Kestrel 5500	Vantage Vue	Vantage Pro2	Modular build to order
	Typical prices $US, £UK	$20 £20	Up to $299 / £200	$190 / £190	$300 / £515	$350 / £300 inc options	$350–420 £405–500	$460/ £350 inc console	$800–1300 £750–1200	> $2500 > £2500
Elements logged	Air temperature	Yes	Yes	Yes	Yes	Yes	Yes	Yes	Yes	Yes
	External humidity (RH)	No	Yes	Yes	Yes	Yes	Yes	Yes	Yes	Yes
	Barometric pressure	No	Yes	Yes	Yes	Yes	Yes	Yes	Yes	Yes
	Rainfall	No	Yes	Yes	Yes	Yes (option)	No	Tipping spoon	Tipping spoon	Tipping bucket
	Wind speed	No	Yes	Yes	Yes	Yes (option)	Yes	Yes	Yes	Yes
	Wind direction	No	Yes	Yes	Yes	Yes (option)	Yes (option)	Yes	Yes	Yes
	Solar radiation	No	No	No	Yes	No	No	No	Option	Yes
	Earth temperatures	No	No	No	No	No	No	No	Option	Yes
Resolution	Temperature resolution °C	0.1	0.1	0.1	0.1	0.1	0.1	0.1	0.1	< 0.1
	Temperature accuracy ±°C	0.5	±1–2	1.0	1.0	0.3	0.5	0.3	0.3	< 0.1
	RH resolution %	X	1	1	1	1	0.1	1	1	< 1
	RH accuracy ±%	X	±10	±5	±5	±3	±2	2	2	2–3
	Pressure resolution hPa	X	1	0.3	0.3	1	0.1	0.1	0.1	< 0.1
	Pressure accuracy hPa	X	±5	±2.7	±2.7	±1	±1.5	1.0	1.0	0.2

Table 3.3 (cont.)

		Basic	Entry-level	Budget	Budget	Budget	Portable	Mid-range	Mid-range	Advanced
	Model	RC-5	Typical generic	WS-2902c	WS-5000	Smart Home Weather Station	Kestrel 5500	Vantage Vue	Vantage Pro2	Modular build to order
	Brand	Elitech	(Various)	Ambient Weather	Ambient Weather	Netatmo	Kestrel	Davis Instruments	Davis Instruments	Campbell Scientific
	Rainfall accuracy %	X	±20%	±5%	±5%	±1 mm/h	X	±4%	±3%	±2% or better
	Wind speed resolution unit	X	1 mph	1.4 mph	1 mph	1 unit	0.1 m/s	1 mph	1 unit	0.01 m/s
	Wind speed accuracy %	X	±10–15%	±10%	±5%	±0.5 m/s	±3%	±5%	±5%	±2%
	Typical prices $US, £UK	$20 £20	Up to $299 / £200	$190 / £190	$300 / £515	$350 / £300 inc options	$350–420 £450–500	$460/ £350 inc console	$800–1300 £750–1200	>$2500 >£2500
Features	Can sensors be separated?	No	No	No	No	Yes, radio linked	No	No	Yes	Yes
	Radiation shield included?	No	Option	Yes (small)	Yes (small)	No	No	Yes (small)	Yes	Various
	Update interval wireless wind/ other	x	Typical 1 min	16 s	5 s	5 min fixed	1 s	Wind 2.5 s, others 10–60 s	Wind 2.5 s, others 10–60 s	< 1 s
	Stated wireless range (m)	x	25–50 m	30–100 m Wi-Fi 30 m	30–100 m Wi-Fi 30 m	Max 100 m	Max 30 m	300 m	300 m	Various
	Display/console included?	Yes	Yes	Yes	Yes	No	Yes	Yes (option)	Yes (option)	No (monitor)
	Logger included	Yes	No Internet only	No Internet only	No Internet only	No Internet only	Yes	No (option)	No (option)	Yes
	Logger capacity	32,000 points (at 10 min = 222 days)					10,000 datapoints (at 5 min = 35 days)	x	About 9 days	Months
	Software included	ElitechLog Win or Mac	Limited, basic	Proprietary	Proprietary	Via Android or Apple app	Yes - options	Console only	Console only	Yes (basic) Options
	Notes					Requires high-speed internet connection		Optional cloud storage	Optional cloud storage	Built to user specification

The following sections are based upon the six product categories suggested in **Table 3.1**. Each category is briefly described, and typical products and suppliers within that category summarised. All specifications are from manufacturer's literature via Internet searches, correct at the time of going to press. However, very few state details of how sensor precision or accuracy were determined, but where *independent* 'real-world' reviews of such products are known (particularly against standard instruments over a period of weeks to months), these are included as references for further reading. Note that product specifications are subject to change over time. Errors and omissions are excepted.

Basic/single element product category

At their most basic, weather monitoring systems may consist of a dedicated standalone datalogger component paired with one or two sensors, temperature and/or humidity sensors being the most common combination, and these devices come in various forms. Standalone USB loggers consist of a suitable sensor paired with a bare-bones datalogger integrated into a convenient USB memory stick format, and are bundled together with basic logging and retrieval software for use on a host computer (see also Chapter 3, *Buying a weather station*). Such devices are inexpensive, and small enough to be used almost anywhere: when placed within a suitable external enclosure or thermometer screen to shield them from sunshine and rainfall they can give surprisingly good results, particularly if the sensor is on a trailing lead rather than integrated into the logger body. Suppliers include Elitech, Lascar Electronics and others (details in **Appendix 3**). Alternatively, an exterior wireless sensor may be paired with an interior display unit which includes basic logging software.

Temperature-only units can take either form. Several manufacturers offer an integrated temperature sensor/logger units in a USB device similar in appearance and size (25 × 80 × 12 mm) to a memory stick. One such inexpensive device is the

Figure 3.2 Elitech 'memory-stick' RC-5 temperature datalogger, with display; this unit is 25 × 80 × 12 mm, and weighs just 25 g. (Photograph by the author)

Elitech RC-5 logger (**Figure 3.2**) available from www.elitechlog.com; a version of this with a trailing-lead thermistor is also available at little extra cost, and remarkably both come with a calibration conformity certificate as standard. These devices are uploaded simply by connecting to a USB-A port on a Microsoft Windows laptop or desktop computer. Such inexpensive devices (starting from $20 to $30) are more commonly found in cold-room stores or in industrial heating and ventilation applications than in meteorology, but when exposed correctly in a suitable radiation shield or thermometer screen (see Chapter 5, *Measuring the temperature of the air*), or even fixed to bicycle handlebars for an urban heat island transect, they can provide a reasonably good record. Some protection from the elements is advisable, as such products are best described as 'splashproof' rather than water-resistant. Some models may also include a humidity sensor in the same unit, and perhaps a small on-screen temperature display (**Figure 3.2**). Typical memory for such single-element loggers may be 30,000 datapoints, which at 5 minute logging intervals (288 data points per 24 hours) will comfortably suffice for three months or more, although the supplied logging/graphing/export software is often basic in the extreme. Inevitably, sensors are likely to be low-cost at this price point – potentially less accurate and subject to drift over time (particularly humidity), but sometimes surprisingly good (**Figure 3.3**). On the occasion shown in this example, which is not atypical, the output from the low-cost logger compared closely against a calibrated professional-grade platinum resistance sensor (PRT) when both sensors were exposed next to each other within a Stevenson screen (**Table 3.4**).

Other than sensor quality and possible calibration drift, the largest disadvantage of such sensors, at least in meteorological applications, is their relatively slow response time when the temperature changes rapidly, a condition known as 'thermal

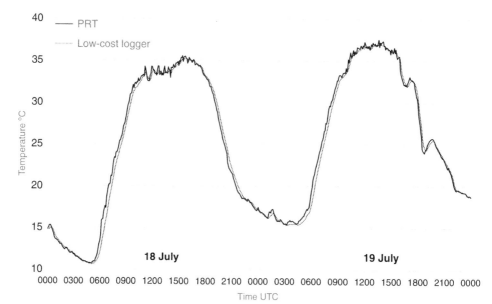

Figure 3.3 A 48 hour temperature record for two very hot July days, comparing a low-cost 'USB temperature logger' (the Elitech RC-5, dotted trace) against a calibrated professional-grade platinum resistance sensor (solid line), exposed next to each other within a Stevenson screen. Southern England.

Table 3.4 *Comparison of mean and extreme temperatures observed in Figure 3.3, with logged times of occurrence (in UTC). The logging interval was 5 minutes for both records.*

Date	Minimum temperature °C		Maximum temperature °C		Mean temperature 00–24 h UTC °C		Mean temp (max + min) /2, °C		Daily range degC	
	Elitech	PRT	Elitech	PRT	Elitech	PRT	Elitech	PRT	Elitech	PRT
18 July	10.6	10.7	35.3	35.7	23.78	24.21	22.9	23.2	24.7	25.0
	0505	*0500*	*1540*	*1530*						
19 July	15.4	15.4	37.0	37.6	26.13	26.25	26.2	26.5	21.6	22.2
	0315	*0310*	*1330*	*1330*						

inertia'. This is evident in **Figure 3.3**, where the low-cost logger can be seen to lag the PRT slightly. As a result, the relatively slow response of the low-cost logger tends to 'smear out' the fine detail of the temperature record, and observed maximum temperatures are slightly reduced, because the rate of short-period temperature changes is normally at its greatest around the time of the daily afternoon temperature maximum on sunny days such as these. Where the rate of change of temperature is reduced, the effects on the record of a slower response time become less evident. A sensor on a trailing lead would probably reduce or eliminate thermal inertia issues entirely.

Other temperature logger products consist of an exterior temperature sensor connected wirelessly to an interior display unit with integrated logger: sometimes a humidity sensor is also included, and such sensors are included in many multi-element budget AWSs. Provided the exterior sensor is adequately protected from sunshine and rainfall with an appropriate radiation shield or thermometer screen (see Chapter 5, *Measuring the temperature of the air*), such devices can provide surprisingly good results. Calibration should be checked carefully prior to use, however, as not all units are necessarily accurate 'out of the box' (see Chapter 15, *Calibration* for details). As with the 'memory stick' unit earlier, the unit's bulk will render it prone to 'thermal inertia lag' when temperatures change rapidly.

For those who seek only temperature logging, and who can provide suitable exterior protection for the sensor, devices such as these can be cost-effective and offer excellent value for money, provided a basic calibration check can be undertaken.

More capable and sophisticated dedicated loggers include the Tinytag range from Gemini Dataloggers (www.geminidataloggers.com, also **Appendix 3**). These include a selection of small, weatherproof temperature and humidity loggers which are ideal for offline meteorological monitoring. Most can be supplied with a sensor calibration certificate on request. While they make ideal temperature and RH loggers in their own right, or can be used as a backup to a larger system, they are perhaps most useful as portable calibration benchmarks to check the accuracy of other sensors (see also Chapter 15, *Calibration*). Their robust casing makes them suitable to check calibration of grass and earth thermometers as well as those within a screen, for example, but it should be noted that they are also 'offline' loggers in that most require physical downloading via USB to a host computer at regular intervals; they do not include real-time communications, nor a display showing current values.

With any standalone sensor, it is best wherever possible to opt for a 'plug-in' rather than a 'built-in' sensor, because the response time of a sensor embedded within the body of the logger itself will be adversely affected by thermal inertia and will be slower to respond as a result. A slow response will mean that the amplitude of measurements (maximum and minimum temperatures, for example) will be reduced, and rapid changes may be missed altogether – see **Appendix 1** for more on response times. For all such devices without a calibration certificate, it is unwise to assume accurate calibration 'out of the box', and accordingly calibration checks against another sensor of known calibration prior to use are strongly recommended.

Rainfall-only units usually consist of a tipping-spoon or TBR connected wirelessly to an interior display console. The capacity per 'tip' is usually between 0.5 and 1 mm (0.02 to 0.04 inches). Most of these devices are display-only – they may store recent values for the last 7–10 days, but usually there is no download/export capability, although an add-on logger can sometimes be wired in. It is important to choose a unit with an open round funnel, not a square or rectangular opening, as distorted airflow over non-circular gauges leads to unpredictable results which vary with wind direction. Avoid any that have any obstructions or grids within the funnel opening (other than insect or debris filters) as these can obstruct airflow and rainfall ingress; they will also catch 'fogdrip' in windborne fog and generate spurious precipitation records.

Given a reasonable exposure (see Chapter 6, *Measuring precipitation*), these little devices can provide a reasonable approximation of rainfall totals: tests by the author on one model (**Figure 3.4**) showed that over a 2 year period the gauge caught a fairly consistent 10 per cent less than an adjacent standard raingauge. However, the relatively coarse (0.5–1 mm) resolution precludes accurate recording of small daily amounts, mostly because minor falls are more likely to evaporate from the tipping-bucket mechanism than to be recorded. Like all tipping-bucket raingauges, they are next to useless in snowfall.

Figure 3.4 Entry-level wireless raingauge. (Photograph by the author)

Entry-level systems

The minimum entry for inclusion in this category is the ability to log two or more weather parameters (such as temperature and pressure) to a computer. Prices for entry-level AWSs vary from under $100 up to about $300, typically 'cheap and cheerful' consumer-boxed products offered by branded electronics stores, online resellers or other retailers. Many include some element of weather forecast options (algorithms normally based upon current values and trends in air pressure and wind direction), but these are not considered further here. Such systems do provide a good starting point for beginning weather monitoring activities, although limited accuracy, functionality, wireless range and durability are more likely at this price point. Almost all include some form of display console and elementary logging capability, but particularly at the lower end of the price range the logging capability is most likely to be on the host computer or remotely (a cloud-based system) rather than built into the system itself (see Box, *What does 'computer connection' really mean?*). Many products in this segment include garish colour displays, perhaps suggesting greater emphasis being placed by the manufacturer on presentation aspects over sensor quality and reliability. System software may be the manufacturer's own or a limited-functionality version of a more fully featured third-party package, and is often very basic: however, higher-quality third-party software is available for some popular AWS models, and upgrading the supplied software may make these products more suitable (or indeed usable). More details are given in Chapter 13, *AWS data flows, display and storage*.

What does 'computer connection' really mean?

Check the manufacturer's specifications carefully before purchasing, as some are (perhaps deliberately) rather vague on exactly what 'computer connection', 'computer link' or 'data recording' actually refers to. Some systems require 'always on' host computer, display console and/or Internet connection (not allowed to drop into 'sleep' mode), to receive and store regular incoming data, and there may be implications for electricity or metered Internet bills as a result.

Systems in the entry-level category typically offer the basic set of sensors (usually air temperature, rainfall, barometric pressure, sometimes humidity, wind speed and/or wind direction), built into a single 'all-in-one' exterior unit (**Table 3.3**). There are many entry-level systems, often originating from no-name consumer electronics companies in China. It is rare to find reliable manufacturer specifications for sensor accuracy in this price range, but experience shows these are often rather poor – typically ± 1–2 degC / ± 3–4 degF for temperature and ± 5 hPa for pressure, for example. Where one is supplied, the raingauge is typically a low-resolution (1 mm / 0.04 inch) tipping-spoon device, or a 'haptic' sensor which attempts to convert the noise of falling raindrops on a small plastic sheet into a rainfall intensity. Both devices are simply too coarse for reliable rainfall measurements.

Systems such as these are almost always limited to the manufacturer's sensors only, and expansion or customisation is rarely feasible. The sensors themselves should be replaceable in the event of malfunction, although this is not always the case: reports of frequent sensor failure on entry-level systems are common (beware also of high prices for replacement sensors). Sensor update intervals are also rather long, typically 1–2 minutes: this severely limits the high-frequency resolution of the records obtainable (particularly gust wind speeds), although depending upon requirements such relatively coarse resolutions may be within acceptable limits for other parameters.

Is it possible, therefore, for products within this category to provide any worthwhile weather measurements? The answer is definitely *yes* – at least for some parameters – provided a few basic principles are followed from the outset.

Firstly, site and exposure. It has already been stated that a basic AWS with a good exposure will often provide more representative results than a top-of-the-range model in a very sheltered position. There is more on site and exposure in the following chapters: paying careful attention to the detail of where the sensors will be located makes an enormous difference to both the quality of the records and, at least as important, how comparable they are with other records, whether they be observations made in the next city or records made 100 years ago. For sites where the exposure is very restricted, the accuracy of the temperature and rainfall records in particular may be limited more by the degree of shelter than by the sensors themselves. It is often difficult to find an ideal exposure for truly representative wind measurements, particularly in urban or suburban areas.

Secondly, it is vital to ensure that the temperature sensor is properly shielded from sunshine and rainfall. Very few entry-level systems include adequate protection for the temperature sensor – some provide none at all. Without such protection from both solar and terrestrial (infrared) radiation, indicated temperatures will be much too high on sunny days and too low on clear nights (and when it's raining). Contrary to popular belief, exposing temperature sensors on a shaded north wall will also give rise to significant errors and is far from ideal. If the supplied radiation shield is poor (or non-existent), allow additional budget for adding one to the system, taking care to ensure that the temperature sensor will fit into the option chosen. (Note that it may be less expensive to specify a higher-spec system which includes a better radiation shield.) Chapter 5, *Measuring the temperature of the air* gives additional details on the requirements for measuring air temperatures.

Finally – it is essential to check accuracy by carrying out a calibration check of at least the temperature and rainfall sensors prior to installation, and every few months thereafter. Basic calibration checks are not difficult to do: more details in Chapter 15, *Calibration*. With a little attention to detail in this regard, it is perfectly possible to reduce the likely 'out-of-the-box' errors significantly.

One of the biggest shortfalls in performance in entry-level wireless systems is in wireless range. A typical quoted wireless range spec for entry-level systems is 25–50 m, but based upon feedback from existing users these rarely seem attainable in practice. The transmission range of one entry-level system, user-reviewed on an online site, was described as 'very poor, typically 5 m or less' despite a specification of ten times that figure. Wireless range will decrease in poor weather, in cold spells and when the transmitter batteries are running down, and loss of record will inevitably follow – sometimes for long periods.

A final factor to consider is longevity. Obviously, and by definition, AWSs need to be able to withstand prolonged exposure to all types of weather. Entry-level systems are built to a price, and cannot be expected to be as robust as higher-specification units: regular replacements of sensors should be expected. It is unrealistic at this price point to expect spares to remain available for many years after purchase, and accordingly the availability of spare parts to keep the system operational may become a problem quite soon after the manufacturer has discontinued production, or replaced it with a newer model. Even with regular replacement of sensors, some exterior system components may not last more than a year or two. Online reviews abound of cheaper systems lasting no longer than the basic warranty period – sometimes not even that. If your supplier is merely an import agent for a manufacturer on the other side of the world, your warranty options may be limited.

Entry-level systems – summary

There are many situations where an entry-level system may perfectly meet the requirements. Provided their limitations in terms of accuracy, capability and lifetime are understood and accepted at the outset, and careful attention is paid to siting and exposure, such systems can represent reasonable value for money for a 'starter' weather monitoring system to explore basic home weather monitoring as a hobby, or those with limited budgets.

Budget systems

Budget systems are a step up from entry-level units, with typical prices starting at $200 or $250 and extending up to about $500 at the time of going to press (excluding local sales taxes, shipping and options). Taking the budget to this level begins to ensure a more capable, functional, accurate and robust system which will meet many user requirements for a basic weather monitoring system. Many of these devices are euphemistically referred to by their manufacturers as 'professional' systems, but don't be misled – none of these meet the requirements of professional weather monitoring systems.

Systems in this price range offer a similar range of sensors to entry-level products, typically comprising exterior temperature and humidity, rainfall, wind speed and wind direction and barometric pressure, all except perhaps the barometric pressure sensor being mounted in a single 'all-in-one' housing such as the Davis Instruments Vantage Vue model illustrated in **Figure 3.5**. Many will also display 'derived' measurements, calculated from the readings from two or more sensors, such as windchill (which is derived from temperature and wind speed readings). As with entry-level systems, such 'all-in-one' systems are convenient for installation (a single module to affix to a post or pole), but are significantly compromised as regards sensor exposure. What may be an ideal siting for one sensor, perhaps atop a 3 m or 10 ft pole for wind measurements, will be highly unsuitable for others – such as the raingauge, which is best located at or near ground level to minimise wind losses.

When compared to entry-level systems, sensors are generally of a slightly higher standard and with improved accuracy (remember that 'accuracy' and 'precision' or 'resolution' are not the same thing). Quoted specifications are typically ± 1 degC / ± 2 degF for temperature, ± 5% for humidity, ± 10 per cent for wind speed and ± 1 hPa for pressure. Brands in this category include Ambient Weather (ambientweather.com, a division of Nielsen-Kellerman who also make the Kestrel range of portable AWSs), Bresser (bresser.de/en/Weather-Time), Netatmo (netatmo.com) and Davis Instruments (davisinstruments.com), part of the AEM group: typical product specifications are listed in **Table 3.3**, correct at the

Figure 3.5 A typical 'all-in-one' sensor housing – the Davis Instruments Vantage Vue model. (Photograph by the author)

time of going to press, and suppliers are listed in **Appendix 3**. As with all other AWSs, local calibration checks are always advisable prior to installation and at six-monthly intervals thereafter. Only towards the higher price points of this category is anything other than a 1 mm / 0.04 inch capacity tipping-spoon or TBR included as standard; as previously suggested, a 0.2 mm / 0.01 inch unit is greatly preferable for climatological monitoring purposes and is a worthwhile add-on if budget permits. Some include a haptic rainfall sensor instead, one which estimates the rate of rainfall by the noise the falling droplets make on a sensitive plate; the author's experience of these is uniformly poor, and they are unlikely to provide reliable precipitation records [77]. Most raingauges are connected by a simple two-wire connection, however, and replacing with a higher-spec model from another manufacturer is usually straightforward (see Chapter 6, *Measuring precipitation*).

Most budget-level systems also include some form of radiation shield for the temperature sensor, although to judge by their visual appearance many of these are unlikely to be particularly effective (some are not even white!). If air temperature measurements with any claim to accuracy are sought, it is essential to budget for a higher-quality radiation shield, or consider higher-spec systems which have more effective units included as standard. NetAtmo systems include a curiously solid block of aluminium as a radiation shield for the external temperature sensor, which field evaluations have shown to give a very slow response to changes in temperature [78]. It should, in any case, be mounted away from solar radiation and precipitation, while remaining within wireless range of the base unit.

Models in this category are also usually limited to the manufacturer's own sensors, with little or no upgrade/expansion/customisation potential other than those. Some systems, including Netatmo, permit limited physical separation of wind and other sensors for better siting of the anemometer and wind vane, but some users may prefer the 'all-in-one' instrument package model used in the

Davis Vantage Vue AWS, which is also included in this price band. A detailed review of the Davis Vantage Vue system, carried out by the author over a 14 month period against professional-standard instrumentation, can be found on www.measuringtheweather.net [79].

Systems in this price range normally include an internal display console as standard, while some form of basic standalone datalogger function can be expected on at least some products in this category (see *Computer-based or separate logger?* in Chapter 2, *Choosing a weather station*), although on the Davis Instruments range the logger is an optional extra. It pays to check specifications very carefully, as some systems imply datalogging capability, which close reading reveals is local computer- or cloud storage–based rather than logger-based. A standalone logger removes the necessity for a permanently connected, always-on computer to collect data from the system; depending upon logging interval and memory, a separate datalogger will allow unattended operation for days or weeks.

As with entry-level systems, software included with budget units may be the manufacturer's own basic package, or a limited-functionality version of a more fully featured third-party component: higher-quality third-party software is available for many popular AWS models. Intended purchasers may wish to take note and allow budget to cover the upgrade costs, or choose an alternative system. More details on AWS data flows, display and storage are given in Chapter 13, *AWS software*.

Sampling intervals on budget systems are an improvement on entry-level systems, and are typically 1 minute or less. More frequent updates are essential to provide a good record of high-frequency elements, particularly gust wind speeds: Chapter 9, *Measuring wind speed and direction* includes more details on the importance of high-frequency measurements when measuring wind gusts. The Davis Instruments Vantage Vue model updates wind data every 2.5 seconds, for example.

Check what fittings are included

Check system specifications for what fixtures and fittings are included, as most AWS packaged systems will include only the basic bolts and brackets for fixing sensors to a 25–35 mm (1–1½ inch) diameter mast. Suitable lengths of sturdy steel tubing suitable for mounting anemometers can be obtained from most hardware stores (think TV aerial/antenna masts). An excellent UK website containing lots of advice on masts and fittings suitable for weather stations, and good prices too, is www.aerialsandtv.com. Builders' merchants can source heavier-duty lengths of galvanised scaffolding pole for more substantial installations, rooftop or gable-end masts or particularly exposed sites; if in doubt, seek professional installation.

More details on siting anemometers are given in Chapter 9, *Measuring wind speed and direction*.

The higher price point of this category also results in an improvement in wireless transmission range. Typical quoted specifications are around 100 m, although, as has been noted in the previous chapter, these are maximum rather than typical ranges, and they will drop sharply where there are obstructions, in heavy rainfall, or when the battery is near exhaustion. However, loss of record owing to wireless transmission dropouts tends to be much lower with budget systems than those at entry level. System longevity, too, is much improved compared with entry-level systems. Although some sensor failures can be expected during the system's lifetime, given typical exposure and occasional maintenance a useful working life of 5 years or more can be anticipated. Availability of spares and replacements may remain problematic for older systems, however.

Site and exposure remain essential ingredients to successful weather measurements, and the importance of both factors stressed earlier in this chapter applies to budget as well as entry-level systems. A good radiation shield is also essential for representative and comparable temperature measurements. As with cheaper systems, it is also essential to check the calibration of at least the temperature and rainfall sensors prior to installation, and every six months thereafter. Methods of doing this are set out in Chapter 15, *Calibration*. With a little attention to detail in this regard, it is possible to reduce the likely 'out-of-the-box' errors considerably.

Budget systems – summary

Budget AWSs comprising 'all-in-one' system units will meet the needs of many users looking for a system that has tolerable accuracy and covers a reasonably wide range of weather parameters. The siting and exposure of such 'all-in-one' systems typical of this category inevitably forces significant compromises in the siting/exposure of the individual sensors, but provided this limitation is understood such systems can provide tolerably accurate weather records over a number of years.

Mid-range systems

Stepping up a notch are the mid-range systems, where differentiation from budget systems arises from a more modular approach to sensor configuration. This not only enables appropriate exposure for different sensors according to their needs, but begins to offer integration options for third-party sensors and system expansion over time. Sensors, where supplied with the system, are generally both higher quality (improved precision and accuracy) and more durable. Inevitably, price points are also higher, typically from $500 or $750 upwards depending upon specification. From this point, specifications and price points begin to overlap with the advanced or professional category.

At this level, products can be relied upon to be reasonably accurate and robust, assuming of course a satisfactory exposure to the elements being measured can be found. A dedicated standalone logger and high-quality reliable software are often included as standard, but it pays to check this detail before purchase as some models opt instead for direct Internet connection with cloud-based logging and data storage instead of 'local' record management. (It is as well to note that Internet and cloud storage options usually incur additional ongoing costs, typically through an annual subscription, and of course depend upon a fixed and reliable Internet connection for unbroken records.) Mid-range AWSs meet many user requirements for a good weather monitoring system: provided the sensors are reasonably well-exposed, and the system is given occasional maintenance (including the odd sensor replacement as necessary), there is every reason to expect high-quality reliable and comparable observations for a decade or more.

The range of sensors offered on mid-range systems is similar to budget systems (typically – exterior temperature and humidity, rainfall, wind speed and direction and barometric pressure, together with 'derived' measurements such as windchill). Sensor accuracy approaches professional standards: typically ± 0.5 degC / ± 1 degF or better for temperature (assuming of course an effective radiation shield is in place), ± 3–4% for relative humidity, ± 0.5 hPa or better for barometric pressure and ± 5 per cent for wind speed: a 0.2 mm / 0.01 inch increment TBR is often a standard component. The mid-range portion of the AWS marketplace has been dominated by Davis Instruments (now part of the AEM group of companies) for many years. At the time of writing, the Davis Instruments Vantage Pro2 AWS was the bestselling product in this category, although both system and software are showing their age (the model was launched in late 2004). System specifications can be found at www.davisinstruments.com/pages/vantage-pro2.

The radiation shield included as standard on this model (**Figure 3.6**) is of reasonable quality and efficiency, and consequently temperature measurements from it are broadly comparable to those obtained in Stevenson screens (see Chapter 5, *Measuring the temperature of the air* and reference [80] for more details). Some expandability is possible – the Davis Instruments Vantage Pro2 unit can be expanded to include earth temperature and solar radiation measurements, for example – although apart from support for one third-party anemometer, at the time of writing the choice of sensors remains restricted to the manufacturer's units. On the Davis Instruments Vantage Pro2 system, the radiation shield and raingauge are combined into one unit, which is far from ideal from an exposure perspective

Figure 3.6 Exterior sensors of a Davis Instruments Vantage Pro2 AWS. (Photograph by the author)

(it is worth asking whether your reseller can supply the radiation shield and raingauge separately, to special order): more usefully, however, the anemometer/ wind vane unit can be separated to optimise exposure, if necessary using an optional separate wireless transmitter to locate the wind sensors in a more suitable elevated spot some distance from the main unit.

The Davis Instruments Vantage Pro2 system includes an interior display console as standard (**Figure 3.7**); data collection via the company's website is the preferred route, although a standalone datalogger is optional at extra cost. A 'local' datalogger avoids the necessity for a permanently connected, always-on computer and Internet connection to collect data from the system. Depending upon logging interval, a separate datalogger permits unattended operation for days or weeks.[*] Software included is of a considerably higher standard than budget-level systems, and as the brand leader, Davis Instruments systems are most likely to be supported by third-party software options where available (see Chapter 13, *AWS software*).

Figure 3.7 The Davis Instruments interior display dashboard, colour model 6313. (Photograph by John Dann, weatherstations.co.uk)

Sampling intervals are a step up from budget systems and come much closer to that required to provide a good record of high-frequency elements, particularly gust wind speeds: the Davis Instruments Vantage Pro2 model updates wind data every 2.5 seconds, for example.

The higher price point on this model delivers a substantial improvement in wireless transmission range. Typical quoted specifications are around 300 m, although, as has been noted in the previous chapter, these are maximum rather than typical ranges, and they will reduce sharply where there are obstructions, or in poor weather conditions. Except where such systems operate at or close to the maximum supported wireless range, accurate measurements show less than 0.25 per cent loss of record owing to wireless transmission dropouts. System longevity, too, shows a marked improvement on systems at lower price points. Of course, some sensor failures can be expected during the system lifetime, but given typical

---

[*] Davis Instruments legacy WeatherLink software builds a local database on the host computer, which usefully expands over time and allows text-based export to spreadsheet and database systems. However, the database capacity is limited to 25 years, after which no new entries can be added unless and until earlier years are deleted to make room.

exposure and occasional maintenance a useful working life of at least a decade can be anticipated, and probably longer. (The author's first Davis Instruments AWS lasted 17 years before being retired, although more recent products have been less reliable, the plastic anemometer sub-system in particular lasting only a few years before requiring replacement.) Availability of spares and replacements is usually good, albeit often pricey relative to the price of a new system. Purchasing spares for key components and sensors, even if not immediately needed, will avoid loss of record while spares are shipped, and can also extend system life in the event of the model and spares becoming obsolete, or a manufacturer going out of business.

A viable alternative to pre-configured systems such as the Davis Instruments model described previously is to consider a system based around a multi-input datalogger of the type sold by Campbell Scientific (www.campbellsci.com), Lambrecht meteo (lambrecht.net/en), Onset (www.onsetcomp.com) and various other specialist companies, then add sensors to suit from a wide range of third-party suppliers. This approach is considered in more detail in the 'professional systems' category later in this chapter, but it is mentioned at this point because opting for a reputable datalogger as the system core with perhaps only two or three sensors at the outset, subsequently expanding the range of sensors as requirements grow, is likely to prove more cost-effective over a 5–10 year timescale.

Site and exposure remain essential ingredients to successful weather measurements, and the importance of both factors stressed earlier in this chapter applies to every system. As with other models, it is essential to check calibration of at least the temperature and rainfall sensors prior to installation, and every six months or so thereafter. Methods of doing this are set out in Chapter 15, *Calibration*. With a little attention to detail in this regard, it should be possible to reduce most 'out-of-the-box' errors to near-professional levels.

Mid-range systems – summary

Mid-range AWSs, whether pre-configured systems or one built around a combination of core datalogger with a few third-party sensors, will meet the needs of many users looking for a system that has acceptably good accuracy across a wide range of weather parameters. Provided careful attention is paid to siting/exposure and calibration, such systems can be expected to provide reliable and accurate weather records over a decade or more. A typical mid-range AWS costing three times as much as an entry-level or 'all-in-one' budget system is likely to provide higher-quality records and to outlast its cheaper rival in a similar ratio. Viewed over a typical 10 year period, mid-range systems therefore represent much better value for money.

Portable weather stations

Before moving on to the 'advanced' or professional AWS category, it is worth side-stepping to consider portable instruments. These are less often used for 'routine' measurements at fixed sites but are invaluable for particular applications, specifically portable calibration reference units and portable AWSs. Handheld anemometers are widely and cheaply available, reasonably accurate (site limitations are usually the larger source of error in wind speeds) and are ideal for spot wind measurements where budget or site considerations do not permit permanent installations, or for fieldwork.

Portable calibration reference units

A portable reference unit is one which can be accurately and professionally calibrated in a laboratory, and then used on-site to check the calibration of other instruments by being exposed alongside for a period of parallel-running measurements. This is most frequently and conveniently undertaken with temperature sensors, and a process for doing this is given in Chapter 15, *Calibration*.

One such suitable product is one of the Tinytag logger range from Gemini Dataloggers (Chichester, West Sussex, UK – www.geminidataloggers.com). Tinytag dataloggers are self-contained, rugged and reliable battery-operated electronic devices for monitoring environmental parameters. Records are quickly and easily transferred to a computer using a fast USB connection, or wireless data loggers are also available. Tinytag temperature (and humidity) loggers can be supplied with a three-point calibration certificate, making them ideal for checking other sensors. (Members of the UK's Climatological Observers Link – see **Appendix 3** for contact details – can borrow one of these units to check the temperature calibration of their own equipment over a few weeks for a nominal fee, plus postage and packing.)

Temperature and/or humidity can be easily and accurately monitored using one of these units, with or without a display. They are small, light, easy to use and very reliable. Tinytag loggers are available with various thermistor options: a thermistor on a short lead will give better results than a logger with a built-in sensor, because the thermal inertia of the logger body slows their response time. Logging times are software-selectable from seconds to days, and their memory capacity is sufficient to run for typically 4–6 weeks between downloads. Temperature sensors will of course require protection from solar radiation and rainfall in order to provide accurate air temperature measurements. If a screen/radiation shelter is available, the Tinytag sensor should be placed in the screen close to the temperature sensor whose calibration is being checked.

Portable AWSs

Portable AWSs are small, light and entirely self-contained. They are therefore particularly useful for fieldwork and field courses, for walkers, for outdoor sports enthusiasts including rowers, glider pilots and the like, together with others who require current on-the-spot wind and weather conditions with the convenience of a handheld unit.

The Kestrel range of handheld weather meters, manufactured and sold by Nielsen-Kellerman (www.nkhome.com) of Boothwyn, Pennsylvania, encompass a surprising number of AWS options in a small device. The top-of-the-range model Kestrel 5500 (**Figure 3.8**) is about the size of a mobile phone and runs on a single AA battery for months. It will even float if dropped into water. This model can measure and log air temperature, barometric pressure, relative humidity and wind speed (current, average and gusts). With the optional wind vane housing, wind direction can also be recorded, although of course if the unit is exposed outdoors by day, incident sunshine and rainfall will bias the unit's temperature and humidity readings. Derived values are available for pressure trend, altitude, heat stress index, dew point, wet bulb temperature, density altitude and wind chill: other models are available to support different user requirements. A built-in datalogger records up to

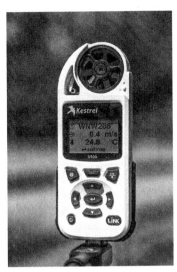

Figure 3.8 The Nielsen-Kellerman Kestrel 5500 portable AWS. (Photograph by the author)

10,000 data points, and at 5 minute resolution the unit can record for about a month before the memory becomes full. Logged data can be inspected or graphed on-screen or downloaded to computer using the optional dongle, or more easily via a Bluetooth link to a smartphone on supported models. Logging intervals from 2 seconds to 12 hours are available. Sensor responsiveness, accuracy and repeatability are reasonably good, particularly where local calibration checks against fixed instruments can be made beforehand: temperature within 0.5 degC / 1 degF and pressure within 0.5 hPa (millibar) is easily attainable in steady-state conditions. Owing to the relative bulk of the unit, however, thermal inertia can be a problem when temperature or humidity changes rapidly: the transition from a warm room or coat pocket to a cold outside environment can take 10–20 minutes for the unit to settle to the ambient conditions, and this must be allowed for in use. A review of the Nielsen-Kellerman Kestrel 5500 can be found on www.measuringtheweather.net, and supplier details in **Appendix 3**.

Portable systems – summary

Portable systems are particularly useful for calibration checks, for fieldwork or for a variety of outdoor users. With a suitable choice of logging interval they can be used for short-term logging or backup system at permanent sites, although battery life and memory limitations mean they are not best suited for permanent installation.

Advanced or professional systems

The definition adopted here for the advanced or professional category is a system consisting of a core multi-function, multi-input datalogger, a range of suitable meteorological sensors from the manufacturer or third parties, and a bespoke logger operating system to configure and run the combined system. Both individual components (datalogger, sensors and software), pre-packaged systems and bespoke configurations are available from numerous manufacturers, resellers and system integrators; leading brands in this category include Campbell

Scientific (www.campbellsci.com), Lambrecht meteo (lambrecht.net/en), Environmental Measurements Ltd (www.emltd.net), Met One Instruments (metone.com/meteorology), Vaisala (www.vaisala.com/en) and others. Such organisations typically supply a range of customers, from individuals or small businesses to regional or national hydrological or meteorological agencies. Pre- and post-sales technical support for configuration options and installation and maintenance services are also part of most contracts. Prices for such systems reflect capability, quality and professional support, but used loggers can sometimes be found on Internet market sites such as eBay for a fraction of the list price. In addition, price need not be a barrier for an individual content to purchase products from a reliable brand and build a system over time by adding sensors as budgets and expertise allow. The author purchased his first Campbell Scientific logger back in 2001, and grew from there; that logger is still in everyday use, although since supplemented by other faster and more expandable loggers from the same stable.

As might be expected from professional-quality products, systems in this category are amongst the best in their field, and the choice of sensors can be highly customised for particular applications. Typical installations can be found not only at observatories, universities, airports and wind energy sites throughout the world, but also in numerous 'serious amateur' sites – many of which have upgraded from initial 'consumer-level' packaged systems. Provided site and exposure requirements are satisfied and regular calibration checks undertaken, such systems can be relied upon to provide accurate, reliable and high-quality weather measurements over many years, for almost all applications, whether in remote or inaccessible locations or in city centres, even in the most hostile of climates (**Figure 1.1** and **Figure 3.9**).

Figure 3.9 Two British Antarctic Survey staff checking and maintaining the automatic weather station at Limbert AWS, 'Site 8' in the British Antarctic Territory at 75.73°S, 58.71°W, 34 m above sea level. (Photograph by Jo Cole and courtesy of Eloïse Chambers, British Antarctic Survey)

Professional sensors need to be of a high standard, not only in terms of accuracy but robust enough to withstand extreme operational environments and climates in remote locations where site visits are likely to be infrequent. Typical professional-level accuracy levels include ± 0.2 degC / ± 0.5 degF or better for temperature, ± 2–3% for relative humidity, ± 0.2 hPa or better for barometric pressure and ± 1 per cent for wind speed. Sensor calibrations traceable to national or international standards are usually available on request. Air temperature measurements are made within Stevenson-type screens, smaller multi-element AWS screens or, increasingly, aspirated screens (see Chapter 5, *Measuring the temperature of the air* for details).

The range of measurements possible is limited only by the availability of sensors: if there is a suitable sensor for that parameter that can be read by the datalogger at the heart of the system, then measurements can be made. A wide variety of additional meteorological parameters can be logged beyond the capability of 'packaged' AWSs – visibility, cloud base, CO_2 concentrations, turbulence measurements, snow depths, atmospheric pollution, lightning detection and warning, rainfall acidity, atmospheric electric fields and present weather are just a few examples. When specifying multi-element systems, care should be taken to ensure the datalogger has both the physical capacity (number of available inputs) and the on-board processing power to manage all of the required or anticipated sensor channels, together with sufficient on-board memory to store days or weeks of data between downloads if necessary. A useful WMO tender guide, regularly updated, is also available [76]. Dataloggers are covered in more detail in Chapter 13, *AWS data flows, display and storage.*

System software is of a high professional standard, and such systems offer almost infinite flexibility for customisation (although sometimes at a cost of a steep learning curve in datalogger coding competency). Sampling and logging intervals can be separately specified, and sampling rates as high as 200 kHz are supported. Although no meteorological variables would require a sampling rate anywhere near this figure, sub-second sampling may be advantageous in some wind and microclimate work, particularly turbulence studies where 10 Hz or 20 Hz is more typical. Some systems are utility powered (with battery backup), but most are run from batteries kept charged using solar cells and/or wind turbines. Communications can be wireless (cellular mobile or bespoke radio links, with ranges to suit most requirements), or even directly to satellite in remote areas or at sea.

As with every other class of weather measurement, the importance of adequate site and exposure remain unchanged. It is worth repeating that a budget-level system in a well-exposed location will normally provide superior measurements to a more expensive AWS which has been poorly sited. No amount of expenditure on advanced or professional systems will bring the required results without paying careful attention to site and exposure requirements: these are considered in more detail in the following chapter.

Advanced systems – summary

Advanced AWSs tend to be custom-built to a specific requirement, whether for the serious amateur or professional installation, and are capable of almost unlimited expansion. Systems in this price range are accurate, robust and capable of measuring a very wide range of elements, but at a price to match. Provided site and exposure requirements are satisfied and regular calibration checks undertaken, such systems can be relied upon to provide accurate, reliable and high-quality weather measurements over many years, for almost all applications and locations, even in the most remote areas or hostile climates.

One-minute summary – *Buying a weather station*

- There are enormous differences in functionality and capability between basic and advanced models. The general rule that 'you get what you pay for' holds true for AWSs as well as most other products, but some systems *are* better than others and it pays to check available products carefully against your requirements to ensure the best fit.

- To simplify selection, this chapter suggests five product and budget categories. Most systems fit within one of these price/performance bands – entry-level systems (single-element, or AWS), budget AWS, mid-range AWS, portable systems, and advanced or professional systems.

- *Entry-level systems.* There are many situations where an entry-level system may perfectly meet the requirements. Provided their limitations in terms of accuracy, capability and lifetime are understood and accepted at the outset, and careful attention is paid to siting and exposure, such systems can represent reasonable value for money for a 'starter' weather monitoring system, or those with limited budgets.

- *Budget AWSs* will meet the needs of many users looking for a system that has tolerable accuracy and covers a reasonably wide range of weather parameters. As with entry-level systems, provided careful attention is paid to siting/exposure and calibration, such 'all-in-one' systems can provide reasonably accurate weather records over a number of years. Some represent good value for money at those price points, while others are best avoided.

- *Mid-range AWSs*, whether pre-configured systems or one built around a combination of core datalogger with a few third-party sensors, will meet the needs of many users looking for a system that has acceptably good accuracy across a wide range of weather parameters. Provided careful attention is paid to siting/exposure and calibration, such systems can be expected to provide reliable and accurate weather records over a decade or more. A typical mid-range AWS costing three times as much as an entry-level or 'all-in-one' budget system is likely to provide higher-quality records and to outlast its cheaper rival in a similar ratio. Viewed over a typical 10 year period, mid-range systems therefore represent much better value for money.

- *Portable systems* are particularly useful for calibration checks, for fieldwork or for a variety of outdoor users. With a suitable choice of logging interval they can be used for short-term logging or backup system at permanent sites, although battery life and memory limitations mean they are not best suited for permanent installation.

- *Advanced AWSs* tend to be custom-built to a specific requirement, whether for the serious amateur or professional installation, and are capable of almost unlimited expansion. Systems in this price range are accurate, robust and capable of measuring a very wide range of elements, but at a price to match. Provided site and exposure requirements are satisfied and regular calibration checks undertaken, such systems can be relied upon to provide accurate, reliable and high-quality weather measurements over many years, for almost all applications and locations, even in the most remote areas or hostile climates.

- AWS specifications are suggested within four very loose 'user profiles' – Starter, Hobbyist, Amateur and Professional – intended as a pragmatic starting point to what is practical and affordable within various budget and site restraints. As an example, with a limited budget it is probably better to concentrate on air temperature and rainfall observations: wind speed and direction (for instance) are more complex and the site requirements more challenging. These and other elements can probably follow at a later stage as budgets (and perhaps an improved site) allow.

PART II

MEASURING THE WEATHER

The following chapters provide brief descriptions of how each weather element is measured. The emphasis is mainly on modern electronic sensors and approaches, although 'traditional' or legacy methods are summarised where appropriate. Following a short explanation of what is meant by 'site' and 'exposure', and a brief summary of exposure details by element, each chapter thereafter follows a similar structure for each variable in turn. International guidelines from the World Meteorological Organization (WMO) on siting and exposure are summarised, followed as necessary by country-specific details for the United States, the United Kingdom and the Republic of Ireland. A book of this size cannot hope to include detailed observational practice for every observed element covering every country in the world, but by setting out the appropriate WMO recommendations followed by references to individual national or state weather services it is hoped that information regarding any variations in observing practice for other countries or regions can be quickly and easily identified.

Guidelines for choosing a representative exposure for each instrument type are given in turn, and methods to ensure compatibility with existing national or international standards and sensors are also suggested. Of course, it is not always possible to follow WMO guidelines in every detail, particularly where site and/or exposure may be limited, and wherever possible tips for obtaining optimum results under such circumstances are set out.

A brief summary of each chapter is given at the end of that chapter; for those looking for a quick overview of each element, this short section ('One-minute summary') summarises briefly the main points covered. For convenience, these short summaries are collected together in Chapter 20, Summary and getting started.

An understanding of how mechanical and electrical instruments function and respond is always beneficial in getting the most out of any measurement system. More technical details on instrumental theory and methods are given in Appendix 1 *and more specialised material listed as references for further reading as required.*

4   Site and exposure: The basics

There's an oft-repeated saying among real-estate agents: that the three most import-ant factors when it comes to property are *location, location, location*. When it comes to setting up instruments to measure the weather, the refrain could be similar: *exposure, exposure, exposure*.

It is certainly true that a well-exposed budget AWS will give more representa-tive and reliable statistics than a poorly exposed top-of-the-range AWS costing as much as a small car. However, a garden the size of New York's Central Park is not a prerequisite to making worthwhile weather observations, because by taking care when siting your sensors and following the advice in this chapter, good results can be obtained from all but the most sheltered locations.

Firstly, what is meant by **site** and **exposure**? The two terms are often used synonymously, but in this book *site* is normally used to refer to 'the area or enclosure where the instruments are exposed', while *exposure* refers to 'the manner in which the sensor or sensor housing is exposed to the weather element it is measuring'.

Exposure to what?

Self-evidently, sensors to 'measure the weather' need to be located where they are exposed to the elements. It is not immediately obvious to those venturing into weather measurement for the first time that the ideal exposure for one sensor can be very far from ideal for another. For example, a World Meteorological Organization (WMO) Class 1 anemometer exposure is one where the sensor is mounted on a 10 m (33 ft) mast and located such that no obstacles higher than 4 m are located within 30 times their height; so a two-storey building, 9 m high, would need to be at least 270 m distant (more on wind instrument exposures in Chapter 9, *Measuring wind speed and direction*). However, siting a raingauge on the mast adjacent to that anemometer would represent a very poor exposure for that instru-ment, because stronger winds and increased aerodynamic turbulence at height would blow more of the rain over and around the raingauge, rather than allow it to fall into the funnel and be measured. A gauge mounted in such a location typically receives a greatly reduced catch when compared to a standard exposure near ground level: the same goes for raingauges mounted on rooftops. The effects are more pronounced with stronger winds, and so are more apparent in windier locations, in winter compared to summer, and on wet days that are also windy.

The following chapters set out preferred site and exposure characteristics for each of the major weather elements in turn, based upon WMO published guidance (the so-called CIMO guide [4], which will be referred to frequently). The single most important reason for setting out standard siting, exposure and observing practices guidelines is to minimise or eliminate instrumental or process differences caused by non-standard practices which might otherwise appear to be genuine climatic variations. Adopting such guidelines, or as near as possible, greatly increases the likelihood that observations made at one site will be directly comparable to those made at another – regardless of whether that site is located 10 or 10,000 kilometres away, or whether the records were made an hour ago or a century ago.

Before considering the needs of each element in turn, however, some general remarks about the siting of instruments can be established.

The ideal location for sensors to measure **air temperature** and **rainfall (precipitation)** is near ground-level on flat ground, or no more than gently sloping terrain, well away from hedges, buildings, trees and other obstructions. The instruments should be mounted above short grass (in areas where grass grows, otherwise natural, low vegetation representative of the locality) and well clear of buildings, areas of tarmac, concrete paving and other artificial surfaces. Sites close to significant expanses of water are also best avoided, unless this is a natural feature such as a lake or coastline. As far as possible, the site chosen should be typical of its locality (whether city-centre, suburban, rural, coastal, mountaintop or whatever). That way the readings obtained are most likely to be representative of the area in which the instruments are located, and accordingly more useful and comparable to other sites, than those which exhibit purely local effects. Many 'official' weather observing sites are today located at airfields which are often a long way from the centre of the nearest town or city whence they derive their nominal location: but which provides the more representative picture of the city's climate – the windswept airfield outside the built-up area, or a carefully sited AWS in a suburban garden?

For **sunshine**, generally the higher above ground level the sensor is located, the better, because horizon obstructions reduce the amount of sunshine recorded by the instrument and make comparison of records with other locations more difficult or even impossible. (Instruments exposed at height also tend to suffer less from dew or frost deposits than those at ground level; such deposits can obstruct low-angle sunshine on many types of sensor.) As stated earlier, **wind** instruments also benefit from being exposed at height, to reduce the frictional effects of buildings, trees and other surface obstructions which affect surface wind measurements (both speed and direction) very considerably. Wind speed is probably the most difficult of all elements to measure reliably in a sheltered suburban park or domestic setting. Rooftop sites are generally not ideal for wind measurements as they can be affected by considerable turbulence, which will itself result in some distortion to measurements, but often such sites may represent the only viable opportunity to obtain wind readings.

Particularly with mast- or rooftop-based measurements, **safety considerations for installation and maintenance are paramount** (see Box, *Important safety considerations for installing and maintaining weather instruments*).

If *wireless sensors* are to be used, ensure the distance to the receiver is no more than about half of the manufacturer's maximum separation distance (reception conditions often deteriorate in poor weather). There should not be any significant 'line of sight' obstructions, such as walls or buildings, that may attenuate or block the signal. For *cabled sensors*, check beforehand that the entire cable length required is supported by the AWS or datalogger and interfacing software: some systems, and some sensor types, will simply not function correctly with long cable runs. Check also that the cabling can be safely installed without incurring safety risks. For example, unsecured cabling should not obstruct a walkway, be strung close to head height in an area where it may not be easily visible or where it may become entangled with other wiring. This applies particularly in public areas and in schools.

An hour's careful site survey prior to installation will be time well spent and may avoid the laborious task of subsequently relocating instruments to a more suitable location at a later date. A good compass and tape measure are essential, and a clinometer is also useful. Note the areas which have best exposure to sunshine, wind and rain; they may be in different locations at the chosen site. The prevailing wind direction in temperate latitudes is between south and west, and in these climate zones a location open to winds from this quarter, but not too windy, is a good start. (The second-most common wind direction in temperate latitudes is generally between north and east, so if possible try to optimise exposure in those directions too.) Other climate zones should optimise exposure with a view to prevailing winds. Avoid positions close to buildings, close to or under trees, or near solid fences or dense or tall hedging. Take care to avoid locations that might be suitable now but may soon become over-sheltered owing to hedge or tree growth in only a few years – rapidly growing trees or hedges to the south will interrupt a sunshine record, reduce wind speeds, affect maximum temperatures and substantially reduce measured rainfall totals within a very few seasons. If the proposed site may be subject to unwelcome visitors, whether curious small children, domestic pets, wild animals or vandals, you may need to consider some form of site access restriction (see *Access restrictions?*).

Important safety considerations for installing and maintaining weather instruments

Installing and running ground-based weather instrumentation should present few health and safety concerns, provided trailing cables and the like are carefully secured. Risks increase when elevated or rooftop sites are used. Modern electronic instruments offer many advantages over conventional instruments, not least that small, low-power sensors (such as those used to measure sunshine, wind speed and direction and solar radiation) can more easily be exposed on a mast or rooftop to provide a better exposure than at or near ground level. With little or no maintenance required, they can be left in place for months or even years. Predecessor instruments which required manual chart records to be changed daily (the Campbell–Stokes sunshine recorder, for example) required safe working access on a daily basis, and this limited the number of sites where the instrument could be deployed.

Rooftops or masts may provide much better exposure for sunshine and wind sensors, amongst others (although they should be used only as a last resort for measurements of temperature and rainfall), but carefully consider the accessibility of the site before attempting to install the sensors. If you have no head for heights, are not comfortable on long ladders, or are in any way unsure whether your do-it-yourself (DIY) skills are up to the job, **DON'T TAKE RISKS**. TV aerial fitting contractors or local builders will often quote to undertake the work required when the requirement is made clear. With appropriate equipment and experience, a job that might be a very risky undertaking for a DIY installation will probably be 'all in a day's work' to a specialist.

Installing equipment in a position where it is difficult to gain safe and easy regular access makes it absolutely essential to ensure the equipment and all its connections are set up and tested thoroughly (including logging requirements) over a period of at least a couple of days prior to installation. Finding out that your new wireless anemometer needs a dipswitch setting changed the day after a builder has installed it on your roof is likely to be both frustrating and expensive. Many instruments require regular maintenance – wireless transmitters need batteries replaced from time to time, for example – and unless risk-free access is available, difficult-to-reach locations are best avoided.

Exposure to the weather will eventually cause most instruments to fail, but if the expected lifetime is measured in years rather than months, then balancing a better exposure against a builder's bill for hire of a scaffolding tower once every 10 years or so may be a fair compromise. Some things remain unpredictable, however. Anemometer bearings may seize up and need lubrication: birds may build a nest around your sunshine recorder: cable clips may snap and leave cables whipping about in strong winds – and may cost money to put right. The Golden Rule is **don't take any risks you are uncomfortable with taking**. The DIY installation and maintenance of sensors exposed at considerable heights falls considerably outside the gamut of most domestic or small-office DIY tasks.

Rooftop or mast installations may increase the risk from lightning strikes. Full lightning protection, such as that afforded to church steeples, rooftop satellite dishes and the like is commercially available from specialist contractors, but the costs of doing so will dwarf expenditure on the instruments themselves. Except in the areas most prone to severe electrical storms, the risks of being struck in any one year are quite small unless the building, mast or tower is particularly tall or very exposed, but even a close strike stands a good chance of writing off sensors, host computer and Internet connection (and quite possibly nearby utility power circuits too). Some form of commercial lightning protection or grounding kit is therefore advisable in vulnerable areas. When possible, isolate equipment from other components during thunderstorms by using a surge protector or physically unplugging it, and take care not to stand near tall instrument masts during electrical storms.

Access restrictions?

Physical access and site security will not normally be concerns in domestic garden or backyard sites, but they must be carefully considered for sites with public access such as federal or local authority parks, schools and similar environments where vandalism may be a problem. Erecting dense high fencing around the instruments may keep the vandals out, but it will probably keep most of the weather out too. The records of some long-established regional climatological sites have deteriorated badly owing to the worsening exposure resulting from installation of vandal-proof fencing.

Especially in domestic locations or schools, observation sites may have to share space with other activities. It is not a good idea to site the raingauge where children may use it as a proxy goalpost, for example, and neither is it a good move to shield the instruments from sight by small trees – because in a very few years those trees may have grown tall enough to overshadow your raingauge, and you will wonder why your rainfall records seem to show a steady (but very local) tendency towards a climate more typical of Morocco than Manchester or Minneapolis. Small children, domestic animals, wild animals or birds and electronic sensors exposed on lawned areas simply don't mix – for their safety and the continuity of your records, plan fencing appropriate to the requirement when setting out an observing location. Crows, foxes, rabbits and squirrels tend to be particularly persistent and creative offenders.

Carefully consider ease of access for maintenance when laying out the site – can the grass be kept tidy around and between the instruments without risk to the instruments or cabling, for example? If the maintenance is being undertaken by an external contractor, in schools or colleges for example, can the required work be undertaken by the contractor without risk of damage to the instruments? Are there any automatic garden sprinklers which may occasionally 'water' the raingauge? Think ahead to consider whether a well-exposed site today may become over-sheltered in just a few years time once that new conifer hedge planted around it becomes established.

Table 4.1 summarises outline site and exposure requirements by element. The following chapters expand and elaborate upon these guidelines and introduce WMO element-specific site classifications. Very few sites rate optimum (class 1) for all sensors and their exposure; most sites are slightly sub-optimal (class 2 or 3), because it is not always possible to find a site which combines optimal instrument exposure with ease of access, maintenance, security, permanence of tenure and so on. Some sites, including some of the long-period sites described in Chapter 1, *Why the weather?*, are sub-optimal by modern standards, but remain in use today because minimising changes to their environment is a greater priority. Two very different sites within the United Kingdom are illustrated in **Figures 4.1a** and **4.1b**.

Table 4.1 *Summary of the main site and exposure requirements for meteorological instruments, by element*

Element	Preferred siting	Access required	More details
Air temperature and humidity	Representative ground-level position, on flat or gently sloping ground, well away from obstructions. Instruments should be mounted in a radiation screen (see Chapter 5 for details) between 1.25 m (4 ft) and 2 m (7 ft) above short grass, and well away from areas of tarmac, etc. Rooftop sites should be avoided.	Maintenance including cleaning of thermometer shelter, regular grass cutting, etc. Hedges and trees should be cut back if growth encroaches. Securely fenced if vandalism may be a problem.	*Temperature*, Chapter 5 *Humidity*, Chapter 8

Table 4.1 (*cont.*)

Element	Preferred siting	Access required	More details
Precipitation	Representative ground-level position, on flat or gently sloping ground well away from obstructions. Some shelter to reduce wind losses is beneficial. Raingauges should be mounted on the ground, preferably above short grass, with their rim at the national standard height. Areas of hardstanding, concrete or tarmac are best avoided, as they may cause insplash in heavy rain. Avoid locations (or access routes) that may be subject to flood inundation after heavy rain. Rooftop sites are not suitable.	Maintenance including regular checking and cleaning of raingauge funnel, regular grass cutting, etc. Hedges and trees should be kept cut back to avoid encroaching growth. Securely fenced if vandalism may be a problem.	*Precipitation*, Chapter 6
Wind speed and direction	Ideal is a 10 m (33 ft) mast in open country. Failing this, rooftop sites will probably provide better records than sheltered ground-level sites, but representative records can be difficult to obtain. In some districts, building or planning permission may be needed for masts, etc.	Some instruments or wireless transmitters require regular maintenance, such as battery replacements or bearing lubrication; where this is so, safe access is essential.	*Wind speed and direction*, Chapter 9
Sunshine and solar radiation	Clear horizon from north-east through south to north-west (in northern hemisphere, temperate latitudes); a rooftop or mast may be ideal for electronic sensors.	Some instruments or wireless transmitters require regular maintenance, such as battery changes or desiccation capsule replacements; where this is so, safe access is essential.	*Sunshine and solar radiation*, Chapter 11
Atmospheric pressure	Sensors can be mounted indoors, provided the building is not sealed. They should not be mounted where they are subject to significant vibration or changes in temperature or airflow (not in direct sunshine, for example, or near heating, ventilation or air conditioning outlets).	Little access or maintenance is normally required.	*Atmospheric pressure*, Chapter 7
Grass and earth temperatures	Open ground-level site freely exposed to sunshine, wind and precipitation – similar to temperature and rainfall instruments – probably co-located with those instruments.	Maintenance including regular grass cutting, etc. Hedges and trees should be kept cut back to avoid encroaching growth. Securely fenced if vandalism may be a problem.	*Grass and earth temperatures*, Chapter 10

(a)

(b)

Figure 4.1 (a) The automatic weather station at St James's Park in central London. This site, located only 400 m south-west of Trafalgar Square, is sub-optimal, with large areas of tarmac and concrete in the immediate vicinity, but is an example of what is often a difficult compromise exposure, without which no direct inner-city central London observations would otherwise be possible. Observations have been made in this park since 1903, although the record is not continuous and there have been several minor site moves in that time. (b) Another automatic weather station in a very different environment – Moor House in Upper Teesdale in northern England. This site, at 556 m above sea level, is extremely remote, located as it is within a nature reserve some 10 km from the nearest small settlement. While urban influence is non-existent, its very remoteness and liability to heavy winter snowfalls mean that access for regular maintenance or emergency repairs is far from easy. As with St James's Park, it is often better to have some observations, representing here the moorland of the high Pennines, than nothing at all. Observations were first made manually in this location in the 1930s, and continued by automatic weather station since the 1990s. (Both photographs by the author)

A brief overview of the main meteorological observing networks in the United States and the United Kingdom

US ASOS: Automated Surface Observing System

The Automated Surface Observing Systems (ASOS) program is a joint effort of the United States National Weather Service (NWS), the Federal Aviation Administration (FAA), and the Department of Defense (DOD). Automated Surface Observing Systems such as those shown in **Figure 1.2** serve as the nation's primary surface weather observing network, designed primarily to support aviation operations and provide surface observations for forecasting models, but also as the primary climatological observing network in the United States. Because of this, not every ASOS is located at an airport; for example, one is located at Central Park in New York City to continue the long weather record there (see **Figure 1.19** in Chapter 1). A key priority for airfield ASOS sites is to measure critical aviation weather parameters where they are most needed, namely at runway touchdown zone(s). Typical ASOS sites report air temperature and dew point, wind direction and speed (mean and gusts), visibility (to at least 15 km or 10 miles), basic present weather information such as the type and intensity of rain, snow, and freezing rain, barometric pressure, precipitation accumulation and sky conditions (cloud height and amount, up to 12,000 feet/4000 m). Some also report nearby lightning strike activity. Observations are updated every minute, 24 hours per day.

At the time of writing, there were 947 ASOS sites within the United States (**Table 4.2**). More information on ASOS sites can be found at www.weather.gov /asos and current weather observations by state are available from www.faa.gov /air_traffic/weather/asos.

USCRN: NOAA's Climate Reference Network

The US Climate Reference Network (USCRN) is a system of climate observing stations developed by the National Oceanic and Atmospheric Administration (NOAA), whose primary goal is to provide reliable and consistent long-term temperature, precipitation, and soil moisture and temperature observations made in stable settings [73, 81, 82, 83]. There are 141 commissioned USCRN sites, 117 within the continental United States, 21 stations in Alaska, two stations in Hawaii and one in Canada. The USCRN provides the United States with a reference network that adheres closely to climate principles endorsed by the National Academy of Sciences (NAS) and WMO's Global Climate Observing System (GCOS). USCRN sites have spatial representativeness and temporal stability as their primary goals, for which sites are chosen in pristine environments expected to remain free of development for many decades. Each station collects independent measurements of temperature, precipitation, and soil moisture at five depths, as well as relative humidity, solar radiation, surface temperature, and surface wind speed, using a standard site configuration plan (**Figure 4.2**). Highly accurate measurements and reliable reporting are critical: instruments are calibrated annually, and maintenance includes routine replacement of aging sensors. A typical USCRN site is shown in **Figure 4.3**. The program is managed and coordinated by the National Centers for Environmental Information (NCEI), a component of NOAA's National Environmental Satellite, Data, and Information Service (NESDIS).

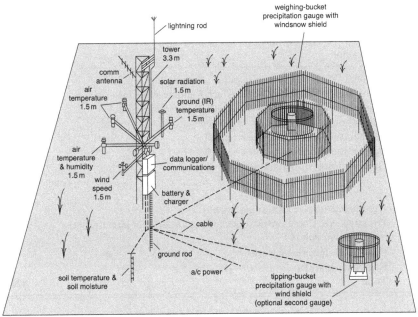

Figure 4.2 The site layout of a typical US Climate Reference Network (USCRN) location. (National Oceanic and Atmospheric Administration/National Centers for Environmental Information)

Figure 4.3 A remote site in the USCRN network (Toolik, Alaska – 68.6°N, 149.4°W, 750 m / 2461 ft AMSL) showing the three aspirated screens which are used at over 100 sites across the United States to obtain parallel and fail-safe air temperature data. (Photograph courtesy of the National Oceanic and Atmospheric Administration/National Centers for Environmental Information NOAA/NCEI)

Data from USCRN sites are used in a variety of climate monitoring and research activities that include placing current climate anomalies into an historical perspective. These sites are rightly distinct from the less stringent requirements of day-to-day

Table 4.2 *United States Federal weather observing network population, early 2023 (includes Alaska, Hawaii, Puerto Rico and US Affiliated Islands). Source: National Centers for Environmental Information [90].*

Type	Description		No of sites
NWS COOP	National Weather Service Cooperative network (COOP):		
	Rainfall-only sites with manual 8 inch standard raingauge		1256
	Rainfall-only sites with manual 4 inch plastic raingauge		267
	Rainfall-only sites – automated gauges		891
	Temperature and rainfall sites – MMTS		3977
	Temperature and rainfall sites – non MMTS		495
	Other – mostly river stations		190
		Total	7076
ASOS	Automated Surface Observing System		947
PLCD	Primary Local Climatological Data (First Order), not already included in ASOS		17
USCRN	US Climate Reference Network		140
Various	Regional networks		88
	Total		**8268**

Note: 426 of the 947 ASOS sites are also included in the NWS COOP site numbers, and the total site count has been reduced accordingly. In addition to these, there are 21,834 CoCoRaHS sites within the US Global Historical Climatology Network (GHCN-Daily); CoCoRaHS is covered in more detail in Chapter 19, Sharing *observations.*

operational meteorology required for aviation and forecasting purposes. The performance of each station's measurements is monitored on a daily basis and any issues are addressed promptly. Each station transmits data hourly to a geostationary satellite. Within minutes of transmission, raw data and computed summary statistics are made available via the USCRN web site [82], which also contains information on the instruments in use and site photographs.

USCRN is an exemplary example of a 'new generation' climatological monitoring network designed from the outset to adopt modern technologies, capable of updating as technologies and measurement systems advance over time, and maintaining excellent digital site metadata (see Chapter 16, *Metadata, what is it and why is it important?*). Similar 'climate reference networks' are being established in other countries, notably Australia (112 locations [84]) and China (672 locations [85]), although in both cases these refer mainly to existing high-quality long-period station records rather than new purpose-built stations in pristine locations using optimal twenty-first-century instrument standards. At the time of writing there remain no plans to set out such a reference network in the UK. The establishment of a truly *global* climate reference network on similar grounds to USCRN would bring enormous advantages at relatively small costs, and has been proposed several times (for example, [86, 87]), but has as yet to find global backing from the UN or WMO.

US MMTS: Maximum-Minimum Temperatures System

The US National Weather Service (NWS) began to update its second-order climatological observing network in the mid-1980s, with a programme to replace traditional liquid-in-glass thermometers and Cotton Region Shelters at thousands of cooperative observer sites across the country [88]. The wooden shelters had become increasingly expensive and difficult to maintain, while NWS was also experiencing difficulties in sourcing high-quality self-registering thermometers at an acceptable price. An ageing corps of volunteer observers was also finding these thermometers difficult to read. Over a period of a few years, about half of the network was migrated to a remote-reading (cabled) temperature measurement system, the so-called maximum-minimum temperatures system (MMTS), sometimes referred to as NIMBUS after the display unit itself (**Figure 4.4**). The system comprises an electrical resistance sensor (thermistor) housed in a specially designed radiation shield (**Figure 4.4a**) mounted at about 1.5 m / 5 feet above ground level. There are almost 4,000 MMTS sites across the United States operating under the NWS co-operative observer program (COOP), **Table 4.2**.

MMTS is a manual system – no logger is included. Daily maximum and minimum are recalled from the remote display unit's memory (**Figure 4.4b**) and noted manually, and the memory then reset, thus directly replicating the observational routine for 'traditional' thermometry. Observations are then transcribed to an electronic or manuscript form and then sent monthly to a regional National Weather Service office, where they are digitised and added to the regional and national weather archives.

The MMTS system is simple and inexpensive: however, its rapid introduction was criticised as giving rise to major discontinuities in many long-term US climatological records [89], although independent temperature measurements using precision aspirated thermometers 'suggested that MMTS measurements were likely closer to truth … than those from the traditional wooden weather shelters' [88]. However, one sustained criticism has been its use of cabling between sensor location and display, and relatively short cable runs at that, the cabling providing both power and data transmission. Where obstacles or lack of resource have made burying the cable impossible, some sensor locations had to be relocated, and some were badly compromised in doing so. Some previously satisfactory observing sites were moved to highly unsuitable positions, being relocated only 2–3 m from buildings, in parking lots, too close to air conditioning outlets and so on. Many have since been resited to more representative locations, although in the interim period the continuity of some long-term temperature records was damaged. Lightning-induced currents in the cables also resulted in frequent damage to early electronics modules, and even started a few fires. More difficult to spot was the change in resistance in the thermistor circuit that accompanied less severe electrical surges, some producing permanent temperature changes of 1–2 degC (2–4 degF). Damage to cables also occurred from rodents and burrowing animals, while (particularly in the southern states of the United States) various unpleasant insects found MMTS radiation shields to be ideal residences.

The total number of observing sites in the United States has declined slightly in 40 years; there are currently about 8,200 operational sites with data reported to the National Centers for Environmental Information (**Table 4.2**), which compares with 11,615 observing sites in 1981 and 10,406 in 2011. Generally, there have been a significant number of COOP station closures over this time period, a move to more automated equipment, and a change for stations to report more than one element.

Figure 4.4 (a) MMTS radiation shield at Elko, Nevada (40.86°N, 115.74°W). (Photograph by Famartin on Wikimedia Commons, CC BY-SA 3.0, https://commons.wikimedia.org/wiki/File:2013-10-14_12_27_21_Nimbus_electronic_temperature_sensor_housing.JPG). (b) MMTS NIMBUS interior display unit. (Photograph courtesy of National Weather Service)

State mesonets

Table 4.2 does not include numerous state and regional weather observation networks operated by non-federal authorities. These so-called mesoscale networks of automated weather stations, or 'mesonets', typically collect data at finer space and time resolution than traditional national networks, make data available in (near) real time, and usually record a wide variety of meteorological elements. Mesonet data serve a multitude of application areas, including emergency warning, forecasting, education, agriculture, leisure, hydrology and research applications in a wide variety of fields. A survey conducted back in 1991 [91] identified 100 such networks and 831 sites in the United States and Canada, and since then many more mesonets have been established. The Kentucky Mesonet, for instance (www.kymesonet.org), operated by Western Kentucky University and the University of Alabama in Huntsville, comprises 72 research-grade meteorological and climatological observation stations distributed at a spacing of 20–30 km across the state which measure air temperature, precipitation, relative humidity, solar radiation, wind speed and wind direction [92, 93]. Other state mesonets exist in New York (126 sites, www.nysmesonet.org/), Oklahoma (120 sites, www.mesonet.org), Missouri (32 sites, agebb.missouri.edu/weather /stations), West Texas (150 sites, www.mesonet.ttu.edu) and Alabama (21 stations, wx.aamu.edu), with a 75 station network in construction in Maryland at the time of going to press. In all, the various current and proposed mesonet sites account for several hundred additional observing sites across the United States.

The UK land surface synoptic observing network

At the time of going to press, the UK Met Office land surface observing network consisted of 264 sites, of which 137 were synoptic stations reporting through the WMO communications channels (most of which provide the full range of synoptic observations, updated hourly in most cases [94]), while another 127 are automated sites with a reduced observational coverage reporting in to the Met Office network. **Figure 4.5** shows a typical synoptic site (Kenley Airfield, on the South Downs south of London). The average separation of stations in this network is about 40 km, sufficient to allow good coverage of synoptic-scale weather features to be observed; satellites and weather radar networks play an important role in observing smaller-scale features such as convective storms in real time.

Observations made at these sites are intended to be representative of the wider area around the station (a circle of 5–20 km radius), and as a result most will be WMO class 1 or class 2 site classification (the details of site classifications follow in subsequent chapters). Measurements made will normally include most if not all of the following: air temperature, relative humidity and dew point at 1.25 m above the

Figure 4.5 The Met Office synoptic station at Kenley Airfield, south of London: WMO number 03781, 51.30°N 0.08°W, 170 m AMSL. (Photograph by Richard Griffith)

ground; grass surface temperature (or, more usually, an artificial turf surface); surface temperature over a concrete slab; soil temperatures (typically at 10 cm, 30 cm and 1 m depths); rainfall amounts; snow depths; wind speed and direction at 10 m above ground, including gust speeds; barometric pressure at both station level and reduced to mean sea level, together with amount and sign of change over the previous 3 hours; visibility; cloud amounts and heights of cloud base (and cloud types, if an observer is present); and past and present weather types.

Almost all measurements are fully automated and are logged every minute. Data are logged at the station, processed to convert the measurement to a standard format, then transmitted hourly to a central collecting system based at Met Office HQ in Exeter before being communicated on the global synoptic observations networks. Although most of the observing process is automated, there remain a few locations where a trained meteorological observer provides additional 'eye' input into the observation, most often at airfields where manual oversight of cloud, visibility, wind, present weather and atmospheric pressure are of particular importance for aircraft safety.

UK climatological and rainfall observing networks

Two supplementary station networks, the 'climatological' and 'rainfall' stations, provide a more limited range of data but with a denser station network to allow coverage of all but the smallest-scale meteorological phenomena. The climatological network consisted of 141 locations at the time of going to press, excluding synoptic sites as detailed earlier [95], most of which also provide climate records. Observations at these sites are made manually: most report once per day at or close to 0900 UTC, and these typically include measurements of air, grass and soil temperatures, daily rainfall, and perhaps sunshine duration, wind speed and direction. Instruments are provided by the Met Office, but sites are located within local authorities, universities, agricultural and research establishments and the like, including a few private individuals.

The rainfall network is denser still, with (at the time of writing) 2,741 precipitation-only sites in the UK (the network consists of 3,146 locations including synoptic and climatological stations), some of which have records extending back 100 or even 150 years. The station density is much greater than for other elements because the spatial variation in precipitation is much greater: across the UK as a whole the average spacing between gauges is about 10 km (**Table 6.1**, page 142), although the number of rainfall sites has fallen by more than half in the past 50 years owing to cost-cutting exercises (also known as 'network rationalisation'). 'Rainfall only' sites usually consist of a standard five-inch (127 mm diameter) raingauge, read at or close to 0900 UTC daily; some sites will have a tipping-bucket raingauge in addition to, or sometimes instead of, the standard 'checkgauge'. (More details of raingauges are included in Chapter 6, *Measuring precipitation*.) The majority of the rainfall network is managed by regional partner agencies – the Environment Agency in England, Natural Resources Wales/Cyfoeth Naturiol Cymru in Wales, and the Scottish Environment Protection Agency (SEPA) – although national data archiving remains the responsibility of the Met Office. The network includes several sub-networks for flood forecasting purposes, not all of which necessarily contribute to the regional and national rainfall archive. Most rainfall sites outside the synoptic and climatological networks are classed as 'voluntary co-operating sites', and many are run by private individuals or other interested bodies. At the time of writing, most rainfall sites report their records once monthly by post, but pilot schemes are being developed which will eventually lead to a much greater number of rainfall records being reported daily using current or future online systems (more details in Chapter 19, *Sharing observations*). Much of the SEPA network already reports at least daily; totals for the most recent 36 hours, daily totals over the last month and monthly totals over the last year can be viewed online at www2.sepa.org.uk/rainfall. The SEPA site also includes details on how interested observers in Scotland can make and contribute their own observations.

One-minute summary – *Site and exposure: the basics*

- *Site* refers to 'the area or enclosure where the instruments are exposed', while *exposure* refers to 'the manner in which the sensor or sensor housing is exposed to the weather element it is measuring'.
- Satisfactory site and sensor exposure are fundamental to obtaining representative weather observations. An open well-exposed site is the ideal, of course, but with advance thought and careful positioning of the instruments, good results can often be obtained from all but the most sheltered locations.
- The ideal exposure for one sensor can be the exact opposite for another. For representative wind speed and direction readings, for example, an anemometer mounted on top of a tall mast is ideal, but this would be a very poor exposure for a raingauge owing to wind effects (more on this in Chapter 6, *Measuring precipitation*).
- Based upon World Meteorological Organization (WMO) published guidelines, this chapter outlines preferred site and exposure characteristics for the most common sensor types. No single exposure will provide a perfect fit for the requirements of all sensors, and some compromises may be necessary, particularly for 'all-in-one' personal weather station equipment where all sensors are located in one module. More details on WMO site classifications for individual instrument types follow in subsequent chapters.
- Rooftops or masts may provide much better exposure for some sensors, but carefully consider the accessibility of the site before attempting to install the sensors. If the proposed site cannot be reached safely, fit appropriate safety measures or find another site. **Do not take personal risks, or encourage others to do so, when attempting to install weather station sensors, particularly at height**.
- Brief summaries are included of the main operational weather monitoring networks in the United States and the United Kingdom, whose functions vary from frequent, broad-scale real-time information intended primarily for aviation and forecasting purposes to more spatially dense precipitation-only sites reporting in slower time.

5   Measuring the temperature of the air

Air temperature is the first element reviewed in this section of the book, as for many this will be the foremost measurement priority. This chapter opens by setting out what we mean by 'air temperature', and precautions needed to obtain measurements that are both accurate and representative, followed by details on recommended siting and exposure of air temperature instruments and the sensors themselves. World Meteorological Organization (WMO) siting guidelines are set out [4], followed where relevant by country-specific details most relevant to the United States, the United Kingdom and the Republic of Ireland. Recommendations on observing practices in other countries can generally be found on that country or region's state weather services web pages on the WMO website [96] (at the time of going to press, this WMO page has links to 187 Member States and six Member Territories). The chapter concludes with a brief summary of the content.

Methods for making *grass and earth temperature measurements* are covered in Chapter 10.

What is meant by 'air temperature'?

In its simplest form, 'temperature' is a measure of how hot or cold an object is. 'Meteorological air temperature' was defined by WMO in 1992 as ' ... the temperature indicated by a thermometer exposed to the air in a place sheltered from direct solar radiation' [4], although this definition cannot be considered comprehensive. (Note that the term 'thermometer' is used throughout this book as convenient shorthand for 'a device capable of measuring temperatures', rather than in the limited traditional sense of a graduated liquid-in-glass sensor. Unless a specific context is given, it should be taken as covering all sensor types used to measure temperatures. Where the narrower sense is meant, this is made clear in the text.)

Starting from first principles, the 'temperature' of a body is a measure of the heat energy of that object, itself a measure of the kinetic energy of the atoms or molecules of which the object is composed. Temperatures are measured with reference to defined fixed scales set out in terms of physical changes in state of various substances, such as ice and water for temperatures within normal meteorological ranges, as currently set out in the International Practical Temperature Scale of 1990 [97]. In national and international meteorological and climatological use, temperatures are normally expressed in degrees Celsius (°C), although the older Fahrenheit scale (°F) is still in general public use within the United States. Temperature *intervals* are expressed in Celsius degrees (degC) or Fahrenheit degrees (degF): 1 degC is also identical to 1 kelvin, a measure of absolute tempera-ture, where absolute zero = 0 k (–273.15 °C). A conversion table from °C to °F is given in **Appendix 6**.

Temperature is one of the most important atmospheric quantities, with societal applications from short-term weather forecasts to long-term climate change. Temperature measurements have been made for almost 400 years, but high accuracy still presents a challenge, for the measurements are themselves significantly influ-enced by many factors, including amongst others the calibration, response time (or *time constant* – see **Appendix 1**) and exposure of the sensor(s) involved, the housing of the sensor, external solar and terrestrial radiation balance, surface albedo and wind speed. Consequently, care is required when exposing air temperature sensors to ensure that, as far as possible, the instrument reading is both accurate and representative, and not unduly influenced by the instrument housing, surrounding vegetation or ground cover, the presence of buildings or other objects. For standard meteorological observations, the objective is normally to seek measurements (air temperatures in this case) which are representative of conditions over a wide area – typically 100 to 1000 square kilometres, perhaps more conveniently thought of as a circle of radius 5–20 km centred on the measurement site, making due allowances for significant topographic and land use variations. Following national and inter-national standards and exposure guidelines for measuring air temperature (such as those published by WMO) as closely as possible helps ensure that temperature records from one locality, country or last month, can be confidently compared with those made in another country, or several decades ago.

Factors influencing air temperature measurements

There are numerous factors which can influence the reading of a thermometer exposed to the air [98, 99]:

- During daylight hours, the sensor element must be adequately protected from both incoming and reflected short-wave solar radiation (sunshine) at all incident angles, and from re-emitted long-wave (infrared) terrestrial radiation from the Earth's surface and atmosphere. Without adequate shielding the sensor will absorb this radiation, and as a result the temperature indicated will be higher, perhaps much higher, than the true air temperature.
- At night, the sensor must be shielded against outgoing terrestrial radiation, from both sky and ground, because outgoing long-wave radiation losses to the sky will cause it to read lower than the true air temperature, particularly under clear skies.
- Air is a very effective insulator, and to ensure changes in air temperature are reflected in the sensor reading the instrument must be in good contact with the air – and preferably well ventilated, so that it quickly takes up and indicates the temperature of the air passing over it and responds quickly to changes. In most conventional thermometer housings, however, this requirement has to be bal-anced against the need for protection against solar and terrestrial radiation, not always successfully.
- The sensor requires protection from precipitation, for a device that is wet will cool below the true air temperature in dry air, owing to evaporative cooling

(this is the principle of the wet bulb thermometer, used to determine water vapour content of the air – see Chapter 8, *Measuring humidity*).

- The thermometer housing should also provide a uniform internal temperature environment which is the same as the true external air temperature. Its time constant to take up changes in air temperature should be as small as possible, preferably no more than about a minute (see **Appendix 1**), although as subsequently set out this is rarely attained in conventional louvred thermometer screens.
- The sensing device used must be sensitive enough to respond quickly to changes in air temperature on timescales of a minute or less, but not so sensitive as to respond to minor second-by-second fluctuations which are largely irrelevant for most meteorological purposes.
- The sensor itself should be robust, stable in calibration, easy to use and capable of deployment and use by non-specialists in different operational environments, some of which may be in challenging climatic conditions or in remote locations. It should also be easy to calibrate, inherently safe, avoid the use of dangerous or toxic materials, and minimise environmental impacts in manufacture or use. An operational life of a decade or more is preferable, both to provide consistency in measurements and minimise sensor changes in the station's climate record.
- As far as is commensurate with other requirements previously listed, the sensor must also be protected from the corrosive effects of air pollution or the weather itself, from the risks of accidental damage, and all too frequently from the attention and destructive influences of thieves or vandals.

Many of these requirements mandate a physical thermometer shelter, often referred to generically as a 'thermometer screen'. Many different types and designs of thermometer screen have been used over the years and in different countries, and many descriptive and comparative analyses have been documented – see, for example [23, 99, 101, 102]. WMO guidelines state clearly that 'In order to achieve representative results when comparing thermometer readings at different places and at different times, a standardised exposure of the screen and, hence, of the thermometer itself is also indispensable' ([4], section 2.1.4.2.1). The main types of thermometer screen, together with their advantages and disadvantages, are covered later in this chapter.

The height above ground at which the temperature measurements are made is another important factor. The surface of the ground can become much warmer in sunny conditions, and much colder on clear nights, than the air just a metre or two above the surface. Large vertical temperature gradients can come and go very quickly. For this reason, air temperatures are usually measured at a height between 1.25 m and 2.0 m (4 to 7 feet) above ground level, with a little variation from country to country. The type of ground surface will also influence air temperatures – readings made above black tarmac will be higher in sunny weather than those measured above short grass, for instance – and measurements made over or near such artificial surfaces are likely to be less representative. These requirements are covered within this chapter, starting with site and exposure and including the details of WMO site classifications for temperature measurement locations.

In order to ensure compatibility with other observing locations, the time period(s) within the day to which measurements relate (such as mean, maximum and minimum daily temperatures) must be consistent between sites. This important topic is covered in more detail in Chapter 12, *Observing hours and time standards*.

The remainder of this chapter sets out three main headings:

- **Site**: the location where the temperature measurements are to be made
- **Exposure**: how the temperature sensors will be exposed to the weather, and finally
- **Sensors**: the device or devices that will respond to and measure changes in air temperature.

Site requirements for representative air temperature measurements

WMO's guidance ([4], paragraph 2.1.4.2.1) is clear and concise:

> The most appropriate site for [air temperature] measurements is, therefore, over level ground, freely exposed to sunshine and wind and not shielded by, or close to, trees, buildings and other obstructions.

An ideal observation site should be located well away from significant obstructions such as buildings, walls, hedges and trees, and areas of non-natural surfaces such as roads, car parks, concrete hardstanding and the like, as well as bodies of water (unless they are natural features, such as lakes or coastlines). More sheltered locations, or those in closer proximity to artificial surfaces, can still provide worthwhile measurements, albeit subject to greater uncertainty, provided the instruments are sited carefully as set out in WMO's temperature site classifications in the following section. Certain locations are best avoided altogether, however, as readings obtained in these situations may bear little comparison to observations made elsewhere under standard conditions:

- Very sheltered positions, with little free airflow or exposure to sunshine, including north-facing walls, obstructions or buildings which will lead to significant local sheltering (such as a tall thick hedge located upwind in the direction of the prevailing wind), or other locations surrounded by buildings or tall fencing or hedging;
- Locations which may result in significant additional reflected radiation, such as immediately to the south of a south-facing wall or windows (in the northern hemisphere), should be avoided because the additional reflected radiation will result in warming by day, while stored heat released during the night may also affect nocturnal temperatures;
- Rooftop or chimneypot sites, house eaves and shed roofs should be avoided altogether, owing to the complex effects of the building itself on the observed temperature;
- Sites with significant topographic shelter – on steep slopes, in narrow valleys or in hollows, where such shelter may induce or enhance stable stratification and enhanced radiational cooling leading to the generation, draining and trapping of cold air. Such sites can be subject to exceptional conditions and may be interesting in themselves, but are unlikely to be representative of the wider area;
- Masts or towers where the screen is significantly sheltered by the mast structure, or where the sensor/screen combination is higher than about 2 m above the ground, may give misleading results, as will locations under overhanging trees or near exhaust gases (for example, close to air conditioning outlets);
- Areas of tarmac or concrete near the site should be avoided, as these can become very warm in sunny conditions, and may lead to artificially high

readings. Many airport or airfield sites are very close to extensive areas of tarmac; where there is little choice of site for operational reasons, if possible, ensure the thermometer screen is located *upwind* of the prevailing wind direction of such surfaces.

If a site open in all directions cannot be found, one allowing the best available exposure to sunshine and wind (particularly the prevailing wind) should be chosen. Any site shelter should not be so dominant in any direction as to make readings difficult to compare with other sites under varying wind conditions.

WMO air temperature site classifications

Table 5.1 sets out the WMO air temperature and humidity site classification criteria (summarised from reference [4], Annex 1D). These classifications were established in an attempt to give some indication of how representative temperature measurements from any given site are with regard to the possible impact of its immediate environment, on the scale of tens to hundreds of metres around the location of the thermometer screen or equivalent. A class 1 site is deemed a 'reference' site, one which is considered broadly characteristic of an area of 5–20 km radius centred on the site; a class 5 site is one where site conditions may impose additional estimated uncertainty of several degrees Celsius in unfavourable circumstances, when compared with records from a class 1 site. The lower the site class, the lower the expected uncertainty of observations made at that site. However, these site classifications recognise and reflect that the real world is not perfect, and that operational, practical, social and logistical considerations (including the availability of power supplies) often place limits on where instruments can be sited. Accordingly, there are few class 1 sites in most national networks. That is not to say that sites with a higher class number are unimportant: many long-period sites (such as some of those described in Chapter 1) do not meet class 1 limits. Where continuity of records in a particular location may assume greater importance than strict conditions of local representativity, and at some urban sites where the location may represent a compromise, such records may represent the 'best available option', where the only alternative may be no measurements at all. All else being equal, however, class 1 or class 2 sites are preferred wherever possible.

The effects of shelter

The site classifications set out in **Table 5.1** are guidelines rather than strict rules and should be interpreted accordingly. As it stands, the table would presumably permit a location with a huge tree immediately poleward of the thermometer screen, or tall, dense hedging immediately south of the screen, to be recognised as a class 4 site rather than class 5, even though such obstacles would make such an environment highly unsuitable for representative air temperature measurements.

Many sites necessarily face compromise in one area or another, most often in the degree of shade or shelter resulting from hedges or trees, or proximity to buildings or artificial surfaces such as tarmac, car parks and so on. It is important to realise that

Table 5.1 *WMO air temperature site classification criteria. All measurements are from the thermometer screen, or the main operational thermometer screen if there is more than one on site.*

WMO temp. Site class	Estimated uncertainty (note 1)	Surroundings	Ground cover	Heat and water sources (note 2)	Shade
1	Reference site	Flat, horizontal land, surrounded by an open space, slope less than ⅓ (19°)	Natural and low vegetation (< 10 cm) ground cover, representative of the region	> 100 m distant, or < 1% within 10 m radius, < 5% within 10–30 m radius and < 10% within 100 m radius	No shade whenever solar elevation > 5°
2		*As above*	*As above*	> 30 m distant, or < 1% within 5 m radius, < 5% within 5–10 m radius and < 10% within 30 m radius	No shade whenever solar elevation > 7°
3	± 1 °C	*As above*	Natural and low vegetation (< 25 cm) ground cover, representative of the region	> 10 m distant, or < 5% within 5 m radius, < 10% within 10 m radius	*As above*
4	± 2 °C	Close, artificial heat sources and reflective surfaces or bodies of water occupying < 30% of the surface within 3 m radius around the screen and < 50% within a 10 m radius			No shade whenever solar elevation > 20°
5	± 5 °C	Sites not meeting the requirements of class 4			

Note 1. The WMO annex from which this is taken states 'This uncertainty is derived from bibliographic studies and comparative tests', although these are not listed or referenced in the document.
Note 2. Heat sources or reflective surfaces include buildings, concrete surfaces, car parks, and the like; similarly for bodies of water, unless representative of the locality such as lakes or coastlines.

the effects of site shelter, or proximity to buildings, tarmac surfaces and the like will vary with time of day, time of year and with weather conditions: they are often most pronounced under conditions of little or no cloud and light winds. Under sunny skies and light winds, a sheltered site (class 3 or higher) will often show higher air temperatures than a nearby 'open' location, as a result of the reduction in heat transport by the wind (advection/forced convection) away from objects warmed by sunshine – including of course the screen structure itself. All else being equal, maximum temperatures in such locations under such conditions during the summer months in temperate latitudes may be 1–2 degC / 2–4 degF above those measured in more open locations nearby, as suggested in column 2 of **Table 5.1**. Even under cloudy and windy conditions, differences can remain substantial – during the summer months a difference of 1 degC / 2 degF is not unusual even under unbroken cloud cover. In subtropical or tropical latitudes, these differences can be expected to be larger, given higher solar radiation receipts.

During the night, effects due to direct solar radiation are obviously eliminated, but other factors come into play. Wind speeds are normally lower at night, and a sheltered location may experience little or no air movement at screen height for several hours, whereas a more exposed site may experience a persistent breeze throughout. This continual stirring of the air may act to keep the temperature at the more open site significantly different (higher or lower) from that in a sheltered location.

A sheltered urban or suburban site which is surrounded by buildings often experiences higher night-time minima (typically by 0.5–2 degC / 1–4 degF) under clear-sky conditions in both winter and summer, the effects being a combination of delayed heat release from the urban infrastructure and a reduction in both outgoing radiation and screen-level ventilation affecting the screen/radiation shelter itself. In fine, settled, hot spells in summer these urban heat-island effects can quickly become substantial, and differences of 5–7 degC (9–13 degF) between suburbs and nearby rural districts just a few kilometres away are not uncommon, particularly early in the night [103, 104]. On the other hand, cloudy, windy, dry nights can be a good time to check calibrations across different types of screens and sensors, because differences should be small – less than 0.1 degC / 0.2 degF, once any altitudinal differences are allowed for – and any significant calibration errors or drift can be more easily identified, assuming of course that the readings of at least one sensor are known accurately. (See Chapter 15, *Calibration* on calibration techniques for details.)

How representative are urban and suburban sites?

How best to measure and represent urban climates is a subject that has generated debate amongst the professional climatological community for decades, for a well-exposed open site in a city centre is, almost by definition, not likely to be typical of the built-up area. Although the biases likely to result from higher class number sites can be comparable with, or even exceed, sensor calibration errors, a sheltered location in itself need not rule out useful local weather measurements. Indeed, many a back-garden or backyard site, with carefully located instruments, may be more typical of the location and provide a more representative picture of the 'true' climate of the town or suburb. It is, however, more difficult with such sites to distinguish between purely site- or instrument-specific characteristics, and those that are truly representative of the urban or suburban character of the area.

Exposing the thermometer: screens and radiation shelters

We have already seen that thermometers need to be protected from the elements, while at the same time ensuring adequate ventilation is allowed for. Such protection is most often provided by a suitable instrument shelter, usually referred to generically as a 'thermometer screen' (*not* a 'temperature screen') or 'radiation shield'. Because the means by which air temperature sensors are exposed has a much greater impact on the observed readings than all but the least accurate sensors, the type and choice of thermometer screen is the most important factor after site characteristics when it comes to making accurate and representative air temperature measurements.

There are many different types and designs of thermometer screens in use worldwide [105, 106], but three types dominate – the louvred or *Stevenson screen* type, still the standard shelter in many countries; smaller plastic *AWS radiation screens* ideally suited for deploying smaller electronic sensors; and *aspirated screens*, which use a fan to provide a constant flow of air drawn from the immediate surroundings over the sensors. Aspirated sensors are standard in many 'climate reference' sites, such as the US Climate Reference Network (USCRN), which was covered in the previous chapter. Descriptions and details of all three main types of screen follow, together with their advantages and disadvantages.

Less expensive alternatives, including home-made shelters, can suffice where high accuracy or comparability with other sites is not required, or for those on a tight budget, and these are also covered briefly. It should be noted at this point that *almost any form of radiation shield will give better results than a bare sensor.*

Louvred screens

Louvred wooden, or more recently plastic, thermometer screens are as close to a worldwide standard as currently exists and are a familiar sight around the world (**Figure 1.6**). The double-louvred shelter now known as the *Stevenson screen* (**Figures 5.1** and **5.2**) was first described by Thomas Stevenson in Scotland in 1864 [26] (see Chapter 1). After various experimental trials in England in the 1870s [27, 107], this type of thermometer screen was adopted as the standard shelter by the UK Meteorological Office about 1872. Slight refinements on the original design were recommended by the Royal Meteorological Society in Britain in 1884 [28] and thereafter quickly adopted in many parts of the then British Empire during the late nineteenth century. Various iterations of the original design remain the standard screen to this day in Great Britain, Ireland, Canada, Australia, New Zealand and many other countries, including the smaller Bilham variant [108]. The *Cotton Region Shelter*, introduced by the US Weather Bureau towards the end of the nineteenth century and still widely used throughout the Americas (**Figure 5.3**), is slightly larger than a standard Stevenson screen but otherwise similar in design and construction

Figure 5.1 A modern plastic-and-aluminium standard size 'Metspec' Stevenson screen. (Photograph by the author)

Figure 5.2 A double-width or large 'Metspec' thermometer screen. (Photograph by the author)

Figure 5.3 A Cotton Region Shelter near Asheville, North Carolina. (Photograph by Grant Goodge, NOAA)

(some are single-louvred, rather than double). Broadly similar designs of louvred screen remain in use in many other countries.

The basic elements of the louvred screen design are similar – a four-sided single- or double-louvred enclosure with overlapping floorboards, topped off with a ventilated rain-proof roof. The ventilated roof, louvres and the overlapping bottom boards allow natural ventilation of the interior of the screen, while preventing the ingress of direct or reflected solar or terrestrial radiation or rainfall (although fine snow or dust does tend to be blown into such screens). One side is hinged as a door to allow access to and observation of the thermometers – normally on the north side in the northern hemisphere, to prevent the Sun shining on the instruments at any time of day while the door is opened. The screen is usually mounted on a metal stand. WMO guidance is that the thermometers be located between 1.25 m and 2 m above ground level (the standard height in the UK and Ireland is 1.25 m ± 0.1 m; in the United States between 4 and 6 feet), although in areas where significant accumulations of snow occur the screen can be mounted on an adjustable stand to keep the thermometer height at roughly the same level above the snow surface as the snow depth varies.

Double-width Stevenson screens (**Figure 5.2**) were originally developed in the early twentieth century to allow autographic instruments, such as thermographs and hygrographs recording on clockwork-driven paper charts, to be installed alongside conventional thermometry [108]. With the reduction in size of electronics-based temperature and humidity sensors, the need for such large screens has since dwindled almost to nothing. Because their response characteristics differ somewhat from the 'standard' screen, wooden screens of this pattern are best replaced by a standard-size plastic screen as they near the end of their useful life, unless there is a compelling reason to replace like with like – perhaps for the purpose of maintaining consistency and homogeneity with long-period records.

External and internal dimensions of typical Stevenson-type thermometer screens are given in **Table 5.2**.

Table 5.2 *Comparative dimensions of the US Cotton Region Shelter and plastic 'Metspec' Stevenson screens, excluding roof panel*

	External W × D × H (mm)	Internal W × D × H (mm)	Internal volume (litres)
'Metspec' Stevenson screen – standard size (Figure 5.1)	570 × 390 × 550	490 × 315 × 430	66
'Metspec' Stevenson screen – large pattern (Figure 5.2)	1225 × 495 × 550	1145 × 420 × 430	207
Cotton Region Shelter, large (Figure 5.3) [109]	770 × 525 × 800	660 × 460 × 690	210

Plastic Stevenson-type screens

Stevenson screens were originally made of wood (and some still are); but since the millennium, aluminium and plastic or fibreglass screens (**Figures 5.1, 5.2**) have replaced many traditional wooden models. Careful side-by-side trials conducted in several locations – by the UK Met Office in particular, who adopted the 'Metspec' plastic screen as standard in 2006 – have shown that differences between plastic and wooden screens are small, typically < 0.1 degC, and so mostly insignificant for operational and climatological purposes [110]. Metspec screens have since become the preferred screen in many other countries.

Plastic Stevenson-type screens possess the enormous advantage of being almost maintenance-free, requiring little more than an annual wash inside and out, particularly in areas with significant levels of airborne pollution or high windborne salt loading – although a regular wipe-down also helps to keep the inevitable resident insect population in check. The bright, shiny white exterior finish on such screens is much more resistant to the elements than gloss paint finish on traditional wooden screens, retaining near-constant radiative properties over decades and more, although plastic can become brittle at low temperatures. The annual chore of sanding down and repainting wooden screens can thankfully be confined to history. For these reasons, the standard Metspec thermometer screen is a 'recommended product' (**Appendix 2**).

In the United States, many wooden Cotton Region Shelters containing conventional thermometers were replaced by electronic sensors housed within small plastic radiation shields as part of the MMTS programme [75, 111], and some are also being replaced by plastic alternatives.

Louvred thermometer screens of this type and size were usually built to accommodate more than one type of liquid-in-glass thermometer, most often maximum and minimum thermometers together with a dry and wet bulb hygrometer, used to determine humidity (see Chapter 8, *Measuring humidity*). Modern electronic sensors, and even loggers, are so much smaller that smaller radiation screens are steadily replacing 'traditional' louvred screens, although an overlap record with existing methods should be made where the length of the existing temperature record is two decades or more. (Note, in passing, that where additional equipment is housed within a thermometer screen, whether that be an old-fashioned thermograph or a modern datalogger, its volume must not be such as to result in significant obstruction to ventilation through the screen.)

The construction of any type of shelter inevitably offers some resistance to natural ventilation: tests have shown that airflow speed through a Metspec standard Stevenson screen averaged just 7 per cent of the wind speed at 10 m above ground (10 per cent of that at 2 m): from this it is evident that air movement through the screen (and thus across the temperature sensor or sensors) is almost non-existent in surface winds below 4–5 m/s [112] . The time constant of both sensor and the combined screen/sensor combination also increases with reducing airflow, as a result of which the indicated air temperatures can lag changes in true air temperature by anything up to an hour in light winds [113, 114] (see also **Appendix 1**). For this reason, temperature records in passively ventilated screens tend to be somewhat damped when temperatures change quickly, in comparison against actively ventilated or 'aspirated' screen/sensor combinations, both by day (**Figure 5.11**) and by night (**Figure 5.12**, page 126). Such damped responses can also result in the under-recording of daily maximum or minimum temperatures, particularly the latter as winds tend to be lighter around the time of minimum temperature.

Installing and maintaining thermometer screens

All louvred-type thermometer screens require a suitable stand, often of aluminium or treated steel, with four legs for stability and wind resistance (**Figures 5.1, 5.2**). The base of the stand should be buried at least 30–50 cm (12 to 18 in) below ground level, depending upon the model, and oriented so that the door of the screen when mounted on the stand will face due north in northern temperate latitudes, to prevent the Sun being able to shine on the sensors at any time (except perhaps near dawn and dusk at midsummer in high latitudes). In the southern hemisphere the door should face south, and in the tropics the screen should either be rotated according to season, or fitted with doors to both north and south, their usage varying with time of year. The screen height should be adjusted so that the air temperature sensor within the screen – not the base of the screen – is at the correct height for the country

(WMO guidelines state between 1.25 and 2 m above ground level: the standard height in the UK and Ireland is 1.25 metres, 4 to 6 feet in the United States, higher in other countries or regions, particularly where there is a high annual snowfall). The soil removed should then be replaced and packed down firmly to ensure the stand cannot move, and the grass cover reinstated.

The screen should be firmly secured to the stand (this requires two people to lift into place) using appropriate fixing brackets and bolts. It is important to ensure it is immovable once fixed to the stand, because screens have been blown off stands in severe gales – with resulting damage to the contents. In very exposed sites, some additional guying of the screen may be required.

Ready-made standard-pattern wooden Stevenson-type screens or Cotton Region Shelters are expensive, and becoming more so as they are increasingly superseded by plastic screens. It is perfectly possible to construct one, as plans and designs are still available [109]: note, however, that reasonable carpentry skills are required! Given occasional maintenance, a new, well-constructed wooden thermometer screen should last 20 or 30 years. Regular care is essential: they should be thoroughly washed, inside and out, at least twice per year (more often in areas of high atmospheric pollution loading) and external surfaces repainted at least every other year (internal surfaces in good condition need repainting less frequently). As with any exterior woodwork, high-quality gloss paint should be used and the appropriate base coats carefully applied to previously well-prepared surfaces. Wooden screens with deteriorating exterior paintwork will warm more than well-maintained gloss-white models in sunshine, and this will gradually affect the temperature readings obtained – differences of 1 or 2 degC between newly painted screens and those in need of repainting have been reported. Minor repairs should be attended to promptly, well before decay becomes established, because if the major structural members start to rot there is often little that can be done to save the rest of the structure. Sanding-down and repainting a wooden Stevenson screen is a major task and will likely involve the loss of one or two days record, so a parallel 'backup' temperature measurement system should be readied in advance to avoid any loss of readings while the work is carried out and the paint allowed to dry thoroughly before instruments are reinstated.

Cheaper self-assembly screens The same installation principles apply – the screen should be firmly secured to a post or small mast, such that it will not be blown over in strong winds. The door should open to the north in the northern hemisphere, and the thermometer/sensor unit should be fixed so that readings are made at or close to 1.25 m / 4–5 feet above ground, preferably above short grass. Fixing to walls, even north walls, is not recommended, as the different infrared response and thermal inertia of the building to which it is attached will significantly affect the readings obtained. The screen should consist of white exterior-quality (UV-resistant) plastic or gloss white painted wood, to minimise any solar heating effects on the sensors within the shelter.

AWS radiation shields or screens

AWS radiation screens come in a wide variety of shapes and sizes: all are physically much smaller than traditional louvred screens (**Figure 5.4**). Most are made of moulded plastic with a gloss white exterior (necessary to minimise any solar heating effects on the sensors within the unit), but some are black inside. The black interior

Figure 5.4 Two plastic 'multiplate' passive radiation shields – the original Gill screen on the left, Davis Instruments model 7714 on the right. (Photograph by the author)

apparently reduces solar radiation penetration, although whether this results from the black finish or merely the use of plastic with better infrared opacity has not been convincingly demonstrated. Beware of low-end models on 'budget' AWS models, some of which are grey rather than white, and too flimsy to protect the contents from overheating in strong sunshine.

Almost all AWS radiation screens are varieties of the 'multi-plate' design (**Figure 5.4**). Assuming high-quality weatherproof and ultraviolet-resistant materials are used, they should prove both durable and maintenance-free for many years and require little more than the occasional wipe down with a damp cloth. Changes in surface albedo or colour over the lifetime of the screen itself are not unknown, particularly but by no means exclusively in budget models: manufacturers' guarantees on these aspects should be sought, for such changes have been reported as resulting in unreliable temperature records within as little as 3 years [115].

Electrical temperature sensors themselves are smaller and less bulky than 'traditional' liquid-in-glass thermometers, and as the sensors are remotely displayed and/or logged the radiation shield does not have to open up every observation to enable an observer to read and/or reset the instruments. Both factors combine to reduce the size of the units, bringing benefits in reduced thermal inertia (which can be significant with louvred screens, particularly in light winds [114, 116]) and thus improved time constants (a faster response to changes in air temperature). Smaller screens are also cheaper, lighter and easier to deploy – usually a single sturdy pole firmly secured or concreted into the ground will suffice in place of a substantial metal stand, the unit affixed using the supplied brackets or other fixings. Direct attachment to (for example) an existing fence or fencepost is not recommended except as a last resort, as the fence will warm in sunshine and thus affect the readings obtained.

As with louvred thermometer screens, a well-designed radiation shield must provide protection against solar and terrestrial radiation and precipitation while permitting good natural ventilation throughput [117]. Some are much better at doing this than others, and some are frankly useless. Some expensive professional models are no better than others a third of their price. Unfortunately, it is not always obvious at first glance which is which – beware of any manufacturer claims regarding performance which are not backed by *independent* comparative side-by-side trials over a period of at least several months, using identical calibrated sensors in each type of screen tested (few online 'reviews' of consumer/budget AWSs test in any detail, or for any length of time). The best-performing units can provide a temperature record almost indistinguishable from the larger and more expensive Stevenson screen (see, for example, a review of Campbell Scientific's Met21 radiation shield, available on www.measuringtheweather.net). Some less expensive units perform almost as well – the Davis Instruments passive radiation screen (on the right in **Figure 5.4**) for example, proving much more effective than units costing several times as much in the author's tests.

'Mixing and matching' sensors and screens is perfectly possible, and it is worth spending a little more on a screen which has been shown to perform well: but pay careful attention to interior screen dimensions and sensor sizes to ensure the chosen sensor will fit comfortably and benefit from unrestricted airflow.

Aspirated radiation shields or screens

Stevenson screens and AWS radiation shields are naturally ventilated, in that air transport through the screen is solely by means of the surface wind (they are referred to as 'passive' radiation screens for this reason). In sunshine, exposed surfaces of thermometer screens or radiation shields will warm as they absorb solar radiation. The effect is slight in wind speeds greater than 3–4 m/s (**Table 5.3**), but in light winds the excess heat is less easily carried away and as a result all passive screens tend to overheat in conditions of strong sunshine and low wind speeds, the excess occasionally surpassing 2 degC.

This excess warming of louvred screens in sunshine and light winds has been known for well over a century (see, for example, [119, 120, 121]), and various instruments have been developed to measure directly the temperature of a stream of external air. The Assmann psychrometer, first described in 1887 [122], consists of a paired dry and wet bulb (two liquid-in-glass thermometers with narrow, fast-response bulbs, formerly mercury, now more usually coloured alcohol), mounted inside concentric polished steel or aluminium tubes (**Figure 5.5**) to minimise radiative effects. The wet bulb sleeve is moistened with pure water prior to operation. The device is then hung vertically in the open air and air drawn through the tubes and over the thermometers for several minutes by a clockwork-driven fan. During this time both thermometers are read every 30 seconds, and values noted once the indicated temperatures stabilise. Correctly used, the psychrometer provides an excellent indication of 'true air temperature and humidity' at ambient temperatures above 0 °C. In its original form it is less suitable for continuous automated measurements owing to the requirement for relatively frequent replenishment of the water reservoir and replacement of the wet bulb sleeve, the latter quickly becoming contaminated with atmospheric aerosol owing to the high volume of forced airflow.

The relatively recent introduction of small remote-reading sensors and low-power miniature fans has made continuously aspirated air temperature measurements both practical and cost-effective. One such device is the RM Young Model 43502 Aspirated Radiation Shield, illustrated in **Figure 5.6** (a 'recommended

Table 5.3 *Average differences (°C) between the air temperature measured in a 'Metspec' Stevenson screen and that derived from a nearby aspirated sensor (in this table, the RM Young Model 43502 Aspirated Radiation Shield), as a combination of 10 m wind speed (m/s) and global solar radiation (W/m^2, lower bound). Positive values indicate Stevenson screen warmer than aspirated value, with mean differences greater than 0.25 degrees highlighted in bold and greater than 0.5 degrees bordered. Based on 49,853 hourly values over 10 years, minimum class size 5 entries (Source: [118])*

		Wind speed m/s										
		0	1	2	3	4	5	6	7	8	9	≥ 10
	0	−0.08	−0.04	0.02	0.03	0.03	0.04	0.04	0.04	0.03	0.03	0.07
	50	−0.14	0.04	0.06	0.05	0.03	0.03	0.04	0.03	0.04	0.05	0.01
	100	−0.12	0.12	0.11	0.07	0.05	0.05	0.04	0.05	0.07		
	150	0.08	0.23	0.15	0.10	0.07	0.06	0.06	0.05	0.06		
	200	**0.29**	**0.32**	0.19	0.12	0.07	0.05	0.04	0.08	0.00		
	250	**0.39**	**0.38**	0.20	0.14	0.07	0.06	0.03	0.03	0.02		
	300	**0.50**	**0.43**	**0.26**	0.15	0.09	0.06	0.04	0.01	−0.02		
GLOBAL SOLAR RADIATION lower bound W/m^2	350	**0.54**	**0.47**	**0.27**	0.15	0.10	0.03	0.02	−0.02			
	400		**0.49**	**0.26**	0.16	0.09	0.05	0.01	0.08	0.03		
	450		**0.51**	**0.29**	0.15	0.11	0.01	−0.09	−0.06	−0.08		
	500		**0.51**	**0.31**	0.15	0.08	0.03	−0.03	0.00			
	550		**0.50**	**0.31**	0.18	0.12	0.05	0.02				
	600		**0.54**	**0.34**	0.17	0.08	0.01	0.00	−0.07			
	650		**0.56**	**0.34**	0.18	0.04	0.05	−0.02	0.00			
	700		**0.60**	**0.34**	0.18	0.07	0.06	−0.07	−0.02			
	750		**0.58**	**0.33**	0.17	0.06	−0.01	−0.08				
	800		**0.56**	**0.32**	0.19	0.13	0.08					
	850		**0.52**	**0.32**	0.16	0.04	−0.02					
	900		**0.58**	**0.40**	0.12	0.07	0.00					
	950			0.19	0.11							
	1000											
	Speed mean	−0.06	0.16	0.17	0.11	0.06	0.04	0.03	0.03	0.03	0.03	−0.01

product', see **Appendix 2**). Air is drawn upwards through the unit using the top-mounted fan, over a fast-response temperature sensor (usually PRT) mounted within coaxial PVC tubes which are thermally insulated from each other. These provide shielding from solar radiation and precipitation, thereby almost entirely eliminating external radiative influences and dissipating any internal heating effects within the body of the unit. Typical performance specifications for aspirated screens are for a radiative heating effect of 0.2 degC or less even under very intense insolation (1000 W/m^2 – a value rarely attained for more than a few minutes in temperate latitudes, even at midsummer – compare **Table 5.3**). Airflow across the temperature sensor in aspirated systems is typically 5 m/s or greater – equivalent to at least Force 3 on the Beaufort Scale – much greater than typical ventilation levels inside a Stevenson screen [112]. Such forced ventilation greatly improves contact between sensor and ambient air, reducing time constants and thus improving response times [113] and so reducing sensor lag on temperature records.

As such, temperatures indicated by aspirated units such as the RM Young shield are as close to 'true air temperature' as is conveniently and cost-effectively attainable by modern methods. WMO 'best practice' in thermometer exposure is achieved by

Figure 5.5 The Assmann psychrometer, first described in 1887, is a simple device which enabled high-accuracy aspirated measurements of air temperature and humidity to be made quickly and easily in almost any location using two calibrated liquid-in-glass thermometers. (Photograph by the author)

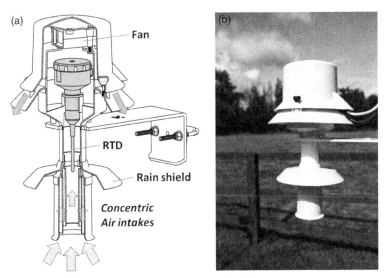

Figure 5.6 (a) Sectional view of RM Young Model 43502 Aspirated Radiation Shield (33 cm high × 20 cm diameter), as described in the text. Sectional diagram courtesy of RM Young Company, Traverse City, Michigan, USA; (b) photograph on the right shows the device in use. (Photograph by the author)

'triple redundancy' aspirated sensors (CIMO guide, [4], sections 2.1.4 and 4.3.3.1) as used, for example, in the exemplary US Climate Reference Network [73] and the US Automated Surface Observation System (ASOS), as described in the previous chapter: but see also Box, *Screens in hot, dry climates.*

Well-designed aspirated systems possess a less obvious 'futureproofing' benefit, namely that temperature measurements from different models or designs of aspirated screens can be expected to be identical – assuming of course that all other factors (such as sensor calibration, time constant and airflow velocity) remain similar (**Table 5.4**). This important advantage applies equally to individual sites as well as to regional or national network managers, for it lessens the risk of inhomogeneity in future temperature records owing to system or network replacement/update cycles (especially so as such changes are likely to occur more frequently than was the case with 'traditional' wooden thermometer screens and liquid-in-glass thermometers).

Table 5.4 *Comparison of air temperatures (mean daily maximum and minimum 0000–2359* UTC *and 24 h mean temperature) together with differences observed between two different aspirated screens, RM Young model 43502 (RMY) and the Apogee Aspirated Radiation Shield, over a 24 month period in southern England (Source: [123])*

Month	Mean max RMY (°C)	Mean max Apogee (°C)	Tmax Apogee minus RMY (k)	Mean min RMY (°C)	Mean min Apogee (°C)	Tmin Apogee minus RMY (k)	Mean temp RMY (°C)	Mean temp Apogee (°C)	Tmean Apogee minus RMY (k)
Jan	8.12	8.14	+0.02	0.35	0.36	+0.01	4.33	4.35	+0.02
Feb	8.39	8.41	+0.02	1.08	1.11	+0.03	4.81	4.82	+0.01
Mar	11.12	11.18	+0.06	2.29	2.32	+0.03	6.77	6.79	+0.02
Apr	14.72	14.76	+0.04	3.99	3.93	−0.06	9.65	9.61	−0.04
May	18.94	18.99	+0.05	6.70	6.73	+0.03	13.09	13.10	+0.01
June	22.62	22.71	+0.09	10.12	10.18	+0.06	16.57	16.61	+0.04
July	25.15	25.25	+0.10	11.92	11.98	+0.06	18.58	18.62	+0.04
Aug	22.49	22.61	+0.12	10.50	10.56	+0.06	16.54	16.60	+0.06
Sept	19.37	19.46	+0.09	7.82	7.92	+0.10	13.61	13.67	+0.06
Oct	16.06	16.12	+0.06	6.43	6.48	+0.05	11.47	11.52	+0.05
Nov	10.87	10.99	+0.12	2.73	2.81	+0.08	7.16	7.25	+0.09
Dec	8.87	8.95	+0.08	1.60	1.66	+0.06	5.58	5.64	+0.06
Year	**15.30**	**15.37**	**+0.07**	**5.32**	**5.36**	**+0.04**	**10.49**	**10.53**	**+0.04**

Screens in hot, dry climates

WMO published the results of a 12 month comparison of 18 different types of thermometer screens (11 'passive', 7 aspirated) at Ghardaïa, Algeria (32.6°N, 3.8°E, 468 m above sea level) in the northern Sahara desert [124]. It might be expected that aspirated screens would provide the most representative temperature readings in these hot, dry desert conditions, but in fact their results were 'disappointing', partly because airborne dust and sand reduced the ventilation efficiency of the units in the trial. Most small passive multi-plate plastic radiation shields performed well. The large Stevenson screens provided 'very good results', although with significant lag.

Not all sites are suitable for aspirated measurements, however, owing to the need for a reliable power supply to operate the fan, drawing typically 6 W from a 12 v source. Where a mains/utility power supply is available nearby this is not usually an insuperable obstacle, but for remote sites the power drain from one or more 'always-on' ventilation fans may exceed the power drawn by all other sensors and datalogger combined. During daylight this power need can often be met from solar cells or wind turbines, but to maintain ventilation during the night, for long periods of dull or calm weather, or in the polar regions during the winter months, a substantial battery-based storage system is essential. The fan mechanism on aspirated screens also requires regular maintenance (and occasional replacement), again making such systems less suitable for exposed or remote sites. During periods of power failure (or fan failure) air temperature readings from aspirated screens quickly become unreliable, particularly under light wind conditions and/or strong sunshine, as natural ventilation for the sensor is often limited, if any. One way to minimise this risk is to adopt a triple-redundancy approach, as in WMO's 'best practice' recommendation, although of course this adds expense and requires a yet larger power supply. A more sustainable approach comes from Apogee Instruments [125], whose aspirated radiation shield draws just 1 W at full speed (at 12 v). Further, to conserve power, fan speed can be reduced to draw 0.3 W at night or when wind speed is greater than 3 m/s without loss of accuracy. In terms of lifetime, when continuously operated at full speed the fan lifetime is rated at 50,000 hours or about 6 years. A comparison of daily maximum and minimum temperatures logged in adjacent RM Young and Apogee aspirated radiation shields over a 2 year period is given in **Table 5.4**. Within the limits of instrumental error, the records from the two shields were almost identical, although there is a suggestion that the lower-power fan in the Apogee unit was perhaps slightly less effective than the RM Young model in summer daytime radiative conditions. A longer comparison period would be needed to confirm this conclusion.

Two minor operational points regarding aspirated temperature records should be mentioned. The first is to note that the temperature sensor exposed in the entry airflow corridor, and the walls and interior of the fan system, can quickly become very dirty owing to deposition of aerosols and particulates from inbound air, particularly evident during episodes of high particulate loading. The resultant darkening of the sensor and the entry chamber increases sensitivity to reflected terrestrial radiation, particularly where the underlying ground surface is dry, and can result in artificially high measured temperatures. The error in a moderately blackened sensor/airflow inlet tube on a Young aspirated sensor can reach half a degree Celsius or more, an error which is not otherwise obvious without a comparison reference value. It is advisable to ensure the sensor and the inlet tube are given a quick wipe-down with a damp cloth every month or so, while to keep dust and dirt accumulations (and insect populations) to a minimum the fan should be switched off and the interior assembly given a thorough clean with a soft brush at least twice per year. Proactive maintenance of these areas will greatly prolong the life of the fan, but it is always advisable to have a spare at hand to avoid delays in sourcing and fitting a replacement because there are normally few if any prior indications of imminent motor failure.

A second 'quirk' of aspirated temperature records relates to occasions of fog or persistent very damp air, such as fine drizzle. In these conditions, inbound airflow will deposit a fine film of moisture on the temperature sensor, making it a near-perfect wet bulb. In conditions of saturation or near-saturation the temperature shown by paired dry bulb and wet bulb thermometers will be identical, or very nearly so (more on this in Chapter 8, *Measuring humidity*), but when the air begins to dry out, as when fog clears suddenly, for example, the wetted aspirated sensor continues to act as a wet bulb for some time, before eventually drying out. At this point the indicated aspirated temperature rises quickly back to the dry bulb temperature. This rapid rise is a known artefact of aspirated methods of measuring air temperature. The delay in reaching the true air temperature (as, for example, shown in the adjacent Stevenson screen record, **Figure 5.7**) occasionally becomes evident when comparing the mean daily temperature between aspirated and screen records.

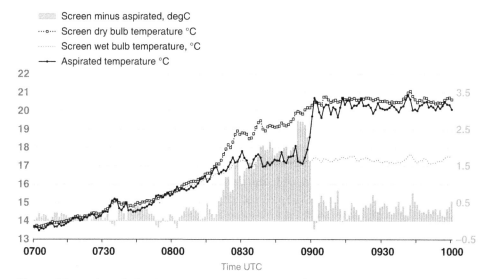

Figure 5.7 A plot of simultaneous Stevenson screen (dry and wet bulb) and aspirated temperatures (°C, left scale) on an occasion when overnight wet fog left the aspirated temperature sensor coated in water and acting as a wet bulb (compare with the plotted screen wet bulb temperature) for some time following fog clearance 07–08 h. The aspirated sensor dried out just before 0900 UTC, quickly rising to match the screen dry bulb temperature after reading nearly 3 degC lower just beforehand (grey columns, right scale) [126].

Why not aspirate a Stevenson screen or Cotton Region Shelter?

Retrofitting entire Stevenson screens with fans to provide aspiration has previously been attempted, but ventilating the much greater internal volume of a Stevenson screen or Cotton Region Shelter versus that of a typical commercial aspirated screen such as the RM Young unit necessitates a much more powerful fan, and poses difficulties with both power supply and the disposal of waste heat from the fan mechanism. A combination of smaller, low-power personal-computer type fans mounted inside the screen and close to small, fast sensors may provide a way forward, however, and research continues.

Hybrid or part-time aspirated systems

One manufacturer markets a version of their AWS which uses a solar panel to drive an aspirated fan system by day, but which reverts to passive ventilation after sunset – and, presumably, on any other occasions when solar energy is insufficient, such as under thick cloud cover, in high latitudes during winter months and so on; accordingly, it is poorly suited to temperate or high-latitude sites. The flaw in the logic is that maximum temperatures tend to be recorded at or close to the diurnal wind speed maximum, when artificial ventilation is of less benefit. This is in contrast to minimum temperatures, which tend to be recorded close to sunrise when wind speeds are often at their lowest and as a result screen + sensor response times can become substantial (tens of minutes), and the recorded minimum temperature may be too high as a result [116]. A record which may or may not be aspirated at particular times of day is not directly comparable with other records which are either passively or actively aspirated. For these and other reasons, hybrid fan-aspirated AWSs are not recommended.

Other types of thermometer screen

In some parts of the world, many nineteenth- or early twentieth-century temperature records were made in 'thatched screens', particularly where wood was scarce. These were large, open-plan shelters with roofs and sides made from local materials, often palm fronds (a good insulator). Very few such screens remain in operation. **Figure 5.8** shows one still in use today at the Hong Kong Observatory, continuing a daily record which began in January 1884. Records from a conventional Stevenson screen are made a few metres away, and have also been maintained in nearby King's Park since 1951 [127].

Figure 5.8 The thatched screen in the grounds of the Hong Kong Observatory, in use for over 140 years. (a) Screen interior: traditional liquid-in-glass thermometers remain in use for historical continuity for the time being, while electronic sensors provide an overlap record. (b) Screen exterior view. (Photographs by the author) (Photographs by the author)

Do different types of thermometer screen give different results?

Yes. Differences are greatest in strong sunshine and light winds. In most cases the readings differ only slightly, but even a few tenths of a degree are sufficient to damage the continuity of a long-term temperature record, for example, or when comparing records across a limited geographical area, as in urban heat island studies. Numerous side-by-side measurements made in different climatic regimes around the world using sensors exposed in louvred screens, small plastic AWS radiation shelters and aspirated screens show that the results obtained can differ wildly in some conditions: such trials and comparative analyses have been documented by WMO [105, 124, 128], within the International Standards Organization [106], in the United States [75, 89, 111, 129], UK [80, 110, 130], Australia [131], Spain [132], the Netherlands [133], Japan [134], and Norway and Sweden [135, 136] amongst others. **Table 5.5** summarises the results from the many and diverse screen trials held around the world, while the following section provides sample comparisons between simultaneous records made within a modern Metspec Stevenson screen and the Young aspirated screen.

Table 5.5 *Thermometer screen types compared. See text for details and references.*

	Advantages	Disadvantages
Wooden louvred screens *Examples*: Stevenson screen, Cotton Region Shelter	Still the current standard measurement benchmark in many countries Ideal housing for manually read traditional thermometry where still in use	Relatively expensive Require regular maintenance including annual painting Overheat in sunshine, particularly in light winds, owing to low ventilation throughput Less responsive than smaller or aspirated screens, owing to considerable thermal inertia (due to bulk) and reduced ventilation Manual instruments require screen to be open for duration of the observation Requires substantial stand
Plastic Stevenson-type screens *Examples*: Metspec standard within UK Met Office and other countries	Results almost indistinguishable from wooden models and thus valid substitute Minimal maintenance requirements Ideal housing for manually read traditional thermometry where still used Little evidence of surface deterioration after two decades	Relatively expensive Overheat in sunshine, particularly in light winds, owing to low ventilation throughput Less responsive than smaller or aspirated screens, owing to considerable thermal inertia (due to bulk) and reduced ventilation Manual instruments require screen to be open for duration of the observation Requires substantial stand
Small plastic AWS radiation shields *Example*: NOAA MMTS, Davis Instruments Vantage Pro2	Much cheaper Lighter – less thermal inertia, more responsive Ideal for small sensors	More responsive than records from louvred screens and so not fully homogeneous with the latter

Table 5.5 (*cont.*)

	Advantages	Disadvantages
Small plastic AWS radiation shields, *continued*	Remote-reading sensors mean screen can be sited away from buildings, etc. No need to open housing to make observation Low maintenance Easy mounting on small mast or tripod	Wide variations in performance – some are dreadful No clear leading design or model to consolidate standards Too small to house conventional thermometers
Aspirated screens *Examples*: USCRN and ASOS networks, RM Young model 43502, Apogee Aspirated Radiation Shield	**WMO Best Practice method** Reliable and cost-effective means to measure 'true air temperature' Highly responsive, particularly when fitted with short time constant sensor(s) Consistent data between different types and models – important for record continuity Ideal for small sensors Low maintenance No need to open housing to make observation	More responsive – thus records not fully comparable with other screen measurements, or homogeneous with existing records made in louvred or radiation screens Requires mains power or solar power/battery combination for 24 hour operation Readings quickly become invalid if power or fan fails Cannot house traditional thermometers May become unreliable in hot, dusty climates

Comparisons of air temperatures within a Stevenson screen and an aspirated shield

This is a fairly major topic, and there is insufficient space in this chapter to go into any great depth. Instead, average hourly temperature differences between a Stevenson screen and an aspirated shield are given, together with three examples (two by day, one by night) showing simultaneous 1 minute mean air temperature records over a 3 hour period; see also **Table 5.3** for a comparison in differing wind speed and solar radiation conditions. All example cases utilise the records from a standard Metspec screen and a nearby RM Young aspirated shield at an observatory site in southern England (51°N, 1°W), using identical sensors (calibrated twice annually) mounted at 1.25 m above ground level.

Figure 5.9 shows average differences by month and hour UTC over a 10 year period: positive values show the Stevenson screen warmer than the aspirated value, and vice versa. The largest differences occur near sunrise and sunset at all times of year, at a time when temperatures are changing relatively rapidly: the slower time constant of the Stevenson screen compared with the aspirated unit results in a slight lag, particularly in the lighter winds around the time of the daily minimum temperature [114, 116]. The most obvious feature is relative warming of the Stevenson screen by day, typically by 0.3 degC during summer afternoons. Peak daytime warming does not coincide with peak solar radiation (around local noon), but instead with maximum warming (peak terrestrial radiation) during the afternoon. **Table 5.3** shows the

Hour UTC

	0000	0100	0200	0300	0400	0500	0600	0700	0800	0900	1000	1100	1200	1300	1400	1500	1600	1700	1800	1900	2000	2100	2200	2300	Mean
Jan	-0.03	-0.03	-0.03	-0.03	-0.04	-0.05	-0.05	-0.04	-0.07	0.00	0.05	0.07	0.11	0.11	0.12	0.08	0.03	0.00	-0.01	-0.03	-0.04	-0.05	-0.05	-0.05	0.00
Feb	-0.04	-0.03	-0.03	-0.03	-0.04	-0.04	-0.04	-0.04	-0.02	0.05	0.06	0.06	0.07	0.09	0.09	0.12	0.13	0.09	0.06	-0.01	0.00	-0.03	-0.04	-0.03	0.02
Mar	-0.06	-0.07	-0.04	-0.05	-0.06	-0.04	-0.06	-0.07	0.02	0.07	0.07	0.07	0.06	0.08	0.09	0.10	0.11	0.10	0.08	0.03	-0.03	-0.03	-0.05	-0.06	0.01
Apr	-0.08	-0.09	-0.10	-0.11	-0.10	-0.11	-0.19	0.01	0.16	0.17	0.15	0.14	0.13	0.14	0.11	0.15	0.14	0.14	0.14	0.13	0.04	-0.03	-0.05	-0.06	0.03
May	-0.04	-0.03	-0.08	-0.06	-0.13	-0.20	-0.13	0.16	0.19	0.20	0.20	0.19	0.18	0.20	0.20	0.20	0.18	0.19	0.22	0.25	0.16	0.07	0.04	-0.04	0.09
June	0.02	-0.02	-0.01	-0.03	-0.07	-0.14	-0.14	0.16	0.20	0.23	0.21	0.19	0.20	0.25	0.24	0.23	0.21	0.22	0.24	0.31	0.25	0.14	0.09	0.04	0.12
July	0.02	-0.02	-0.02	-0.02	-0.05	-0.15	-0.21	0.20	0.23	0.21	0.22	0.21	0.22	0.25	0.25	0.25	0.24	0.24	0.25	0.34	0.28	0.14	0.07	0.03	0.13
Aug	-0.03	0.00	-0.01	0.00	-0.03	-0.04	-0.17	0.05	0.15	0.22	0.20	0.20	0.20	0.22	0.24	0.24	0.24	0.23	0.19	0.23	0.18	0.10	0.04	0.01	0.11
Sept	-0.03	-0.06	-0.04	-0.06	-0.04	-0.05	-0.06	-0.06	0.11	0.23	0.22	0.22	0.23	0.25	0.27	0.27	0.30	0.25	0.24	0.16	0.04	0.02	0.01	-0.02	0.10
Oct	-0.02	-0.04	-0.05	-0.02	-0.01	-0.02	-0.03	-0.03	0.05	0.16	0.17	0.18	0.17	0.21	0.22	0.25	0.27	0.19	0.09	0.03	0.00	0.00	-0.01	-0.01	0.07
Nov	-0.02	-0.02	-0.02	-0.01	-0.02	-0.03	-0.02	0.00	-0.02	0.03	0.07	0.12	0.12	0.14	0.16	0.19	0.12	0.06	0.02	0.01	0.02	-0.01	-0.01	-0.01	0.04
Dec	-0.01	-0.02	-0.01	-0.02	-0.01	-0.01	0.01	0.00	0.00	-0.01	0.01	0.11	0.11	0.11	0.13	0.17	0.09	0.03	0.01	0.00	0.02	0.01	-0.01	-0.01	0.03
Mean	-0.03	-0.04	-0.04	-0.04	-0.05	-0.07	-0.09	0.03	0.09	0.12	0.13	0.15	0.15	0.17	0.18	0.19	0.18	0.15	0.13	0.12	0.08	0.03	0.00	-0.02	0.06

Figure 5.9 Average hourly differences (degrees Celsius) between air temperatures in a Stevenson screen and those in an adjacent aspirated screen, by month, at a site in southern England over a 10 year period (87 608 valid observations, 99.93 per cent data availability). Positive numbers indicate screen warmer than aspirated, and vice versa [137].

degree of warming as a function of solar radiation and wind speed: differences increase with solar radiation and decrease with wind speed, occasionally reaching almost 3 degrees Celsius. Perhaps surprisingly, some of the greatest differences occur very occasionally during the *winter* half-year, at times of low solar elevation sunshine in light winds.

The three examples which follow are fairly typical; each plot is 3 hours duration. **Figure 5.10** shows a midwinter day with unbroken sunshine and light winds. Near-normal incidence solar radiation at low elevation (just 15 degrees at noon) warmed the louvres on the south side of the screen. A light southerly wind, 0.1–0.2 m/s at 3 m above ground, advected air warmed by passage over the louvres into the screen interior, resulting in the observed screen temperature rising to about 2 degC higher than the aspirated value (grey columns, right-hand scale). A little after 1220 UTC, the light southerly wind backed to north-easterly for about 40 minutes, maintaining a similar speed. From this point onwards, air entering the screen was drawn over the north exterior face of the screen, which was cooler because it received no direct sunshine. The excess warming rapidly ebbed away, only to reappear once the light southerly wind resumed shortly after 1300 UTC, the difference quickly attaining 2 degC once more. The maximum temperature recorded in the Stevenson screen was 4.4 °C at 1338 UTC, but in the aspirated screen 2.7 °C two minutes earlier, a difference of 1.7 degC. It is perhaps surprising to note the magnitude of the warming of the south face of the screen, even in mid-latitude midwinter, and its direct dependence upon wind direction; some of the largest differences between screen and aspirated temperatures occur with low solar angles in the winter months.

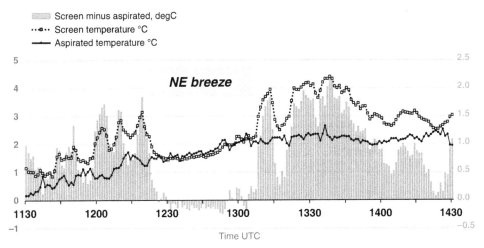

Figure 5.10 A comparison of 1 minute mean temperatures recorded in a Stevenson screen and a nearby aspirated screen around noon on a sunny midwinter's day in southern England. Unbroken low-elevation sunshine warmed the south side of the screen, warmer air then being advected into the screen on a light southerly wind. Between 1221 and 1305 UTC the surface wind backed to north-easterly, removing the source of warm air and temperatures within the screen fell back, to rise once more as the southerly wind resumed thereafter. The plot shows both screen and aspirated temperatures (°C, left scale) and the difference between the two (columns, degC, right scale); for details, see text [138].

Figure 5.11 shows a daytime example from midsummer. In strong sunshine and light winds (typically 2–3 m/s at 10 m above ground), the aspirated temperature sensor is clearly much more responsive to near-surface turbulent mixing processes than the Stevenson screen record. Although the latter's temperature remains typically 0.5–1 degC above the aspirated value (grey columns, right scale), it is notable that the maximum aspirated temperature during this 3 hour period (33.63 °C) fractionally exceeded the equivalent Stevenson screen value (33.59 °C) as a result of its lower (faster) time constant [138].

Finally, **Figure 5.12** shows a midwinter nocturnal example. On clear nights with light winds, radiational cooling results in ground surface temperatures falling more quickly than air immediately above the surface, resulting in stratification of layers of air

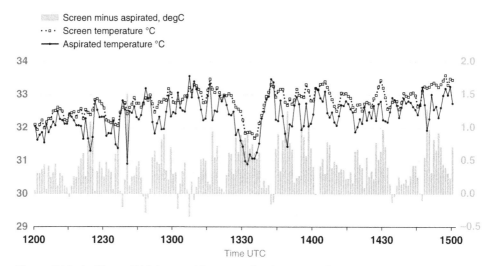

Figure 5.11 As Figure 5.10, but a midsummer daytime example.

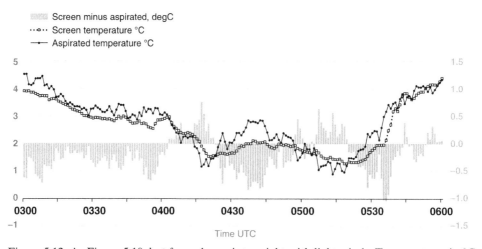

Figure 5.12 As Figure 5.10, but for a clear winter night with light winds. Temperatures in °C (left scale): grey columns show the difference between the two (right scale, degC) [138].

nearest the ground. The cooling gradually extends upwards towards screen level by continued radiational cooling and gentle turbulent mixing, but if the top of the cooled layer is gently ruffled by a temporary increase in wind speed and/or change in wind direction, then colder air may be stirred upwards. Such disturbances may last for only a few seconds. If the 'ruffled' air layer is close to screen height, the brief change in temperature may last only long enough to be sampled and detected by a temperature sensor with sufficiently fast response (the aspirated unit). In near-calm conditions, airflow within the Stevenson screen is close to zero and external air exchange very slow. Lack of ventilation, and the resulting lengthy sensor response time, ensure that the response of the temperature sensor within the screen is muted, if any. Such conditions are not uncommon at any time of year, and occasionally result in significant differences in minimum temperature between the two records. On this occasion, the minimum air temperature by the aspirated sensor was 0.89 °C at 0513 UTC, 0.38 degC below the Stevenson screen minimum (1.27 °C at 0518 UTC), despite the aspirated temperature being *above* the Stevenson screen value for the majority of the period plotted in this example (133 minutes out of 180).

Which type of thermometer screen is best?

Although WMO 'best practice' recommendation for measuring air temperature is to adopt an aspirated method, it is not always feasible to do so (especially where power supplies are an issue). It is also evident from the previous section that aspirated measurements, although certainly closer to 'true air temperature' than any alternative at a similar price point, are unlikely to be perfectly homogenous with existing long-period records made using traditional screens and instruments. Any significant change of air temperature instrumentation (for example, migration from liquid-in-glass thermometers within a louvred screen to logged electronic sensors using an AWS or aspirated radiation shield) should be recorded meticulously in site metadata and 'side-by-side' comparisons made for a period of at least 2 years in line with WMO recommendations (CIMO guide, [4], section 4.3.2), to quantify any impact on existing records. For sites with more than five decades of temperature records, an overlap period of 5 years or 10 per cent of the existing record length is preferable. Where compatibility with existing standards and/or existing records is important or in doubt, seek guidance from the national or regional weather service or climatological network operator prior to the planned change.

Some entry-level and budget weather stations do not include any form of radiation screen for the temperature sensors, while those that do will not normally provide any choice of radiation screens (and the component included can be expected to accommodate only the manufacturer's proprietary sensors). Very few systems below 'professional' level (Chapter 2, *Choosing a weather station*) will offer optional aspirated units which can be fitted in place of their standard build. Some budget AWS radiation shields are very small and exhibit good ventilation and minimal thermal inertia, and thus offer a fast response, but shielding from direct or reflected solar radiation is demonstrably insufficient because they overheat badly in sunshine (and become too cold on clear nights), and as such cannot be recommended. That said, however, the best patterns of AWS radiation screens can provide both a faster response and a closer approximation to 'true air temperature' than traditional louvred screens, owing to lower thermal inertia and better natural ventilation. Unfortunately, mere build and appearance are little help in determining their real-world performance characteristics, although those that are a little larger (typically constructed of a minimum of six or seven 'inverted saucers') tend to provide better results. Very few have been evaluated alongside conventional 'standard' screens to assess their effectiveness; exceptions are the Vantage Vue and Vantage Pro2 unit from Davis Instruments and Campbell Scientific's Met21 screen [79, 80].

New Stevenson-type louvred screens, whether wooden or plastic, are not cheap, but with occasional minor maintenance they should last for decades and can be viewed as an investment. Louvred screens can also usually contain one or more conventional thermometers, useful for those who maintain records using existing instrumentation and who wish to run an AWS alongside conventional thermometry, or who retain liquid-in-glass thermometers for now as a backup or as a calibration check on electronic sensors. Check beforehand whether the AWS sensors or legacy thermometers can fit (and be securely affixed into) the unit chosen.

For those looking for a less expensive way to get started, cheaper alternatives are available where compatibility with existing standard methods is less important – a home-made shelter made from white-painted flower-pot bases is better than nothing at all. Relatively inexpensive self-assembly wooden screens, and various self-build patterns, can be found on the Internet – for example, www.weatherfor schools.me.uk/html/weatherboxes.html. Some equipment suppliers sell simple wooden screens, in ready-made or build-it-yourself kit versions. Records from such screens will not be directly comparable with those made in a Stevenson-type louvred screen, but they are a fraction of the price and will provide reasonable protection from radiative effects (both solar and terrestrial) and from rainfall. Most contain ample room for a couple of electronic/digital sensors. If considering the purchase of an entry-level or budget AWS where no radiation shield is included, one of these little screens, properly exposed, will significantly improve the measurement of ambient air temperatures. An appropriate pole or stand is also required to expose the screen at the correct height above the ground surface. Avoid mounting on a fencepost, or on a north wall.

Third-party suppliers of thermometer screens, including wooden and plastic Stevenson screens and Cotton Region Shelters, are listed in **Appendix 3**.

Temperature sensors

Almost any physical property of a substance which is a function of temperature can be used as the basis for indicating temperature. Traditionally, the expansion of mercury or alcohol within liquid-in-glass thermometers fitted with a graduated scale has been the main thermometric mechanism, but statutory restrictions on the use of mercury following the UN Minamata Convention in 2013 and stringent WMO guidelines [5] resulted in the termination of manufacturing and supply of all mercury-based instruments (including meteorological thermometers and barometers). As a result, most countries have withdrawn such instruments from use, although the continued use of such legacy instruments as are already in place may be sanctioned by local legislation (although breakages cannot then be replaced). Extension of legacy instruments for a few years is particularly important where overlapping periods of record between 'traditional' and modern instruments are undertaken to assess record homogeneity. Operational details covering legacy liquid-in-glass thermometers, and potential alternatives, are covered briefly in **Appendix 4**.

For almost all current purposes, however, electrical temperature sensors or resistance temperature devices (RTDs), displayed and/or logged by electronic means, now represent the dominant meteorological temperature measurement mechanism, and these are covered in some detail in the remainder of this chapter.

Automating existing manual weather stations

It is often the case that changes arise owing to the cessation of manual observations, for whatever reason, and the automation of the observing site. Unfortunately, it is also often very evident that the withdrawal of daily observer visits results in a decrease in record quality and reliability. Minor faults – such as a blocked raingauge, or instruments becoming buried by snowfall – can remain unnoticed and uncorrected for days or even weeks at a time, while 'local' maintenance such as grass cutting and fence repairs are often reduced or withdrawn altogether by the host authority. Enclosure and instruments can quickly become unkempt or overgrown (**Figure 5.13**). Unfortunately, the cumulative effect is often damage to the continuity, reliability and quality of the observational record, and the risk of this happening should be carefully considered when proposing to automate an existing site.

Resistance Temperature Devices (RTDs)

Metals or semiconductors whose electrical resistance varies with temperature are particularly useful as temperature sensors, because Ohm's law can be used to determine resistance given accurate measures of voltage and current in an electrical circuit; the measurements and calculations required lend themselves well to remote logging applications. The sensors themselves can also be made very small, much smaller than a traditional liquid-in-glass thermometer, for example, and as stated previously this confers a very considerable advantage in much lower (faster) time constants (see also **Appendix 1**). Because of their small size, they do not necessarily require relatively large louvred thermometer screens, and housing in a smaller screen (particularly when aspirated) will improve responsiveness, although for reasons of consistency and homogeneity of record many countries continue to expose electrical sensors within Stevenson-type screens.

There are two types of electrical sensor in common meteorological usage – *the platinum resistance thermometer*, or PRT, and the thermally sensitive resistor or *thermistor*. For more detailed technical information on these – and other – sensor types, the reader is referred to more specialist works [139, 140].

Platinum resistance thermometers (PRTs)

The resistance of platinum (Pt) varies significantly with temperature in an almost linear fashion over a very wide range in temperatures, and this property makes it a popular choice for electrical temperature sensors. Professional-quality AWS systems use PRTs manufactured to repeatability tolerances better than traditional liquid-in-glass thermometers, better than ± 0.1 degC over a typical range of temperatures (see Box, *PRT classes*). The sensors themselves are constructed from platinum wire or more usually a thin-film chip (platinum deposited on an alumina

Figure 5.13 The climatological station at Kew Gardens in west London. This is a well-exposed and representative site, established in 1981 following the closure of the nearby Kew Observatory. Until 2019 this site held the record for the highest air temperature reliably measured in the British Isles, namely 38.1 °C on 10 August 2003. Manual observations ceased when the site was automated in 2007, and the site maintenance quickly deteriorated, as evident from these two photographs, taken before automation (a) and following automation (b). (Photographs by the author)

substrate). Such sensors are small (2 mm thin-film chip, **Figure 5.14**), both to minimise radiative effects [141] and optimise their time constant [113]. They are usually mounted within a narrow steel sheath (**Figure 5.15**), using conducting thermal paste or resin, to provide some physical protection, although doing so does degrade their time constant somewhat. Most have four leads for datalogger connection, allowing for nulling out the resistance of the connecting leads. Calibration is easier for PRT probes than for other sensors, usually requiring only one fixed point (see Chapter 15, *Calibration*). Calibration stability is usually good – although

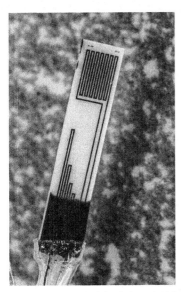

Figure 5.14 A typical 'bare' thin-film platinum resistance temperature sensor, the sensitive 'grid' being 2 mm across. Sensors of this type are normally enclosed within a narrow steel sheath, typically 3 mm diameter or less, to provide some degree of physical protection as the base chip and its connections are otherwise quite fragile. (Photograph by the author)

Figure 5.15 A 3 mm diameter platinum resistance thermometer (PRT) mounted vertically within a Stevenson-type thermometer screen. Mounting the sensor vertically upwards minimises the risk of condensation on the sensor cable running downwards and accumulating as a drip on the tip of the sensor sheath, which would then act as a wet bulb. (Photograph by the author)

calibration checks should be undertaken at least annually, preferably twice per year, as slow calibration drift may not become obvious until it has reached several tenths of a degree, by which time months or even years of records may have been damaged beyond repair.

Sensor size and mounting

The time constant of a PRT (see **Appendix 1**) is primarily a function of the diameter of the probe (smaller sensors = lower time constant and thus faster response) together with the speed of ventilation across the device (greater ventilation rates = lower time constant) [113]. Careful measurements of airflow within Stevenson-type louvred screens show that ventilation is reduced to about 7 per cent of the external wind speed at 10 m (10 per cent of that at 2 m), when external wind speeds are above 1 m/s [112]. Such low ventilation rates are not conducive to optimal time constants. For the time constant of a PRT to attain the WMO CIMO guideline of 63 per cent response within 20 s, the sheathed PRT should be no more than 3 mm in diameter, and preferably 2 mm or less. Larger sensors do not meet the time constant guidelines (a 6 mm diameter sensor typically takes four times as long as a 3 mm diameter sensor to respond to the same change in temperature [113]), while smaller sensors tend to be insufficiently robust enough for current operational deployment, although research into smaller, faster sensors is ongoing.

Within a louvred screen, the dry bulb PRT should be mounted vertically, with the sensor at the top, as close to the middle of the screen volume as is practical to maximise ventilation within and through the screen (**Figure 5.15**). Vertical mounting avoids condensation on the PRT cable running down the sheath and forming a drip, which may take some time to evaporate: in the meantime, the sensor will act as a wet bulb rather than a dry bulb and thus indicate a lower reading than the true air temperature. A vertical mounting also allows rapid convective loss of any minor self-heating of the sensor itself owing to the measurement current.

If the PRT is logged by a datalogger, whether connected directly or via a powered amplifier, the datalogger and/or amplifier must remain *outside* the screen to avoid any waste heat from the electronics affecting the sensors.

PRT standards

International standard IEC 60751, revised in 2022, sets out the requirements and temperature/resistance relationship for industrial platinum temperature sensors, of which meteorological sensors are but one sub-group (see Box, *PRT classes*). The standard covers temperature responses from –196 °C to 660 °C, although sensors intended for meteorological applications typically cover a more limited range appropriate to atmospheric conditions.

When logged using a datalogger, maximum and minimum temperatures over any specified period will normally be extracted using software – usually with the time of occurrence – obviating the need for three separate traditional thermometers. Platinum resistance thermometers are typically used as dry bulb thermometers but are easily configured as wet bulbs for accurate determinations of atmospheric humidity (see Chapter 8, *Measuring humidity*).

PRT classes

IEC 60751:2022 specifies a standard RTD to have an electrical resistance of 100.00 Ω at 0.0 °C and a temperature coefficient of resistance defined by formula of 0.00385 Ω/°C between 0 and 100 °C. Many dataloggers are set up by default to accept such 'standard' PRTs, which are known as 'Pt100' sensors (500 Ω and 1000 Ω variants are also available), and the standard nature of such devices and interfaces renders them easily interchangeable. By measuring the PRT resistance, the sampled temperature can be quickly and easily determined by calculation within the datalogger and output in relevant units; for example, the temperature of a Pt100 sensor whose resistance is known to be 100.132 Ω is 34.29 °C (0 °C + 0.132/0.00385).

Four resistance tolerances for PRT RTD thermometer devices are set out in IEC60751:2022, essentially defining the confidence in the resistance versus temperature characteristics for the sensor type – Classes AA, A, B and C. The larger the element tolerance, the more the sensor may deviate from the 'standard' resistance characteristic curve, and the more variation possible between sensors. Within high-accuracy meteorology applications, it is important to be sure one sensor can be swapped out for another without introducing significant calibration errors (although the calibration should always be checked, and if necessary adjusted by datalogger offset, when sensors are installed). Sensors will often be characterised as a fraction of these classes – for example '1/3 B' would indicate a sensor whose tolerance was 1/3 of the Class B specification, namely within 0.03 degC. It is important to note, however, that IEC 60751:2022 specifies different temperature ranges for each class according to whether the sensor is wire-wound or thin film construction, but whereas wire-wound sensors are valid for all temperatures between –50 and 250 °C, thin-film sensors in class AA are only specified over the range 0 to 150 °C; Class A thin-film sensors cover the range –30 to 300 °C, and Class B –50 to 300 °C. Thus meteorological sensors will normally be chosen from Class A Pt100 PRTs, no more than 3 mm in diameter, for which maximum errors remain within 0.2 degC between –20 and +20 °C. The actual error can be reduced to within 0.05 degC by individual calibration of each sensor prior to installation, but should be checked at least annually thereafter as set out in Chapter 15, *Calibration*.

The action of applying a voltage across a PRT will cause self-heating of the sensor element, and this can quickly exceed expected tolerances set out in **Table 5.6**. For this reason, datalogger programming must limit voltage excitation of PRTs to very short low-voltage pulses, typically no more than 2500 mV (often much less) and a measurement time of 250 µs, followed by a settling time of a few milliseconds before the pulse is repeated. Doing so ensures self-heating remains negligible.

Table 5.6 *IEC 60751:2022 platinum resistance thermometers class specifications*

IEC 60751:2022 PRT class	Tolerance °C	Range error at –20 °C	Range error at +20 °C
Class AA	± 0.1 + range error (0.0017 Δt)	*Not supported*	± 0.1 + 0.034
Class A	± 0.15 + range error (0.002 Δt)	± 0.15 + 0.04	± 0.15 + 0.04
Class B	± 0.3 + range error (0.005 Δt)	± 0.3 + 0.1	± 0.3 + 0.1
Class C	± 0.6 + range error (0.01 Δt)	± 0.6 + 0.2	± 0.6 + 0.2

Note that the IEC PRT class does not imply any particular time constant of the sensor, and for accurate meteorological work this should be evaluated to ensure it is fit for purpose.

Semiconductor resistance thermometers (thermistors)

A thermistor is a semiconductor device whose resistance varies significantly with temperature. Their resistance, and the variation of resistance with temperature, is considerably higher than PRTs. They are somewhat smaller than PRTs, and considerably cheaper, and as a result are used as the temperature sensor in almost all AWS systems below advanced and professional models. They offer almost all of the advantages of PRTs given earlier but are slightly less accurate (a typical error for a high-quality thermistor being ± 0.5 degC / 1 degF over the –10 °C to +35 °C / 15 °F to 95 °F range) and less stable in their calibration, although both can be optimised with regular calibration checks every couple of years.

Not surprisingly, less expensive thermistors tend to have larger errors and/or less stable calibrations, but again regular calibration checks can reduce this to manageable levels: very often site or exposure deficiencies will result in larger temperature errors than those originating from the sensor (see the WMO recommended site guidelines in **Table 5.1** for indicative values). Because of the variability between units, there is no 'standard thermistor' in the same sense as the Pt100 PRT, and because of this calibration is less repeatable between devices. It may not hold outside narrow limits, and additional corrections may be required at extremes of temperature.

As with PRTs, metal sheathed thermistor sensors should also be mounted vertically upwards where it is possible to do so, for the same reasons.

Logging requirements

Air temperature does not change as quickly as other weather elements (wind speed and solar radiation, for example): very rapid changes, of more than a few degrees Celsius within a couple of minutes, are uncommon in most climates. In any case, as noted earlier and in **Appendix 1**, time constants of the screen/sensor combination for all except aspirated screens are likely to be measured in minutes rather than seconds, and thus very frequent sample rates are largely a waste of datalogger processor and bandwidth. Sampling air temperature at 10 second intervals, and logging 60 second running averages every minute (the average of the previous 6×10 second samples), meets all WMO recommendations (see Box, *Measuring responsiveness*). Even 5 minute averages are sufficient for many climatological requirements, but hourly averages are insufficiently granular to provide accurate daily maximum and minimum temperatures.

Where supported by the logger and software, short-period running averages can be very useful to remove minor electrical noise or smooth out high-frequency natural random fluctuations (which are almost certainly faster than the sensor's ability to respond fully in any case).

Observation times

For observations to be comparable between different locations, it is preferable that the sampling and logging intervals, the times at which observations are made, and the time period covered by the daily maximum and minimum temperatures, are as nearly identical as possible. This latter topic is covered more fully in Chapter 12, *Observing hours and time standards*.

Measuring responsiveness

Enhanced responsiveness is desirable, up to a point (unlike wind speeds, for example, there is little benefit in sampling air temperature every second), but too sensitive a system will simply generate slightly higher maximum and slightly lower minimum air temperatures than those recorded by traditional instruments in a louvred screen, for no reason other than differences in instrumental responsiveness. It is for this reason that it is good practice, where supported by the logger functionality, to sample the air temperature frequently but to take a running average over a short period. Doing so also helps to iron out minor stray electrical noise or sensor/ logger resolution artefacts. However, sensors and screens cannot react instantaneously to changes in temperature. Responsiveness is quantified using a measure known as the *time constant*, which is the length of time an instrument (or screen, in this case) takes to respond to a certain fraction of a step change in a variable (this topic is covered in more detail in **Appendix 1**).

WMO CIMO guidelines ([4], Annex 1A) suggest 1 minute mean temperatures be adopted (and the highest and lowest of these logged as the period maximum and minimum temperature, respectively), as this provides some measure of compatibility with the time constant of traditional liquid-in-glass thermometers. Such a mean temperature can be considered as the average of 60 × 1 s samples, or more typically the average of 6 × 10 s spot values. (The lower frequency reduces still further self-heating of the sensor by the measurement device.) How does the time constant affect the derivation of mean temperature over a 60 s interval?

Both sensor and screen have their own time constants, the latter normally being considerably larger for traditional louvred screens. Both are very dependent upon ventilation, so temperature time constants are expressed in terms of the time taken to show a response to a fraction of an instantaneous change in temperature – usually 63 per cent – at a given ventilation speed. Response time theory, outlined in **Appendix 1**, shows that a 95 per cent response to a change in input – temperature in this case – requires three times the 63 per cent sensor time constant. From this it is easy to appreciate that, to obtain a 95 per cent response within 60 seconds for reliable 60 s means, the time constant of the temperature sensor should be 20 seconds or less.

As stated previously, for small cylindrical sensors, typical of platinum resistance temperature devices used in meteorological applications, time constants are primarily a function of both sensor diameter and ventilation rate [113, 141]. Taking a typical commercial meteorological platinum resistance temperature sensor of diameter 3 mm, laboratory tests show that a 20 s time constant can only be attained with a ventilation rate of at least 1.5 m/s. (For larger sensors, much greater ventilation is required to achieve the same time constant – a 6 mm sensor would require an unrealistic 16 m/s ventilation to respond as quickly, hence small sensors are preferred for air temperature applications.) Airflow across the temperature sensor within a typical aspirated radiation shield is 3–6 m/s, meaning the time constant for a 3 mm sensor will be less than 20 s, and thus

95 per cent changes in temperature will be fully attained within 60 s. Unfortunately, such ventilation rates are rarely attained within louvred screens.

A recent study [112], based upon 3 months comparisons between external wind speeds and internal ventilation rates within a standard Metspec Stevenson screen, showed that internal screen airflow averaged just 7 per cent of the wind speed measured nearby at 10 m above ground (or 10 per cent of the wind speed at 2 m), at wind speeds of 1 m/s or more. Over the entire comparison period, the average ventilation rate within the screen was just 0.2 m/s, implying a 63 per cent time constant of about 65 s for a 3 mm PRT. The consequence here is that the sensor will take upwards of 3 minutes to respond to 95 per cent of a step change, even before the larger time constant of the screen itself is factored in [116], and therefore logged 60 s mean temperatures will inevitably lag behind the truth when temperatures rise or fall rapidly.

From the above, it is evident that to attain the WMO guideline 20 s 63 per cent time constant, either an aspirated sensor is required, or a particularly responsive (small) sensor. At the time of going to press, only one commercial PRT had demonstrated conformance to the WMO guideline at typical in-screen ventilation rates. In reality, of course, screen-based temperature sensors continue to be widely used across the world: the demonstrated limitations of their relatively slow time constant resulting from greatly reduced ventilation rates within the screen can be offset to some extent by the explicit specification of small sensors (maximum 3 mm diameter), but in all cases 60 s mean temperatures from screen-based sensors are by definition subject to lag, and may slightly underestimate the true daily range in temperature. Recent work in Australia examines the implication of this for both climatological means and extreme conditions [142, 143], making clear that what may appear to be a 'spot' temperature reading at any instant is necessarily an integral over a period of time dictated by the time constant of the sensor in use.

In the United States, the adopted preference is for a 5 minute running average in the ASOS system, and fixed 5 minute periods in USCRN [73, 74, 81]. This is not just an academic concern, as it affects the acceptance – or not – of weather extremes. A good example is accorded by two recent heatwave events. The first affected the southern and eastern states of America in summer 2011. Dodge City, Kansas, has one of the longest continuous temperature records in the United States, commencing in 1875. The hottest day on its long record stood at 110 °F (43.3 °C). On 26 June 2011 the highest 1 minute temperature observed was 111 °F (43.9 °C). However, the value (logged on an ASOS system) was not accepted as a new record because ASOS takes the maximum temperature as the highest 5 minute running mean, which was 110 °F. Thus, the official high by the US method was 110 °F, tying rather than exceeding the previous record: by the WMO recommended method the maximum was 111 °F, which would have set a new record.

The second event related to the extreme heatwave which affected western Canada and the north-western United States in late June 2021, when the long-standing maximum temperature extreme for Washington state of 118 °F (47.8 °C) was apparently exceeded in several places. A detailed investigation conducted by the State Climate Extremes Committee subsequently examined records from a large number of reporting sites in Washington state, some official NOAA sites and others run by other bodies. It concluded that the maximum temperature of 120.0 °F (48.9 °C) recorded in a small AWS screen at site H100F within the Hanford Mesonet was valid and acceptable as a new state extreme, despite the value in question being an 'instantaneous' peak value (deriving from 5 s samples) rather than a mean over several minutes [144]. It begs the question whether the '5 minute rule' applies only to ASOS sites, and whether *ipso facto* extremes at other sites cannot be regarded as strictly comparable. It appears strange that the WMO guideline (namely 1 minute means) was not adopted, which would otherwise avoid the likelihood of similar uncertainty regarding future extremes. (I am indebted to Christopher C. Burt, US weather historian, for drawing my attention to both events.)

Many wireless-display in/out temperature displays and AWSs use sensors which are encased in a much larger block containing batteries, electronics and the like, and this is particularly so with 'all-in-one' systems (see Chapter 3, *Buying a weather station* and Chapter 4, *Site and exposure*). Owing to their relatively bulky nature, these systems can be very slow indeed in their respond to sudden changes in air temperature; the temperature sensor on the popular NetAtmo consumer AWS has a time constant calculated as 12.7 *minutes* [78], compared to the guideline 20 *seconds* or less for professional systems. It is a fact of physics that any two sensors whose time constants differ, no matter how carefully calibrated, will display or log different readings when the temperature changes – as it does almost continuously in meteorological applications. This will become evident as a lag in indicated temperature of the less responsive unit, more so when the temperature is changing rapidly (see also **Figures 3.3** and **5.9**). Lag can result in under-recording extremes – the maximum temperature being under-recorded and the minimum over-recorded – the magnitude of the effect depending upon the rate of change of temperature at the time of the extreme. Days with short but intense spells of sunshine often see an under-recording of maximum temperatures by 'slow' sensors. Although **Figures 5.11** and **5.12** were prepared using fast-response sensors, it can be appreciated that a sensor with a larger lag time would be even less likely to record peaks and troughs accurately.

One-minute summary – *Measuring the temperature of the air*

- Air temperature is one of the most important meteorological quantities, but it is also one most easily influenced by the exposure of the thermometer. Great care needs to be taken in exposing temperature sensors to ensure that, as far as possible, the instrument measures a true and representative value, which is not unduly influenced by the instrument housing, surrounding vegetation or ground cover, the presence of buildings or other objects.
- WMO guidelines specify preference for an open site on level ground, freely exposed to sunshine and wind and not shielded by, or close to, trees, buildings

and other obstructions, but recognise within the classification of sites from 1 to 5 that it is rarely possible to find a 'perfect' site. The expected degree of error that may occur in unfavourable circumstances is set out in these class definitions. Certain locations, such as hollows or rooftop sites, are best avoided, as readings obtained in these situations may bear little comparison to observations made elsewhere under standard conditions.

- Some form of shielding for the temperature sensor(s) is essential to provide protection from direct sunshine, infrared radiation from Earth and sky, and from precipitation. The main screen types – louvred (Stevenson screen, Cotton Region Shelter), small plastic radiation screens (typical of AWS systems) and aspirated screens – are covered in some detail, because the thermometer housing (or lack of it) is likely to have the largest impact upon the observed temperature. Almost any form of radiation shelter will provide better results than a bare sensor exposed to sunshine. If the AWS model chosen does not include an effective radiation screen, allow budget to purchase a suitable third-party one and use that instead.

- Traditional louvred screens can accommodate both legacy liquid-in-glass thermometers (where they remain in use) and small electronic sensors, but small AWS radiation shields will hold only the smaller electronic sensors. Aspirated units are preferred by WMO as providing a cost-effective means of measuring 'true air temperature'; they are fast in response and largely free of radiative effects, but they provide a slightly different temperature record from other standard methods. Next-generation climate monitoring networks, such as the US Climate Reference Network USCRN, have adopted multiple-redundancy aspirated methods of measuring air temperature from the outset. Aspirated radiation shields can also provide consistent data between different types and models, important for historical continuity as instruments are changed or updated over time.

- To avoid significant vertical temperature gradients near the Earth's surface, thermometer(s) to measure air temperature should be exposed between 1.25 and 2 m above ground level. In the UK and Ireland, the standard height is 1.25 m; in the United States, between 4 and 6 feet.

- Sites that have long current records of temperature made with legacy instruments in traditional louvred thermometer screens (Stevenson, Cotton Region Shelter) should not substitute an alternative method of measuring temperature (for example, electronic sensors in an aspirated screen) without a substantial overlap period, because doing so risks destroying the continuity and homogeneity of the long record. The overlap period should be a minimum of 12 months, or one-tenth of the existing station record length, whichever is the longer.

- Most air temperature measurements are now made using resistance temperature devices (RTDs), which have superseded traditional liquid-in-glass thermometers. The main types of sensors in use today are the *platinum resistance thermometer* and the *thermistor*. The former is more accurate and more repeatable, but more expensive. Both can be made very small and thus highly responsive.

- Logging intervals of 1 to 5 minutes, with shorter sampling intervals (typically 5 to 15 seconds), are sufficient for most air temperature measurement applications. Running means can be used to smooth out any stray electrical noise together with very short-period temperature fluctuations, which are of little significance in climatological measurements.

- Sheltered sites can introduce significant measurement errors, but with some care given to siting the screen and sensor(s) reasonable air temperature measurements can be made in all but the most restricted locations. Temperature records from suburban sites, even those with limited exposures, can often provide more numerous and perhaps more representative climate records for a town or city than those from more distant sites such as airfields, although the latter may have near-perfect exposures.

6 Measuring precipitation

Determining a rudimentary indication of the amount of precipitation that falls (the term includes rain, drizzle, snow, sleet, hail and so on as well as – conventionally – smaller deposits from dew, frost or fog) is not difficult: almost any bucket left out in the rain will suffice. Obtaining accurate, consistent and comparable measurements does, however, require a little more care and sophistication in technique. This chapter sets out how precipitation is measured and covers instrument types, siting and exposure (based upon World Meteorological Organization (WMO) CIMO guidelines on siting and instruments [4]), and discusses some of the pitfalls involved. Making precipitation measurements in snowfall is also covered. A brief 'One minute summary' completes the chapter.

The *measurand* – what is being measured?

'Precipitation' is defined by WMO ([4], section 6.1.1) as 'the liquid or solid products of the condensation of water vapour falling from clouds, in the form of rain, drizzle, snow, snow grains, snow pellets, hail and ice pellets; or falling from clear air in the form of diamond dust.' The more usual term *rainfall* is frequently used in a more general sense to include all types of precipitation, including the water equivalents of frozen precipitation. Contributions from deposits of fog, dew, rime and hoar frost are not included in the formal WMO definition of precipitation, but minor amounts resulting from these processes are conventionally included in precipitation measurements.

The total amount of precipitation which reaches the ground at a particular location in any given period of time is specified in terms of the depth of water (or water equivalent in the case of solid forms) that accumulates on a horizontal surface, disregarding losses through evaporation, percolation or run-off: this is the measurement given by a raingauge. Precipitation depth is normally expressed in millimetres of liquid water, although inches are still the preferred unit for public communication in the United States. A rainfall depth of 1 mm is dimensionally equivalent to 1 litre of (pure) water per square metre, or 1 kg of water per square metre, and rainfall values are sometimes more conveniently expressed in these units. Daily precipitation measurements are normally made to a precision of 0.1 or 0.2 mm, or 0.01 inches. Rainfall intensities are expressed as millimetres (or inches) per unit time, for example mm/hour.

Precipitation is one of the most variable of all weather elements, in both space and time. For this reason, most countries have a greater density of precipitation measurement locations than is the case for any other meteorological element. In recent decades, remote sensing techniques (especially by radar and satellite) have enhanced the spatial coverage of precipitation measurements, but these are not always available and necessarily rely upon accurate 'ground truth' observations for calibration and quality control. Precipitation measurements are very sensitive to exposure – particularly to the wind – and it can also be difficult to derive records representative of an area from point values provided by a ground-based instrument, particularly in areas with complex topography. For obvious reasons, precipitation measurement networks tend to be densest in well-populated areas of gentle terrain rather than in remote mountainous areas with complex topography and harsher weather. Unfortunately, the latter are often the areas with highest annual average precipitation and/or snowfall, and may well be the prime source of a city or region's water resources, requiring careful and consistent long-term measurements to monitor and maintain supplies.

How do we measure precipitation?

At this point it is pertinent to refer to a key distinction in precipitation instruments, namely 'catching' versus 'non-catching' methods. 'Catching' instruments, as the name implies, catch precipitation for subsequent measurement, and many are referred to simply as 'raingauges'. 'Non-catching' or 'optical' sensors typically infer precipitation type and intensity using algorithms based upon attenuation of one or more laser light pathways by falling precipitation. The latter class of instruments are widely used as 'present weather' detectors, and they can cope as easily with frozen precipitation as with rain, drizzle or even wet fog. The advantage of a non-catching precipitation sensor lies in its ability to sample precipitation in free air, rather than relying on varying collection efficiencies of a raingauge funnel. In theory at least, 'non-catching' sensors have the potential to provide 'ground truth' precipitation measurements from first principles, although their cost and complexity continues to restrict them to research or professional, operational meteorological monitoring (such as detecting the onset of freezing precipitation at airports). Precipitation measurements from such instruments remain subject to considerable uncertainties and at present are usually regarded as incompatible with measurements from networks of 'catching' gauges, although field trials have been conducted and are continuing [145, 146]. The remainder of this chapter should be taken as referring to 'catching' instruments unless otherwise stated, and the term 'raingauge' is used as a convenient general-purpose term to refer to precipitation monitoring devices.

For reasons of history and politics as much as differences in climate, there remain many different types of raingauge in use around the world, and as a result precipitation measurements are not strictly comparable between countries. The analysis of precipitation data is greatly simplified where common standards of equipment, siting and observation times are followed, such as those within WMO's CIMO guidelines [4] and national and international standards (for example, [145, 147]). This chapter contains WMO site classification guidelines together with country-specific details for the United States, the United Kingdom and the Republic of Ireland. Reference to the websites of individual national or state weather services [96] will provide links to current precipitation measurement practices and policies for other countries.

A global habit

The measurement of surface precipitation is the most common form of meteorological measurement made globally – a WMO survey in 1989 identified more than 150,000 manually read raingauges then in use [148]. Since then, much rationalisation of rainfall networks has taken place (more often for cost-cutting reasons than network optimisation, it has to be said), while at the same time most networks have been at least partially automated. The net result has been a general reduction in spatial density but an increase in near real-time precipitation coverage using various communication methods including Wi-Fi, wireless and radio, dedicated telephone lines or cell phone systems, or even direct-to-satellite circuits for the most remote locations. Without another WMO survey similar to the 1989 exercise, it is impossible to put a reliable figure on the current number of raingauges in use worldwide, but if both professional and consumer sites are included it is probably in excess of one million. The degree of standardisation has also decreased since 1989, which is an increasing headache for long-term homogeneity of rainfall records. Many are still read only once daily, but an increasing proportion of the global network consists of automated instruments which can provide a record of rainfall against time, typically hourly resolution or better.

Inevitably, any site counts can only represent a snapshot at any time, but at the time of writing over 7,500 sites reported rainfall into the NOAA network in the United States (**Table 6.1**), the vast majority through NOAA/NWS's voluntary

Table 6.1 *Precipitation measurement sites around the world*

Country	Network	Land area (km^2)	Population 2023 (millions)	No of raingauge sites	km^2 per raingauge	Average gauge separation (km)	Raingauges per million people
USA	NOAA networks	9,629,091	336	7547	1276	40	22
	CoCoRaHS	9,629,091	336	21 834	441	24	65
Australia	Bureau of Meteorology	7,692,024	26.3	6652	1156	38	253
United Kingdom	Met Office	242,900	68.8	3146	77	10	46
Republic of Ireland	Met Éireann	70,273	5.1	518	136	13	102
Switzerland	MeteoSwiss	41,277	8.8	618	67	9	70
France	Météo France	640,294	65.7	5520	116	12	84
Netherlands	KNMI	37,354	17.2	370	101	11	22
Germany	DWD	357,114	84.5	1868	191	16	22
India	India Met Department	3,287,263	1416	5296	621	28	4

Land areas and population (2023 estimates) from online sources. Raingauge site numbers are from enquiries to national met services and are inevitably a 'snapshot' as of March 2023 [155]. The UK numbers from the Met Office include those administered by devolved country agencies (see text); see also **Figure 6.1**. The US site numbers are taken from **Table 4.2**. No reply to the enquiry was received from Météo France, and these numbers are those supplied for the first edition in 2012. More countries are given in reference [148], although the information in that 1989 WMO report is now more than 30 years out of date.

cooperating observer programme COOP. In addition, the Community Collaborative Rain, Hail and Snow Network (CoCoRaHS) [149, 150, 151, 152, 153], founded in 1998 (see Chapter 19, *Sharing your observations*), includes more than 25,000 observation locations across 50 US states, Canada, Puerto Rico, the US Virgin Islands, the Bahamas and Guam, and data from almost 22,000 of these sites are included in the Global Historical Climate Network (GHCN-Daily) dataset (**Table 4.2**). In Australia, the raingauge network operated by the Bureau of Meteorology consisted of 6,652 sites in early 2023. Most are run by volunteers, who provide daily rainfall totals for the impressive Bureau of Meteorology weather and climate website [154]. In the United Kingdom, there were 3,146 sites reporting rainfall measurements to the UK Met Office (including synoptic and climatological sites) in early 2023 (**Figure 6.1**), of which just over half (1,618) were automated gauges. The majority (94 per cent) of the 2741 rainfall-only sites report via partner organisations (England's Environment Agency, Natural Resources Wales/Cyfoeth Naturiol Cymru in Wales, or the Scottish Environment Protection Agency SEPA). In the Republic of Ireland, there were 518 rainfall-recording sites reporting to Met Éireann at the time of writing, around 90 per cent of which are run on a voluntary basis.

Switzerland has probably the densest rainfall network in the world (**Table 6.1**), with an average of one gauge per 67 km^2, or an average gauge separation of about 9 km. The United Kingdom is close behind at one gauge per 77 km^2: for England alone, the figure is about one per 60 km^2. In Australia's wide-open spaces, the figure rises to one gauge per 1156 km^2 (about 40 km between gauges), although Australia

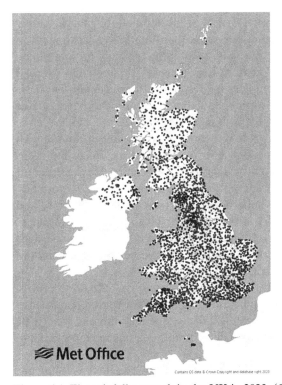

Figure 6.1 The rainfall network in the UK in 2023. (© Crown copyright 2023, the Met Office)

has five times the number of raingauges per head of population as the UK or the United States. In contrast, CoCoRaHS has 20 observers in the (extremely rural) Liberty County, Montana, area 3748 km^2 but population just 1,946, which *pro rata* equates to over 10,000 raingauges per million people ... if the raingauge density was the same in London (population 2022 9.541 million), there would be over 98,000 raingauges within the UK's capital city.

Many sites have long and homogenous rainfall records, which are extremely valuable for long-term climate change studies. Probably the oldest current rainfall record in the United States is that for Charleston, South Carolina, which has precipitation records dating back to 1738 [156], although with gaps to 1830. Surprisingly, continuous records were kept there throughout the American Civil War. In 2023 there were an impressive 1,927 raingauge sites in Australia with more than 100 years record (compared with just 223 in the UK) [157]. In Tasmania, the site at Hampshire, Neena Road has a record commencing in 1835, while records from Adelaide in South Australia and Grafton in New South Wales began in 1839 and continue today. At Sydney Observatory (**Figure 1.21**), temperature and rainfall records began in July 1858 [70]. The oldest same-site rainfall record in the British Isles is in Oxford, where monthly records at the Radcliffe Observatory site date back to January 1767, with an unbroken daily record from January 1828 (see also Chapter 1) [57].

It is easy to forget, of course, that 70 per cent of the planet's area is ocean, and that the only viable methods of assessing precipitation over oceans are satellite-based, with large measurement uncertainties.

Standard methods of measuring precipitation

There are two basic types of 'catching' raingauge, and both are described in this chapter. *Manual gauges*, often known as 'storage' or 'accumulation' gauges, simply collect liquid precipitation using a funnel, for subsequent manual measurement in a measuring cylinder or similar. Most are read once daily, typically around 9 a.m. local clock time, although in remote locations high-capacity versions may be read weekly or monthly. These are the simplest in design and construction, with no moving parts and needing no power supply, and they comprise the backbone of both current and historical rainfall networks in most countries.

Automated or *recording gauges* provide a record of the amount of rainfall with time, using a variety of mechanical, electronic or optical sensors. Recording rain-gauges are often co-located with a manual storage gauge. Because of losses inherent in mechanical recording gauges, the 'standard' rainfall measurement is normally taken from the manual raingauge where both are co-located. In the UK and Republic of Ireland the standard gauge is usually referred to as the 'checkgauge' for this reason. Sub-daily measurements from the recording gauge or gauges, such as hourly totals, are conventionally adjusted to agree with the period total from the checkgauge using a simple linear adjustment factor.

Many 'automated' gauges transmit their readings in near real time to a central collecting location using suitable communications links, and/or log their measurements using a datalogger, and at such sites the checkgauge may be read weekly or monthly rather than daily. **Figure 6.2** shows such a 'remote reading' gauge site in the English Lake District.

Site and exposure requirements are common to both gauge types and are described first.

Figure 6.2 The checkgauge and tipping-bucket raingauge at Honister Pass in the English Lake District. Rainfall records are sent by communications link every 15 minutes to an Environment Agency control centre, and hourly and daily records summed from these. Honister Pass is one of the wettest locations in the British Isles for which daily records are available, with an annual average rainfall close to 4000 mm (160 inches) per annum. In December 2015, this gauge measured 341 mm in one 24 hour period, the highest such fall on record anywhere in the British Isles [158]. (Photograph by the author)

Site and exposure requirements for precipitation measurements

As with most other meteorological observations, the purpose of precipitation records is to obtain a measurement representative of the area where the raingauge is sited, and thus the choice of site (and its documentation) is of particular import-ance. An ideal site needs to be open to the weather, but not too exposed, nor too sheltered. As most other meteorological instruments are best exposed in sites that are as open as possible to the elements, particularly the wind sensors, it is sometimes necessary to site the raingauge or raingauges in a more sheltered spot, perhaps even some distance from the other instruments. WMO guidelines suggest an accuracy of 5 per cent is attainable in precipitation measurements, but to achieve representative and comparable measurements which are consistent over time and space, a number of important factors need to be considered, the most important of which are as follows:

- *Over-exposure to wind* Very exposed sites are rarely ideal for making rainfall observations. Locations on headlands or cliffs, on windy moorlands, exposed lighthouses and even some airfields can be particularly troublesome. Locations on a slope, and particularly rooftop locations, should also be avoided. The physical presence of a raingauge causes eddies over and around the gauge, carrying away droplets that are then lost to the catch, and of course this is particularly evident in very exposed situations. Loss of catch by wind is the greatest single factor in precipitation measurement uncertainty, especially in snow, and can result in very significant under-estimation of mean annual pre-cipitation values, particularly where snowfall is a main contributor [159, 160], but it is likely that almost all precipitation gauges under-read by some amount owing to wind losses. In an attempt to reduce these, the fitting of windshields to reduce eddies around the body of the gauge is standard practice in some countries, particularly where annual snowfall is significant (these are covered later in the chapter). Much research has been carried out in recent years in attempts to make raingauges more aerodynamic and thereby partially overcome some of the worst effects of wind losses, and these are also considered later in this chapter.

- *Shelter* Conversely, too much shelter is likely to reduce the catch of a raingauge: it can be difficult to achieve a balance between shelter and over-exposure to wind. Some shelter is beneficial to reduce wind losses, but the site chosen should, as far as possible, be clear of buildings, trees and other obstructions (including other meteorological instruments) for some distance around the raingauge. The effects of shelter can vary with wind direction and speed, over time (as trees or hedges grow in the vicinity of the raingauge, for example, or if new buildings are erected nearby) and seasonally (particularly near deciduous vegetation). Shelter can affect gauge catch from perhaps 10 per cent above to as much as 80 per cent below the 'true' value [159, 160, 161, 162, 163, 164]. A slow increase in shelter over a period of years or even decades, perhaps from the growth of trees around the site, can be difficult to identify until the value of the record has been seriously affected.

- *Height above ground* Wind speed increases with height, so raingauges exposed well above the ground (whether mounted on the ground with a high rim height, or exposed on rooftops or masts, for example) will almost invariably catch less precipitation than an identical gauge nearer the ground. This was first described by William Heberden in 1769 following trials on the roof of Westminster Abbey [165], although the true cause was not correctly identified for decades thereafter. Differences increase with height and with wind speed (the latter, of course, is highly variable on all timescales). Theoretically at least, the ideal height for a raingauge is flush with the ground surface (**Figure 6.3**), but such exposures are impractical to establish and often difficult to maintain, and they quickly become useless in snowfall. Unless some form of splash-proof surround is provided, such as illustrated in **Figure 6.3**, raingauges exposed at or close to ground level are vulnerable to insplash or surface water ingress in heavy rain. ('Splash height' can extend to 1 m above impermeable surfaces in intense rainfall, although splash heights are considerably reduced over grass and bare soil, and accordingly gauges are best placed in and around such surfaces.) In WMO's 1989 survey across more than 100 countries [148], the most common height of the raingauge rim was between 50 cm and 150 cm (18 inches to 5 feet) above ground level, tending to be greater in countries where substantial snowfall occurs because gauges at lower heights can become buried by heavy or drifting snow: in such districts the standard raingauge height is normally set well above the mean annual maximum expected snow depth. Ironically, the recent adoption of large-capacity weighing gauges with rim heights considerably in excess of standard gauges has led to concerns that their increased rim height itself results in wind-induced undercatch [166].

Figure 6.3 Ground-level or 'pit' raingauge; gauges are exposed at ground level within a strong metal mesh which reduces turbulence and prevents insplash. (Photograph by the author)

- *Level* The raingauge collecting surface must be set, and maintained, absolutely level. Slight errors in the level of the funnel rim can result in significant under- or over-catch, the reduction averaging about 1 per cent for each degree of tilt [164]. Unless firmly fixed in the ground and regularly checked, it is very easy for slight errors of level to occur – movement of the gauge in dry soil, even minor knocks from a lawnmower, can affect the gauge and are easily overlooked. Raingauges located on masts are not only subject to greatly increased wind effects, but from experience can prove almost impossible to keep level: they are discouraged for these reasons.
- *Observation times* To ensure that rainfall measurements from differing sites are comparable, a common period within the day to which the measurement refers should be specified. Many synoptic or automatic weather stations will provide near real-time rainfall measurements at hourly, three-hourly or six-hourly inter-vals, but most manual raingauges are still read once daily, usually at a convenient morning observation time. The standard rainfall observing time in most countries is between 7 a.m. and 9 a.m.
- The *design, construction and materials* of the raingauge itself can influence the amount of measured precipitation. A deep, round funnel is important, to avoid outsplash in heavy rain, and to help retain solid precipitation (particularly snow) in windy conditions. Shallow funnels are liable to lose precipitation catch due to outsplash and are completely ineffective in snowfall. Square or rectangular funnels can create turbulent eddies over the gauge, the effects of which will vary with wind direction, and should be avoided. The material of the gauge itself can affect the catch, particularly where amounts are small. Different materials have differing wetting characteristics – droplets react differently to well-weathered metals, such as copper, and shiny plastic surfaces, for example, while painted surfaces are especially poor. Where a particular surface favours the formation of near-spherical droplets on the funnel, which then remain in place rather than quickly running into the gauge itself, those droplets may

evaporate and thus be lost to the record. Plastic gauges are particularly prone to heavy dewfall (occasionally sufficient to register as 'precipitation', which can increase the apparent frequency of small amounts of precipitation), most probably because the conduction of heat upward from the ground in copper or steel gauges reduces the likelihood of condensation in the gauge funnel. The surface characteristics of the catching surface can also change over time, affecting runoff and thus gauge catch. Where the gauge is designed to store liquid water for subsequent measurement, it is vital to ensure that the design of the unit minimises evaporation from the storage container to avoid losses between the time of precipitation and the next measurement, particularly if the gauge is not read daily.

- Finally, and perhaps surprisingly, *observer bias* has been identified as a cause for concern. A study using US Cooperative Observer Program (COOP) datasets found significant and widespread biases in the way COOP observers measured daily precipitation. The two most obvious biases were under-reporting of light precipitation events (daily totals of less than 0.05 in., or 1.3 mm), and over-reporting of daily precipitation amounts evenly divisible by five- and/or ten-hundredths of an inch, that is, 0.10, 0.25, 0.30 in. and so on (2.5, 6.3, 7.6 mm, etc.) [167]. Observer biases were found to be highly variable in space and time, and all observers should be aware of the risk of unconscious bias in measuring and noting rainfall measurements.

For all these reasons, the measurement of precipitation is, perhaps more than any other element, closely defined by standards.

WMO precipitation site exposure guidelines

Raingauges are best exposed where a moderate degree of shelter can be found, whether that be natural (a line of hedges forming an effective windbreak, for example) or artificial (porous plastic sheeting, or a site at some distance from low buildings, or perhaps a suitable windshield as covered later in this chapter). In contrast to preferred 'WMO class 1' sites for representative temperature and wind observations, suburban sites can often prove more suitable for precipitation measurements, and voluntary observers with suitable secure sites are welcomed in regional and national precipitation networks in many countries.

Table 6.2 sets out WMO site class guidelines for precipitation measurements. Note that the prevailing wind direction is deliberately not taken into account, as heavy convective rainfall may not be wind direction dependent. Determining the wind direction(s) which produce the majority of rain in your location is itself an interesting project – see Chapter 18, *Making sense of the data avalanche* for how to do this.

The 'class 2 or better' siting guideline ('no object should be within twice its height above the rim of the gauge') has been adopted by most state weather services around the world, and class 3 or higher sites (including rooftop sites) are normally considered only where there is no viable alternative nearby site at ground level. Objects closer than twice their height above the gauge rim are likely to intercept precipitation, particularly in windy conditions, while rain or droplets of wet fog from very close objects may also drip or be blown into the raingauge (even when it is not raining). That being so, however, some raingauge sites appear so sheltered at first glance as to make finding the raingauge itself quite difficult, but may produce results indistinguishable from

Table 6.2 *World Meteorological Organization classes for precipitation sites, summarised from WMO CIMO guidelines ([4], Annex 1D). 'Uniform height' implies those where the ratio between top and bottom is 2 or less. The higher class applies where either surroundings or obstacles criteria are not met.*

WMO pptn site class	Estimated uncertainty from siting	Surroundings	Obstacles (those with an angular width of 10° or more)
1	Reference site	Flat, horizontal land, surrounded by an open space, slope less than ⅓ (< 19°)	Low obstacles of uniform height, distant minimum 2–4 times their height above raingauge rim; or, for a gauge fitted with windshield, obstacles minimum 4 times their height above raingauge rim
2	± 5%	*As above*	Obstacles distant minimum 2 times their height above raingauge rim
3	± 15%	Open area, slope less than ½ (< 30°)	Obstacles distant at least their height above raingauge rim
4	± 25%	Steeply sloping land, > 30°	Obstacles distant at least half their height above raingauge rim
5	± 100%		Obstacles closer than half their height above raingauge rim

All measurements are from the raingauge position, or the main operational raingauge if there is more than one on site. Measurements from sites rated 3 or higher are unlikely to be representative and are not normally accepted unless there is no viable alternative location.

neighbouring gauges: conversely, other apparently perfect sites may give precipitation totals which simply do not fit in with the local pattern, and these may point to an instrument fault, or perhaps a mismatch between funnel size and measuring cylinder calibration. The only real test is to set up the equipment and make observations over a period of at least several months, and compare results with local stations whose observations and equipment are known to meet appropriate national standards for equipment and exposure – see also Chapter 19, *Sharing your observations*. It is important that sites used for comparisons are (as far as possible) reasonably close, in similar terrain and at a similar altitude above sea level, because large differences in rainfall can result from sometimes seemingly minor changes in topography or aspect, particularly in hilly areas. A comparison period of at least several months, rather than days or even weeks, is essential as rainfall patterns can vary substantially over short periods. The inclusion or otherwise of a single heavy rainfall event at a single site can distort conclusions.

The easiest way to determine whether and where the 'twice the height' condition is best satisfied is by making a site plan. (Making a dated site plan is good practice in any case, and is useful for documenting the site details and location of other instruments, if any – see Chapter 16, *Metadata: what is it, and why is it important?* – and documenting hedge and tree growth around the site.) Measure the distance to all significant obstructions – buildings (including outbuildings), fences, hedges and trees and any other nearby meteorological instruments – and draw up a site map approximately to scale. Next, using a clinometer, take accurate elevation bearings on each significant object from the proposed or existing site of the raingauge (the measurement should be made from the height of the raingauge rim, although this is easier said than done from gauges which are close to ground level, as in the United Kingdom: it may be easier to take the measurement from normal eye height, and adjust accordingly). The height of the object h is given by $\tan \theta \times D$, where θ is the observed clinometer angle (in degrees) and D is the distance to the object. Values of tangents can be obtained from standard tables, on the web or in smartphone calculator apps, and within Excel. An example is given here.

Determining 'safe' distances for objects around a raingauge

Example: the top of a tree 17 m away subtends an angle of 12 degrees when measured at eye height (**Figure 6.4**). How high is the tree? Is it far enough away for the proposed site to meet WMO class 1 raingauge exposure criteria?

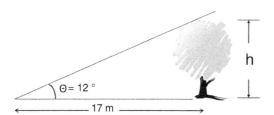

Figure 6.4 Determining raingauge site exposure.

- Tan 12° is 0.212; the tree is therefore 0.212 × 17 m tall = 3.6 m above eye height. Assuming the clinometer is read ± 1 degree, the error is therefore ± 0.3 m, and the tree height 3.6 m ± 0.3 m.
- If eye height is 1.7 m, then the tree is 3.6 + 1.7 m tall, 5.3 m ± 0.3 m.
- If the raingauge rim is at 30 cm above ground (UK standard), then the tree is 5.0 m above its rim; as the tree is 17 m distant, the height multiple is (17 / 5) = 3.4 ± 0.1

... which exceeds the minimum recommendation of twice its height above gauge rim. Assuming no other obstacles are nearer than twice their height, and there are other low obstacles of uniform height surrounding the gauge fairly equally, then the site would rate 'WMO precipitation class 1'.

Working the equation backwards will show the *minimum* distances that objects of a certain height need to be above the raingauge rim to meet at least 'WMO class 2' guidelines. For example:

- For a US standard raingauge with rim at 4 feet, an outbuilding 11 feet high should be at least 2 × (11 − 4) ft = 14 ft distant: a building or tree 52 feet high should be at least 2 × (52 − 4) ft = 96 feet distant.
- For a UK standard raingauge with rim at 30 cm, an outbuilding 2.5 m high should be at least 2 × (2.5 − 0.3) m = 4.4 m distant: a building or tree 12 m high should be at least 2 × (12 − 0.3) m = 23.4 m distant.

Simple trigonometry will show that the WMO recommended minimum height multiple of 2 corresponds to an elevation angle of 27 degrees, and thus a quicker method is simply to check with a clinometer whether any objects in the vicinity subtend an angle of 27 degrees or more from the raingauge rim.

How high should the rim of the raingauge be?

The standard *height of the raingauge rim* varies by country. Rim height is an inevitable compromise between several factors – low rim heights reduce wind errors but increase the risk of insplash in heavy rain, while higher rims increase wind losses significantly but are less likely to be buried by snowfall. In addition, the physical size of raingauges renders near-ground level exposure impractical except in research facilities (**Figure 6.3**), although such pit gauges are preferable where practical. However, pit gauges or windshields are not necessary in the average garden, park or university campus. The national standard for raingauge rim height will normally be defined by the state weather service, and for observations to be comparable the instrument types, exposure and rim height should follow national standards as closely as possible. Most countries specify a gauge height between 50 cm and 150 cm (18 inches to 5 feet) above ground level. Varying climatic factors, together with the need to maintain consistency and homogeneity in a country's long-term precipitation records, the expense of re-equipping or re-siting significant networks and inevitable international politics, render a fixed worldwide common standard unlikely.

In the UK and Ireland the standard height of the raingauge rim was established at 30 cm (1 foot) above ground level following a number of comparative trials in the 1860s and 1870s (**Figures 6.5** and **6.6**): in the United States, the standard height is usually between 3 and 4 feet (90 to 120 cm) above ground (**Figure 6.7**).

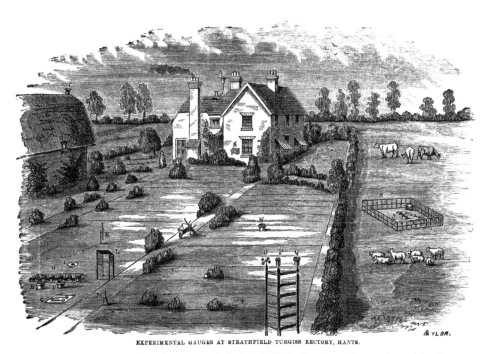

EXPERIMENTAL GAUGES AT STRATHFIELD TURGISS RECTORY, HANTS.

Figure 6.5 The Frontispiece from *British Rainfall 1868*, showing 42 experimental raingauges in the grounds of Strathfield Turgiss (now Stratfield Turgis) rectory in north Hampshire, England. Detailed comparisons of the different types and exposures were published in *British Rainfall* by the observer, the Rev. Charles Griffith. These tests were largely responsible for establishing the UK's standard raingauge type, height and exposure in 1874, which remains in place today [27, 168]

Figure 6.6 Standard UK 'five-inch' (127 mm) diameter funnel raingauge with rim at 30 cm above ground. See also Figure 6.10 for details of construction. (Photograph by the author)

Figure 6.7 US National Weather Service standard eight-inch raingauge (SRG). (Photograph courtesy of Steve Hilberg, CoCoRaHS)

Particularly in domestic or school situations, ideal sites will not always be available and the exposure of a raingauge is often a compromise. Even if the site is sheltered, and the exposure less than perfect, careful consideration of the available options may allow a rainfall record that is sufficiently close to standard neighbouring gauges to permit useful comparisons to be made.

A raingauge exposed to meet defined national standards of instrument type, exposure (WMO site class 1 or 2), observing time, site and rim height can be regarded as providing a 'standard' rainfall record, fully comparable both with other records made under similar conditions, and with the historical rainfall record.

The gauge itself should be set firmly into the ground, or bolted to a suitable surface, with its rim level and at the 'country standard' height above ground level. Mounting the gauge within an area of short grass or a surround of gravel or shingle will reduce the risk of insplash in heavy rain, particularly if the gauge rim is quite close to ground level. A length of metal garden edging strip or similar around the gauge (**Figure 6.8**) can usefully enclose a small area of shingle to reduce insplash while also providing some protection against bumps from lawnmowers or over-zealous strimming, but note the gauge should be fixed into the ground and not merely supported on this loose surface as otherwise it will be difficult to keep upright and level. Hard surfaces such as concrete or tarmac are best avoided to minimise insplash. If using a CoCoRaHS-type four-inch plastic raingauge (or similar) [150] mounted on a post (**Figure 6.9**), avoid mounting it on a fence post where the fence is impervious to the wind, as turbulence caused by the fence structure itself will affect the raingauge catch.

Figure 6.8 Surrounding a raingauge with loose stones within an edging strip reduces insplash and provides useful protection from lawnmowers and the like. (Photograph by the author)

As with all external meteorological instruments, security and access to the site should also be considered carefully: in public areas pay particular attention to site security to reduce the risk of vandalism, theft or sometimes unwanted 'additional contributions'.

Types and choice of 'catching' raingauges

As outlined previously, there are two main types of raingauge: manual gauges that simply store the water from liquid precipitation (melted in the case of solid precipitation) for subsequent measurement, and those that provide a record of the amount of rainfall with time. Both types are known as 'catching' gauges because they 'catch' precipitation for subsequent measurement.

Manual raingauges

The first known measurements of precipitation were made in India in the fourth century BCE ([6], chapter IV): the basic principle has changed little since. Most manual raingauges are deceptively simple, consisting of a funnel of known diameter (and thus surface area) which collects rainwater which is held in a suitable container for subsequent measurement using a graduated vessel. The measuring vessel is typically a graduated glass or transparent plastic measuring cylinder whose surface area is considerably smaller than that of the collecting funnel. The ratio of the two surface areas provides scale magnification, making small amounts easy enough to read to a precision of 0.1 mm or 0.005 inches. It is essential to ensure that the measuring cylinder is correctly paired with the gauge funnel diameter, as even slight differences in funnel size will make substantial differences in the measurements made. Chapter 15, *Calibration*, shows how to calculate the volume or weight of water for a given depth of rainfall and a specified funnel diameter, and thereby to check that the calibration is accurate.

There are many and varied types and designs of manual raingauge in use around the world: the 1989 WMO report [148] illustrated 54 main types. Many more have appeared since.

Measuring small amounts of precipitation

A rainfall amount below the lowest graduation on the measuring cylinder – usually 0.05 mm or 0.005 inches – is conventionally noted in the records as 'trace'. Where it is known from personal observation that precipitation has fallen, but there is none in the gauge, it is also acceptable to enter 'trace' (or perhaps '< trace' to distinguish such events).

When the measured amount results from condensation from dew, fog or frost, the entry can be made as 'trace – dew', 'trace – fog' or 'trace – frost', as appropriate. Occasionally fog or dew can result in as much as 0.1 or 0.2 mm / 0.01 inches in the gauge, in which case the entry should read 'Dew 0.2' ('Dew 0.01' for inch measurements), or similar. 'Dew – trace' and similar terms should be entered only where there is some water in the gauge, and not merely when dew is seen on the grass, for example. Where there is a trace in the gauge resulting from precipitation, and a further trace subsequently results from (say) dew or fog, only 'trace' should be entered.

The US standard raingauge

The most common pattern of manual raingauge in use in the United States, known simply as the Standard Rain Gauge, or SRG, consists of an aluminium (aluminum) metal cylinder with a copper, aluminium or plastic funnel on top and a plastic measuring tube in the middle [169]. The standard funnel diameter is eight inches (203 mm) and includes an accurately turned knife-edged rim. The body of the gauge itself is typically about 2 feet (600 mm) long, and the gauge is normally fixed to a small metal stand or tripod, itself bolted into the ground or a suitable heavy object such as a block of concrete, to bring the rim height normally to between 3 and 4 feet (90 to 120 cm) above ground level, as shown in **Figure 6.7**. In areas subject to heavy snowfalls, gauges can be mounted on supports at greater heights to remain sufficiently clear above snow accumulation levels. Windshields may also be fitted to gauges in such districts, as described later in this chapter.

Water collected in the deep funnel is measured by eye in a measuring cylinder located inside the gauge unit, readable to a precision of 0.01 inches. This can hold up to 2 inches or 50 mm of rain before it overflows into the larger outer cylinder. If rainfall overfills the tube, the excess is caught in the outer overflow can. If this occurs, the overflow from the outer can is poured back into the measuring cylinder after the 2 inch measurement is noted, and incremental overflow measurements are added to this to obtain a daily total.

This type of gauge has been in use for more than 100 years, and most are read daily. The majority of long-term US precipitation records are derived from gauges of this type. Differences between (mostly unshielded) raingauges in the US COOP programme and the shielded tipping-bucket gauges used in the US Climate Reference Network USCRN (see Chapter 4, *Site and exposure* for details) are discussed in references [73, 83].

The US plastic raingauge

There are more plastic raingauges in the United States than SRGs – currently more than 25,000 in the CoCoRaHS network alone (see **Table 6.1** and Chapter 19, *Sharing your observations*). The most common model (about one-tenth of the price of the standard metal gauge) is shown in **Figure 6.9**. To be approved by CoCoRaHS, the gauge has to conform to the minimum requirements set by the National Weather Service, to be within ± 4 per cent of the SRG in all types of weather. The gauge is made of clear, tough butyrate and has a capacity of 11 inches (275 mm) of precipitation. The internal measuring tube is graduated to 0.01 inch and has a capacity of 1 inch. Precipitation greater than 1 inch overflows into the outer cylinder and is measured by pouring into the measuring tube. Digital/electronic raingauges are not approved for CoCoRaHS use.

Figure 6.9 Four-inch plastic raingauge, as used in the US CoCoRaHS network; this version has a scale in millimetres for use within Europe. As with the larger US SRG, the receiving funnel and measuring tube are removed for collection of snow (see text). (Photograph by the author)

Metric versions of the CoCoRaHS gauge are available from European resellers. A comparison by the author and others against UK standard 'five-inch' raingauges showed very good agreement, and as such this gauge makes an excellent and long-lasting budget alternative to the more expensive standard copper gauge, and deserves to be a 'recommended product' (**Appendix 2**). Some regional hydrology authorities within the UK now use this gauge, or very similar alternatives, to increase station density within their networks.

The UK and Ireland standard raingauge

The British Rainfall Organization encouraged a number of comparative raingauge tests in the 1860s and 1870s (**Figure 6.5**), and in 1874 the deep funnel copper 'Snowdon' storage raingauge was defined as the UK and Ireland standard. This remains the case 150 years later, enshrined in British Standard BS 7843 [147]. WMO's 1989 analysis revealed that, at that time, the Snowdon gauge was in use at around 18,000 sites around the world, and was standard in 29 countries [148]. This simple, inexpensive and robust instrument is also known as the 'five-inch' raingauge, as the diameter of the funnel is exactly 5 inches (127 mm). The 'Met Office Mk II' or 'splayed base' raingauge (**Figures 6.6** and **6.10**) is identical to the Snowdon pattern with the inclusion of the outer splayed base, which makes the raingauge more stable when dug into the ground – a Snowdon gauge can work loose over time, particularly in sandy soils or dry weather, and requires occasional checks to ensure it remains upright and level. The splayed base also provides additional overflow capacity in the event of exceptional rainfall.

The standard British raingauge is traditionally made of copper, which with care will last for many decades, although in recent years cheaper stainless steel variants have been introduced by the UK Met Office (stainless steel had not been invented at the time of the British Rainfall Organization raingauge trials in the 1860s and 1870s and was therefore not one of the materials originally considered). Changing the material from which the standard gauge is made probably has little effect on the resulting measurements, but it is unfortunate that no side-by-side trials were undertaken beforehand by the UK Met Office to compare copper and stainless steel gauges (at a few sites, over a year or two) to assess whether any systematic effects resulted from the change.

The basic principle is the same in all 'five-inch' storage gauges. Precipitation falls into the deep funnel, rimmed with a substantial brass knife-edge to define the area of the catchment orifice very precisely, and then falls through the connecting spout into the collecting bottle. The deep funnel largely eliminates insplash and outsplash in heavy rain and hail and is an essential feature. (Some UK weather-equipment resellers sell copper five-inch gauges with shallow funnels, but whatever the description quoted, these are *not* standard raingauges and should be avoided at all costs. Gauges with deep funnels are only slightly more expensive but offer greatly superior performance in conditions of heavy rain, hail or snow.) The connecting spout has a narrow cross-sectional area and fits the neck of the collecting bottle snugly; this together with the fact that the bottle is mostly underground and largely unaffected by extremes of surface temperature both minimises evaporation and almost eliminates the chance of the contents freezing in cold weather. The collecting bottle is surrounded by an outer container or can which serves as an overflow in extremely heavy rainfall, allowing for about 150 mm capacity in the standard daily versions of this instrument, while the splayed base in the Mk II gauge acts as a second-level overflow container, further expanding the capacity by about another 100 mm. (Higher-capacity versions are available for weekly- or monthly-read gauges of this pattern, allowing for up to about 1500 mm monthly in the wettest UK locations.) It is important to ensure that the seams on the outer case remain sound, otherwise groundwater may leak into the outer container, or overflow may leak away.

At the nominated daily observation time, the collected water is carefully poured into a 10 mm capacity measuring cylinder (**Figure 6.10**) and read by eye, to the *bottom* of the water meniscus, to a precision of 0.1 mm. Amounts of more than 10 mm are summed from successive fillings of the measuring cylinder (taking care not to drop or spill the raingauge bottle while doing so, of course). The measuring cylinder is tapered at the bottom to facilitate the measurements of small amounts.

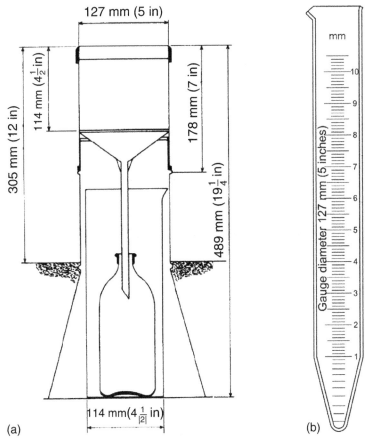

Figure 6.10 (a) UK splayed base five-inch 'Mark II' raingauge (© British Crown copyright, Met Office). (b) Measuring cylinder, 10 mm capacity, for a standard British five-inch raingauge. The lowest graduation, 0.05 mm, distinguishes between 'trace' and 0.1 mm. The measuring cylinder is 300 mm in length.

Other types of manual raingauge

The many different types of manual or storage raingauge in use around the world all use a similar principle, but funnel diameters, rim heights and measurement details vary considerably from country to country (sometimes within a country). It is difficult to be sure that rainfall statistics are comparable between different gauge types or countries or are consistent in time where instruments or methods have changed. As a modern continuation of the British raingauge trials of 150 years ago depicted in **Figure 6.5**, WMO continues to conduct international intercomparison field trials of meteorological instruments, including raingauges. **Figure 6.11** shows part of the test array at a intercomparison of 31 rainfall intensity gauges hosted at the Centre of Meteorological Experimentations of the Italian Meteorological Service: detailed results of this trial were published by WMO [77]. The results of other national and international raingauge intercomparison trials have also been published [170, 171, 172, 173], covering both manual and recording instruments and the measurement of snowfall in precipitation gauges.

Figure 6.11 International raingauge intercomparison by WMO of 31 raingauges held in Italy. (Photograph courtesy of the World Meteorological Organization, Geneva, from reference [77])

How much should a raingauge hold?

Reliable raingauge records of extreme rainfall events are of enormous scientific and civil engineering value. The capacity of a manual or storage raingauge should be sufficient to capture (at least) a 'once in 100 years' daily rainfall event for the locality or region, and preferably a 1000 year occurrence. For gauges which are read less frequently than daily, the gauge capacity should be in proportion. While the chances of exceptional short-period rainfall events are very remote at any one point, it would be frustrating in the extreme to lose the record of a once-in-many-lifetimes event simply because the raingauge had overflowed!

Within the UK and Ireland, a manual (storage) gauge should hold at least 150 mm of rain, particularly if the site is not visited daily (and this may include amateur observers' gauges whilst absent on holiday, or schools over holiday periods, for example). A standard Snowdon gauge will hold about 150 mm, a MkII gauge about 250 mm. Falls in excess of 100 mm in a few hours have been recorded in almost all parts of the British Isles, and are not particularly uncommon in the wetter mountainous districts, and as such 150 mm should be the minimum capacity of any gauge in regular use. The highest 24 hour rainfall total yet reliably recorded in the British Isles, 341 mm, was recorded by a logged and telemetered tipping-bucket raingauge (TBR) at Honister Pass in Cumbria over 4–5 December 2015 [158] (**Figure 6.2**). Great difficulties attend the measurement of very intense short-period rainfall events, not least enormous volumes of floodwater, but examples include a raingauge record of 193 mm in about 2 hours at Walshaw Dean Reservoir near Halifax, West Yorkshire on 19 May 1989 [174, 175], while up to 300 mm may have fallen in 4 hours near the centre of the storm responsible for the Boscastle flood in Cornwall on 16 August 2004 [176]. Despite their location in the driest part of England, two well-attested falls of 240 mm in 4–5 hours at East Wretham in Norfolk on 16 August 2020 [177], and 180 mm in under 2 hours at Brettenham in Suffolk on 25 July 2021 [178, 179], clearly underline the importance of both gauge capacity and design, particularly as regards insplash and outsplash characteristics.

Within the United States, the expected maximum 24 hour fall varies from below 18 inches (450 mm) in Montana to in excess of 38 inches (close to 1000 mm) in Texas and Louisiana [180]. A standard US raingauge will hold 20 in (500 mm). The greatest 24 hour rainfalls on record anywhere in the United States are 49.69 in (1262 mm) at Waipa Garden on the island of Kauai, Hawaii on 14–15 April 2018, and 43.0 in (1092 mm) at Alvin, Texas on 25–26 July 1979 [180, 181]. While it would be unrealistic to implement as standard a raingauge network that could cope with such extreme falls without difficulty – the Hawaii fall would amount to 41 litres, or almost 11 U.S. gallons, of collected water in a standard eight-inch gauge – a minimum raingauge capacity of 20 in / 500 mm is advisable. Note that daily falls in excess of 600 mm have been recorded as far north as New Jersey and Iowa.

Reliably calibrated and maintained automated gauges, particularly tipping-bucket gauges, are capable of recording such extreme events (provided, of course, that they remain unaffected by surface flooding or storm debris), thus avoiding the requirement for a physically very large storage container as part of the gauge itself. In February 2007, a 400 cm^2 TBR near the summit of Cratère Commerson on La Réunion island in the Indian Ocean (21.60°S, 55.65°E, elevation 2310 m / 7,579 ft) successfully recorded new world record three- and four-day falls of 3929 mm (154.7 inches) and 4869 mm (191.7 inches), respectively, during the passage of Tropical Cyclone *Gamede* [182].

Raingauge sites which are themselves susceptible to flooding – either the site of the gauge itself, or access to the gauge – should be avoided, as the gauge site itself may become inaccessible during and after a heavy rainfall event. Worse still, the gauge itself may have been flooded above its rim, or even swept away in floodwater.

Recording or automated raingauges

Most 'catching' recording raingauges can be classified in one of four categories:

- Tipping-bucket raingauges (TBRs),
- Float gauges,
- Weighing gauges, and
- Hybrid weighing/tipping-bucket instruments.

All four are discussed briefly in the remainder of this chapter: each has its advantages and disadvantages. Non-catching gauges, such as present weather detectors and the like, are not included, and the reader is referred to references [145, 146] for more information.

Tipping-bucket raingauges (TBRs)

By far the most common type of recording raingauge used in modern AWSs is the tipping-bucket type. Ironically this is one of the oldest of today's meteorological instruments, first described (in 'tipping spoon' form) by Christopher Wren, the architect of London's St Paul's cathedral, in 1663 [14, 16]. This type of sensor is ideally suited to digital logging – Sir Christopher was clearly around 350 years ahead of his time.

The principle of the TBR is simple and robust (**Figure 6.12, left**). Rainwater from the collecting funnel falls through a spout and into one of two counterbalanced 'buckets' mounted on a pivot. One bucket fills with water until its weight exceeds that of the counterbalancing weight, at which point it 'tips' forward, out of the path of the incoming flow of water from the funnel, emptying as it does so. The other bucket quickly pivots into place underneath the collecting tube. As the bucket tips, a magnet attached to the bucket mechanism swipes over a reed switch, making and breaking a brief electrical contact. The pulse thus generated represents one increment of rainfall. The second bucket then fills until it also tips and empties in turn, generating another pulse as the magnet again activates the reed switch, at which point the original bucket takes its place under the funnel once more. The cycle repeats itself as long as water continues to flow into and through the funnel. The gauge *capacity* is unlimited, as it is self-emptying in operation. The instrument is sensitive to level, and unbalanced buckets or an off-level gauge will lead to irregular tip behaviour and unreliable measurements.

Bucket and funnel capacities are paired to achieve a specific increment, usually 0.1, 0.2, 0.5 or 1 mm, or 0.01 to 0.10 inches of rainfall. Some manufacturers offer 'mix and match' TBR funnel diameter and bucket capacities, which reduce unit costs. An increment of 0.1 or 0.2 mm (or 0.01 inches) is required for accurate climate measurements, including AWSs, although a 1 mm tip may be suitable for flood warning systems or for remote areas with a high annual average rainfall. Note, however, that the 1 mm units typically found on entry-level and some budget AWS models are too coarse for accurate daily or sub-daily rainfall records.

Most TBRs are connected to a display counter, or datalogger of some description, or both: the logger may be a small dedicated unit or, more often, one channel of an AWS multi-element system. Pulses are counted as they occur, and the accumulated rainfall over time is simply derived from the number of pulses (tips) × bucket capacity. Short-period estimates of rainfall *intensity* can be obtained if the number of

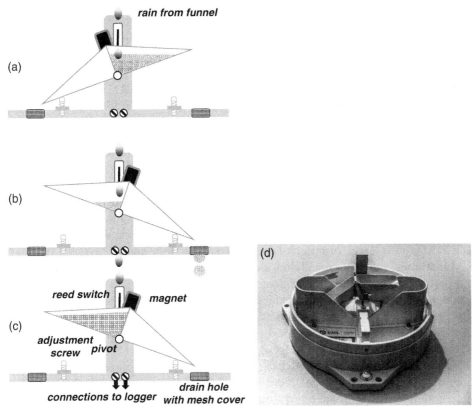

Figure 6.12 The operating principle of a tipping-bucket raingauge (TBR). At (a), one bucket is filling as rain enters from the gauge funnel; at (b), the bucket has tipped on its pivot and has emptied, briefly swiping the magnet over the reed switch and generating a switch closure (pulse) signal as it did so – the second bucket is then in place to receive incoming rainwater from the funnel and begins to fill; at (c), the second bucket is almost full and about to tip, and in doing so will swipe the magnet over the reed switch once again as it empties, ready to repeat the cycle starting at (a) once more. The adjustment screw stops enable the capacity and thus the calibration of each bucket to be precisely adjusted, as set out in Chapter 15, while the mesh cover over the drain holes deters insect entry to the gauge body. For simplicity, the counter-balancing weights are not shown. (d) A modern, 0.1 mm resolution tipping-bucket gauge, EML's ARG314 model with the 314 cm^2 funnel removed. The reed switch is housed inside the pillar visible behind the bucket mechanism, and the activating magnet is in the small arm attached to the bucket structure. (Photograph by the author)

tips in a given period is known, or if the tip times are also logged (see *Event-based logging* below and in Chapter 13, *AWS data flows, display and storage*). However, standard TBRs are poorly suited to the measurement of the *duration* of rainfall, particularly persistent light rain or drizzle, owing to their relatively coarse resolution. A count of the frequency of hours with 0.2 mm or more is a useful climatological statistic in itself, which although not a true measure of 'rainfall duration' is both quicker and more objective than manual methods of determining rainfall duration from paper chart records. It is also much more easily derived from digital summaries, and is becoming more widely adopted.

Event-based logging capabilities are particularly suited to tipping-bucket precipitation measurements, where a short sampling interval is necessary to provide adequate information on high-intensity rainfall events on timescales as short as a few seconds or tens of seconds, but where logging at such short intervals would quickly generate a huge and unwieldy log file. The method was first described more than 30 years ago [183], but it is only within the last couple of decades that datalogger processing power and memory has been able to operate at the sampling and logging intervals required to make the proposition both practical and economic. Here the logger is programmed to log the exact time of each tip (pulse) from a TBR, to the nearest second, but *only once a tip occurs*, regardless of whether these occur one second or several weeks apart (see also Chapter 13, *AWS data flows, display and storage*). The system remains capable of coping with lengthy dry spells and thus provides a scale extension impossible to attain with older ink-and-paper chart-based devices. The logger software may also be configurable to generate hourly and daily rainfall totals for standard climatological analyses, or this can be undertaken by post-processing using standard spreadsheets such as Microsoft Excel, or bespoke analysis packages, as required.

Operating in this manner makes much more efficient use of datalogger memory and battery life, as can easily be demonstrated by considering a fixed time-based datalogger logging a TBR at 1 second resolution. Such a logger would generate $60 \times 60 \times 24 \times 365$ or 31.5 million data points per gauge per year. At a site with an annual rainfall of 1000 mm, that is, 5,000 tips measured with a 0.2 mm tipping-bucket raingauge, 99.98 per cent of the logged values would be zero. Contrast this with an event logger, requiring only 5,000 data points over the same period, from which sub-minute intensities can easily be determined (as in the example in **Table 6.4**), without interruption to conventional hourly and daily total log files. Event logger capability therefore provides an extremely efficient way to capture very high-intensity rainfall events, allowing for precise determination of the times of heaviest precipitation to within a few seconds. Multiple events within the minimum sampling time (usually 1 second) do not pose any difficulty, allowing ample headroom for intense rainfall events – at least in theory. It should be borne in mind, however, that even a single 0.2 mm tip in 1 second amounts to a rainfall rate of 720 mm/hr, a rate at or beyond the capability of most TBRs. In addition, unless the instrument's calibration at high rainfall intensities (above about 30 mm/h) has been accurately determined, uncertainty in the derived rainfall rate is likely to be greater than at normal rates of fall, and as such certainly does not justify precision in the derived intensity better than 1 mm/h. Suitable calibration methods are described in Chapter 15, *Calibration*.

Any resolution tipping-bucket gauge will suffice, but 0.1 mm or 0.2 mm tips permit greater precision than larger bucket capacities. The rainfall intensity I (in mm/h) in the interval between the two tips is simply determined by

$$I = R/t \times 3600$$

... where R is the bucket capacity of the TBR in mm, and t is the interval between tips in seconds. For example, for a 0.2 mm bucket capacity, two tips 50 seconds apart indicate an average rainfall intensity during the period between the two tip times of 14 mm/h. A worked example spreadsheet is given on www.measuringtheweather.net. Note that the incremental nature of the tipping-bucket record means that there is effectively a lower resolution limit, as shown in **Table 6.3:** a 0.2 mm resolution gauge with tip events 60 seconds apart cannot distinguish between nil and 12 mm/h rainfall rates ($I = 0.2/60 \times 3600 = 12$ mm/h).

Table 6.3 *Minimum intensity increments (mm/h) for various resolution tipping-bucket raingauges*

TBR resolution	Logged at 1 minute	Logged over 5 minutes	Logged over 60 minutes
0.5 mm	30 mm/h	6 mm/h	0.5 mm/h
0.2	12	2.4	0.2
0.1	6	1.2	0.1
0.01	0.6	0.12	0.01

Table 6.4 *Example output from an event-based rainfall logger using a 0.1 mm TBR, for the event shown in Figure 6.27: central southern England, October [184]*

Date and time of tip, UTC	TBR total (mm)	Seconds since last tip	Interval intensity (mm/h)
30 Oct 13:16:46			
31 Oct 17:11:14	0.1	100 468	0
31 Oct 17:11:58	0.1	44	8
31 Oct 17:12:32	0.1	34	11
31 Oct 17:12:56	0.1	24	15
31 Oct 17:13:29	0.1	33	11
31 Oct 17:13:55	0.1	26	14
31 Oct 17:14:10	0.1	15	24
31 Oct 17:14:21	0.1	11	33
31 Oct 17:14:27	0.1	6	60
31 Oct 17:14:40	0.1	13	28
31 Oct 17:14:52	0.1	12	30
31 Oct 17:15:07	0.1	15	24
31 Oct 17:15:19	0.1	12	30
31 Oct 17:15:32	0.1	13	28
31 Oct 17:15:49	0.1	17	21
31 Oct 17:16:12	0.1	23	16
31 Oct 17:17:06	0.1	54	7
31 Oct 17:17:36	0.1	30	12
31 Oct 17:17:44	0.1	8	45
31 Oct 17:17:54	0.1	10	36
31 Oct 17:18:07	0.1	13	28
31 Oct 17:18:31	0.1	24	15
31 Oct 17:45:33	0.1	1622	0

To smooth the incremental nature of the record at lower rainfall rates, it can be helpful to sum increments over longer periods; over a 5 minute summation period, for example, one 0.2 mm tip can reliably indicate rainfall rates down to 2 mm/h (moderate rainfall). Reliable intensity measurements at 0.1 mm/h (light drizzle) require weighing TBR gauges with an effective resolution of 0.01 mm summed over 5 minute increments.

Space precludes a longer example, but **Table 6.4** and **Figure 6.13** illustrate the event-based approach using as an example a short but briefly intense October shower logged at a site in southern England. The record was made with an EML ARG314 raingauge (0.1 mm resolution) and logged by a Campbell Scientific

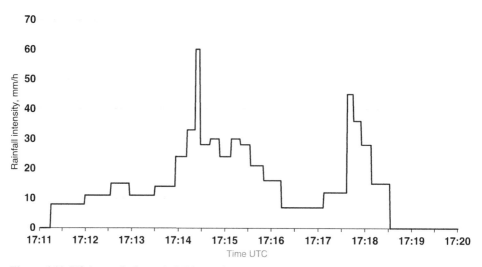

Figure 6.13 High-resolution rainfall intensity record of a brief spell of heavy showery rainfall (7 minutes duration) derived from the 'tip times' logged record of a 0.1 mm resolution TBR; Berkshire, southern England, October. Time is in UTC. See Table 6.4 for logged data.

CR1000 logger in event mode. The table shows a short burst of heavy showery rainfall lasting just 7 minutes, with one short pulse peaking at 60 mm/h (0.1 mm falling in 6 seconds) ending at 1714:27 UTC, followed by a secondary pulse reaching 45 mm/h just over 3 minutes later, and the abrupt cessation of rainfall less than a minute afterwards. Considerable detail and insight can be obtained using this simple approach. (Caution: although this method can also be used with high-resolution weighing TBRs such as the Lambrecht rain[e] unit described subsequently, it is *not* recommended, because reducing the effective resolution to 0.01 mm impacts both logger memory and processing time and results in large and potentially unwieldy data tables, of only limited hydrological relevance.)

Tipping-bucket raingauges are very widely used, but although highly reliable they are not infallible, nor are they quite as accurate as manual gauges. Many external factors, ranging from leaves or bird droppings blocking the funnel to snowfall or a rare defective reed switch or electronics module, can partially or completely damage or destroy the record, while calibration can sometimes fluctuate for no very apparent reason. *For these reasons it is not recommended that TBRs be used as the sole precipitation measurement device.* Without a manual checkgauge sited close by to provide 'reference' period totals – even if this is read only occasionally, perhaps weekly or monthly at unmanned or remote sites – the record will be incomplete in the event of any instrument malfunction. Exact agreement with the manual gauge is unlikely (see Box, *Should my raingauges agree exactly?*), but significant or increasing differences should be investigated promptly as indicative of possible impending failure. Some TBRs used in the United States retain the rainfall passing through the buckets for subsequent manual measurement, acting as a useful double-check on the instrument's calibration, although some 'wetting loss' remains inevitable.

Should my raingauges agree exactly?

Probably not. For a number of reasons outlined in the text, similarly exposed manual and automated gauges will normally give slightly different rainfall totals. Larger differences can be expected with solid precipitation, if the gauges are exposed some distance apart or if one of the gauges is exposed differently (mounted considerably higher above the ground, on a mast or rooftop, for example).

The key word here is *slightly*. A correctly calibrated and well-exposed TBR can be expected to agree with a standard manual gauge to within about 2–5 per cent over a period of a month or longer (daily totals will vary more than this, particularly small falls): normally the automated raingauge will be the lower. Where period totals differ by more than about 5 per cent, the reasons for the discrepancy should be investigated – it may be that the TBR calibration is adrift (see Chapter 15, *Calibration* for ways to check and correct this), that the TBR funnel has become blocked, that the gauge is no longer level, or that the magnetic reed switch mechanism is defective (easily checked with a multimeter). Some field trials have suggested that the accuracy of TBRs with small funnels (and thus bucket capacity), such as the unit included as standard with the Davis Instruments Vantage Pro2 AWS, may also vary significantly with temperature [80], probably owing to temperature-related changes in water's viscosity rather than density, but this is not a general conclusion.

Sometimes it is simply not possible to identify the reasons for significant discrepancies, or the discrepancy may be irregular in nature. Assuming both gauges are level and the TBR calibration has been checked and adjusted as necessary (and this should be repeated every year or so), continual slight calibration tweaking simply to force agreement with the checkgauge should be avoided. The two gauges can be expected to differ slightly, and the pursuit of perfect agreement is ultimately unnecessary.

Like most instruments, tipping-bucket raingauges benefit from occasional maintenance. To increase the chances of obtaining an uninterrupted precipitation record, wherever it is feasible to do so *two* TBRs can be used alongside a standard raingauge, so that in the event of one becoming defective the record from the second unit should still be available.

Common problems to watch out for include the following:

Blocked funnel, or obstructions to bucket mechanism. Most TBRs include a mesh filter on the funnel exit pipe, but these can become blocked by insects, windborne seeds and leaves, and especially bird droppings. Small birds seem to find raingauge rims to be excellent toilet perches, and the fitting of bird spikes – **Figure 6.14** – helps to discourage this without harming the birds. However, it is not always obvious at a glance whether the funnel is obstructed. Sometimes the first indication will be a period of no record during a known period of rainfall, or (more likely) a period of apparently very steady even 'rainfall' accumulation, which may also continue long after the rain has ceased. Such events arise because inbound rainwater held up by the blockage slowly seeps through into the buckets. If this happens, there is no

Figure 6.14 Bird spikes fitted around the rim of the gauge will deter most birds, but ensure water collecting on the exposed area of the spikes cannot drain into the funnel. (Courtesy of Davis Instruments)

remedy – the record is irrecoverable unless there is a backup gauge providing a parallel data source. It is thus best practice to check the funnel and filter visually for blockages frequently (every day if possible), and thoroughly flush out the funnel, the collector pipe and the filter every month, whether or not obvious obstructions are present. Some types of insects also find TBRs irresistible; no record from the TBR after a dry spell may mean that spiders or earwigs have taken up residence and obstructed the pivot mechanism.

Unattended period totals from tipping-bucket raingauges should present few difficulties (other than the finite memory capacity of the attached logger), although in practice the likelihood of loss of record owing to mechanical blockages or obstructions increases with time – typically, no more than 4–6 weeks between site visits is advisable, less if the gauge is part of a flood warning system.

Snowfall. Unfortunately, running two or more parallel tipping-bucket raingauges will not guarantee a record in snowfall, because TBRs – like most raingauges – are largely useless in such conditions (**Figure 6.15**). Models with internal heaters are available which will melt snow falling into the funnel as it falls, and such precautions are advisable in climates where significant falls of snow are a regular feature of the winter months. The heating strips are controlled by a temperature sensor within the gauge unit itself to minimise evaporative losses (and electrical power consumption), but they require a substantial power supply to keep pace with heavy snowfall, and as such may be impractical or infeasible in remote locations or with solar-powered or battery-driven systems. See also *Measuring* snowfall, later in this chapter.

Evaporation from partially filled buckets. One reason why tipping-bucket raingauges typically under-record compared with standard raingauges is that the receiving bucket is likely to remain partially filled at the end of a period of rainfall. If more

Figure 6.15 Manual and tipping-bucket raingauges buried following heavy snowfall. (Photograph by the author)

rain falls soon afterwards and the humidity remains high, little water will evaporate, but on occasions when small amounts of rainfall are separated by spells of dry weather, the contents can evaporate entirely and thus be lost to the record. This tends to be more of an issue in the summer months or in hot, dry climates, and with small buckets where the relatively small amount of water in the bucket is more easily lost to evaporation. Under such circumstances, the total rainfall indicated over any particular period will therefore be too low.

Errors increase where the bucket resolution is relatively coarse, particularly where small amounts of rain (1 mm or less) are much more frequent than larger falls, and this is often the reason for poor performance within budget AWSs where the tipping-bucket capacity may be 1 mm (0.04 in) or similar. **Table 6.5** shows an example of how this can lead to significant errors. The values here are in millimetres, but the principle applies equally to inch measurements.

Here the hypothetical 'actual fall' is shown to 0.05 mm for the sake of example. All the gauges are assumed to be accurately calibrated (i.e., tip at the nominal capacity shown), while an evaporation rate of 0.2 mm/day is assumed for the tipping-bucket gauges.

The table shows that the 0.05 and 0.1 mm tipping-bucket raingauges follow the 'actual' and the manual gauge closely, recording the correct number of days with 0.1 mm or more of precipitation (7 in this 10 day example).[*] The 0.2 mm TBR shows only 5 days with rain, missing out the slight falls on day 1 and day 6 but under-recording several of the falls by 0.1 mm. These then evaporate, leaving the period total 4.2 mm (9 per cent below the manual gauge value). The 1 mm capacity bucket is simply too coarse a resolution to record small amounts accurately, and it records rainfall on only 2 days. Because some rainwater remains in the bucket from previous days, both days are slightly higher than the total from the manual gauge. However, the period total is only 3 mm, only two-thirds of the manual gauge total. Clearly this

---

[*] This is perhaps over-generous to a 0.1 mm unit, as small amounts such as 0.1 mm are easily lost in wetting the sides of the funnel, and in evaporation thereafter, particularly where the funnel has previously been heated in sunshine – immediately preceding a light shower, for example.

Table 6.5 *Example comparison of a daily rainfall sequence between various raingauges; see text for the assumptions made. Values in millimetres.*

Day number	Actual fall (mm)	Manual gauge total (mm)	Tipping-bucket capacity			
			0.05 mm	0.1 mm	0.2 mm	1.0 mm
1	0.15	0.2	0.15	0.1	Nil	Nil
2	Nil	Nil	Nil	Nil	Nil	Nil
3	0.30	0.3	0.30	0.3	0.2	Nil
4	Nil	Nil	Nil	Nil	Nil	Nil
5	0.70	0.7	0.70	0.7	0.6	Nil
6	0.10	0.1	0.10	0.1	Nil	Nil
7	1.90	1.9	1.90	1.9	1.8	2
8	0.50	0.5	0.50	0.5	0.6	1
9	0.90	0.9	0.90	0.9	0.8	Nil
10	Nil	Nil	Nil	Nil	Nil	Nil
PERIOD TOTAL	**4.55 mm**	**4.6 mm**	**4.55 mm**	**4.5 mm**	**4.2 mm**	**3 mm**
		100%	100%	98%	91%	65%
Days with 0.1 mm or more	7	7	7	7	5	2

sensor would hardly present an accurate and reliable climatological account of this spell of weather.

Loss in high-intensity rainfall events. TBRs perform best at rainfall rates between about 0.5 mm and 30 mm (0.02 and 1.2 inches) per hour. During periods of very heavy rainfall, a tipping-bucket gauge can under-read by far more than normal compared to a standard raingauge. This is often the result of two factors – losses in the tipping process or so-called 'continuous tipping' or 'bucket bounce', both of which become much more likely above a particular threshold, typically 100–150 mm/hr (4–6 inches/hr). At these rainfall rates, the inflow of water from the raingauge funnel is so rapid that the smooth mechanical operation of the tip mechanism is disrupted or even ceases, as a result of which the bucket tip rate slows down or stops altogether. At a rainfall rate of 500 mm/hr (20 in/hr), for example, it takes only 1.4 seconds to fill a 0.2 mm capacity tipping bucket, compared with less than 0.1 s for the reed switch to actuate and 0.1–0.2 s for the tipping process to complete. The bucket may empty and bounce back almost immediately, or the empty bucket may not be able to move into position because of the rapid flow of incoming rainwater, and the full bucket therefore remains in position (no further tips) until the rain rate eases off. The effects are probably non-linear with rainfall rate (and the threshold will be lower with smaller buckets) and can be very difficult to assess retrospectively in exceptional storms. Chapter 15, *Calibration* provides guidelines on calibrating tipping-bucket gauges at various user-defined intensities. Field trials such as those operated by WMO [77] suggest that, as long as the tipping mechanism continues to function, empirical calibration at high rainfall intensities in addition to more usual rates of fall is particularly valuable. Field verification kits are available, simple mechanical devices which can be set up to simulate different flow rates (and thereby rainfall intensity) [185]. Using such a kit, a field engineer can quickly assess whether a TBR can either be (re-)certified as being within the required calibration limits, or requires replacement or return to a laboratory for re-calibration.

Making accurate records of very intense short-period rainfall records is one of the most useful areas in which relatively dense networks of 'personal' weather stations can contribute to the scientific understanding of severe convective storms [177, 179], but without an adjacent manual gauge to give a good estimate of 'ground truth', TBR records always remain subject to some doubt over their accuracy. A good example occurred in the series of severe thunderstorms which caused catastrophic flooding in Boscastle, Cornwall in south-west England on 16 August 2004. The nearest TBR to the centre of the storm, at Lesnewth, recorded 155.2 mm, 19 per cent below the adjacent manual raingauge total of 184.9 mm [176]. With the manual gauge reading known (from a well-maintained gauge, read by a conscientious observer), it was possible to make assumptions of the likely losses sustained by the TBR and thus estimate the peak rainfall intensity during the storm – which approached 500 mm/hr (20 in/hr) at this site. The adjusted record from the Lesnewth TBR at the height of the Boscastle storm is shown as **Figure 6.16**. This remains the highest resolution automated record yet obtained of any major rainfall event in the British Isles.

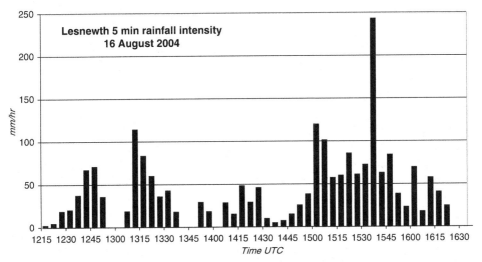

Figure 6.16 Rainfall record from Lesnewth in Cornwall, south-west England, during the 'Boscastle storm' of 16 August 2004 (from reference [176]).

In extremis, however, all TBRs are likely to fail to produce an accurate record in the most extreme rainstorms, the ultimate limit being the flow rate through the pipe leading from the funnel (including a particle filter, where fitted). Too narrow a pipe or too restrictive a filter will cause 'backing up' at unacceptably low rainfall rates, smoothing out the true intensity profile (this also happens during hailfalls), while in dry conditions too generous a diameter may permit the passage of insects, seeds or leaves into the gauge mechanism and also increases evaporation losses from the buckets. A realistic upper limit for a well-designed operational TBR is a minimum of 500 mm/hr (20 in/hr). Fortunately, such intensities are rare and fleeting in most locations.

Raingauge suppliers

Manual gauges. In the United States, standard and plastic raingauges are available from the suppliers listed in **Appendix 3**. It may be worth a call to your local National

Weather Service office, because if you live in a gap in the existing rainfall network, they may loan you equipment free of charge. The same applies in the UK, and your local Environment Agency office in England (and their equivalents in Wales and Scotland) may also be happy to provide equipment if your proposed site is suitable. Alternatively, standard 'five-inch' copper Snowdon or MkII gauges are available from several UK suppliers (see **Appendix 3**): remember to include a 10 mm measuring cylinder with your order. A well-made copper raingauge is not expensive, can be expected to last a lifetime, and is a 'recommended product' (**Appendix 2**).

Tipping-bucket raingauges can be purchased as standalone instruments, with or without a simple single-channel logger, or as part of multi-element AWSs. In most cases, a sub-standard AWS system 'bundled' TBR can be easily replaced by a higher-quality and/or higher-resolution unit, as the connection is a simple two-wire one (ensure the logger calibration can be adjusted to reflect the upgraded unit).

Recommended products: EML's miniature TBR, their 0.2 mm tip Kalyx model (**Figure 6.17**), is based on the dimensions of the traditional 'five-inch' Snowdon raingauge. This gauge has proved easy to install, well designed, very reasonably priced and reliable. One step up is EML's ARG314 model, with a 200 cm^2 funnel and choice of 0.1 or 0.2 mm resolution, itself widely deployed within UK hydrological and meteorological agencies (**Figure 6.17**). Either is ideally suited for siting alongside a 'five-inch' checkgauge, and both gauges offer the added benefit of EML's unique and proven aerodynamic profile to improve catch efficiency by reducing wind eddies around the gauge structure. More details in **Appendix 2**.

Stray pulses. TBR records from any logger-based system can be very susceptible to spurious pulses. These may be mechanical in origin (the gauge being rocked by high winds, perhaps, or accidentally bumped while the grass is trimmed around it – resulting in one or more spurious tips) – or a result of stray electrical impulses. The latter are most likely where long runs of unscreened cable connect sensor and logger, particularly following close lightning strikes and especially if the cable is coiled anywhere along its run. The use of screened cable throughout (see *Will it be cabled or wireless?* in Chapter 2, *Choosing a weather station*) is therefore strongly recommended. All such spurious tips known not to result from genuine precipitation, including those from calibration or maintenance tests, should be removed from the archived record, and station metadata annotated accordingly.

Figure 6.17 Two of EML's range of tipping-bucket raingauges – left, the 0.2 mm resolution Kalyx gauge, with funnel diameter 127 mm; right, the 0.1 mm or 0.2 mm resolution ARG314 unit, funnel diameter 203 mm. (Photograph by the author)

Float raingauges

In this type of recording raingauge, rainwater collected using a funnel falls into a chamber containing a float attached to an indicating mechanism, usually a pen arm, which marks a clock-driven chart. As rainwater accumulates in the float chamber, the float rises and the pen with it, the slope of the trace being directly proportional to rainfall intensity. The design of such instruments factors in a suitable ratio of the surface area of the catchment funnel versus float chamber to achieve sufficient scale magnification, and a very sensitive and responsive instrument is possible. Because the record would otherwise be limited by the capacity of the float chamber, a reliable and automatic mechanical siphoning mechanism to empty the float chamber (and return the pen to zero) once a certain amount of precipitation has been reached is required for a practical instrument. The float chamber must also be relatively small, both to maximise instrument sensitivity and minimise physical size.

Over the years, many ingenious designs of float gauges were developed (the earliest from as long ago as 1782), but most such instruments have now been superseded by TBR/logger combinations. Significant disadvantages of such instruments include the necessity for a manual observer to change the paper chart every day, the need for regular maintenance to keep the instrument in good working order, and the freezing of the float chamber in cold weather. Manual analysis of the resulting chart records is also very labour-intensive, although it is possible to automate some chart digitisation [186]. Most requirements for accurate and reliable automated rainfall recording can be more efficiently met with a TBR and datalogger combination than with paper charts from a float gauge, and as such few now remain in use.

Weighing raingauges

The basic principle of weighing raingauges is very straightforward: precipitation falls through an orifice of known surface area and accumulates in a collecting chamber, the mass (weight) of which is measured digitally by a load cell or strain gauge at regular intervals. After allowing for the weight of the collection container, changes in system mass are directly proportional to precipitation accumulating within the chamber. Such gauges are (at least in theory) equally effective for rain, hail or snow, as there is no funnel to obstruct solid precipitation. Their large, open apertures can result in evaporation of the gauge contents (some manufacturers recommend a thin film of oil to reduce this, and antifreeze to prevent freezing, although both may have undesirable ecological consequences). The collecting chamber is emptied manually, at intervals from weeks to months, which tends to make them more suitable for remote or mountainous areas. In some areas, their large physical dimensions offer some advantages against vandalism or tampering, but also makes installation complying with standard rim height guidelines much more difficult – a rim height of 1 m or more is common (**Figure 6.18**), much greater than the UK standard of 30 cm. Ironically, this latter factor results in additional losses in precipitation receipts through wind effects [166], and their use within standard networks has been limited as a result.

Weighing raingauges, such as the OTT Pluvio model shown in **Figure 6.18**, use sensitive strain gauges, load cells or a vibrating wire (whose frequency is a function of the applied weight) connected to a datalogger to provide

Figure 6.18 An OTT Pluvio weighing gauge; the rim is 100 cm above ground level. (Photograph by the author)

a continuous record of the mass of the large water container which holds accumulated precipitation (not to mention an interesting insect collection after a few months). Because they have no moving parts and do not depend upon liquid water for successful operation, they can provide precipitation records in snowfall as easily as in rain, and this is a significant advantage in locations where snowfalls contribute significantly to annual precipitation totals, such as in the Rocky Mountains: see also *Measuring snowfall* later in this chapter. For this reason amongst others, weighing gauges of one type or another are more widely used within the United States and Canada, where in some regions they form the backbone of the recording raingauge network.

Combination weighing tipping-bucket raingauges

One of the biggest disadvantages of a 'traditional' TBR has always been its incremental nature: the operating principle of the instrument means that precipitation amounts are counted only once the bucket capacity (typically 0.2 mm) has been reached and the mechanism then 'tips', generating a pulse. Although tip *times* can be determined accurately by the datalogger, the time taken to *fill* the bucket is unknown and may represent more than one period of rainfall. As a result, intensity estimates are poorly constrained, particularly in light precipitation, unless the effective resolution of the instrument can be improved tenfold or more. However, the practical lower limit to TBR capacity is about 0.1 mm, beyond which surface tension and viscosity effects in the decreasing volume of water render tipping less reliable in small bucket sizes.

The answer has come about by combining the operating principles of both tipping-bucket and weighing gauges, resulting in a completely new generation of small, reliable and very precise precipitation instruments with unlimited capacity. Such 'weighing tipping-bucket' devices (WTBRs) continuously measure the mass of the entire tipping-bucket mechanism using a sensitive load cell to fill in the 'between tips' detail[*], permitting an effective resolution of 0.01 mm or less with a 200 cm^2 funnel. (In practice, the surface tension of small droplets acts to resist passage through the funnel and its filter, especially when surfaces are dry: only once droplets reach a certain size will gravity outweigh surface tension and the droplets drip into the buckets and be weighed. Manufacturer claims of 'single raindrop' capability are far-fetched, but 'a few raindrops' is certainly achievable.) Reductions in mass as the tipping bucket empties are automatically allowed for by onboard electronics, incidentally permitting a larger bucket capacity without sacrificing precision. The larger bucket is also less liable to loss or malfunction during intense rainfall.

One of the first of these new instruments, and still the class leader, is the rain[e] digital raingauge from Lambrecht meteo in Germany (**Figure 6.8**) [187, 188]. This device is equally capable of both real-time digital second-by-second rainfall intensity output, ideal for such applications as flood warning or real-time 'ground truth' calibration input to radar rainfall networks, to logging daily and monthly totals to a high and consistent degree of accuracy within an existing gauge network. Field calibration is also quick and easy. The precision of the digital output stream has also been used to derive objective automated assessments of rainfall duration, comparing favourably with time-intensive manual methods based on paper chart records from float gauges. Two examples are illustrated by real-world data. **Figure 6.19** shows peak 10 s rainfall rates

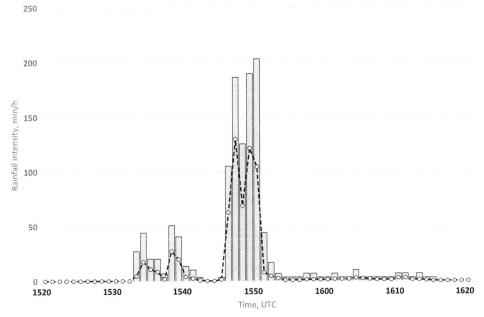

Figure 6.19 Example of Lambrecht rain[e] gauge output showing peak 'running 10 s' rainfall rates minute by minute (grey columns) and average 1 minute rainfall intensity (dashed line) minute by minute during a heavy October thunderstorm in southern England: rainfall intensity reached 203 mm/h at 1549:09 UTC, a 'once in 5 years' rate for the site. Only a little over 10 mm fell in this storm, but most of that within 5 minutes. The records shown here were available in real time. [189]

---

[*] The desirability of both measures was recognised very early on. Robert Hooke's weather clock, in 1679, recorded both the number of tips and 'shewed what part of the bucket is fill'd', according to William Derham (*The Philosophical Experiments and Observations of Dr Robert Hooke*, London, 1726).

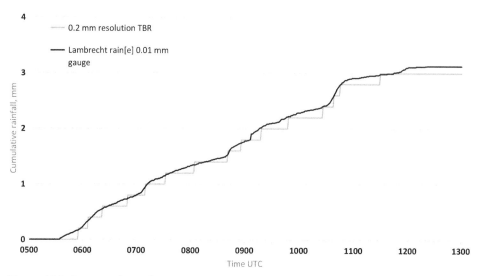

Figure 6.20 A comparison of cumulative rainfall totals (mm) over an 8 hour period in January at a site in central southern England; the high-resolution (0.01 mm) Lambrecht rain[e] gauge is shown by the darker line, the output from a conventional 0.2 mm resolution TBR by the stepped grey trace. On this occasion, rainfall was almost unbroken between 0535 and 1058 UTC [190]

by minute over a 60 minute period during a heavy thunderstorm in southern England, while **Figure 6.20** compares the output from the rain[e] unit with a standard 0.2 mm resolution TBR. On the latter occasion, rainfall was continuous for a period of 5½ hours, amounting to just over 3 mm; the higher resolution rain[e] record clearly identifies the steady and unbroken character of the rainfall event, whereas the record from the 0.2 mm TBR has insufficient detail to determine whether the fall was continuous or intermittent in nature over the period shown.

Although relatively new to the market, Lambrecht's rain[e] gauge is already and deservedly becoming widely adopted within hydrological and meteorological networks in several European countries, and is a 'recommended product' (**Appendix 2**).

Precipitation detectors

Rainfall sensors, or more accurately precipitation detectors, have numerous uses, including determination of leaf wetness in a growing crop, for example, or automated sensors for opening and closing greenhouse vents, but perhaps surprisingly they have seen little use in meteorological applications, probably because of a lack of common working standards. In themselves, they can only provide a 'dry/wet' signal, rather than indicate precipitation intensity or accumulated depth. Such sensors can usefully help determine the time of onset and cessation of a period of precipitation, particularly light rain or drizzle where it may take an hour or more to accumulate 0.1 mm and thus trigger a pulse from a tipping-bucket gauge.

One such innovative instrument is EML's SW120 R rain sensor (**Figure 6.21**) [191]. The device consists of three electrodes on a plane surface mounted on a 5 W ceramic resistor, the latter being gently heated to ensure the exposed surface

Figure 6.21 EML's SW120 R precipitation detector. (Photograph by the author)

remains a few degrees warmer than ambient air temperature. Droplets of precipitation landing on the sensor reduce its electrical resistance for as long as the surface remains wet. A datalogger connection evaluates the sensor resistance at regular intervals (typically once per minute); if the resistance drops below a given threshold, the datalogger sets a binary flag indicating 'wet' and the duration is incremented by the sampling interval. To discourage precipitation droplets remaining on the surface once rainfall has ceased, the unit is mounted at an angle of about 30° to the horizontal, as shown in **Figure 6.21**, while in addition the gentle warming of the substrate by the ceramic heater aids evaporation of surface water. Correctly installed and configured, this little sensor provides a reliable indication of the duration of liquid precipitation which can be compared with the duration of rainfall estimated by other means. The only drawback of this unit is its power requirement, which renders it unsuitable for sites without utility power source to provide its 12 v / 5 W supply, which will quickly drain AWS battery configurations.

Reducing the effects of wind on raingauges

As outlined earlier in this chapter, turbulent airflow due to the presence of the raingauge is the largest source of error in precipitation measurements. Various methods have been tried over the years in an attempt to reduce such errors in precipitation measurements, which can be particularly significant in snowy climates or in very exposed locations. Such options may include fencing or hedging, aerodynamic gauge shapes, ground-flush or 'pit' gauges, windshields, or a turf wall exposure.

Fencing or hedging?

The erection of windbreaks at appropriate distances from the gauge can provide effective shelter, especially in the direction of the prevailing wind, and these may consist of artificial materials or natural vegetation. Impervious fencing should be avoided, as it may increase rather than decrease turbulence across the site, but open fencing fitted with semi-permeable plastic mesh can provide a simple and cost-effective solution. Hedging with natural vegetation will take longer to achieve the same effect, and requires regular maintenance to ensure it does not grow to a level where it results in over-shelter, but may be preferable on ecological or aesthetic grounds.

Aerodynamic raingauges

Perhaps the simplest approach to reduce lost precipitation catch owing to wind effects is to redesign the shape of the gauge to minimise disruption to the wind flow around and over it. Much research has been undertaken, both with theoretical models using computational fluid dynamics and practically in wind tunnels, to reduce the impact of the gauge structure on the flow of air around it [192], resulting in a more streamlined, tulip-shaped gauge (examples shown in **Figure 6.17**). Comparison trials in windy locations have demonstrated that such gauges collect more precipitation than co-located standard raingauges, the conclusion being that standard gauges are losing some catch to wind losses [172, 193]. Effects vary not only with wind speed but with rainfall intensity [194].

Ground-flush or 'pit' raingauges

If the body of the raingauge itself does not intrude into the airflow by being at ground level, then – at least in theory – precipitation losses owing to wind should be negligible. This is the reasoning behind the ground-level or 'pit' raingauge, defined by the World Meteorological Organization (WMO) as its standard reference instrument for measuring liquid precipitation ([4], Annex 6A): similar recommendations and detailed specifications appear in both International and British Standards documents [147, 195]. Such gauges are mounted flush with the ground surface within a strong plastic or metal anti-splash grid (**Figure 6.3**). Although in reality the catch varies with the exposure and surroundings of the gauge, they generally record 5–10 per cent more precipitation than gauges whose rim sits above ground level, the difference increasing with rim height as expected [166]. However, such installations are large (around 2 m × 2 m), often impractical to install and maintain, ill-suited to widespread adoption, and perform poorly in snowfall. Such operational difficulties mean that pit gauges are usually found only in observatories or similar research establishments.

Raingauge windshields

By reducing turbulence due to wind, windshields such as the Nipher and Alter models can improve precipitation measurement efficiency, particularly in windy sites and more especially in snowfall. The Nipher windshield (**Figure 6.22**, left), first described in the United States in 1878, consists of an inverted trumpet-shaped cone surrounding the gauge, which acts to deflect the wind downwards. Although used with some success in mountainous areas of the western United States for many years, they are prone to fill with snow and can block the gauge completely when this happens.

The Alter shield (**Figure 6.22**, right, and **Figure 6.23**), dating from 1937, consists of a ring of metal strips which hang loosely around the gauge. The deflection of these

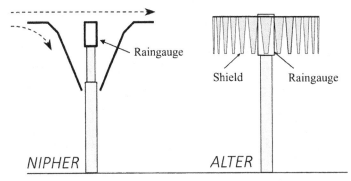

Figure 6.22 (Left) Nipher windshield (right) Alter windshield. (Courtesy of Ian Strangeways)

Figure 6.23 Alter windshield around a UK standard 'five-inch' raingauge with its rim at 100 cm above ground level: University of Reading Atmospheric Observatory. (Photograph by the author)

by the wind results in a less turbulent airflow over the gauge, improving its performance particularly in strong winds and snowfall. More substantial wooden fence structures around the gauge or gauges have been shown to effect greater improvements in snowfall catch, and are outlined briefly in *Measuring snowfall* later in this chapter.

The 'turf wall'

Windshields are only practical for raingauges exposed as standard at 1 m or more above ground level, because the size of the shields renders them unsuitable for gauges with a lower standard rim height, as is standard in the UK and Ireland amongst other countries. For this reason, very few raingauges are fitted with windshields in the UK. Instead, the recommended method of reducing wind eddies around a standard raingauge in exposed locations in the British Isles is the 'turf wall' (**Figure 6.24**).

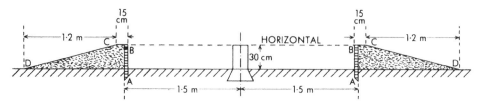

Figure 6.24 A 'turf wall' as used in the British Isles and the Republic of Ireland to provide shelter for raingauges on exposed sites. A-B represents the vertical retaining structure, B-C the level portion of the turf wall and C-D the sloping portion of the wall (gradient 1 in 4). (© British Crown copyright, Met Office)

The turf wall method was first described in 1933 following a multi-year series of trials conducted at an exposed moorland location in northern England [196]. The design of the shallow turf wall surrounding the gauge reduces the effects of turbulence over the gauge structure in a similar way to the taller windshields described earlier. The method is not perfect, however, as the gauge itself (and sometimes the turf wall) may be buried under snowdrifts; in areas where the gauge is visited only infrequently such effects may still lead to serious undercatch in the winter months. The inundation and filling of turf wall enclosures by floodwater or flood-borne debris has occasionally been reported during spells of intense rainfall. The turf wall structure also requires regular maintenance, while in remote areas its very existence can unwittingly highlight the presence of the gauge to casual passers-by, sometimes resulting in vandalism or theft.

Raingauge accessories required

The only accessories required with a manual raingauge are a standard, calibrated measuring cylinder suitable for the funnel diameter (**Figure 6.10b**). Check the calibration of new measuring cylinders before use by carefully pouring a known amount of water into it (see Chapter 15, *Calibration*). A spare raingauge bottle or outer collecting can is also useful as a spare in case of accidents, and for swapping over quickly when melting snowfall. Raingauge collecting bottles can be made of glass or plastic, as long as the material used is transparent or at least translucent.

Where the measuring cylinder is not built-in to the gauge structure, as in the US SRG or the CoCoRaHS models, it should be securely stored nearby, upside down (to allow any remaining rainwater to drain fully) and firmly held by clips or similar to minimise the risk of breakage or movement due to accidental damage or to strong winds. If there is one co-located at the raingauge site, a nearby Stevenson-type thermometer screen provides a convenient storage location.

Recording requirements

Raingauges combined with loggers (and often real-time or near real-time communications links) are now widely used for measuring precipitation – in fact, in many countries they now form the default network, at least for sub-daily measurements. They are ideal for remote sites where automated gauges can be left to record hourly or daily climatological measurements for perhaps a month or more, and for providing real-time, fixed-interval, high-resolution precipitation records invaluable in the study and forecasting and of severe storms, for tracking active fronts or squall lines, or for flood warning systems. However, whatever type of recording raingauge is in use, its calibration must be accurately known for set up within the datalogger (the period rainfall total will be number of pulses × bucket capacity). It is advisable to ascertain the sensor calibration in both normal and intense rates of rainfall, as they may differ considerably. The calibration(s) must be carefully checked, and adjusted if necessary, before bringing the unit into use – it is astonishing how many tipping-bucket raingauges are used 'out of the box' without any calibration checks being undertaken. Details on how to do this are given in Chapter 15, *Calibration*; mobile calibration kits for network field engineers are also available [185].

Short-period rainfall measurements in intense rainfall events

High-frequency rainfall measurements (1 minute or less) are particularly valuable in intense convective rainstorms, and it is surprising how often even a 'normal' heavy shower can produce short bursts of intense rainfall, often lasting less than a minute. The combination of a TBR and datalogger in 'event mode' as described earlier in this chapter can easily provide fine detail on storm intensity and evolution at high time resolution (down to a few seconds) very efficiently. Another method is best suited for high-resolution TBRs, those with resolution 0.1 mm or better, preferably 0.01 mm, such as the Lambrecht rain[e] combination weighing/TBR referred to previously. Logged data from such a gauge can be used to derive minute-by-minute rainfall intensities as in the example that follows. To do so, the gauge is sampled every second by the logger, and the logger programmed to sum rainfall totals over any particular user-selected short period – the author uses four periods, namely 10 seconds, 1 minute, 5 minutes and one hour. (It is unlikely that rainfall rates for periods less than 10 s duration are hydrologically significant, unless part of a sustained fall.) The rainfall intensity at any point in time is derived from this total: as an example, 0.23 mm of rainfall in 1 minute amounts to 0.23×60 mm/h = 13.8 mm/h, rounded to 14 mm/h (precision to better than 1 mm/h is unwarranted). The logger code derives rainfall intensity for each of the periods chosen, and outputs the maximum intensity value for every chosen time period every logged interval (typically every minute). By this means, high-resolution average and maximum rainfall intensities can be routinely logged with minimal logger processing or memory overhead. **Figure 6.19** illustrates 10 s maximum intensity and 1 minute average rainfall intensity from a Lambrecht rain[e] gauge using this method.

It hardly needs stating that the datalogger clock accuracy is critically important to ensure accurate event timing, but unless the tipping-bucket mechanism fails or datalogger clock drift is significant, the *relative* duration and thus intensity of episodes within the storm will be faithfully recorded.

Measuring snowfall

Four hundred years of invention and evolution have produced many fine, accurate, precise, robust, reliable and easy-to-use meteorological instruments. But in one area, namely the accurate and representative measurements of precipitation in snowfall, progress has been slower – a graduated wooden rule is still much the best low-tech method of obtaining a manual record of snow depth, as indeed it would have been in Galileo's time.

Methods to improve accuracy and consistency of snow depth measurements, and the determination of equivalent precipitation amounts, have been published by WMO and by many national meteorological services [197, 198, 199, 200, 201, 202]. This section provides a summary of these, although the detail varies somewhat from country to country – for obvious reasons, places where snowfalls are rare have different processes to countries where metres may accumulate in a few days – and individual country guidance should be consulted for details as appropriate. The CoCoRaHS website at cocorahs.org has a wide range of excellent presentation-quality training materials and videos covering all aspects of precipitation measurement, including snowfall (more details on CoCoRaHS in Chapter 19, *Sharing your observations*).

Manual methods for measuring snow depth

Measure the total depth of snowfall using a graduated rule held vertically. Choose a location free from drifting or scouring by wind. Take several measurements at different places (five will usually be sufficient, unless the values vary widely, in which case take ten spot readings across the range) and note them down. Disregard the highest and lowest readings of the set, then take the average of the rest as the snow depth. (If using a short 30 cm / 1 foot rule to make the measurements, do not forget to allow for the short gap between the end of the ruler and the zero mark when you make your measurement, and ensure the ruler does not pierce the grass or other ground surface beneath the snow: either will give a false reading.)

Note also the maximum and minimum depths within an area representative of the observing site – between drifted areas and parts scoured of snow by strong winds, for example. Measurements should be made in appropriate units – centimetres (record anything less than 0.5 cm as '< 0.5 cm') or inches, according to national standards. The US standard is to report the depth to the nearest inch.

As far as possible, routine measurements of snow depth should be made at or close to the same time as the raingauge is read – typically between 7 and 9 a.m. If it is snowing heavily at the time, precipitation measurements may be impossible or even dangerous to undertake; in such circumstances, the snow depth should be measured and the precipitation measurement delayed until a more opportune time as soon as possible thereafter.

Note that an increase in depth on successive days may not fully reflect the depth of a new snowfall, owing to melting/sublimation or compaction of the previous snowpack. The best method of measuring 'fresh snow' is by placing a wooden 'snowboard' (typically a white-painted board some 600 mm square) level on the snow surface at each observation and subsequently measuring 'fresh snow depths' using the board as the base level. It should obviously be re-laid level on the snow surface at each observation once the measurement has been made. Measurements are normally made at 3 or 6 hour intervals.

Automated snow depth measurements

Some AWSs are fitted with snow depth sensors, which work on the same principle as radar – a short ultrasound pulse is fired from the sensor, the time of return of the pulse from the underlying surface is measured by onboard electronics and converted into a height. These are sensitive enough to measure grass growth (indeed, all such sensors at UK Met Office AWSs are routinely deactivated during the summer months to prevent false readings), but they suffer from a very limited field of view; if the snow is drifting around the AWS, or being blown away from it, the measurement may be unrepresentative. At remote sites the errors may not be obvious to data users, and several such sensors in different locations may be required to obtain multiple samples, particularly where the accurate measurement of snowpack is vital for hydrological balance research.

Measuring snowfall equivalents of precipitation

The relationship between snow depth and water equivalent is very variable for fresh snow, between about 5 and 20 (sometimes even higher). In general, a ratio of 10 or 12 to 1 (i.e., 10 cm of snowfall will produce about 10 mm rainfall equivalent / 10 inches of snow to 1 inch rainfall equivalent) is typical for many snowfalls in temperate latitudes [203]. However, this 10-to-1 snow to liquid ratio varies significantly with air temperature, and 'fluffy' dry snow falling at temperatures well below freezing can result in much higher ratios – 20-to-1 or more (i.e., 20 cm of snow would melt to provide 10 mm of equivalent rainwater). At the other extreme, heavy wet snow falling at or just above 0 °C, particularly if it turns to rain at times, can produce a snow-to-liquid ratio of 5-to-1 or less (i.e., 5 cm of snow would melt to provide 10 mm of water). If the air temperature varies during the snowfall, the type of snow can and does vary over time.

Wherever possible, a more objective measure of 'snow water equivalent precipitation' should be attempted: the procedures for doing so vary somewhat by country.

Using a standard US eight-inch raingauge. During the winter, the observer should remove the funnel and inner measuring cylinder and allow snow to collect in the outer tube. The snow should then be melted using a known (measured) amount of warm water, and the meltwater measured in the same manner as for liquid precipitation, remembering to subtract the amount of warm water added to melt the snow. Measurements of liquid and solid precipitation are normally identified separately on US precipitation returns.

For observers in the British Isles, and in light to moderate snowfalls, light winds. On such occasions, the funnel of a standard raingauge will be partially filled with snow. Before the observation, prepare approximately 500 ml of warm water (not hot – about 30–40 °C) in a suitable container. At the raingauge site, fill the 10 mm measuring cylinder almost full with the warm water, and note this quantity using the measuring cylinder graduations in the usual manner. Then carefully pour the warm water onto the snow in the raingauge funnel, taking care to melt as much of the snow as possible. This may need to be repeated several times to melt all the snow: note down the amount of warm water added each time. Then carefully remove the funnel and measure the water content in the raingauge bottle. The rainfall equivalent is the measured amount of water in the bottle less the amount of warm water added. If a spare raingauge funnel and bottle is available, it may be easier to swap both over and bring the snow-filled units inside to melt. Stopper the bottle to avoid evaporation or sublimation losses.

During heavy snowfalls, or snowfalls accompanied by strong winds. On these occasions, a raingauge at the standard rim height of 30 cm may become partially or completely buried by drifted snow, strong winds may sweep most or all of the snow out of the gauge, or the snow may simply exceed the funnel capacity: any of these will result in the funnel contents being unrepresentative of the general precipitation level (**Figure 6.15**). Provided the gauge is not completely buried, it is worth first attempting the method previously described.

A more reliable method in such cases is the 'snow core' approach, which gives best results in fine, dry snow (heavy wet snow, or snow followed by rain followed by further snowfall, may produce misleading results). Assuming it is not snowing at the time of the observation, after measuring any snowfall contained in the funnel, insert the inverted funnel (or a spare) vertically into a representative area of lying snow, avoiding drifts or areas where snow has been removed by strong winds, to obtain a 'snow core' sample down to ground level. As far as possible, ensure all the snow in the area enclosed by the raingauge funnel is collected in the funnel, then melt and measure the snow sample using the warm water method as described earlier. Repeat this three times in locations several metres apart and take an average. (This method will obviously include any existing lying snow in the total, but when used for successive snowfalls the previous day's snowfall equivalent measure should be subtracted from the total to obtain the incremental amount.)

In spells of severe weather, or where significant additional snowfall is expected before the next observation, the process can be simplified by 'snow coring' down to the surface of a snowboard at subsequent observations, as described previously. A thin cane inserted next to the board will assist in finding it after a snowfall event. Measuring the water content of snow cores daily provides essential information on the hydrology of the snowpack [203].

Automated measurements of snowfall

WMO's Solid Precipitation Intercomparison Experiment (SPICE) evaluated the performance in snowfall of a wide variety of both instruments and windshield methods at 20 field sites located in 15 different countries between 2012 and 2015 [199, 200, 201]. Precipitation catches within larger, multiple fenced windshields such as the Double Fence Automated Reference or DFAR (**Figure 6.25**) were evaluated against smaller fence configurations, single shields (such as the Alter windshield) and unshielded sensors. Adjusted relative catches varied from 54 to 123 per cent of the reference DFAR total. As expected, windier sites saw a greater loss relative to the reference (54 to 83 per cent), while the less windy sites tended to record a slight gain (102 to 123 per cent). The conclusions were that transfer functions to mitigate wind bias in solid precipitation measurements were desirable, especially at windy sites and for unshielded gauges, but that the variability in the observed performance metrics between sites suggests that any such functions must be applied with caution.

The standard windshield structure used by the US Climate Reference Network is the Small Double Fence Intercomparison Reference (SDFIR) – see **Figures 4.3** and **4.4**. At about 8 metres in external diameter, it is smaller than the DFAR and thus more practical to use at many sites, yet it attained collection efficiencies in high-wind snow events similar to the larger DFAR structure.

A more detailed account of the experiment is outside the scope of this chapter, and the cited references should be consulted for detailed summaries and recommendations, but in general, for weighing gauges and tipping-bucket gauges, double-shields were preferred recommended over single-shields, and single-shields over

DFAR - DOUBLE FENCE AUTOMATIC REFERENCE

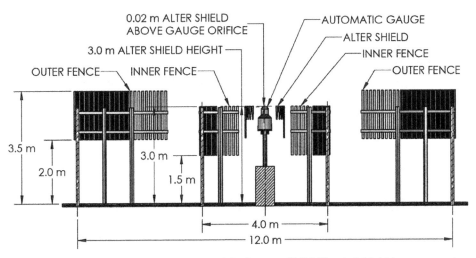

Figure 6.25 The Double Fence Automated Reference (DFAR) windshield in cross section. (Courtesy of World Meteorological Organization, Geneva: from reference [200], figure 3.4: original artwork by Jeffery Hoover, Environment and Climate Change Canada)

unshielded configurations. However, the benefits in terms of improving catch efficiency of the larger double-shield configurations relative to single-shield patterns were less clear, and results varied significantly with different climate regimes and their characteristic snow type(s) and wind conditions.

With few exceptions, unless the measuring device is fitted with heating elements to melt falling snow as it falls, most instrumental records will be unreliable at best, and probably useless. The larger *weighing gauges* can provide a good record of snowfall accumulation where fitted with a suitable windshield and gauge neck heater; more comprehensive details are provided in the main WMO SPICE report [200]. Snow melting in a TBR funnel will produce a series of tips as the snow melts, but unless the snow is melting as quickly as it falls or the gauge has a substantial heater attachment, the record will bear little resemblance to the actual rate of fall. After a heavy snowfall, the melting of snow in the funnel may not take place for some time (days, possibly weeks) after the snowfall event, and to avoid uncertainties in subsequent records it is best to scoop out the snow from the funnel soon after the snow has stopped falling. Note this in the observation record.

If sub-daily totals are required (hourly rainfall equivalent estimates, for example) where the record is unavailable owing to heavy snowfall, often the only method will be to apportion the rainfall-equivalent total for the snowfall period (from melted snowfall in the gauge) using eye observations and/or intermediate observations of snow depth, if such exist. Otherwise, rainfall radar evidence, if available, may provide some indication of precipitation onset, intensity and cessation.

In all cases where records have been lost owing to snowfall, a note should be made in the station records to indicate this. Where the records have been completed using estimates, the source and basis for the estimates should be clearly stated.

Estimates are not ideal, but reasonable 'best efforts' estimates are always better than gaps or 'nil or missing entry due to snowfall' in the record.

Accuracy versus precision in precipitation measurements

Two well-maintained and reliably calibrated standard raingauges exposed adjacent to each other should agree to within better than 5 per cent (see Box, *Should my raingauges agree exactly?* earlier in this chapter). Errors due to shelter (too much or too little), incorrect exposure, poor levelling and so on can easily double or treble these differences. Small differences are next to impossible to spot without regular comparisons with another well-exposed gauge, and yet even over as little as a few months they can be enough to destroy the value of any rainfall record.

While a quoted precision of 0.1 mm for daily falls makes sense, quoting monthly or annual totals to this precision is certainly unjustified in terms of their accuracy. The greatest mismatch of precision and accuracy comes from one leading brand of AWS, who quote 'highest rainfall intensity' rates to 0.1 mm/hr – even when that rate is over 100 mm/hr – for the rate quoted is probably no better than ± 20 per cent at best under such circumstances.

Access and security

The guidelines given in the previous chapter with regard to site security apply equally to rainfall; in many cases both temperature and precipitation instruments will be co-located on the same site. Any security fencing should not be of a size or nature itself to shelter the raingauge, but even in a domestic or school situation some protection should be considered to avoid unwanted attention by young children, young children's curious friends, or pets (for obvious reasons in the case of the family dog). Bird spikes fitted around the rim of the gauge will deter most birds, but ensure the exposed area of the spikes cannot drain into the funnel. Consider carefully the operation of any automatic garden sprinklers, and avoid planting fast-growing vegetables, crops, flowering plants or hedges anywhere near raingauges. Remember to allow sufficient clearance around the gauge and any associated cabling to permit easy grass cutting. Copper is widely used for raingauges – it is a soft metal, easily formed and soldered, but sheet copper dents very easily. If the maintenance is being undertaken by an external contractor, in schools for example, can the required work be undertaken by the contractor without risk of damage to the instruments or cabling? More than one shiny new raingauge on a golf course has been turned into mangled copper strips by a gang mower whose driver didn't know it was there! Stainless steel gauges are increasingly common and are more resistant to bumps and knocks, but care is still needed to avoid accidental damage or upset of the gauge level. Where the gauge is sited at or close to ground level, a small surround about 20 cm around the gauge base edged with metal garden edging strips and filled with loose gravel or stone chippings and treated occasionally with weedkiller (**Figure 6.8**) will help deflect the attention of lawnmowers and strimmers while also reducing insplash in heavy rainfall.

Measurement and observing standards

Keeping daily records with a standard raingauge, and perhaps one or more recording raingauges to indicate the timing, duration and intensity of precipitation, should present few difficulties to most observers, provided the raingauges are kept in good condition and the funnels are checked regularly for blockages. Snowfalls can make observations more difficult, however, and special measures are required to obtain accurate precipitation measurements (see *Measuring snowfall* earlier in the chapter).

Observation times – and 'throwing back'

For observations to be comparable between different locations, it is important that the times at which observations are made are as similar as possible: this particularly applies to the time period covered by daily rainfall totals.

For rainfall measurements made once daily during the morning, the convention is to 'throw back' the reading to the previous day, since the majority of the 24 hour period since the previous measurement occurred on the day prior to the measurement being made. This applies even if it is known from personal observation that all of the rain in the gauge fell in a shower 2 minutes before the measurement was made. When observations are made at other times, the date applied should also be the one in which the majority of the observing period falls. This important topic is covered more fully in Chapter 12, *Observing hours and time standards*.

One-minute summary – *Measuring precipitation*

- The term 'precipitation' includes rain, drizzle, snow and snow grains or snow pellets, sleet and hail; minor contributions from dew, frost or fog are also conventionally included in precipitation measurements. Precipitation is highly variable in both space and time, and precipitation measurement networks are usually denser than for other meteorological elements to maximise spatial coverage. There may be as many as 1 million raingauges operating globally, although standards vary from country to country.
- Precipitation measurements are very sensitive to the exposure of the gauge itself – particularly to the wind – and the choice of site is very important to ensure comparable and consistent records. Choose an unsheltered (but not too exposed) spot for the raingauge(s) – loss of catch through wind effects is the greatest single error in precipitation measurements, particularly in snow. A site on short grass or gravel is preferable. Wherever possible, obstructions (particularly upwind obstructions in the direction of the prevailing rain-bearing winds) should be at least twice their height away from the raingauge. Rooftop sites are particularly vulnerable to wind effects and should be avoided. The site should also be secure, but accessible for maintenance (grass cutting, etc.) as required.
- The gauge should be exposed with its rim at the national standard height above ground – in the UK and Ireland, this is 30 cm; in the United States, between 3 and 4 feet (90 to 120 cm). Most countries define a 'standard rim height' as

between 50 cm and 150 cm above ground. Take care to set the gauge rim level, and to maintain it accurately so.

- Manual raingauges should have a round, deep funnel to minimise outsplash in heavy rain (shallow funnel gauges should not be used) and should have a capacity sufficient to cope with at least a '1-in-100 year' rainfall event – a minimum of 150 mm in the UK and 500 mm (20 inches) in most parts of the United States. The gauge must be paired with an appropriately graduated and calibrated glass measuring cylinder.

- Most manual raingauges are read once daily, usually at a standard morning observation time, typically between 7 a.m. and 9 a.m. local time. The morning reading should be 'thrown back' to the previous day's date.

- To obtain records of the timing and intensity of rainfall, one or more automated raingauges are often sited alongside the manual storage raingauge. The record from the manual (storage) gauge should be taken as the standard period total and sub-daily records (hourly totals, for instance) taken from the automated gauge adjusted to agree with the daily total taken from the manual gauge, where there is one.

- The preferred resolution of a tipping-bucket raingauge is 0.1 or 0.2 mm; 1 mm capacity devices are too coarse for accurate measurements of small daily amounts. Recording raingauges should be logged at 1 minute or 5 minute resolution (higher frequencies are possible using an event-based logger or modern weighing tipping-bucket gauges). They should be regularly inspected for funnel blockage or any obstruction to the operating mechanism, which will result in the complete loss of useful record if not quickly corrected.

- Snowfall is difficult to measure accurately with most types of raingauge, and without some form of windshield most raingauges will lose 50 per cent or more of the 'true' catch through wind errors introduced by the presence of the gauge, which interferes with the flow of the wind over it, causing a loss of some of the catch.

- Procedures for measuring snow depth and the water equivalent of snowfall are set out.

7 Measuring atmospheric pressure

This chapter covers the measurement of atmospheric or barometric pressure (often abbreviated to 'air pressure' or simply 'pressure'), and its importance.

Air pressure is one of the most important of all meteorological elements. Fortunately, it is also the easiest of all to measure, particularly with modern sensors, and even basic AWSs, household aneroid barometers or smartphones can provide reasonably accurate readings. It is also the only instrumental weather element that can be observed indoors, making a barometer or barograph – analogue or digital – an ideal instrument for weather watchers living in apartments, or those who for whatever reason are unable to site weather instruments outside. To ensure consistent and reliable readings for professional users, particularly aviation, it is essential that pressure sensors are correctly exposed: World Meteorological Organization (WMO) recommendations on exposure and instrument accuracy ([4], section 3.1.4) are included.

Great accuracy is not required for casual day-to-day observations, as very often the trend of the barometer, whether it is rising or falling, and how rapidly, provides the best single-instrument guide to the weather to be expected over the next 12–24 hours, in temperate latitudes at least.

Where accurate air pressure records are required, the observed barometer reading needs to be adjusted to a standard level, usually mean sea level (MSL). Air is a compressible gas, as a result of which pressure decreases rapidly with altitude. Uncorrected readings simply reflect the height of the instrument above sea level, the absolute values varying also with the movement of weather systems, air temperature and humidity, and other factors. For meteorological purposes, pressures are corrected to MSL to eliminate differences due to variations in the altitude and temperature of individual observing locations. MSL values from all observation sites within a given area are plotted or analysed by computer to derive the familiar isobars (lines of equal pressure) on a weather map which delineate centres of low or high pressure. This chapter explains how to correct or 'set' a barometer to mean sea level. Accurate records also require the calibration of the pressure sensor to be checked regularly to avoid calibration drift, which can become substantial if not corrected. Methods for doing this are explained, with examples.

The *measurand* – what is being measured?

Barometric pressure refers to the force per unit area exerted by a column of air extending from the Earth's surface out to (at least in theory) the outer limits of the

Figure 7.1 Household barometer legends. (Photograph by the author)

atmosphere. Air is a compressible fluid acted upon by the gravitational attraction of the Earth, and so the mass of the atmospheric column (and thus the air pressure) decreases upwards above any point on the surface. The atmosphere is therefore densest at the Earth's surface. The outer limit of the atmosphere is rather arbitrary, but if we take it as a point where the pressure has fallen to one thousandth of that at sea level, then it is about 50 km above the Earth's surface. About half of the mass of the atmospheric column lies below about 5 km.

We often refer to something as being 'as light as air', and yet the weight of the air all around us is very substantial. At sea level and at typical atmospheric pressure and temperature, the weight of the column of air above a 1 metre square surface is around 11 tonnes. (This forms the basis of a great general knowledge quiz question: 'Which is heavier, a cubic metre of air, or a 1 kg bag of sugar?' The answer is air – at sea level and at 20 °C, 1 m³ of dry air weighs 1.2 kg.) We do not notice this great weight or pressure because the pressure within our bodies is the same, but very few humans adapted to life at or near sea level are able to function without prolonged acclimatisation at altitudes above 3000 or 4000 metres where the pressure is 30 per cent or more lower than at sea level. Our bodies cannot sense barometric pressure directly, nor anything but the most rapid changes in pressure (such as 'ear popping' experienced by aircraft passengers in the first few moments of a flight, for example), yet it has been known since the seventeenth century that relatively small fluctuations in atmospheric pressure are often closely linked to significant changes in the weather in temperate latitudes – hence the familiar legends adorning many household aneroid barometers (**Figure 7.1**). Such annotations were probably first suggested by Robert Boyle as long ago as 1660 [204]; the earliest surviving reference to English instruments being inscribed thus dates back to 1688 [205].

Standard methods of measuring pressure

The earliest form of the barometer was invented by Evangelista Torricelli in 1644 [206] and consisted simply of an inverted glass tube in a bowl of mercury. Torricelli correctly reasoned that the weight of the mercury column in the inverted tube exactly counterbalanced the weight of the atmospheric column of air on the mercury reservoir. As the weight of the mercury column was directly proportional to its height, so the earliest units of barometric pressure were expressed as the height of a column of mercury, measured in millimetres or inches of mercury (mmHg or inHg, respectively), or often simply mm or inches (' ... of mercury' being assumed, although there were numerous country-specific versions of 'inch' [207]). The earliest surviving records of barometric pressure were those made by Vincenzo Viviani and Alfonso Borelli in Pisa in northern Italy, covering the period November 1657 to May 1658 [32], barely a decade after Torricelli's invention of the barometer. Daily barometric pressure records exist, with only a few short gaps, for Paris from 1670, and for London from 1685 [33, 34, 208].

In 1914, the standard unit of pressure became the millibar (mbar). This is numerically identical to the preferred unit in the international SI system of units, the hectopascal (hPa): thus 1 mbar = 1 hPa = 100 Pa (1 Pa = 1 Newton per square metre): 1 hPa = 0.75 millimetres of mercury, or mmHg. Such equivalents are strictly valid only at 'standard conditions', normally defined as 1000 hPa pressure at a temperature of 0 °C and gravitational constant 9.806 65 m/s^2. 'Inches of mercury' are still used on some household barometers and in some public weather communications within the United States: 1 inHg = 33.86 hPa. For the remainder of this chapter, the hPa unit is used. The International Civil Aviation Organization (ICAO) 'Standard Atmosphere' defines the barometric pressure at sea level as 1013.25 hPa (29.92 inHg) at 15 °C.

Mercury barometers

Mercury was the traditional working liquid in barometers for three reasons. Firstly, its high density (about 13.5 times that of water) made for an instrument of practical size. The height of a column of mercury at average atmospheric pressure is about 760 mm (0.76 m); if the measuring fluid was water, the column would be about 10 metres high, the height of a two-storey building. Secondly, under normal atmospheric conditions mercury is an opaque silvery liquid, which makes it easy to read the height of the liquid column. Finally, mercury has a vanishingly small vapour pressure at room temperature, which means that in properly constructed instruments the vacuum at the top of the barometer does not deteriorate over time owing to the evaporation of the barometric fluid into it, which would be the case with (say) water or alcohol.

However, mercury is a known and dangerous neurological toxin, and its compounds are also known to accumulate in the environment and in food chains. The United Nations Environment Programme (UNEP) Minamata Convention on Mercury, signed by 128 countries in 2013 and which came into force in 2020, prohibits the manufacture, import and export of mercury-based instruments – including barometers and thermometers. Safety concerns have rightly hastened mercury's replacement in modern scientific instruments, and accordingly mercury barometers were mostly replaced by electronic sensors as part of WMO directives during the 2010s. Aside from a few mercury barometers retained in use for historical, legacy or parallel-running climatological overlap purposes, few remain in use today, although it is important to note that the Convention does not restrict the continued *use* of existing mercury-based instruments, notably antique mercury barometers. However, replacements or repairs are now impossible to obtain.

For this reason, this chapter does not cover mercury barometers, although notes on legacy instrument procedures are included in **Appendix 4**.

Aneroid barometers and barographs

The aneroid (from the Greek a-nēr(ós), 'without fluid') barometer consists of a partially evacuated closed metal capsule, prevented from collapsing under the influence of atmospheric pressure by an internal spring. Constant fluctuations in atmospheric pressure cause the distance between the two faces of the capsule to vary slightly. This movement can be amplified using a system of levers (as in a household aneroid barometer or barograph – see **Figure 7.2**), with a direct-reading micrometer (as in a portable precision aneroid barometer, often used as a travelling calibration standard), or electronically.

Electronic pressure sensors

Electronic pressure sensors use a variety of sensor types – typically variable capacitance circuits or piezo-electric substances – to generate an electrical output as the aneroid capsule flexes in response to changes in atmospheric pressure. Many 'pressure chips' are so small and inexpensive that most smartphones now include a pressure sensor.

Modern electronic pressure sensors (**Figure 7.3**) are, when correctly calibrated, almost as accurate as mercury barometers and have the great advantages of being small, robust (mercury barometers are very vulnerable to damage in transport or with careless handling) and are relatively insensitive to ambient temperature variations. Their electrical output signal also makes such sensors easy to include in computerised logger-based data acquisition systems such as AWSs. All such sensors

Figure 7.2 Aneroid barograph. This consists of a stack of aneroid pressure capsules (visible to the right of centre) connected via a lever mechanism to a pen arm. The lever mechanism magnifies the small changes in volume of the aneroid capsules with changes in atmospheric pressure. The pen marks a paper chart which rotates using a clock-driven drum. The charts are usually changed weekly. (Photograph by the author)

Figure 7.3 Modern electronic barometric pressure sensor; this unit is about 60 mm square. (Photograph by the author)

are, however, prone to calibration drift and they require occasional checking against a reference instrument or the local synoptic pressure field over at least a few days. How to do this is described later in this chapter.

Siting air pressure instruments

Barometric pressure sensors, whether aneroid or electronic, are easy to expose, because they are normally sited indoors. Avoid placing them in locations subject to significant variations in temperature, such as near a heating or air conditioning system, and keep them away from draughts or vibration, and especially out of direct sunlight. Electrical noise (whether from the sensor power supply, or nearby electromagnetic sources such as computers or wireless computer/telephone equipment) can also be a problem, and the equipment should be sited appropriately to reduce or eliminate these effects. Professional systems often have little choice but to locate the pressure sensor in the external logger enclosure, although this is not ideal – transient 'spikes' caused by wind eddies, the greater risk of condensation on sensor components and large temperature ranges affecting instrument calibration can all reduce the accuracy and reliability of pressure data under such circumstances.

It may be difficult to obtain reliable pressure readings inside an air-conditioned building, owing to the differential pressure created by such equipment, and an external connection is sometimes required to obtain satisfactory readings. Barometric pressure sensors can usually be fitted with a length of flexible tubing connected to a static port on an outside wall to achieve such an external connection. In windy locations it can be difficult to find a suitable location for a static port that is reliable in all wind directions, and several alternatives may need to be tried before a satisfactory position is found ([139], section 2.5).

Types and choice of sensors

Most budget and mid-range AWS systems include the barometric pressure sensor within the interior display console, where one is provided (**Figure 3.7**). Standalone sensors are available for advanced systems, with simple electrical connection to a datalogger (**Figure 7.3**). Considering the considerable difference in cost, there is surprisingly little difference in day-to-day performance between 'budget' and 'professional' pressure sensors. A modern electronic pressure sensor, once correctly calibrated, should provide barometric pressure readings of comparable accuracy to a good mercury barometer, without the disadvantages of the traditional instrument; even the much cheaper sensors used in most smartphones can give readings within a few hPa. However, calibration drift remains one of the largest potential sources of error with electronic sensors, particularly so with budget sensors. Drift results from long-term changes in sensor sensitivity, or settling-in of the sensor components: it therefore tends to be more pronounced with new sensors, reducing somewhat with time. Despite this, step jumps in calibration can occur – sometimes for no obvious reason – and it is best to check outputs frequently (ideally at least once per month) against another sensor of known accuracy, or against other local observations on days of light winds, to identify any sudden changes. Frequent checks will not eliminate calibration errors but will reduce their duration and impact on the records when they are identified and corrected promptly. Ways to do this are covered later in this chapter.

Electrical sensors can be logged remotely, and as frequently as required, enabling a continuous record of barometric pressure to be made without the necessity for frequent manual observations or for the tedious manual analysis of weekly paper barograph charts. Traditional barographs – aneroid barometers making a record on a paper chart, normally changed weekly (see **Figure 7.2**) – remain popular for display and aesthetic purposes. Barograph charts can be expensive, particularly for those with an expanded scale such as the model illustrated in **Figure 7.2**, while pens, charts and supplies for older instruments can become difficult to source. Most electronic sensors can provide finer and more accurate graphical records of barometric pressure than those from a barograph: pressure trends can be monitored at least as easily as with the more traditional instrument, and a digital barometer/barograph can be built quite cheaply using off-the-shelf components [209].

Logging requirements

Barometric pressure changes continuously, although not often very rapidly. Hourly readings are sufficient for many purposes, although to examine the fine detail of pressure changes accompanying individual showers or thunderstorms, marked frontal passages or the occasional very rare event such as that shown in **Figure 7.4**, more frequent sampling and logging intervals are preferable. WMO recommends 1 minute logging when using electronic sensors and dataloggers, while obtaining accurate daily or monthly maximum and minimum pressure values requires logging at no more than 5 minute intervals in order to capture extremes accurately.

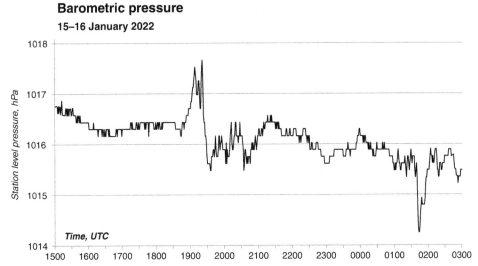

Figure 7.4 Changes in station-level barometric pressure in central southern England over a 12 hour period during the passage of the first 'outbound' and 'reflected' airwaves following the violent eruption of the Hunga Tonga–Hunga Ha'apai volcano in the Pacific Ocean on 15 January 2022. The records used were logged every minute, as the average of 6 × 10 s spot values immediately prior to the logged time. Less frequent observations would have been insufficient to document the rapid variations in atmospheric pressure during this and other similar events [211].

Even within buildings, wind effects can cause significant short-period fluctuations in pressure. To avoid 'noise', whether arising from power supply, electromagnetic induction or wind gusts, a suitable solution is to log running means of sampled pressures – logging the mean of the most recent 6 × 10 second samples every minute, for example, as with outside air temperature.

Figure 7.4 shows high-resolution (1 minute) barometric pressure records from the author's AWS in southern England, using a Setra model 278 electronic pressure sensor logged by a Campbell Scientific logger, following the explosive eruption of the Hunga Tonga–Hunga Ha'apai volcano in the Pacific Ocean at 0415 UTC on 15 January 2022. This colossal eruption resulted in pressure waves which crossed and recrossed the globe several times, in the largest such event since those resulting from the Krakatoa eruption in August 1883 [210, 211, 212, 213, 214, 215, 216] and thus falling into the 'extremely rare' category. The arrival of the initial 2.3 hPa amplitude pressure wave from the eruption at 1846 UTC, 14 hours and 31 minutes after the eruption, is clearly shown; this pulse travelled 16,549 km in great circle distance from the volcano at an average speed of 317 m/s. The first 'reflected' wave from the antipodean point in south Algeria just over 6 hours later at 0106 UTC is also evident, 1.6 hPa in amplitude, having travelled 3456 km at an average speed of 327 m/s. In all, four such 'outbound' and 'reflected' airwaves could be identified over a period of four days following the first arrival. Clearly, once-hourly observations (and analogue barograph records) are insufficient to resolve such transient phenomena, which are invaluable for revealing details of storm dynamics, atmospheric gravity waves, frontal structures and the like, particularly when combined with high-resolution wind, temperature and rainfall records.

Correcting barometer readings for altitude

Atmospheric pressure decreases rapidly with height, much more so than with horizontal distance. Because of this, where barometric pressures from different places are to be compared (for instance, in compiling national or international weather charts, or for setting aircraft altimeters) the observed barometric pressure at each point needs to be corrected to a standard reference height. In the meteorological context, this is normally mean sea level (MSL) – hence the two terms 'station-level pressure' or SLP, and 'mean sea level pressure' or MSLP:

Mean sea level pressure MSLP = station-level pressure SLP + height correction Δp

The details which follow apply only to electronic and aneroid sensors; different procedures apply for legacy mercury barometers, for which several smaller additional terms are needed to correct the observed temperature of the mercury column to a standard temperature, and for variations in gravity on a non-spherical Earth: details are given in **Appendix 4**.

Station level pressure SLP

For most electronic pressure sensors, as the name implies the station-level pressure is simply the pressure value output by the device, corrected where necessary for any calibration or offset reading. The output from electronic sensors is normally well compensated across a typical range of interior temperatures. As a result, corrections arising from changes in sensor temperature are normally insignificant and can be safely ignored, although they may be required if the sensor is exposed externally (in an AWS enclosure, for instance) where the range in temperatures will be considerably larger.

The height correction Δp

The correction to be applied depends upon a number of factors, of which the two largest are the height of the barometer above mean sea level, and the external air temperature.

Four methods of deriving a height correction for a pressure sensor are described, including the method prescribed by WMO ([4], section 3.7 ff). Which one is used depends not only upon the accuracy sought, but also upon the accuracy of the sensor in use – there is little benefit in using a high-precision method with a sensor whose accuracy is no better than 1 hPa, for example, as will be the case with most entry- or budget-level AWS systems (see **Table 3.3** on page 63). Methods 1, 2 and 3 are suitable for use at low elevations; stations above about 200 m (650 ft) above sea level, or where accurate readings are required for safety reasons (such as aviation requirements) should refer to method 4.

Aviation pressure reporting – Q codes

For aviation purposes, barometric pressure is reported slightly differently. There are three main 'Q codes' denoting various aviation standards for reported barometric pressure, as follows:

QFE

Pressure at airfield level; set on an aircraft (pressure) altimeter when height above local aerodrome level (strictly the official threshold elevation) is required.

QFF

Pressure at mean sea level (reduced according to actual/mean temperature). The same as MSLP in the meteorological context.

QNH

Pressure at mean sea level (reduced according to ISA profile); set on an aircraft (pressure) altimeter when height above local mean sea level is required.

In all cases, units are millibars, hPa or inches of mercury (inHg), by convention.

MSL pressure corrections – method 1

This is a quick and easy method, which does not require knowledge of the height of the site above mean sea level. It is accurate to only about 1 hPa at low elevations, and greater errors will occur at greater heights and at the extremes of pressure. This level of accuracy will suffice for many purposes (or for inexpensive sensors, where measurement errors will probably be greater than this).

Most state weather services publish hourly weather observations for a selection of stations on their websites, and these will normally include MSL pressures given to 1 hPa – for example, the US National Weather Service pages at www.weather.gov/forecastmaps (select town or city), the UK Met Office site at www.metoffice.gov.uk (select town or city, click on 'past 24 hours' to see nearest hourly observations including pressure), or Met Éireann at www.met.ie/latest-reports/observations.

Note the reading of your barometer each hour on the hour for 2–3 hours and write down the readings. Then, from the Internet, check the current weather observations at the site or sites nearest to your location (you may be lucky and have an observation point quite close, or you may be between two or more listed locations) and note down the pressures at those sites. It is best to do this on a day when the pressure is fairly steady and winds light, as stronger winds imply tighter pressure gradients (the horizontal variation of atmospheric pressure). Days with a nearby anticyclone (high pressure area) dominating the weather situation are ideal, although comparisons should be made at both low and high pressures, and at different times of day.

If there is an observing location quite close (within say 15 miles/25 km or so), then use the pressure given for that site. If you are between two or three locations with available observations, then take the average of the pressures at those sites, being sure to include observations north as well as south, east as well as west, to avoid biasing the average towards one direction. Compare your barometer readings with the MSL pressures from the official reporting stations, averaging or weighting inversely by distance as necessary. Ensure you compare the observations at the same time as your own readings (allowing for summer time if necessary). Your MSL pressure correction will be the amount you need to add to or subtract from your barometer reading to give approximately the same reading as the website observations. Most AWS systems will allow you to enter either a 'MSL pressure' or a fixed offset to ensure that your barometer readings are thereafter always approximately corrected to MSL by this amount.

Repeat this exercise over several days, at different times within the day, particularly with different wind directions (avoiding windy or very showery days), and at different pressures too. Average the corrections. Check and repeat every 6 months or so to identify and correct for any calibration drift in the sensor, or seasonal temperature differences. The correction obtained should be reliable to within 1 or 2 hPa.

MSL pressure corrections – method 2

This simple calculation is WMO's recommended method for sites below about 50 m above mean sea level ([4], section 3.7). A knowledge of the height of the site above mean sea level is required (see the following for how to obtain this), together with an estimate of the mean annual air temperature (to within 1 degC or so is sufficient).

Correct the observed station-level pressure SLP p to MSL by adding the value C, where

$$C = p \times \frac{h}{29.27 \times T_V}.$$

p is the station-level pressure (in millibars or hectopascals), h is the height of the barometer above MSL (in metres) and T_v is the mean annual virtual temperature at the site (in kelvin – add 273 to the Celsius value). The virtual temperature of moist air is the temperature at which dry air at the same pressure would have the same density as the moist air. To a reasonable approximation, and at a wide range of mean annual air temperature and humidity values, T_v will be about 1 degC above mean air temperature T (the exact value does not affect the result significantly).

Example: for a site at 35 m above mean sea level with a station-level pressure of 1005 hPa and a mean annual temperature of 10 °C (283 K), the correction C will be as follows:

$$C = 1005 \times (35 / (29.27 \times 284)) = 4.2 \text{ hPa}.$$

This correction should be added to the station-level pressure, either manually or automatically using the sensor/logger software. The MSL pressure in this example is therefore 1005 + 4.2 = **1009.2 hPa**.

This calculation is easily set up in a spreadsheet to produce a small barometer correction table (**Table 7.1**) – this spreadsheet can be downloaded from www.measuringtheweather.net and customised as required.

The correction does, of course, vary a little with the observed station-level pressure, but at altitudes of about 20 m above MSL or less the variation across a typical range of pressures (970 to 1030 hPa) is only ±0.1 hPa – within the uncertainty of all but the most accurate of pressure sensors. Even at 50 m the correction varies by only ± 0.25 hPa across this pressure range. Unless great accuracy is required, a single value of C for the MSL correction will suffice for all but the most extreme values of station-level pressure at heights below 50 m or so above MSL.

Table 7.1 *An example of a simple barometer correction table, for sites at or below 50 m above MSL*

Station height h (metres above MSL)	**35** (Valid only for sites 50 m or less above mean sea level)
Mean annual air temperature °C	**10**
Station-level pressure (hPa)	Correction to be added (hPa)
960	4.0
970	4.1
980	4.1
990	4.2
1000	4.2
1010	4.3
1020	4.3
1030	4.3
1040	4.4
1050	4.4

MSL pressure corrections – method 3

This is an extension of method 2, which applies up to about 150 m above sea level. It can be used at greater altitudes, but uncertainties in the method increase rapidly with height thereafter.

The first step requires an accurate determination of the height above MSL of your barometer. This is best obtained from local detailed topographical maps (in the United States, the US Geological Survey local maps: in the British Isles, the Ordnance Survey/ Ordnance Survey Ireland maps – online or hardcopy – at 1:25,000 scale, which include contour lines at 5 m vertical intervals). Google Earth can also provide a height measure digitised to a GPS overlay, although this may not be accurate enough for this purpose. (As barometric pressure at low levels decreases by roughly 1 hPa for every 10 m increase above sea level, a 5 m error in height will result in roughly 0.5 hPa error in barometric pressure, so an accurate determination of height is essential for precise work.) Remember also to allow for the height of the barometer within the building, or datalogger enclosure if outside – if it is in a first-floor room, for example, it is likely to be an additional 6 metres or so above ground level, and that needs to be added to the ground height given from the base map. (A good barometer will easily show the difference in pressure between ground and first floors in a building.)

The method also makes the initial assumption that the pressure sensor has no calibration errors across the normal pressure range (say 950 to 1050 hPa). As this is extremely unlikely, if the calibration errors are known these should be applied to the observed reading before the MSL correction is added,[*] or added to the calculated MSL correction as described below.

---

[*] Few pressure sensors other than those intended for professional standard AWSs will come with a calibration certificate. To determine any calibration errors, obtain the MSL pressure correction as outlined in the rest of the chapter and use that to derive a 'first-pass' corrected MSL pressure. To determine sensor error, compare the 'first pass' readings over a couple of weeks with neighbouring synoptic network observations as described in the section *Checking calibration drift on barometers* in Chapter 15; any calibration error greater than a few tenths of a hectopascal should become apparent. Note that the error may vary with barometric pressure, so determine sensor errors over as wide a range of pressure as possible – in temperate latitudes, the winter months have the largest range in pressure. Repeat every six months or so. Keep a note of corrections applied – this will indicate whether there is continued sensor drift over time.

Table 7.2 *Barometric correction to mean sea level, in hPa, for various heights and external
temperatures, for station-level pressure (SLP) 1000 hPa. From* Handbook
of Meteorological Instruments *[218], Table LVI, pages 446–7*

| | External air temperature | | | | |
Height (m)	−10°C	0 °C	10 °C	20 °C	30 °C
10	1.3	1.3	1.2	1.2	1.1
20	2.6	2.5	2.4	2.3	2.3
30	3.9	3.8	3.6	3.5	3.4
40	5.2	5.0	4.8	4.7	4.5
50	6.5	6.3	6.0	5.8	5.6
60	7.8	7.5	7.3	7.0	6.8
70	9.1	8.8	8.5	8.2	7.9
80	10.4	10.0	9.7	9.4	9.0
90	11.7	11.3	10.9	10.5	10.2
100	13.1	12.6	12.1	11.7	11.3
110	14.4	13.8	13.3	12.9	12.5
120	15.7	15.1	14.6	14.1	13.6
130	17.0	16.4	15.8	15.2	14.7
140	18.3	17.6	17.0	16.4	15.9
150	19.6	18.9	18.2	17.6	17.0
160	21.0	20.2	19.5	18.8	18.2
170	22.3	21.5	20.7	20.0	19.3
180	23.6	22.7	21.9	21.2	20.5
190	24.9	24.0	23.2	22.4	21.6
200	26.3	25.3	24.4	23.6	22.8
250	33.0	31.7	30.6	29.5	28.5
300	39.7	38.2	36.8	35.5	34.3
350	46.4	44.7	43.1	41.6	40.2
400	53.2	51.2	49.4	47.7	46.1

For most purposes **Table 7.2** will be sufficiently accurate to correct an electronic sensor to within 1 hPa for locations below about 150 m above sea level [217]. Above about 150 m above sea level the table can still be used, but corrections become substantial (20 hPa or more), and the accuracy of the MSL correction less reliable as a result, particularly at low temperatures. Corrected MSL pressure readings from high-altitude sites are inherently rather less accurate than those from low-level sites because the assumptions and approximations involved in the corrections rapidly become substantial and are very dependent upon the treatment of the external air temperature.

All corrections to mean sea level are positive (add them to the observed barometer reading) for locations above sea level. An observed pressure of 1000 hPa is assumed: other corrections are in proportion – that is, the value for 980 hPa will be 0.98 × the 1000 hPa value given in the table. Interpolations between the cells shown are in proportion, thus the correction for a site at 83 m above sea level would be the value at 80 m plus 3/10 of the difference between the values for 80 m and 90 m.

Method 2 and **Table 7.2** show that, to within a reasonable margin of error (1 hPa or so), and close to sea level, a single, average sea level correction value is 'close

---

Correcting a barometer to mean sea level: example using average values

Using **Table 7.2**, for an observing site at 65 metres above sea level, barometric pressure 1020 hPa, external air temperature 15 °C, the correction would be obtained from the table as follows:

— Height correction for 10 °C and 1000 hPa would be + 7.9 hPa (midway between the values for 60 m and 70 m above sea level)
— Height correction for 20 °C and 1000 hPa would be + 7.6 hPa (midway between the values for 60 m and 70 m above sea level)

Thus at 15 °C and 1000 hPa, the correction is +7.75 hPa (interpolating between the values for 10 °C and 20 °C derived above).

Finally, as the observed pressure is 1020 hPa, the correction to be applied is $(1020/1000 \times 7.75)$ hPa = 7.9 hPa.

Thus the corrected MSL pressure for this site given the observed temperature and pressure is 1020 + 7.9 hPa = 1027.9 hPa.

---

enough' for many purposes. At 100 m above sea level, for example, the average correction at an outside air temperature of 10 °C (a reasonable figure for temperate mid-latitudes) is 12.1 hPa; this varies by less than 1 hPa on either side between –10 °C and +30 °C, so for a barometer accurate only to ± 1 hPa an average correction will be sufficient. Most budget and mid-range AWS use this 'average' MSL correction method. Note though that at very high or very low temperatures, particularly at altitudes greater than about 100 m above sea level, this assumption departs somewhat from the truth (calculated MSL pressures will be too low in winter, too high in summer – at 200 m above sea level ranging about 2 hPa between winter and summer). Different 'average' corrections for summer, winter and the equinoxes should be used at greater altitudes.

For electronic sensors with an accuracy better than about 0.5 hPa, a more accurate site-specific barometer correction table can easily be prepared; see **Table 7.3**. (Note that this simplified correction table is *not* valid for mercury barometers, which require several additional corrections to be included – see **Appendix 4**.) This simple Excel spreadsheet can be downloaded from www.measuringtheweather.net and customised as required. Enter the height of the sensor above sea level (in metres – remember to include the height of the barometer above ground level if necessary) and, if known, any sensor calibration errors at specific pressures. The spreadsheet will then generate a site-specific sea level correction table, to 0.1 hPa precision, for a range of external air temperatures and observed pressures. The table can then be printed and used as required. This needs to be done only once, and the table will remain valid unless any changes in calibration become apparent (see *Checking calibration drift on pressure sensors* in Chapter 15, *Calibration*), or if the station height changes (the barometer is moved). Advanced loggers can be programmed to use the same calculation method to correct station-level pressures to MSL during logging operations, using actual sampled air temperature.

Table 7.3 *Example of a site-specific barometer correction table. This spreadsheet can be downloaded from www.measuringtheweather.net and customised as required. Small cell intervals minimise the interpolation required, which makes the table easier to use.*

<Name of site>		From *The Weather Observer's Handbook* (second edition) by Stephen Burt

Barometric pressure correction table

Altitude above MSL	**65.0** metres	Add the hPa correction below to the 'as read' barometer reading

Station-level pressure	Outside air temperature, °C																										
hPa	−15	−10	−8	−6	−4	−2	0	2	4	6	8	10	12	14	16	18	20	22	24	26	28	30	32	34	36	38	40
950	7.6	7.4	7.4	7.3	7.3	7.2	7.1	7.1	7.0	7.0	6.9	6.9	6.8	6.8	6.8	6.7	6.7	6.6	6.6	6.5	6.5	6.4	6.4	6.4	6.3	6.3	6.2
955	7.6	7.5	7.4	7.3	7.3	7.2	7.2	7.1	7.1	7.0	7.0	6.9	6.9	6.8	6.8	6.7	6.7	6.6	6.6	6.6	6.5	6.5	6.4	6.4	6.3	6.3	6.3
960	7.6	7.5	7.4	7.4	7.3	7.3	7.2	7.2	7.1	7.1	7.0	7.0	6.9	6.9	6.8	6.8	6.7	6.7	6.6	6.6	6.5	6.5	6.5	6.4	6.4	6.3	6.3
965	7.7	7.5	7.5	7.4	7.4	7.3	7.3	7.2	7.2	7.1	7.1	7.0	7.0	6.9	6.9	6.8	6.8	6.7	6.7	6.6	6.6	6.5	6.5	6.5	6.4	6.4	6.3
970	7.7	7.7	6.5	7.5	7.4	7.4	7.3	7.2	7.2	7.1	7.1	7.0	7.0	6.9	6.9	6.8	6.8	6.8	6.7	6.7	6.6	6.6	6.5	6.5	6.4	6.4	6.4
975	7.8	7.6	7.6	7.5	7.4	7.4	7.3	7.3	7.2	7.2	7.1	7.1	7.0	7.0	6.9	6.9	6.9	6.8	6.8	6.7	6.7	6.6	6.6	6.5	6.5	6.5	6.4
980	7.8	7.6	7.6	7.5	7.5	7.4	7.4	7.3	7.2	7.2	7.1	7.1	7.1	7.0	7.0	6.9	6.9	6.8	6.8	6.7	6.7	6.6	6.6	6.6	6.5	6.5	6.4
985	7.8	7.7	7.6	7.5	7.5	7.5	7.4	7.3	7.3	7.2	7.2	7.1	7.1	7.0	7.0	7.0	6.9	6.9	6.8	6.8	6.7	6.7	6.6	6.6	6.5	6.5	6.5
990	7.9	7.7	7.7	7.6	7.5	7.5	7.4	7.4	7.3	7.3	7.2	7.2	7.1	7.1	7.0	7.0	6.9	6.9	6.8	6.8	6.8	6.7	6.7	6.6	6.6	6.5	6.5
995	7.9	7.8	7.7	7.7	7.6	7.5	7.5	7.4	7.4	7.3	7.3	7.2	7.2	7.1	7.1	7.0	7.0	6.9	6.9	6.8	6.8	6.7	6.7	6.7	6.6	6.6	6.5
1000	8.0	7.8	7.8	7.7	7.6	7.6	7.5	7.4	7.4	7.3	7.3	7.3	7.2	7.2	7.1	7.1	7.0	7.0	6.9	6.9	6.8	6.8	6.7	6.7	6.6	6.6	6.6
1005	8.0	7.9	7.8	7.7	7.7	7.6	7.6	7.5	7.5	7.4	7.3	7.3	7.2	7.2	7.1	7.1	7.0	7.0	6.9	6.9	6.9	6.8	6.8	6.7	6.7	6.6	6.6
1010	8.0	7.9	7.8	7.8	7.7	7.7	7.6	7.5	7.5	7.4	7.4	7.3	7.3	7.2	7.2	7.1	7.1	7.0	7.0	6.9	6.9	6.8	6.8	6.8	6.7	6.7	6.6
1015	8.1	7.9	7.9	7.8	7.7	7.6	7.6	7.5	7.5	7.4	7.4	7.3	7.3	7.2	7.2	7.1	7.1	7.1	7.0	7.0	6.9	6.9	6.8	6.8	6.7	6.7	6.7
1020	8.1	8.0	7.9	7.8	7.8	7.7	7.7	7.6	7.6	7.5	7.4	7.4	7.3	7.3	7.2	7.2	7.1	7.1	7.1	7.0	7.0	6.9	6.9	6.8	6.8	6.7	6.7
1025	8.2	8.0	8.0	7.9	7.8	7.8	7.7	7.6	7.6	7.5	7.5	7.4	7.4	7.3	7.3	7.2	7.2	7.1	7.1	7.0	7.0	6.9	6.9	6.8	6.8	6.8	6.7
1030	8.2	8.0	8.0	7.9	7.8	7.8	7.7	7.6	7.6	7.6	7.5	7.5	7.4	7.4	7.3	7.3	7.2	7.2	7.1	7.1	7.0	7.0	6.9	6.9	6.8	6.8	6.8
1035	8.2	8.1	8.0	7.9	7.9	7.8	7.8	7.7	7.7	7.6	7.6	7.5	7.5	7.4	7.4	7.3	7.3	7.2	7.2	7.1	7.1	7.0	7.0	6.9	6.9	6.8	6.8
1040	8.3	8.1	8.1	8.0	7.9	7.9	7.8	7.7	7.7	7.7	7.6	7.5	7.5	7.4	7.4	7.3	7.3	7.2	7.2	7.1	7.1	7.0	7.0	7.0	6.9	6.9	6.8
1045	8.3	8.2	8.1	8.0	8.0	7.9	7.9	7.8	7.8	7.7	7.6	7.6	7.5	7.5	7.4	7.4	7.3	7.3	7.2	7.2	7.1	7.1	7.0	7.0	6.9	6.9	6.9
1050	8.4	8.2	8.1	8.1	8.0	8.0	7.9	7.8	7.8	7.7	7.7	7.6	7.6	7.5	7.5	7.4	7.4	7.3	7.3	7.2	7.2	7.1	7.1	7.0	7.0	6.9	6.9

MSL pressure corrections – method 4

Accurate corrections of barometric pressure to MSL are required for many purposes, particularly aviation briefings and climatological averages, where precision and accuracy to 0.1 hPa are essential. WMO provide a general reduction formula suitable for sites up to about 750 m above MSL in the WMO CIMO guide [4], section 3.7. Readers with requirements outside the scope of methods 1 to 3 above are referred to the WMO CIMO guide for more details.

Synoptic observing locations in mountainous areas above about 1500 m usually report pressure readings corrected to a different level, such as the 850 hPa surface, because of the very large corrections that would otherwise be needed to correct to sea level. Because of the requirements of aviation forecasts, these methods are defined by international agreement, and can be found in the various WMO publications already referred to.

Calibration

Unless access to a nearby or 'travelling standard' portable reference barometer is available, a comparison with neighbouring synoptic stations using the method outlined in Method 1, but working to a precision of 0.1 mbar, offers the best method of checking the calibration of barometric pressure sensors and evaluating calibration drift over time. The method is described more fully in Chapter 15, *Calibration*.

Precision versus accuracy

For operational, aviation or climatological purposes, precision to 0.1 hPa and accuracy to within 0.3 hPa are mandated by WMO. For many other purposes, accuracy to within 0.5 hPa will be sufficient. However, regular checks for calibration drift should be made to ensure the accuracy of the sensor remains within this range. For stations above about 150 m above sea level, uncertainty in the value of the MSL correction itself will sometimes be greater than ± 0.5 hPa instrumental errors, particularly at extremes of external air temperature.

Hours of observation

When the weather is settled and the pressure fairly constant, the *diurnal cycle* of barometric pressure (or more accurately, the *semi-diurnal cycle*), sometimes known as the *atmospheric tide*, will often be evident on a pressure graph or barograph trace – a twice-daily peak and trough caused by tidal movements within the atmosphere. In tropical latitudes, the amplitude can be as much as 5 hPa, although in temperate latitudes more typically 1–2 hPa. In mid-latitudes such diurnal movements are often obscured by much larger changes in pressure resulting from the day-to-day movement and change in intensity of large-scale weather systems, but are clearly visible when examining hourly pressure means over a period of even a few days in fine weather (**Figure 7.5**). Note how the morning minimum is earlier, and the afternoon minimum later, in July compared to January because of the greater solar heating cycle in the summer months, and how the highest hourly means are close to 0900 UTC.

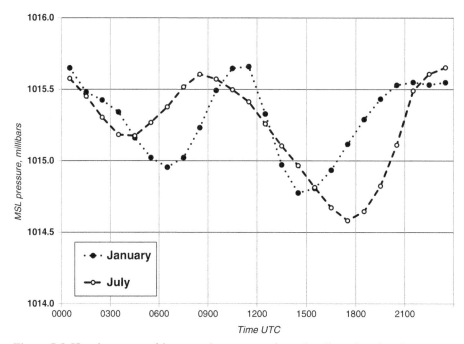

Figure 7.5 Hourly means of barometric pressure show the diurnal cycle of pressure very clearly. The curves here are for January and July and are averages by hour of MSL pressure in hPa over the 20 year period 2001–2020, from the author's own records in central southern England.

Because of the expected diurnal variation, it is therefore important to state the hour or hours at which barometric pressure observations are made regularly, or for which averages are quoted. Long-term pressure means are often quoted for one or more fixed hours of the day, often 9 a.m. and 3 p.m. local time. Sometimes 24 hour means are stated, calculated from hourly or three-hourly observations made throughout the 24 hour civil day and thus averaging out the diurnal cycle. In the UK and Ireland, pressure means are most commonly quoted for 0900 UTC, largely for reasons of historical convention. AWSs can easily provide a true 24 hour mean from sub-daily observations and are gradually replacing published averages for specific observation hours.

Extremes of barometric pressure, where quoted, should always refer to the full 24 hour civil day (i.e., midnight to midnight local time, excluding any summer time adjustments), as an exact coincidence of any particular observation time with the day or month's maximum or minimum pressure will be only fortuitous. Maximum and minimum pressures over any time periods based upon a single daily observation – usually a morning reading – will therefore significantly under-represent the true range of barometric pressure in any given time period.

One-minute summary – *Measuring atmospheric pressure*

- Atmospheric pressure is the easiest of all of the weather elements to measure, and even basic AWSs, household aneroid barometers or smartphones can provide reasonably accurate readings. It is also the only instrumental weather element that can be observed indoors, making a barometer or barograph – analogue or digital – an ideal instrument for apartment dwellers.
- The units of atmospheric pressure are hectopascals (hPa) – a hectopascal is numerically identical to the more familiar millibar. Inches of mercury (inHg) are still used for some public weather communications within the United States – one inch of mercury is 33.86 hPa.
- Pressure sensors must be located away from places that may experience sudden changes in temperature (direct sunshine, heating appliances or air conditioning outlets) or draughts, which will cause erroneous readings.
- Great accuracy is not required for casual day-to-day observations, as very often the trend of the barometer, whether it is rising or falling, and how rapidly, provides the best single-instrument guide to the weather to be expected over the next 12–24 hours, in temperate latitudes at least.
- Where accurate air pressure records are required, the observed barometer reading needs to be adjusted to a standard level, usually mean sea level (MSL), because air pressure decreases rapidly with altitude. A variety of approaches exist to correct or 'set' a barometer to mean sea level are described in this chapter. The choice of method depends upon accuracy sought (and the accuracy of the sensor) and height above sea level. Downloadable Excel spread-sheets are available to simplify the production of site-specific sea level correction tables where desired.
- The calibration of all barometric pressure sensors, particularly electronic units, should be checked regularly to avoid calibration drift. More details are given in Chapter 15, *Calibration*.
- Because of the twice-daily diurnal cycle of barometric pressure, the hour of observation should always be stated when presenting averages. Automatic weather stations can easily provide 24 hour means, which average out diurnal inequalities in atmospheric pressure.

Measuring humidity

This chapter describes what we mean by atmospheric humidity, explains the various humidity terms, and how they are related to each other. Surface-based instruments and sensors used to measure the amounts of atmospheric water vapour are described, together with a brief summary of the advantages and disadvantages of each. Recommendations and guidelines on humidity instruments, siting and standard measurement techniques from the World Meteorological Organization (WMO) CIMO guide are also included ([4], section 4). The chapter closes with a one-minute summary of the key points covered.

The *measurand* – what is being measured?

The term 'humidity' refers to the amount of water vapour in the air. The fascinating physics of water vapour is one of the main components of the atmospheric heat engine which produces our weather. As a result, humidity measurements are an essential requirement for operational meteorological analysis and forecasting, for climate studies, hydrology, agriculture and many other areas of human activity and comfort. In the meteorological context, the terms *relative humidity* (RH) and *dew point* (T_d) are most often used in specifying atmospheric water vapour content, but other terms are also used.

Humidity terminology

Water vapour is a colourless gas and is ubiquitous in Earth's atmosphere. Under normal atmospheric conditions, and at constant pressure, its maximum concentration depends entirely upon the temperature of the sample or 'parcel' of air – warm air is capable of holding a greater concentration of water vapour than is cold air. Various terms are used for expressing the amount of water vapour in the air – each can be converted to any of the others (see the example that follows), so knowing any one together with the air temperature (the '**dry bulb**') enables the others to be found. Most dataloggers intended for meteorological applications will include a variety of subroutines enabling calculation of the various humidity measures from input sensor readings.

The '**wet bulb temperature**' is that indicated by a thermometer or sensor whose sensing area is covered by a thin cotton sleeve or wick which is kept permanently moistened with pure water. Its reading shows the lowest temperature to which the ambient air can be cooled by evaporation. The difference between the readings of the paired dry bulb and wet bulb thermometers decreases as the humidity increases: when the air is saturated, two correctly calibrated thermometers will read the same temperature.

The **dew point temperature** T_d is the temperature at which a parcel of air just reaches saturation when cooled at constant pressure. The *dew point depression* refers to the difference between the air temperature (the dry bulb temperature) and the dew point – the larger the difference, the lower the humidity.

Vapour pressure e – in meteorology, this refers to the partial pressure of water vapour in air: the units are hectopascals (hPa), numerically identical to millibars (mbar). The saturation vapour pressure is the vapour pressure at the dew point temperature, at which the parcel of air is just saturated – that sample of air is then holding as much water vapour as it can at that temperature (and its **Relative Humidity**, or RH, is therefore 100 per cent). Any cooling will lead to condensation, that is, removal of water vapour from the air parcel. Its variation with temperature is shown in standard meteorological tables, in various online calculators and smartphone apps [219] and in simplified form in **Table 8.1**. Vapour pressure varies by more than an order of magnitude across the normal range of observed air temperatures. There are two forms – the saturation vapour pressure with respect to water, e_w, and with respect to ice, e_i. The difference between the two is small, but crucial to many atmospheric processes.

Accurate calculations of vapour pressure require the ventilation speed over the sensors to be taken into account. For this reason, there are different psychrometric formulae and tables for sensors exposed in a passively ventilated shelter such as a Stevenson screen [112, 220] and for those in a forced airflow, such as an aspirated psychrometer (as described in Chapter 5, *Measuring the temperature of the air*). There are also slight differences in the method of calculation for temperatures below 0 °C, owing to differences in the saturation vapour pressure over liquid water and ice surfaces. Such details are beyond the scope of this chapter, but additional information can be found in references [4, 99].

Relative Humidity or RH (sometimes rh%) – is defined as the observed vapour pressure expressed as a percentage of the saturation vapour pressure at that

Table 8.1 *The variation of saturated vapour pressure, humidity mixing ratio r and specific humidity q with air temperature. From this it can be seen that saturated air at 20 °C holds almost four times the amount of water vapour as saturated air at 0 °C. Calculations assume pressure 1000 hPa. Source: [221]*

Variations of various humidity parameters with temperature			
Air temperature (°C)	Saturated vapour pressure (hPa)	Humidity mixing ratio r (g/kg)	Specific humidity q (g/kg)
−15	1.9	1.18	0.54
−10	2.9	1.81	0.64
−5	4.2	2.62	0.72
0	6.1	3.82	0.79
5	8.7	5.46	0.85
10	12.3	7.75	0.89
15	17.1	10.82	0.91
20	23.5	14.97	0.94
25	31.8	20.43	0.95
30	42.6	27.68	0.96
35	56.4	37.18	0.97

temperature (and pressure); that is, $e / e_w \times 100\%$ (or $e / e_i \times 100\%$ below 0 °C). Where the two are the same, the RH is 100% and the air is said to be *saturated*.

Vapour pressure is directly related to the specific humidity q (the amount of water vapour in a sample of moist air, in grams of water vapour per kilogram of air, g/kg) and to the humidity mixing ratio r (the amount of water vapour in a sample of dry air, in grams of water vapour per kilogram of dry air, g/kg): $q = r / (1 + r)$. The derivation of the various measures is outside the scope of this chapter, but interested readers are referred to the materials in reference [221].

In surface operational meteorology, the dew point is the most quoted measure; in upper-air measurements, specific humidity or mixing ratio: in climatology, RH.

Several humidity parameters are given for a range of temperatures in **Table 8.1**.

Example: using humidity parameters

An observation shows that the air temperature (dry bulb temperature) is 25 °C and the RH is 39%. What is the vapour pressure and the dew point temperature?

Using psychrometric tables or an online/smartphone calculator app [219], the vapour pressure is found to be to be 12.3 hPa (assuming ventilation levels appropriate to a Stevenson screen or similar). From tables (such as Table 8.1) or an online calculator, this corresponds to the saturation vapour pressure at 10 °C – therefore the dew point is 10 °C.

Alternatively, the observation parameters could have been stated as – air temperature 25 °C and dew point 10 °C. What is the RH and vapour pressure?

Using Table 8.1 or an online calculator, we can see that the saturation vapour pressure at the dew point temperature of 10 °C is 12.3 hPa. From Table 8.1, or the online calculator, we find the saturation vapour pressure at the air temperature of 25 °C is 31.8 hPa. The RH is then 12.3/31.8 = 39%.

For this example with the air temperature at 25 °C, we could therefore specify the observed humidity as any or all of the following parameters:

RH	39%
Wet bulb	14.6 °C
Wet bulb depression	10.4 degrees Celsius (degC)
Vapour pressure	12.3 hPa
Dew point	10 °C
Dew point depression	15 degrees Celsius (degC)
Mixing ratio	7.7 g/kg

Standard methods of measuring relative humidity

There are three main instrument types used to measure the amount of atmospheric water vapour, or humidity. These are the chilled mirror method, electronic humidity sensors, or the paired dry and wet bulb combination (or *psychrometer*): each instrument is briefly described in turn, with their relative advantages and disadvantages. Each tends to work best under specific conditions, and while these optimal conditions overlap to a great extent, no humidity instrument can be said to work perfectly in all circumstances. For a more comprehensive review of modern methods of measuring humidity, see reference [222].

Older instruments using strands of human hair to sense changes in humidity, such as the hair hygrograph, are not particularly fast-reacting, reliable or accurate, and have been almost entirely superseded by more modern methods.

The chilled mirror method

Chilled mirror sensors are often known as dew point sensors, and the clue is in the name. The principle is extremely simple: a light beam is bounced off a highly silvered mirror surface whose temperature is controlled and maintained within very tight limits by onboard electronics. Ambient air is drawn over the mirror by a fan, and the mirror is progressively cooled using a high-precision Peltier cell. When the mirror's surface temperature falls to the dew point of the ambient air, a fine film of condensation forms on the polished surface, thus altering the intensity of the reflected light beam. This change is detected by a light sensor, and the mirror's surface temperature (the dew point temperature) quickly measured by a rapid-response platinum resistance sensor. The mirror is then gently warmed a little above the observed dew point temperature to allow the induced condensation to evaporate, at which point the cooling cycle starts again and another dew point temperature is measured. This method measures the dew point temperature directly, and thus provides an absolute measure, in contrast to the other two methods which measure humidity and from which the dew point temperature is then derived by calculation. For this reason, chilled mirror sensors tend to be the device of choice in metrology and calibration laboratories. Precise knowledge of the dew point temperature, together with the air temperature, enables the other humidity measures such as relative humidity, mixing ratio and so on to be derived with optimal accuracy.

Although deceptively straightforward in principle, chilled mirror systems require delicate and expensive electronics to operate and for this reason tend to be expensive. In practice they also tend to be difficult to maintain, especially in remote or unmanned environments, because the mirror surface must remain highly polished for optimal operation: dust, atmospheric aerosols or pollution loading can cause rapid deterioration, and without frequent maintenance instrumental accuracy suffers accordingly. In addition, they tend to exhibit a slow response (minutes between readings, although not very different from the response time of electronic sensors), require a significant power supply and can be unreliable in very dry air or when ambient conditions change quickly. Co-locating the assembly used to warm and chill the mirror alongside sensors used to determine air temperature can also result in errors to the latter. The US ASOS systems (see Chapter 5, *Measuring the temperature of the air*) originally used chilled mirror sensors for dew point determination, but these were subsequently replaced with simpler but more reliable capacitance sensors.

Electronic humidity sensors

Arising primarily from the requirements of balloon-borne temperature and water vapour sensors for routine upper-air measurements, small and reliable electronic humidity sensors have been developed which use little power and provide an output signal proportional to relative humidity. Such sensors now generate the majority of automated humidity measurements at both surface and upper levels in the atmosphere: units range between inexpensive devices for use in consumer smartphones and budget AWS units and more sophisticated professional sensors with high reliability and stable calibrations. A typical device consists of a polymer foil sandwiched between two gold foil electrodes to form a capacitor, whose electrical impedance varies with relative humidity, but other methods based on semi-porous ceramics or thin film technologies are also used. Onboard electronics render signal processing, linearisation and calibration to provide output as a direct measurement of relative humidity between 0% and 100%, as either analogue voltage or digital code. In general, dedicated humidity sensors are to be preferred over combined temperature/humidity units, for although a combined probe simplifies exposure and wiring, takes up less space and is less expensive than separate instruments, replacing the humidity sensor mandates replacing the temperature sensor too, which may have a different calibration from the one it replaces. The presence of an RH sensor on a combination temperature/RH unit also makes ice bath temperature calibration checks (see Chapter 15, *Calibration*) impossible, because immersion in water will damage or destroy the RH sensor.

High-quality humidity sensors can be expected to perform reliably in most conditions, while normally remaining reasonably stable in calibration (typically 2–3 years can be expected between recalibration). They tend to work best at low or very low humidity levels [223], and their performance is unaffected by air temperatures below 0 °C, when both chilled mirror and psychrometer systems can become unreliable or unresponsive. On the other hand, response tends to be very slow when the humidity changes only slowly, especially near saturation, and once they have reached saturation they can take some time to 'dry out' (**Figure 8.1**). This is particularly marked if they have been in very wet air for long periods, such as after a foggy night or in persistent hill fog, and can introduce significant bias into, for example, mean hourly humidity analyses. They can also spuriously indicate RH values slightly in excess of 100%, although that is easily taken care of with suitable code in a programmable logger. Less easily managed is the tendency in some instruments for readings to 'plateau' at typically 97–98%, never actually attaining 100% even in saturated air. The non-linear response just below 100%, especially when combined with sensor hysteresis (see **Appendix 1**) and possible wetting of the sensor owing to condensation, often makes it difficult to be confident of typical nominal 2 per cent accuracy readings from electrical sensors in the 95–100% RH range. Such sensors require a protective micro-pore filter, as direct contact with water or airborne aerosols may damage the sensitive element. The presence of the filter significantly increases sensor lag, particularly where ventilation is limited (below about 2 m/s, as is almost always the case inside a Stevenson screen [112]), and response time is often slow as a result. Aspiration (forced ventilation) can improve the response time of humidity sensors, particularly at high humidities, but appears to offer little benefit in other circumstances [224]. Atmospheric dirt deposition and loading on the sensor in forced airflow is much

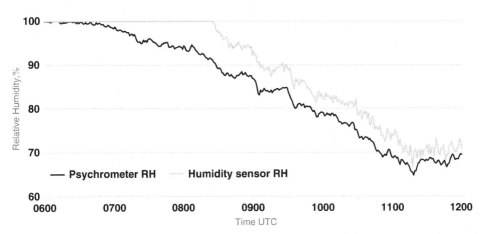

Figure 8.1 Electronic humidity sensors are less responsive in conditions near saturation, and their response time can be slow in such conditions. This plot compares logged relative humidity at 1 minute intervals from an electronic sensor and a dry and wet bulb psychrometer, showing the typical 'overhang' of the former following a long spell of near-saturation, together with the subsequent frequent and rapid minor variations typical of a daytime well-mixed boundary layer [225].

greater than within a more typical Stevenson screen exposure and will eventually cause sensor failure unless regularly checked and cleaned.

Humidity sensors are also prone to calibration drift, of which more below, and can have a fairly limited working life. Useful lifetime is quite variable and not easily predictable, but is rarely more than a few years, occasionally just a few months, especially in entry-level and budget systems or in areas with high air pollution (particularly sulphur dioxide, which degrades the polymer used in many sensors) or plentiful airborne salt particles. The sensor should be replaced if its readings become erratic or the calibration becomes unstable. The readings from a failing sensor will quickly bear little resemblance to changes in atmospheric humidity, and to avoid loss of record a deteriorating sensor should be replaced at the first signs of trouble.

The dry and wet bulb psychrometer

The 'traditional' method of measuring humidity was with a matched pair of mercury thermometers, known individually as dry bulb and wet bulb thermometers and in combination as a dry and wet bulb psychrometer. Fortunately, as mercury-based thermometers have been progressively withdrawn from service, the arrangement is easily replicated using two matched resistance temperature devices, whether thermistors or platinum resistance thermometers (PRTs), mounted adjacent to each other within a thermometer screen. Both 'dry' and 'wet' sensors are then continuously logged using a datalogger. This approach is often used where accurate measures of humidity are required or where strict continuity with existing measurement methods is preferred, or simply to provide a calibration check on a humidity sensor if one is also fitted.

As the names imply, the 'dry bulb' sensor remains 'bare', while the 'wet bulb' is kept permanently moist using a thin, close-fitting cotton sleeve or wick: the latter draws water from an adjacent container by capillary action. The sleeve or wick should extend at least 30 mm along the steel sheath beyond the sensor element to minimise errors due to conduction along the sheath or the connecting cable. The wet bulb is cooled by evaporation, and the difference between the dry bulb and wet bulb temperatures provides a measure of the humidity of the air. The lower the water vapour content of the air, the greater the difference – at saturation (and assuming accurate calibrations) both will read the same temperature. If, in saturated air and when temperatures are changing only slowly, the two thermometers do not read the same temperature, then the calibration of both sensors should be checked. Both thermometers should be read (or logged) simultaneously to a precision of 0.1 degrees Celsius or better, adjusting for any calibration offsets: then, using tables, an online calculator or formulae, or the calculation routines available within meteorological dataloggers, the relative humidity (or any of the other humidity measures) can be quickly and easily determined.

For accurate readings, the wet bulb must be carefully maintained using only pure water (distilled or de-ionised, not tap water). It is also essential that the covering of the wet bulb be as thin as possible commensurate with maintaining an adequate supply of water, and it must be kept clean – a dirty wet bulb will read higher than it should, and as a result the indicated RH will be higher than the true value.

Aside from continuity with legacy thermometers, one key advantage of the logged dry and wet bulb method is its greater sensitivity at high relative humidities. A well-maintained and carefully calibrated psychrometer using carefully matched and calibrated PRTs is capable of reliably distinguishing small changes in relative humidity close to saturation (say from 90% up to 100% RH and back again), a range across which both chilled mirrors and humidity sensors are slow to respond. Humidity sensors in particular will tend to 'stick' at or close to 100% for some time after being exposed to saturated or near-saturated conditions, often showing a sudden fall only once the ambient air has begun to dry out (see **Figure 8.1**). At sites where overnight saturated or near-saturated conditions occur frequently (many temperate climate locations), a humidity sensor will be liable to under-sample the true range in humidity levels. On the other hand, the response time of a psychrometer combination is constrained by the lagging and insulation effect of the wet bulb wick around the sensor itself. Laboratory tests indicate that this is responsible for increasing the response time of a wet bulb sensor by about a factor of three, particularly within the typical low-airflow environment within a Stevenson screen [112]. In the real world, this means that if and when the air temperature changes rapidly, in either direction, the lag of the wet bulb sensor will result in derived humidity values being incorrect for some time after that change, perhaps 10–15 minutes or more. Occasionally, spurious temporary values of RH above 100% (wet bulb reading higher than the dry bulb) will result. In previous times, when dry and wet bulb thermometers were read manually at hourly intervals or less frequently, such circumstances were statistically infrequent, but with the increasing frequency of 1 minute data, they are simply more obvious in the record.

Another disadvantage of the dry and wet bulb psychrometer is that it is difficult to maintain a good wet bulb at temperatures below freezing (an 'ice bulb'), particularly if the air is dry, because capillary action will cease and the wet bulb will quickly dry out [226]. In such circumstances, a humidity sensor will often provide more

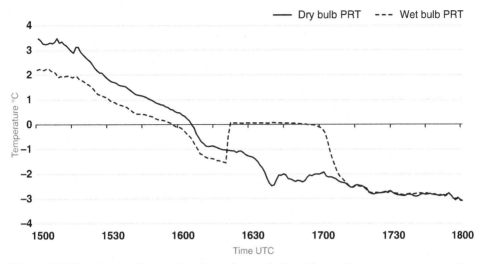

Figure 8.2 The release of latent heat from the wet bulb wick as it freezes shows as a sudden rapid rise in the wet bulb temperature to 0.0 °C at 1620 UTC [227]

reliable records. One consequence of the change of state as the wet bulb freezes is the release of latent heat, which is frequently observed with 1 minute logging (**Figure 8.2**). In the example shown, the wet bulb continued to fall in line with the dry bulb until it reached –1.6 °C at 1620 UTC, whereupon it began to freeze, and the release of latent heat caused the observed temperature to rise quickly to +0.06 °C where it remained until all available water had frozen, thereafter falling back in line with the dry bulb temperature by 1710 UTC. With no liquid water now available to evaporate, dry and wet bulb temperatures remained identical until the following morning (**Figure 8.3**), when the change of state was reversed as the temperature reached 0 °C. At this point, the wet bulb wick began to thaw, and the latent heat involved in the change of state was extracted from the wet bulb wick, keeping it at 0.0 °C until thawing was complete. On the occasion shown, this took a little over 50 minutes, but it can take considerably longer if the rise in temperature is less rapid. During this period, with the wet bulb reading restrained at about +0.05 °C, the depression of the wet bulb below the dry bulb increased to artificially high values, and as a result computed RH and dew point values fell to unrealistically low values (on this occasion the apparent RH derived from the dry and wet bulb values, the thin grey line on **Figure 8.3**, fell below 60%, and the apparent dew point close to –5 °C, until the wet bulb thawed). The wet bulb temperature began to rise only once the thawing process was complete (after about 1130 UTC in the example shown), and only some time thereafter did the humidity record from the psychrometer and the electronic sensor return to close agreement. Under circumstances such as these, humidity parameters from the dry and wet bulb psychrometer should be disregarded, and records from an electronic humidity sensor – if there is one – used in preference.

During occasions of very low humidity (below about 30% RH), humidity measurements from humidity sensors tend to be more accurate and reliable than those from psychrometer measurements. This is partly due to capillary action becoming less effective, or even ceasing altogether, as the wick itself begins to dry

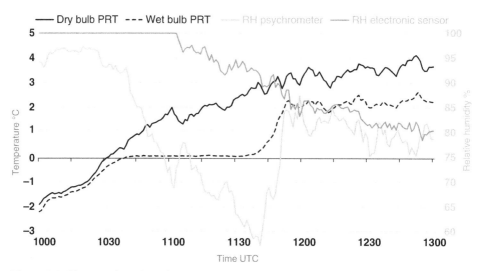

Figure 8.3 The morning after Figure 8.2. As the temperature reached 0 °C once more and the wet bulb wick began to thaw, latent heat involved in the change of state was extracted from the wet bulb wick, keeping it at 0.0 °C until thawing was complete. Circumstances such as these (which occur on almost every occasion of air frost) lead to considerable temporary errors in humidity parameters derived from a dry and wet bulb psychrometer [227].

out, and partly due to heat conduction along the external sensor sheath from areas outside those enveloped by the wick, or along the cabling itself. At 10% RH and air temperature 25 °C, for example, the difference between the dry and wet bulb readings is about 13 degrees Celsius: both effects will act to reduce this difference, in which case the calculated humidity will be higher than the true value. At very low levels of humidity, an error of only a few tenths of a degree Celsius can result in strikingly different RH and dew point calculations [226].

Aspirated psychrometers

Aspirated psychrometers can produce reliable and repeatable humidity measurements. One such device, the Assmann psychrometer, was described in Chapter 5, *Measuring the temperature of the air*.

Humidity, comfort and tracking airmasses

Humans are sensitive to humid air because the human body uses evaporative cooling as its primary mechanism of regulating temperature. When the humidity is high, the rate at which perspiration evaporates on the skin is less than it would be if the air were less humid. Because humans perceive the rate of *heat transfer* from the body, rather than *temperature* itself, we feel cooler when the air is dry rather than when it is humid.

But what is it that gives the best comfort measure? The percentage relative humidity alone is a poor indicator, as a cold winter fog (100% RH at 2 °C) is certainly a lot colder than a humid summer's day (75% RH at 25 °C). The dew point temperature is a much better indicator of comfort levels, although the level of sensitivity depends upon acclimatisation. In general, however, a dew point temperature above 17 °C (63 °F) in southern England will start to see people feeling uncomfortable with 'the humidity', while at 20 °C (68 °F) the majority will be so. For citizens of New York or Washington, DC, with acclimatisation the comfort thresholds are shifted upwards a few degrees. The highest dew points in the world occur near very warm bodies of water such as the Red Sea and the Persian Gulf. Assab in Eritrea, on the coast of the Red Sea, boasts an unenviable average dew point of 29 °C (84 °F), while dew points as high as 35 °C have been recorded in the Persian Gulf. One consequence of the greater capacity of warm air for water vapour is that, contrary to popular perception, fogs are densest at higher temperatures rather than at lower. Visibility in a mountain fog at 20 °C, particularly when it is sustained by a strong breeze, can be extremely poor, and the fog very 'wet' indeed, with copious condensation on any surfaces even slightly below the dew point. (Fog at dew points of 35 °C does not bear thinking about, however.)

Various 'heat index' formulae have been devised to reflect the combined cooling effect (or lack of it) of differing temperature and humidity levels – for example, the US Heat Index [228] and the Canadian Humidex index [229]. These and other similar indices are useful in weather forecasting models to predict and help communicate occasions when heat stress is likely to affect vulnerable sections of the population. Some AWS models can be configured to calculate and display current humidity index values, or even to sound an alarm when particular thresholds are reached.

The dew point value is also of extreme importance in operational meteorology as a means of identifying and tracking airmasses. Unless the water vapour of a sample of air changes (by water vapour evaporating into it, or by cloud droplets condensing out into precipitation), the dew point value remains more or less constant, even if the air is warmed. It is therefore a good conservative indicator of the properties of a sample of air, even when that sample of air has travelled thousands of kilometres horizontally from its source or has been raised vertically by forced ascent over a mountain range. The passage of fronts in the cyclonic systems of temperate latitudes are often more easily identified by changes in dew point temperature than in air temperature, particularly during the summer half-year.

Site and exposure requirements

Humidity sensors are normally exposed alongside temperature sensors in a thermometer screen (**Figure 8.4**), either as a dry and wet bulb PRT pair or as a combined temperature/humidity probe for a logged AWS, or perhaps both. Exposure requirements are the same as those for thermometers. Direct solar radiation will not directly affect the humidity value obtained from the sensor, but if the sensor or the radiation screen in which the sensor is exposed becomes warmer than the ambient air temperature (for example, if it becomes unduly warm in sunshine), then the indicated humidity will be lower than the true value. Restricted airflow through the screen or shelter will lead to a very sluggish response from the humidity sensor, particularly if saturated or near-saturated air persists for many hours [112]. The problem is more acute in sheltered locations, where surface wind speeds are low anyway, and at night, when wind speeds tend to be lower than during the day.

In coastal locations, and even occasionally some distance inland after gales, airborne salt can be deposited on temperature and humidity sensors. As salt is

Figure 8.4 'Traditional' internal sensor arrangement within a standard Stevenson screen. The two horizontal thermometers are the maximum (top) and minimum (bottom). The two vertical thermometers are dry bulbs (the bulb of the right-hand vertical thermometer normally sleeved as a wet bulb). Two 3 mm PRTs are exposed alongside the two vertical thermometers, the dry bulb sensor on the left and the wet bulb sensor within the cotton sleeve on the right. A digital humidity sensor is mounted immediately to the left of the central pillar.

hygroscopic, it will absorb moisture from the air, resulting in erroneously high humidity readings. Regular checking and the occasional wipe-over with a damp cloth will normally keep the problem in check. Some humidity sensors include a micropore filter to keep out dust and salt, but at the expense of increased response times. Aspirated radiation shields (Chapter 5, *Measuring the temperature of the air*) are ideal for accurate temperature measurements, and when fitted with humidity sensors will generally give more representative RH values too, but the greater volume of air movement over humidity sensors tends to exacerbate dust and salt ingress problems and ultimately shortens their working life [224].

Calibration and calibration drift

There is no single device or instrument capable of reliable and economical measurements of surface atmospheric humidity for every possible combination of surface atmospheric air temperature and humidity, although electronic humidity sensors can meet most typical requirements. Humidity measurements derived from calibrated and logged dry and wet bulb PRTs as a psychrometer pair are probably more useful at sites where high humidity is frequent and/or persistent, although the requirement for occasional maintenance (to ensure regular replacement of the wet bulb wick, and top-up the water supply) should be allowed for.

Calibration drift is a particular problem with humidity sensors, particularly so in budget systems where it can exceed 5 per cent per year and where non-linearity of response is common. Whatever type of equipment is used, regular checking over a range in humidities is essential if reasonably accurate long-period humidity measurements are sought. Calibration checking is best carried out annually, or more frequently if spot checks indicate the sensor regularly differs more than about

5 per cent from independent instruments, such as a logged dry and wet bulb psychrometer: the process is covered in more detail in Chapter 15, *Calibration*.

Logging requirements

Logging requirements for humidity are the same as for air temperature, although sampling intervals can be less frequent: once per minute is ample (as recommended in WMO CIMO guidelines [4], section 4.1.3), for, as stated previously, the response times (time constant) of most if not all standard commercial humidity sensors, including psychrometers, are similar (and probably considerably slower if the temperature or humidity is changing rapidly). Depending upon sensor and logger combinations, the output is most often given as RH and dew point, although, if required, other humidity parameters can be easily looked up from tables, calculated directly using dedicated algorithms within a programmable datalogger, or determined from standard formulae ([4], section 4 and Annex 4A; and [140], chapter 6) in a suitable spreadsheet or data processing routine.

Accuracy versus precision

The errors inherent in the measurement of humidity – whichever method be used – necessarily imply that derived RH is accurate only to ±2 per cent at best in mid-range : variations in level of this order within short periods of time are in any case common in a well-mixed convective boundary layer. This level of accuracy is also about what can be expected from the individual calibration errors of two thermometers used as a paired dry and wet bulb, and meets WMO 'working standard' requirements ([4], Table 4.2 therein). Quoting RH to one or more decimal places is therefore unjustified except under strictly controlled laboratory conditions. At high humidities response will be slow, while at low humidities and low temperatures, errors increase and the accuracy falls off further. The same goes for dew point – although it is often quoted to a precision of 0.1 degrees Celsius, in reality the measurement uncertainty is often ± 0.5–1 degree when derived from humidity measurements, with still greater uncertainty at low temperatures and humidities. Such accuracies are permissible for most synoptic and climatological applications, provided calibration drift is watched for and corrected promptly.

One-minute summary – *Measuring humidity*

- 'Humidity' refers to the amount of water vapour in the air, a vital component of the weather machine.
- Various measures are used to quantify the amount of water vapour in the air – relative humidity and dew point being the two most commonly used. Knowledge of any two values can be used to derive other humidity parameters. The amount of water vapour that the air can hold varies significantly with temperature – saturated air at 0 °C holds only a quarter of the amount that saturated air at 20 °C can hold.
- The traditional method of measuring humidity is by using a pair of matched sensors, known individually as dry bulb and wet bulb thermometers and in combination as a dry and wet bulb psychrometer. The wet bulb sensing element

is kept permanently moist using a thin close-fitting cotton cap or sleeve. The wet bulb is cooled by evaporation, and the difference in temperature between dry bulb and wet bulb thermometers is a measure of the humidity of the air. Using tables, an online calculator or formulae, the relative humidity (or any other humidity measure) can be quickly and easily determined from simultaneous readings of the two thermometers.

- Electronic humidity sensors provide an alternative method of measuring humidity, which can be used alongside or instead of a traditional psychrometer. Modern sensors are small, economical on power, more reliable at temperatures below freezing and datalogger-friendly, although subject to greater uncertainty and slower response times at high humidities.
- Establishing and maintaining reasonably accurate calibration can be difficult; even the best humidity sensors are no better than ± 2 % RH. Calibration drift is a problem (regular calibration checks are essential) and working lifetimes can be limited. Combined temperature/RH sensors are popular, but can become expensive and inconvenient if the relatively short working lifetime of the humidity component mandates replacement (and recalibration) of the temperature sensor too. The combination of the two sensors also precludes ice-bath calibration checks being made on the temperature sensor (see Chapter 15, *Calibration*).
- Humidity sensors are normally exposed alongside temperature sensors in a thermometer screen (Stevenson screens or similar, AWS radiation screens or aspirated units).
- Logging intervals should be the same as those for temperature observations, although sampling intervals can be reduced (once per minute is ample) as response times for humidity sensors are greater than those for temperature.

9 Measuring wind speed and direction

Wind is the most variable of all the weather elements. The speed of the wind can double, or halve, within a few seconds. Its direction can, and occasionally does, change by 180 degrees within a minute, and can make several rotations around the compass within an hour or two. Wind direction and speed both vary continuously with a time period measured in seconds, about a mean value which itself changes on a minute-by-minute, hour-by-hour, day-to-day and month-to-month basis (**Figures 9.1** and **9.2**) in fractal-like fashion. The wind may be so feeble as to be barely perceptible, or strong enough to destroy forests and buildings. Measuring and summarising such a fickle element poses considerable challenges, not least in requiring physically robust fast-response sensors along with rapid sampling and logging capabilities. Wind instruments must respond quickly and accurately in the lightest of breezes, yet also be capable of surviving winds in excess of hurricane force.

Observations of surface wind speed and direction are also some of the most important measurements in operational meteorology and in aviation forecasting, and of course consistent and standardised exposures of wind instruments themselves become essential for accurate, reliable and comparable results. World Meteorological Organization (WMO) CIMO exposure guidelines [4] stipulate that wind instruments should ideally be sited on level terrain at 10 m above ground level well away from any obstacles or buildings, but with the rider that, in practice, ' . . . it is nearly impossible to find a location where the wind speed is representative of a large area.' For this reason, high-quality wind records can be the most difficult to obtain of all the more common weather elements, especially in an urban environment where a 'perfect' exposure is almost impossible to achieve. The necessarily elevated nature of the sensors can pose significant safety issues for access, installation and maintenance, while continual exposure to the elements at height (rain and snow, ice and frost, sunshine and solar radiation and possibly lightning, in addition to buffeting by the wind itself) takes its toll on sensor reliability, longevity and electrical connections. In a windy location, even the best sensors may last only a few years before replacement becomes necessary.

Despite these significant obstacles and requirements, it is possible to make useful automated observations of wind speed and direction even without a handy airfield-sized plot of land, although they may be rather more site-specific than is the case with other measured parameters. This chapter describes methods to 'measure the wind', suggests suitable instruments and how best to expose them, and outlines some common pitfalls.

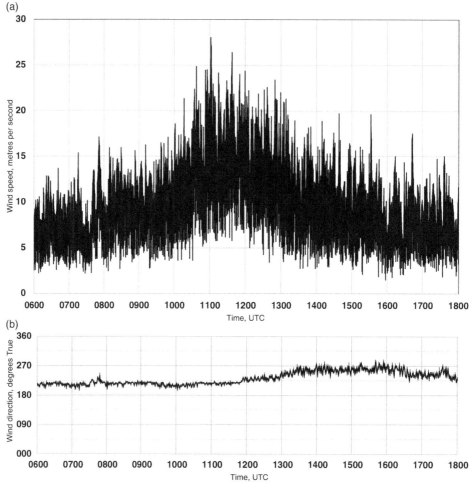

Figure 9.1 A 12 hour record of wind speed (upper graph a, metres per second) and wind direction (lower graph b, degrees True) from a site in central southern England during storm *Eunice* on 18 February 2022, showing 1 minute logged 10 m gust and lull speeds, demonstrating typical rapid variations in both wind speed and direction with time, both on a minute-by-minute and hour-by-hour basis.

Those starting out in weather measurement, or on a tight budget, may find it easier at least initially to prioritise air temperature, rainfall and pressure records, as covered in the preceding chapters, before tackling the more complex territory of automated wind measurements at a later date. Manual estimates of wind speed (using the Beaufort Scale – covered later in this chapter) and wind direction, or 'spot' wind speeds obtained from inexpensive handheld anemometers, will be sufficient for many purposes.

The *measurand* – what is being measured?

'Wind' is the continual movement of air over the surface of the Earth – air currents resulting from differential heating of the planet by the Sun. The Earth's wind systems are vast three-dimensional heat-exchange engines which distribute heat and

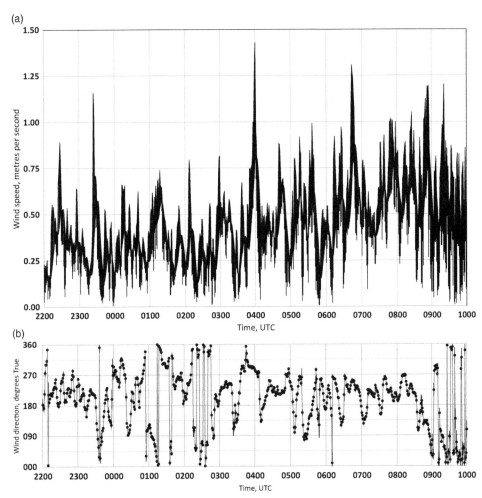

Figure 9.2 As Figure 9.1, but for a prolonged spell of light and variable winds (21–22 January 2023). Both this and the previous figure were logged from the output of a Gill Windsonic sonic anemometer; this figure shows the remarkable sensitivity of the instrument, continuing to operate well below WMO's definition of 'calm' (0.2 m/s, or about 0.4 kn). On this occasion, both alternative cup anemometers on site (Vector Instruments A100 L and Davis Instruments VP2 anemometer) remained unresponsive throughout.

moisture around the planet. We are familiar with surface gusts and lulls – turbulent effects caused by friction in the so-called boundary layer, the lowest layer of the atmosphere in contact with the Earth's surface – but we are probably less familiar with the intricate and continually changing structures of the winds above our heads. Wind speeds are normally lowest, and most variable, close to the Earth's surface; the greater the height of surface obstacles, the greater the short-period variations in wind speed and direction at ground level, which we experience as gustiness. Winds at sea are generally stronger but steadier (in both speed and direction) than on land because frictional effects are much lower: winds in built-up city-centre environments, with many, high and varied surface obstacles, are notoriously variable. Winds are much stronger in the upper atmosphere, where at 10 km or so above the Earth's surface they sometimes blow for days at speeds in excess of 100 metres per second (200 knots, 230 mph) in jet streams.

Mathematically, wind is expressed as a vector quantity – one which has both direction and speed. In this context, *wind speed* and *wind velocity* have different meanings – 'wind speed' (more correctly, 'scalar wind speed') refers to distance covered in a specified time ('a 10 metres per second wind'), whereas 'wind velocity' includes both speed and direction ('a 10 metres per second northerly wind'). Strictly speaking, wind vectors are three-dimensional in nature, namely x, y and z components rather than just x and y, but since the vertical component of wind speed z is normally small near the Earth's surface, at least in comparison to the horizontal component, it is usually disregarded in conventional meteorological measurements. The horizontal 'vector mean wind' is a useful way to combine wind speed and direction records to come up with a 'resultant', or 'averaged', wind direction and speed, and is covered in more detail later in this chapter.

Wind speed and wind direction can be measured directly with separate instruments (namely anemometer and wind vane, sometimes mounted together in one housing), or increasingly with more sophisticated one-piece sensors such as a sonic anemometer, which measures wind as a true vector and resolves it into direction and speed components. Operational meteorology uses 'raw' wind vector information or converts reported x and y components back into vector form, but for ease of handling wind speed and wind direction are most often treated as separate quantities in climatological summaries.

Units of wind speed and wind direction

Wind speed is dimensionally expressed in terms of distance and time – for example, the wind speed could be stated as so many metres per second (m/s or m s^{-1}) or miles per hour (mph). In meteorology, the knot (nautical miles per hour, or kn, but not the tautological 'knots per hour': 1 kn = 1.15 mph) is still the preferred unit in many countries and for aviation purposes. This reflects the preferences of the earliest users of wind speed observations, Britain's Royal Navy in the eighteenth and nineteenth centuries, where coherence with the units of measure of sailing ship speed was essential. Depending upon preference, miles per hour or kilometres per hour are also sometimes used. Conveniently, 1 m/s is very nearly 2 knots; the exact conversions are given in **Appendix 6**.

Whichever units are chosen, it is important to ensure that they are noted alongside instrument and exposure details in the site metadata, as the unit will not be obvious from the record itself (visual inspection of tabular temperature data from an AWS would quickly reveal whether it was in °C or °F, for example, but the distinction between knots and miles per hour would be impossible to determine without additional information).

Two measures of scalar **wind speed** are important – the *mean wind speed* (usually expressed over a defined period of time, most often 10 minutes) and the *gust speed*. Because wind speeds can vary enormously within a few seconds, gusts are defined by WMO as 'the highest mean wind speed over a 3 second period' – see also section *Measuring wind gusts* later in this chapter. There are many and varied effects which influence recorded gust speeds, amongst them being the sensor type and height, sampling interval frequency and the processing applied to the samples. Regardless of site or shelter, for wind gust records to be in any way comparable and meaningful, the sampling interval has to be short enough to 'catch' transient gusts (see *How often is the information updated?* in Chapter 2, *Choosing a weather station*). If accurate records of wind speed – both means and gusts – are an important requirement, as clearly they will be at an airport, or monitoring crosswinds on an exposed railway viaduct, for example, then a short sampling period – no more than a second or so – is needed. For applications where only mean wind speeds over a period of, say, minutes to hours are required, a short sampling period is less important. More details on logging and sampling intervals are given later in the chapter.

By convention, in meteorology **wind direction** is defined as the direction *from which the wind is blowing*, relative to true north, not magnetic north. Thus a south-westerly wind blows *from* the south-west *towards* the north-east.

Compass points have been used to define wind directions for hundreds of years, but for more precise weather measurements the direction is specified using the 360 degrees of the compass, starting from north and working clockwise (technically 'the veer from north'), so that 90° represents an easterly wind, 180° a southerly wind, 270° a westerly wind and so on (**Figure 9.3**). By convention, north is represented as 360° rather than 0°, as '0' is reserved to indicate calm (the absence or near absence of wind – see Box, *How calm is calm?* in this chapter) in both wind direction and speed. In meteorological reporting, wind directions are usually given to the nearest 10 degrees, the final digit being omitted, thus wind direction '23' is understood to mean '230 degrees', or south-westerly. To avoid misinterpretation, leading zeroes are usually quoted, so that '09' refers to '090 degrees' (easterly) and not '009 degrees' (just east of north).

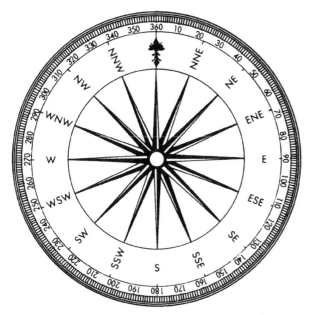

Figure 9.3 Compass points on the 360 degree compass. (© British Crown copyright, Met Office)

Standard methods for measuring wind speed and direction

Wind sensors are located above ground level, usually in an open, level location mounted on an open mast or tower 10 m above ground level (**Figure 9.4**), to minimise frictional effects near the ground. Turbulent effects from obstructions such as trees, buildings or other obstacles can extend downwind to 12 or 15 times the height of any obstacle, and their presence makes measurement of the 'undisturbed' wind flow more difficult. However, 'perfect' sites are few and far between, particularly in urban or suburban areas, a limitation acknowledged by WMO. It may be possible to adjust observed records of mean wind speeds to correspond more closely to those from sites conforming to WMO site standards by the use of empirical 'transfer functions' ('fudge factors'); these are outlined briefly in the wind instrument siting classifications, which are set out in more detail later in this chapter.

The following sections describe the various types of instruments used to measure wind direction and speed, followed by details regarding the exposure, installation and logging of wind sensors.

Measuring wind speed

Wind speed records are particularly sensitive to the type and response of the sensors themselves, together with logger sampling and logging intervals. The response time of the sensor itself, together with the sampling time of the logging or display system, determines whether the instrument(s) in question are capable of recording short-period gusts (**Figures 9.1** and **9.2**). Records from instruments with different response times or sampling intervals will give different values for both peak gusts and gust ratios, as explained subsequently.

Figure 9.4 Anemometer and wind vane on 10 m masts in an exposed, open location; Valentia Observatory in south-west Ireland. (Photograph by the author)

The cup anemometer

Probably the most familiar wind speed instrument today is the cup anemometer invented by Thomas Romney Robinson, an astronomer at Armagh Observatory in Northern Ireland, in 1846; see **Figure 9.5** [6, 52]. Robinson's original design used four large cups; modern instruments such as the Vector Instruments model illustrated in **Figure 9.6** use three cups mounted 120 degrees apart on a vertical shaft with low-friction bearings. The drag coefficient of the open face of the cup is greater than that of the smooth conical or hemispherical opposite face, and this difference causes the shaft to rotate as the cups spin in a breeze. The speed of revolution of the shaft is (very nearly) proportional to the speed of the wind, although cup anemometers tend to speed up in a gust a little faster than they slow down in a lull, and so slightly overestimate true wind speeds. Modern low-power digital cup anemometers such as the Vector Instruments model shown in **Figure 9.6** convert revolutions of the shaft into a distance

Figure 9.5 An early model (1870s) of Robinson's recording cup anemometer, still in working order at Armagh Observatory in Northern Ireland. (Photograph by the author)

Figure 9.6 A modern cup anemometer (nearest the camera) and wind vane by Vector Instruments. (Photograph by the author)

measurement by breaking a beam of light every 0.1 m or so, the 'breaks' then being counted by a pulse counter logger. Older analogue instruments used a small dynamo to generate a voltage, measured by a recording voltmeter, or mechanical gearing to rotate a distance-measurement counter, similar in mechanical principle to a car odometer. Both suffered greater frictional losses than the modern 'light chopper' designs, and so tended to have a higher starting speed (as detailed later in this chapter). Pulses are also more reliably transmitted over long cable lengths than the relatively small analogue voltages generated by the small generators used in anemometers.

The design of the instrument is simple and has been refined over the years, and with modern materials and electronics the sensors when new are both sensitive (low starting speed) and robust (low maintenance, high maximum wind speed capability). Performance tends to degrade over time as bearings wear, particularly so with 'budget' models, leading over time to decreasing sensitivity (higher starting speed) and thus reducing response in light winds.

The Vector Instruments A100 cup anemometer illustrated in **Figure 9.6** – also known as the 'Porton anemometer', after Porton Down in Wiltshire, England, where it was developed in the 1970s – is widely used in professional weather monitoring around the world. It has a starting speed when new of around 0.25 m/s (0.5 knots) – a barely perceptible flow of air – yet is rated up to 75 m/s (over 150 knots), twice hurricane force, with a stated accuracy of 1 per cent ± 0.1 m/s up to 56 m/s (108 knots). The instrument has been shown to produce spurious high gusts in certain atmospheric conditions, however, particularly high potential electric fields in or near convective systems [230].

Cup anemometers are progressively being replaced by sonic anemometers, which are considered in more detail in the following subsection.

Sonic anemometers

Although the principle of the sonic anemometer was first outlined as far back as the 1960s [231], until relatively recently they remained largely confined to research establishments. The increasing capability and processing power of modern dataloggers has broadened their suitability and appeal, and as prices have come down they have become readily available, and it is a welcome sign to see them starting to be included as options in some budget AWSs. Sonic anemometers feature several significant advantages, including reliability (no moving parts to wear out) and high sensitivity.

The principle of the 2D sonic anemometer (**Figure 9.7**) is very straightforward. The speed of sound in air is equal to the speed of sound in still air, plus the speed of the air. Using two small sound emitter/receiver pairs at 90 degrees to each other, the difference in the speed of an ultrasonic sound pulse across the unit is measured very accurately (the temperature is also measured, as the speed of sound is temperature-dependent). Using sophisticated onboard electronics, the instrument derives both wind speed and direction readings, wind direction being calculated as a vector from the measurement of the air speed across two axes at right angles to each other. Commercial sonic anemometer units are available for both two-dimensional (2D) and three-dimensional (3D) monitoring of wind speed and direction, 3D units being particularly useful in turbulence, flux and dispersion boundary layer research projects because of their rapid response times.

Tip: before or during installation, fit bird spikes securely to the 'roof' of sonic anemometers of the type illustrated in **Figure 9.7**, because this elevated platform is otherwise attractive as a resting post for birds, and the accumulation of bird droppings that results can foul the instrument's sensors.

Figure 9.7 Gill Windsonic sonic anemometer. (Photograph courtesy of Gill Instruments)

Propeller or 'windmill' anemometers

A less commonly encountered form of anemometer is the propeller or windmill variant. In this design of instrument, the bladed propeller element is kept facing into the wind by being mounted on a wind vane, thus making the sensing head direction-ally sensitive. As the wind blows through the rotor, differential drag forces across the blades, together with lift from the blade aerofoil itself, causes the blades to spin. In low wind speeds the wind vane may not turn into the wind and the propeller blades may therefore be at an angle to the wind, rendering the instrument unresponsive. A propeller anemometer needs a wind that is strong enough to turn the wind vane into the wind – in contrast to cup or sonic anemometers, which are largely insensitive to direction and therefore benefit from a lower starting speed.

At higher wind speeds, particularly in sites with less-than-ideal exposures such as those typical of urban or suburban areas, turbulence often results in large, rapid and erratic changes in wind direction, resulting in the sensing head again being more often than not at an angle to the wind, reducing the wind speed measured by the instrument. Propeller anemometers perform best in medium to strong winds where the direction does not vary rapidly, and are somewhat less prone to the build-up of ice or rime than cup anemometers. For both reasons, they are often the instrument of choice at exposed sites such as mountain summits (**Figure 9.8**).

Handheld anemometers

There are very many handheld anemometers available on the market, at various prices, and inevitably specification and performance are related to price. Some models offer just a single measurement (instantaneous display of wind speed), while more sophisticated models can display or even log mean speeds and highest gusts over user-determined time periods, perhaps with other measurements such as temperature and humidity. Nielsen-Kellerman's Kestrel 5500 unit (**Figure 3.8**) can even be fitted with an (optional) tripod and wind vane to monitor wind direction (**Figure 9.9**). Instruments such as this are very affordable and reasonably accurate: their greatest source of error is likely to result from exposure at ground level or in a relatively sheltered location. They can often provide 'good enough' indications of wind speed when budget or site limitations preclude more

Figure 9.8 Automated snow, wind and weather AWS in the Swiss Alps, run by the Swiss Institute for Snow and Avalanche Research. The location is Vallée de la Sionne Windstation 'Crêta Besse' (4VDS1), in the Swiss Canton of Valais, at 2696 m above MSL. (Photograph courtesy of WSL-Institut für Schnee- und Lawinenforschung SLF, Davos, Switzerland)

Figure 9.9 Kestrel 5500 handheld AWS fitted with wind vane to sample wind direction and speed. Be aware that temperature and humidity measurements using this device will of course be unrepresentative if the unit is exposed in sunshine, or in rain. (Photograph by the author)

sophisticated automated wind logging equipment, or for temporary or portable field use, although wherever possible their calibration should be verified against a calibrated instrument before use.

The Beaufort wind scale

Reasonably accurate estimates of mean wind speeds can also be made using the Beaufort Scale. Devised by Admiral Sir Francis Beaufort in 1806 (his original manuscript outlining the scale can still be seen in the UK Met Office Archives), the Beaufort wind scale has been adapted somewhat over the years but is still the most frequently used guide when making eye estimates of mean wind speed. Admiral Beaufort's original scale referred only to the effects on sailing ships at sea, but descriptions were later extended to land-based observations. **Table 9.1**

Table 9.1 *The Beaufort wind scale, for use on land, with mean speeds shown (knots, mph and metres per second). The derived empirical relationship between Beaufort Force* B *and 10 m wind speed* v *is* $v = f \sqrt{B^3}$, *where the factor* f *is 1.625 when* v *is in knots, 0.836 for m/s and 1.87 for mph. Source: WMO CIMO guide [4], Table 5.1. Note that CIMO Table 5.2 describes wind speed equivalents for arctic areas and others where there is no vegetation.*

Beaufort Force	Description	knots mean	knots range	mph mean	m/s mean	Effects on land
		WIND SPEED				
0	Calm	0	< 1	0	0	Calm; smoke rises vertically
1	Light air	2	1–3	2	0.8	Direction of wind shown by smoke drift but not by wind vane
2	Light breeze	5	4–6	5	2.4	Wind felt on face, leaves rustle, wind vanes respond
3	Gentle breeze	9	7–10	10	4.3	Leaves and small twigs in constant motion; wind extends light flag
4	Moderate breeze	13	11–16	15	6.7	Dust, leaves and loose paper raised by the wind, small branches move
5	Fresh breeze	19	17–21	21	9.3	Small trees in leaf begin to sway; crested wavelets (whitecaps) form on inland waters
6	Strong breeze	24	22–27	28	12.3	Large tree branches in motion, whistling heard in telephone wires; umbrellas used with difficulty
7	Near gale	30	28–33	35	15.5	Whole trees in motion; difficult to walk against the wind
8	Gale	37	34–40	42	18.9	Twigs and small branches are broken from trees; progress generally impeded
9	Strong gale	44	41–47	50	22.6	Slight structural damage occurs (chimney-pots and slates removed)
10	Storm	52	48–55	59	26.4	Seldom experienced inland; trees uprooted; considerable structural damage occurs
11	Violent storm	60	56–63	68	30.5	Extensive and widespread damage
12	Hurricane	≥ 64	≥ 64	≥ 73	≥ 33	

is the current version, with equivalent wind speeds at 10 m shown. Consistent estimates of wind speeds can be produced with only a little practice. It is important to remember that the scale reflects *mean* wind speeds, and not gusts; twigs may be removed from trees with *gusts* to Beaufort Force 8 ('Gale'), but this does not necessarily mean that the *mean* speed has attained gale force. Simultaneous estimates of *wind direction* should be made, preferably using a wind vane, but failing that on the direction shown by chimney smoke or dropped leaves, grass stems and the like. The direction of low-level clouds should *not* be used, as this may differ significantly from the surface wind direction.

Starting speeds

One of the most important specifications for wind instruments, particularly in areas with low mean wind speeds, is the 'starting speed' (sometimes referred to as 'threshold speed'). As the term implies, this is the speed at which either the anemometer begins to respond (and thus at which measurements commence), or (for wind vanes) the speed at which the vane just turns into the wind.

Almost all modern wind sensors have a starting speed of 1 m/s (2 knots) or less, a huge improvement on older, heavier cup anemometers, some of which had a starting speed of 3 m/s (6 kn) or more [232]. Any anemometer with a starting speed that is a significant fraction of the true mean wind speed will inevitably produce a very distorted wind climatology, particularly an unrealistic frequency of 'calm'. (See Box, *How calm is calm?*)

A selection of starting speeds for various common anemometers and wind vanes is given in **Table 9.2**. The starting speeds on mechanical anemometers and wind vanes tend to increase as the instruments age, particularly cup

Table 9.2 *Starting speed specifications for various common anemometers and wind vanes. Sources are as indicated.*

Anemometers	Gill Windsonic sonic anemometer	0.01 m/s (0.02 kn)	Manufacturer specification
	Vector Instruments A100 cup anemometer	0.2 m/s (0.4 kn)	Manufacturer specification
	Davis Instruments Vantage Pro2 AWS cup anemometer	0.7 m/s (1.4 kn) when new; > 1.5 m/s (3 kn) within 5 years	[233] and author's tests [80]
	RM Young four blade helicoid propeller combined anemometer/wind vane	1.0 m/s (2 kn)	Manufacturer specification
	Met Office Mk 4A cup anemometer	3 m/s (6 kn)	[232]
Wind vanes	Gill Windsonic sonic anemometer	0.01 m/s (0.02 kn)	Manufacturer specification
	Vector Instruments W200P wind vane	0.6 m/s (1.2 kn)	Manufacturer specification
	Davis Instruments Vantage Pro2 AWS wind vane	≥ 1.0 m/s (2 kn)	Author's tests [80]
	RM Young four-blade helicoid propeller combined anemometer/wind vane	1.0 m/s (2 kn)	Manufacturer specification

anemometers as a result of mechanical wear on the bearings. Budget cup anemometers are particularly prone to this, particularly those used on the Davis Instruments range of AWSs which – based upon experience – tend to seize up within a few years and become entirely unresponsive in light winds (see **Figure 9.2**). Sonic anemometers, with no moving parts, are less liable to mechanical ageing: indeed, the manufacturer's specification for the unit shown in **Figure 9.7** is for a mean time between failure (MTBF) in excess of 15 years.

The 'stopping speed' of mechanical anemometers tends to be slightly lower than the starting speed, owing to inertia and reduced 'rolling' friction.

The Gill Windsonic 2D unit has a wide speed range (the manufacturers quote 0–60 m/s, with a lower threshold of just 0.01 m/s) and high accuracy (± 2 per cent) [234]. The sensor consists of a tough corrosion-free polycarbonate body (an anodised aluminium body is also available), and with no moving parts it is ideally suited to harsh environmental conditions or exposure in 'awkward to reach' locations. This type of sensor does require a DC power supply of 5–30 v. It is a 'recommended instrument' (**Appendix 2**).

How calm is calm?

Reliable statistics on the true incidence of very light winds are hard to come by, and are often distorted by changes of anemometer type, site and exposure over the years (see *Starting speeds* earlier in the chapter). Until the introduction of lightweight electronic sensors and loggers starting in the 1980s, standard anemometers often had starting speeds of 2–3 m/s (4–6 kn) or higher: winds lighter than this were often estimated when the chart record was subsequently manually tabulated, based upon the degree of 'mobility' of the wind vane record (although often even the wind vane was insensitive below about 1 m/s or 2 kn). For this reason, statistics of 'calm' tended to be the catch-all for anything below about 2 m/s and were, as a result, higher than reality by varying degrees.

The author's tests on three different anemometer types (a Gill Windsonic, Vector Instruments A100 L and a Davis Instruments VP2 anemometer) mounted adjacent to each other on a standard 3 m meteorological mast – but avoiding mutual shadowing – showed huge differences in each instrument's ability to measure light winds. WMO's definition of 'calm' ([4], section 5.1.2) is 'an average wind speed at or below 0.2 m/s' (0.4 kn), a barely perceptible drift of air. Over a period of 7 months, the percentage of hourly mean wind speeds logged below 0.2 m/s ranged from 2.4 per cent by the sonic anemometer to 23.6 per cent by the A100 L and 37.7 per cent by the Davis Instruments VP2 instrument (**Figure 9.10**). Scaling up to a 12 month period, these figures would amount to just over 200 hours below the 0.2 m/s threshold for the sonic anemometer in a year, to over 2100 hours for the VP2 sensor (averaging almost 6 hours per day). A second 12 month comparison conducted on a 10 m mast, where winds are normally somewhat stronger than at 3 m, showed even greater differences, although no simultaneous Davis Instruments anemometer record was available at this site. Over the 12 month period, the sonic anemometer logged only 19 hours (0.23 per cent) with mean wind speeds below 0.2 m/s, the lowest hourly mean being 0.13 m/s (0.26 kn), while the equivalent figure below 0.2 m/s for the Vector A100 L instrument was 942 hours or 11.5 per cent; 384 hours (4.7 per cent) recorded flat calm (0.00 m/s). Bear in mind also that the Davis Instruments anemometer will output only to 1 mph precision (0.87 kn, or 0.45 m/s), which renders the analysis of low wind speeds rather moot.

True 'flat calm' – wind speed at 10 m 0.0 m/s, 'smoke rising vertically' on the Beaufort scale – is distinctly uncommon in most temperate latitudes. A closer examination using data logged every minute over the same 12 month period revealed that the wind speed recorded by the sonic anemometer at 10 m never actually fell to zero: the longest period of mean winds below 0.05 m/s (0.1 kn) was just 5 minutes, early one September morning. It is also important to appreciate that the sonic anemometer is capable of providing wind *directions* down to such low levels of wind speed (**Figure 9.2**), whereas a conventional wind vane, such as the Vector Instruments W200P used in these comparisons, will not respond below about 0.6 m/s (1.2 kn), even when new (**Table 9.2**). From the preceding, it is clear that the logged duration of 'calm' or very light winds depends to a much greater extent upon the type of anemometer in use than the local climate characteristics. The less sensitive the anemometer, the higher the starting speed and the greater the frequency of calms. Gradual wear on mechanical anemometer or wind vane bearings will also result in a slow increase in starting and stopping speeds, and thus a gradual year-on-year increase in the frequency of calms as the instrument ages.

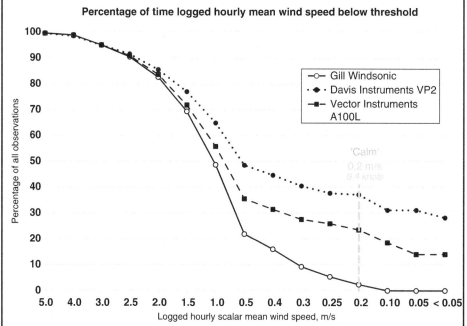

Figure 9.10 A comparison of the percentage of hourly mean wind speeds below a given threshold, for three anemometers mounted at 3 m above ground, logged over a 7 month period at the author's test site. WMO's definition of 'calm' (0.2 m/s) is marked on the plot. Note non-linear x-axis.

Measuring wind gusts

The WMO CIMO guide ([4], sections 5.1.1 and 5.8.2) defines a wind gust as 'the maximum observed wind speed over a specified time interval', typically over a 10 minute or hourly reporting period. Individual wind gusts are defined as the maximum 3 second mean in any given period. To obtain accurate gust measurements, wind speed samples must therefore be made at or less than 3 seconds apart. Where samples are taken at longer intervals, the gust speed will be averaged over a longer time period and individual gusts will be 'smeared out' into a lower speed and longer duration measurement.

Why 3 seconds? The WMO recommendation for a 3 second gust period originated from an analysis [235, 236] published in 1987, which reasoned that, in strong wind conditions, a 3 second gust would possess dimensions typically 50 to 100 m (25 m/s wind × 3 seconds = 75 m), sufficient to engulf typical urban or suburban structures and expose them to the full wind loading of a potentially damaging gust. Gusts of shorter duration are of insufficient scale to engulf complete structures in this way.

AWSs and loggers differ in their sampling intervals, from 1 second or less to a minute or more. **Figure 9.11** and **Table 9.3** illustrate how wind gusts averaged over different sampling periods from 1 second to 60 seconds (open circles) vary in comparison with the standard 3 second mean (solid circle). The figures presented here are based upon the author's observations from a single test site, and will inevitably vary somewhat according to terrain, land use, anemometer height and so on, but are likely to be broadly indicative of all but the most sheltered or most exposed sites. 'Gust' speeds from an anemometer sampling at different intervals from the standard 3 second running mean will vary from an average of 21 per cent above the 3 second value for a sampling time of 0.25 seconds, to 30 per cent below for a 60 second sampling interval. Where wind speeds are sampled at shorter intervals than 3 seconds, the logger should be programmed to calculate 3 second running means from the shorter-period samples. So if the samples were at ¼ second intervals, then a running mean of 12 consecutive samples would become the 3 second value: the highest gust reported would be the highest of the 3 second means, not the highest individual ¼ second sample.

The fine-scale structure of individual gusts, particularly in windy conditions, is such that exact minute-by-minute agreement on wind speeds, particularly gust speeds, is simply not achievable on adjacent instruments where these are more than a few metres apart.

Table 9.3 *Variation of wind gusts with sampling time, as a fraction of the standard 3 second running mean.* '*Observed' – based upon observational data over one calendar year at a site in central southern England (51°N, 1°W): 'Modelled' – based upon the logarithmic profile shown in the dashed line in Figure 9.11.*

Sampling time (seconds)	0.25	0.5	1	2	3	5	10	20	30	60
Observed			1.05	1.03	**1.00**	0.96	0.90	0.82	0.78	0.70
Modelled	1.21	1.15	1.08	1.02	**1.00**	0.94	0.88	0.82	0.78	0.72

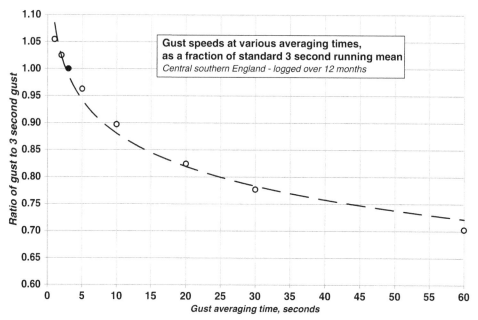

Figure 9.11 Sampling intervals: gust speeds, as a fraction of the standard 3 second gust speeds, for sampling intervals from 1 second to 1 minute. Plotted points are actual observations, and the dashed line is a logarithmic curve fitted to the observations. Based upon the author's observations over a 12 month period at a test site in central southern England; see also Table 9.3.

The **gust ratio** is the ratio of the gust speed to the mean speed over any given interval: for example, an hour with a mean wind speed of 14 m/s and a highest gust of 21 m/s would have a gust ratio of 1.5. Gust ratios are higher over land than over sea or at coastal sites, higher by day and in turbulent or unstable airmasses, and higher in 'cluttered' anemometer exposures more typical of urban sites. Very open exposures – such as a standard 10 m mast site on an open airfield – and stable conditions at night generally record lower gust ratios. Analysis of gust ratios can be useful in airmass stability modelling, while comparisons of gust ratios from a relatively dense network of anemometers across cities or complex topography can provide useful indications of relative turbulence, a benefit to pollution dispersal modelling or the architectural design of city buildings.

The reduction in *average* wind speeds as a result of friction and resulting turbulence is greater than for *gust* speeds, for a variety of physical reasons. An anemometer at 2 m above ground typically records mean wind speeds about 30 per cent lower than those of one at the standard 10 m above ground, but gust speeds might be only 10 per cent less on average – occasionally they may even exceed that of the higher instrument. For this reason, gust speeds should *not* be corrected for height.

Measuring wind direction

The principle of a mechanical wind vane is identical to that of any church spire 'weathervane' – the force exerted by the wind on a vane causes a counterbalanced arrow or pointer to swing into the wind. The two most important characteristics of a mechanical wind vane are that it should turn on its bearings with the minimum of friction, and that it must be balanced. If the unit is not balanced, or is mounted slightly off vertical, it will come to rest in a preferred neutral position, thus biasing wind direction frequencies.

A common sensing element on modern mechanical wind vanes is a potentiometer grid located underneath the vane – the position of the pointer is sensed using a grid of magnetically operated reed switches linked to a small bank of resistors, the measured resistance then being converted into a digital signal, sampled and displayed/logged as required. Other vanes use an array of infrared light-emitting diodes (LEDs) and photodetectors ranged around the wind vane shaft to encode a binary direction code.

The principles of sonic anemometers were outlined in the previous anemometer section. Using orthogonal sensors, north and east wind components are derived directly, and wind direction calculated from frequent (sub-second) data points using on-board electronics or datalogger processing capability. As referred to previously, sonic anemometers operate down to very low wind speeds (< 0.1 m/s), and as a result can accurately record wind directions in very light winds, conditions well below those in which a traditional mechanical wind vane would cease to respond (**Figure 9.2**).

Choosing wind sensors

As with anemometers, modern materials and electronics have driven the development of very sensitive and accurate wind direction sensors which consume very little power and are ideal for digital logging applications. Most budget or 'all-in-one' AWSs include the manufacturer's proprietary wind speed and direction sensors – most often a cup anemometer and potentiometer or LED-based wind vane, although sonic anemometers are available as standard or options on some systems. As wind sensors are the most exposed and are therefore likely to receive the greatest pounding from the weather, failures are most likely here. Accordingly, it is certainly worth considering at the outset whether to purchase a more expensive AWS or optional higher-spec sensor to avoid the greater risk of premature sensor failure and related replacement/reinstallation costs: see also *How robust does the system need to be?* in Chapter 2, *Choosing a weather station*. Where the wind sensor (anemometer or wind vane, or combined unit) may require regular access, such as for battery replacement cycles, then ready availability of access should also be taken into account, for if a 'cherry picker' or scaffolding is required to reach the unit every time a battery fails, then servicing costs over the lifetime of the unit will be very high. On the other hand, sensors not requiring batteries and with 'no moving parts' operating principles, such as sonic anemometers, may have a higher initial purchase cost but are ideally suited for optimum exposure on exposed structures such as masts, and are likely to require far fewer interventions during their operating lifetime. A typical lifetime of professional-quality mechanical wind vanes is around 50 million revolutions, equivalent to 10 years' typical exposure: budget systems or particularly windy locations can be expected to see a shorter mean time between failure.

Typical accuracy of wind direction sensors is ±1–2° obtainable in steady winds over 5 m/s, with a resolution of ±0.2°. The variability of wind direction is normally far greater than these specifications, and the absolute accuracy of the direction will in any case be determined by how accurately the sensor is aligned to true north during installation.

Severe weather performance

The operation of most wind sensors can suffer in severe wintry weather, and they may cease to operate altogether. Heavy wet snow can build up on cup anemometers and wind vanes; repeated melting and refreezing cycles can result in the instruments becoming literally frozen solid for long periods unless access is possible to clear the snow and ice away (this may itself be difficult or dangerous in severe weather, of course). Riming (the build-up of frozen windborne water or ice particles in sub-zero conditions, particularly in cloud) is also a problem for many wind instruments; as well as affecting the measurements themselves, the weight of the accumulated rime can damage or even destroy the sensors and supporting mast or tower (**Figure 9.12**).

Winter riming is a particular problem on many exposed mountain sites in temperate and polar latitudes around the world, particularly in maritime regions

Figure 9.12 Severe riming on a tower at the summit of Mt Washington, New Hampshire (1917 m / 6288 ft) in early May. The small building on the left was the original Mt Washington Observatory, which held the record for the world's highest measured wind speed until 1996. (Photograph by the author)

such as the mountains of the British Isles. Two early AWS models were established by Heriot-Watt University and the then Institute of Hydrology on the summit of Cairn Gorm in Scotland (1245 m, 57°N, 3°W) in 1976 [237, 238, 239], recording wind speed, wind direction and temperature (**Figure 9.13**). To combat the effects of heavy riming, the Heriot-Watt instruments are housed in a heated cylinder, which is exposed to sample the weather for only 3 minutes every half hour, 48 observations per day. The station built up over the years a unique set of observations in the UK's most severe climate, including the highest surface wind speed yet recorded in the British Isles – a gust of 79 m/s (153 kn or 176 mph) at 1148 UTC on 3 January 1993 [240]. The mean wind speed on the summit is 15 m/s, and the average annual temperature +0.5 °C.

Figure 9.13 An early model automatic weather station on Cairn Gorm summit in Scotland (1245 m, 57°N, 3°W) in the 1980s, showing the problems of rime icing. (Photograph by Ian Strangeways)

Exposure of wind sensors

The 'ideal site'

Because wind *speed* increases quickly with height as a result of frictional effects near the ground, the standard exposure for wind instruments is at 10 m above ground level: the sensors are normally mounted at the top of an open mast or tower (**Figure 9.4**). Other instruments benefiting from an elevated exposure, such as sunshine or solar radiation sensors, may sometimes also be mounted on

Life at the top: a challenge for any anemometer

The high-altitude permanently staffed observatory at the summit of Mount Washington in New Hampshire (1917 m / 6288 ft) describes its climate as 'The worst weather on Earth' [241, 242] – the observatory has even trademarked the phrase 'Home of the world's worst weather'. The observatory is supported by NOAA, the National Science Foundation and the University of New Hampshire alongside commercial partners and tourist income, and records have been taken here continuously since 1932. Instruments (and observers) at the observatory are tested well beyond normal extremes in its cold and windy climate [243]. Since early in its history, the observatory has operated and maintained equipment for research, testing and environmental monitoring purposes at its facility on the summit and in Bartlett, New Hampshire in the Mount Washington Valley. Anemometers in particular need to be tough – both to cope with riming (which occurs here even in the summer months – see **Figure 9.12**) and because of the very high wind speeds at this site. Gusts in excess of 67 m/s (130 kn or 150 mph) have been recorded in every month but June and August. Until 1996, Mount Washington also held the world record for the highest surface wind speed measurement, 103 m/s (201 kn, 231 mph), recorded on 12 April 1934. In February 2023, the observatory recorded its lowest-ever air temperature of –43.7 °C (–46.7 °F) while wind speeds averaged close to 45 m/s (around 85 knots or 100 mph); the resulting windchill of –78 °C (–109°F) became the lowest such value ever reliably recorded in North America [244].

The figures support the observatory's tagline – the mean annual wind speed (1991–2020 normal) is 15.6 m/s (30.3 kn or 34.9 mph), the mean annual temperature –2.2 °C (28.0 °F) and the average annual precipitation water equivalent 2317 mm (91 inches) [245]. If visiting New Hampshire, the site is easily accessible during the summer months by the Cog Railway or the auto road and the observatory is well worth a visit – but don't forget to take appropriate clothing, because conditions on the top are often very different from those in the valley! There are also occasional intern vacancies at the Observatory for those who would seek a closer relationship with weather observing at this exceptional location.

Extreme wind speeds present a challenge for most wind measuring instruments, whether or not accompanied by riming. When hurricane *Andrew* hit Florida in August 1992, a Davis Instruments AWS anemometer in Miami registered a gust of 95 m/s (184 kn or 212 mph) [246] shortly before part of the owner's house was destroyed, along with the anemometer itself. Subsequent wind-tunnel tests on similar instruments indicated that the peak gust was probably closer to 79 m/s (154 kn or 177 mph), rather than the 95 m/s originally logged, but the performance for this class of instrument in such severe conditions is noteworthy.

The strongest surface wind speeds on Earth occur in strong tornadoes, but the forces involved are much too great for anemometers to survive. The highest recognised surface wind speed yet reliably recorded by an anemometer, namely 113 m/s (253 mph, 220 kn) occurred on 4 April 1996 at Barrow Island, Australia (20.67°S, 115.38°E, elevation 64 m), during the passage of Tropical Cyclone *Olivia* [247]. The instrument was a heavy-duty three-cup anemometer mounted on a mast 10 m above ground level, sited towards the centre of the island about

4 km from the coast to the south-east and about 7 km inland from the south-south-west, the direction of the strongest wind gusts. The instrument was well exposed in all directions, in good working order and was regularly inspected with comparisons made against a hand-held anemometer.

The peak wind gust measurement was one of five extreme gusts during a series of 5 minute time periods. Gusts of 102, 113 and 104 m/s (199, 220 and 202 kn) were followed by a series of four lower values which were then followed by two more extreme gusts of 96 and 83 m/s (187 and 161 kn) in the subsequent 5 minute periods. The maximum 5 minute mean wind speed was 49 m/s (95 kn).

the mast, although care is needed to ensure that no instrument shields another. In open terrain, the change of wind *direction* within 10 m of the ground is so small as to be disregarded for most purposes. To ensure wind measurements are representative of an area of at least a few kilometres around the site, wind instruments should ideally be sited in a level, open area with few obstructions or obstacles anywhere near the anemometer mast. In practice, of course, perfect sites are rarely available, and thus the World Meteorological Organization (WMO) sets out five classes of wind instrument exposure, along the same lines as those established for other meteorological instruments as covered in previous chapters.

WMO surface wind site classifications

Five classes are set out as shown in **Table 9.4**, which is based directly on the WMO CIMO guide ([4], section 5.9.2 and Annex 1D, section 4). A critical measure here is the so-called 'aerodynamic roughness length z_0', which can be calculated using concurrent wind measurements from two or more heights over a period of weeks or months, or which can be assigned approximately from descriptions given in **Table 9.5**.

As examples, therefore:

- A site with instruments at 10 m with no obstacles higher than the mast within 300 m (including one obstacle 4 m high and 5 m wide located 90 m distant), in open flat terrain (roughness class 3), would be a class 1 reference site;
- Another location with instruments at 10 m with an obstacle 5 m high and 30 m wide located 400 m from the mast in an area of scattered obstacles (x/H width to height ratio typically 15–20) would be a class 3 site;
- A third site, where the instruments were located at 20 m above ground with obstacles up to 6 m high and 5 m in width located 35 m distant (angle subtended 8°), would just fail to meet class 4 criteria and thus would be a class 5 site.

More typical urban or suburban wind sites

It is, of course, almost impossible to find any such class 1 or class 2 site in an inner-city park or typical university or college campus, far less in a suburban or domestic

Table 9.4 *WMO CIMO classification of surface wind observing sites, based upon site and exposure. In this table, 'obstacles' are defined as objects with angular width ≥ 10 degrees, while 'thin obstacles' (such as instrument masts, narrow trees, lamp posts and suchlike) subtend < 10 degrees. The 'roughness class' is set out in the following Table 9.5.*

Class	Distance between sensor(s) and surrounding obstacles of height h	Distance between sensor(s) and 'thin obstacles' of height > 8 m and width w m	Roughness class	Ignore single obstacles below x m height	Uncertainty owing to siting
1	≥ 30 h	≥ 15 w	≤ 4	$x = 4$ m	Reference site
2	≥ 10 h	≥ 15 w	≤ 5	$x = 4$ m	Up to 30%, corrections possible
3	≥ 5 h	≥ 10 w		$x = 5$ m	Up to 50%, corrections cannot be applied
4	≥ 2.5 h	No obstacle with angular width > 60° and height > 10 m within 40 m distance		$x = 6$ m, if measurement at ≥ 10 m	More than 50%, corrections cannot be applied
5	Not meeting requirements of any other class				Undefined

Table 9.5 *Terrain classification expressed in terms of roughness length z_0 (from WMO CIMO guide [4], Annex 1D, section 4.2), where x is a typical upwind obstacle distance and H is the height of the corresponding major obstacles.*

Roughness class	Short terrain description	z_0 (metres)
1	Open sea, fetch at least 5 km	0.0002
2	Mud flats, snow; no vegetation, no obstacles	0.005
3	Open flat terrain; grass, few isolated obstacles	0.03
4	Low crops; occasional large obstacles, $x/H > 20$	0.10
5	High crops; scattered obstacles, $15 < x/H < 20$	0.25
6	Parkland, bushes; numerous obstacles, $x/H \approx 10$	0.5
7	Regular large obstacle coverage (suburb, forest)	1.0
8	City centre with high- and low-rise buildings	≥ 2

setting. WMO guidelines recognise this, and admit 'surface wind measurements without exposure problems hardly exist ... The requirement of open level terrain is difficult to meet, and most wind stations over land are perturbed by topographic effects or surface cover, or by both' ([4], section 5.9.4).

Urban and suburban sites usually contain many unevenly distributed obstacles to free wind flow, the frictional impact of which is to reduce average wind speeds and increase turbulent eddies (and thus gustiness). Some compromise is therefore almost always necessary in finding the best available site for wind observations, and because of this the records obtained may be rather more site-specific than is the case with other measurements. It is therefore very important that site metadata (see Chapter 16, *Metadata, what is it and why is it important?*) provides details on the type and exposure of the wind sensors, particularly the height of the anemometer and full details of all surrounding objects such as buildings and trees (a scale drawing is best to show this information, accompanied by photographs around 360°), the instrument type(s) or model(s) in use, the unit used for wind speed readings, and whether any corrections have been applied to the readings. Date the site plan and photographs, because trees have a tendency to grow over time. Without this it is difficult to make meaningful comparisons with records from other locations, or to attempt any meaningful correction of mean wind speeds to *approximate* those made under standard exposure conditions.

Accepting that compromises will often be necessary, the best exposure to the wind will usually be obtained by exposing both anemometer and wind vane in as open a position as possible, as high as possible, commensurate with both safety and accessibility for installation and maintenance (see Box, *Safety aspects of installing and maintaining weather instruments* on the following pages). Where location permits, the mast should be well above the level of surrounding obstructions to the wind flow. For a typical suburban or urban area with buildings 10–15 m in height, the mast should ideally be around twice as high as the surrounding obstructions – rarely feasible, of course.

Installation and maintenance of wind sensors

Wind sensors are most easily installed when they are a single small unit (most ultrasonic wind sensors), or when both anemometer and wind vane are either combined in one unit, as in the Davis Instruments weather stations, or where separate anemometer and wind vane sensors are mounted on one cross-piece which holds both and is then itself fixed to a mast (**Figure 9.14**). Mounting a single cross-piece to a mast is easier than handling two separate instruments and will also minimise the risk of one sensor shielding the other. Ensure that the wind sensor or sensors are located at the top of the pole or mast used, a minimum of 500 mm / 18 in apart where a separate cup anemometer and wind vane are used, so that the body of the pole or mast itself does not shelter the instruments from the free wind flow – any sheltering effects can be minimised by mounting the cross-frame perpendicular to the prevailing wind direction. Ensure that the sensors are in a secure position, and one where they cannot easily be vandalised.

It is vital to ensure that mechanical anemometers and wind vanes are mounted absolutely vertically, and firmly secured in place. If a cup anemometer is not exactly vertical, its rotation will be lopsided at low speeds, and the starting speed will be higher as a result – affecting the quality of the wind records obtained. If the wind vane is not vertical, it will settle into a preferred neutral position, leading to a bias in wind direction statistics. Ultrasonic anemometers have a little more tolerance with regard to level but should also be mounted as accurately level as possible. Bird spikes may be found useful to prevent wind records being distorted by perching birds and their droppings.

On cabled systems, always leave some slack in cables, as it may be necessary to move or adjust the sensor position for optimum exposure at some later date. Ensure any slack cable is firmly secured to prevent fraying or movement in strong winds.

Before installing wind sensors in their intended location, particularly if access is difficult or if installation is to be undertaken by a contractor, it is advisable to install the sensors at ground level, making sure everything works for at least a few days. Better to find out a wireless anemometer transmitter is not working at ground level than on the end of a ladder. Pay particular attention to connectors for cabled systems, as wind-driven rain is much more penetrating in exposed locations.

Exposed and elevated locations make wind instruments much more vulnerable to weather-related failure than sensors mounted at ground level. Weather-related exposure problems can include snow, ice or rime accumulations (see *Severe weather performance* earlier in this chapter), large temperature ranges, component deterioration owing to ultraviolet exposure, wind-driven rain penetration of connectors or electronic circuitry, and of course damage due to wind loading – whether due to a single strong wind event or resulting from long-term component fatigue of the instrument or its mountings due to repeated wind vibration. Heated anemometers are available to help mitigate rime, ice and snow build-up, although the substantial power requirements generally rule out their use at locations without access to a nearby permanent utility power supply.

Lightning can also pose risks to exposed, elevated sensors. For all but the tallest masts in high-risk, open sites such as open moorland or exposed airfields, or in areas with a high frequency of electrical storms, the chances of an instrument being directly struck by lightning are small, although an electrical surge from a nearby lightning strike (even one several hundred metres distant) can damage or even destroy electronic sensors and datalogging equipment. A good earth connection for the mast or tower is essential to reduce the build-up of static electricity in thundery conditions. However, the cost of full lightning protection may be higher than the cost of the equipment being protected, and a 'self-insure' policy may be more cost-effective where domestic or property insurance specifically excludes exterior equipment such as TV aerials or anemometer masts. At the very least, where the sensors are connected to the datalogger using cables, ensure the datalogger–computer connection is via an optical link rather than an electrical connection if possible. That way a direct lightning strike may fry sensors and perhaps the logger too, but hopefully the damage will be contained at that point rather than endangering downstream electrical connections. It is, of course, good practice to switch off and unplug vulnerable electrical equipment during close lightning storms.

Safety aspects of installing and maintaining weather instruments

Rule number 1: Never take risks with personal safety when installing any weather sensors at height.

If the proposed location for the instruments cannot be reached safely, take appropriate safety precautions, hire an experienced professional installer – or choose another site.

Remember also that all instruments will need occasional maintenance – wireless transmitters will need batteries replaced occasionally (use a solar-powered unit if possible, which itself will need cleaning from time to time), cup anemometer bearings can seize up and may need an occasional squirt of penetrating oil to free them, birds may build nests in the most inconvenient places, the mast fittings or cable run may need checking or tightening after a gale – so make sure they can be reached safely if and when needed.

In domestic or suburban locations, the best available position may be on a short mast attached to a chimneystack or on a mast projecting above the roofline. Try to locate the instruments at least 2 m above the roofline or any other obstructions to the free wind flow, to reduce turbulence and eddying in strong winds (**Figure 9.14**). If mounting near a chimneystack, ensure that hot flue gases will not affect the instruments. If a rooftop site is not possible, find the best exposure available – perhaps mounting the instruments on a tall pole or mast in a garden (see *Do I need zoning or planning permission?*). Lightweight and weatherproof aluminium masts or towers in various heights are available from instrument suppliers. Some are hinged to allow the mast to be tilted over to near ground level to permit maintenance access to the instruments. Towers or masts must be firmly installed (concreted in) to avoid damage in strong winds, and guying may also be required in windy areas or sandy soils.

If possible, optimise the exposure of the instruments for the direction of the prevailing wind – usually between south and west in temperate latitudes. Don't forget that winds between north and east are the second-most common directions here, so try to ensure a good fetch from those directions too. In sub-tropical latitudes in the northern hemisphere, north-easterly or easterly surface winds will dominate.

If (like the author) you are not entirely comfortable with tall ladders and crawling about on rooftops, TV aerial (antenna) installation companies will often agree to undertake the work. Explain clearly what is required (mounting an anemometer/ wind vane set on a mast attached to a chimney stack, with a cable run into a loft, is very similar to the installation procedure for a TV antenna) and ask for a quote. Local builders can often provide similar services – a builder may be a better choice if there is any doubt about the ability of the structure to which the mast is affixed to take the weight and additional windborne stress of the instrument package involved, or if a more substantial mast is required (such as the one shown in **Figure 9.14**). If in any doubt, ask for a pre-installation site inspection and quote, and of course allow sufficient budget to cover the installation costs.

Rule number 2: Always follow Rule No. 1.

Figure 9.14 Vector Instruments cup anemometer and wind vane set mounted on a mast attached to the gable end of a house; the anemometer is 11 m above ground level and 2.7 m above the apex of the roofline. Such an exposure is far from perfect but may be the best available option. The mast also provides a suitable site for a Kipp & Zonen sunshine recorder. This photograph was taken from the south-west (the prevailing wind direction). (Photograph by the author)

Do I need zoning or planning permission?

Most countries operate some form of planning law, the extent of which varies from country to country. Planning law may impose legal constraints on what can or cannot be built within a region of a town or city or within a property's curtilage, and these may include restrictions on the erection of masts or towers to support weather instruments. If your property is rented or leased, then of course the permission of the landlord must also be obtained before commencing installation work.

Within the UK, unless you live in a conservation area, a listed building or restrictive covenants covering your property forbid TV aerials and the like, and with few exceptions, planning permission is not normally required for roof-mounted wind sensors or 'weather vanes', and the erection of an unobtrusive anemometer/wind vane sensor package should not result in a letter from your local authority's planning department – TV aerials, satellite dishes, solar panels and wind turbines all represent more visually obtrusive extensions to a roofline. Planning permission is also not required if the installation replaces or reinstates existing equipment that has been in place for seven years or more.

Within the United States, zoning laws vary by state and city, and local planning authorities should be consulted in advance of any planned installation work.

Unless you happen to live on a remote farmstead, it is of course always advisable to discuss any proposals informally with neighbouring properties and landowners well before any planned installation. Building a 10 m tower in your backyard, even putting up a tall pole, may prompt complaints or contravene local planning regulations. Policies do vary widely, however. If in any doubt, seek professional local advice prior to installation.

Calibration, accuracy and precision

In all but the very best-exposed of sites, the accuracy of the wind *speed* measurements obtained will depend more upon the limitations of instrument exposure than upon the absolute accuracy of the anemometer itself. Very few anemometers at or below those included in advanced-level AWSs will come with a calibration certificate, but the quoted accuracy of mid-range AWS anemometers is around ± 5 per cent (**Table 3.3**, page 63), considerably less than the reduction in wind speed that can be expected from an imperfect exposure. With entry-level and budget system anemometers, reliability, longevity, high starting speeds and slow sampling intervals are likely to result in greater uncertainties in record quality than the absolute accuracy of the sensor itself.

Mean wind speeds are usually quoted only to 1 kn (0.5 m/s) in synoptic or operational environments. For climatological applications, greater precision (to 0.1 kn or 0.1 m/s) is desirable, although this may not be fully justified by the likely errors resulting from imperfect siting or instrument calibration.

The main source of error in wind *direction* measurements is likely to come from the alignment of the sensor (see the following for details on how to align wind vane sensors), although high starting speeds and high-friction bearings in mechanical wind vanes will also result in slow or damped responses in light winds. In addition, wind vanes that are not perfectly level will preferentially settle upon a particular direction, biasing wind direction records in light winds. The fixings holding the sensor in place should be checked occasionally, as they may work loose over time, causing changes in vane alignment in the horizontal or even vertical plane.

Things to avoid

Growing trees and hedges will gradually (and often significantly) reduce recorded wind speeds over time, even when they are tens or even hundreds of metres distant. The erection of new buildings nearby often has a more immediate impact. Keep hedges or trees cut back – not always possible if it is a tree on a neighbouring plot that is becoming overgrown, of course. Take a set of site photographs throughout the full 360 degrees around the instrument every two years to assess or document slow changes in the exposure of wind instruments. The technique is covered in more detail in Chapter 11, *Measuring sunshine and solar radiation*.

Correcting wind speed readings made at a non-standard height

The correction of observed mean wind speeds to emulate those observed under 'standard' conditions is a complex area, and only brief comments are possible in the space available. Guidance, based on the so-called roughness length z_0 (**Table 9.5**), is given in WMO CIMO guidelines ([4], section 5.9.4), but it should be noted that it simply may not be possible to obtain satisfactory and consistent corrections in a location with multiple and unevenly distributed obstacles, such as a typical urban site. However, some general principles regarding the variation of wind speed and direction with height can be stated.

Mean wind speeds increase with height. An approximate correction factor for *mean* wind speeds observed at heights other than 10 m is given in **Table 9.6**. This is derived from the reciprocal of the expression ([232], pages 4–15):

$$V_h/V_{10} = 0.233 + 0.656\log_{10}(h + 4.75)$$

where V_h/V_{10} are wind speeds at height h and at 10 m respectively and h is the height of the anemometer, in metres.

The table shows that, for example, mean wind speeds measured at 3 m above ground need to be *increased* by about 22 per cent to approximate those at 10 m above ground; at 20 m placement, they would need to be *reduced* by 13 per cent.

It must be stressed that these factors apply only to *mean* speeds in an open location over relatively long periods (days or months), and that the variation of

Table 9.6 *Typical variation of mean wind speeds with height*

Anemometer height h, metres	1	2	3	4	5	6	7	8	9	10	15	20
Correction factor to 10 m equivalent	1.37	1.29	1.22	1.18	1.13	1.10	1.07	1.04	1.02	1.00	0.92	0.87

wind speed with height will vary with atmospheric conditions minute by minute, hour by hour and day by day. Individual spot readings over short periods, and observations made in obstructed or sheltered sites, may depart significantly from these averages. The correction factors *do not apply to gust speeds*, which vary much less with height and which should be reported *without* applying any corrections.

Where corrections are made to any wind speed reading, archive both uncorrected and corrected values so that the original data are always available for subsequent (re-)analysis if required.

Table 9.6 can also be used to estimate the so-called *effective height* of the anemometer, which is defined as 'the height above open level terrain in the vicinity of which mean wind speeds would be the same as those actually recorded by the anemometer' [248]. Compare mean wind speeds over an extended period – several months at least – with a neighbouring site having an unobstructed exposure and similar weather types (comparing a hilltop site with a valley site, for example, would be unrealistic owing to topographical and climatic differences) and note the mean differences: the 'effective height' of the anemometer can be estimated by comparison with the factors given in this table. (Differing levels of obstruction around the compass may provide different values of effective height by compass bearing, which greatly complicates any possible applied corrections.)

Example: compare Station A, where the anemometer is located 6.5 m above ground level in a high-clutter suburban location, with Station B a few kilometres distant in a rural area of otherwise similar terrain, where the anemometer is located at 10 m above ground in an unobstructed site. Over a period of 2 years, Station A's mean wind speeds were 19 per cent below Station B. **Table 9.6** shows that a 19 per cent difference corresponds to an effective height of about 3.75 m. So, although the anemometer is actually located at 6.5 m above the ground, the sheltered surroundings and local urban infrastructure are reducing wind speeds at the actual anemometer height to what might be expected to be found at about 3.75 m above ground in 'open level terrain in the vicinity'. In this case, to approximate the 'open level terrain' wind speeds observed in the vicinity, the mean wind speeds from Station A's anemometer should be increased by 1.19 (the correction for an 'effective height' of 3.75 m) rather than the 1.08 that would be suggested from the actual anemometer height of 6.5 m.

For best results, the determination of 'effective height' should be performed from several different comparison sites, at different compass points. The main difficulty here will be scarcity of unobstructed anemometer records (and perhaps the difficulty in getting hold of data from those instruments).

In practice, it is much better to locate the anemometer and wind vane as high as possible at the outset, than to locate them in a sheltered exposure at only a couple of metres above ground and rely on correction factors which may not be appropriate for your site. In any case, no correction factor can provide a representative wind speed if the low-level anemometer shows calm. Particularly with 'budget' systems, operational factors related to anemometer reliability and high starting/stopping speeds may prove more troublesome than height corrections.

Correcting wind directions made at a non-standard height

The variation of wind *direction* with height is insignificant in the lowest few metres of the atmosphere. Assuming that there are no gross obstructions to the flow, the mean direction at lower levels will not be significantly different from that observed at 10 m, although the variability will almost certainly be greater, owing to greater turbulent effects nearer the surface.

In sites where there are significant obstructions, turbulence and eddies around obstacles may increase the variability of the wind direction even further, resulting in rapid and extreme second-by-second fluctuations. Such situations may benefit from the introduction of a datalogger averaging routine which acts to damp 'noisy' direction samples over a 5–10 second interval, thus improving the representation of true wind conditions.

Aligning the wind vane

Setting up a wind vane correctly is (perhaps surprisingly) one of the trickier aspects of installing an AWS. It is very important that it be accurately aligned not only to the compass points, but as stated previously the rotational axis of vane-type sensors must be mounted as close to true vertical as possible to avoid settling into a preferred neutral position.

To obtain an accurate display and record of wind direction, the wind vane must be correctly oriented to true north (see *Finding true north*). There are various ways to do this, and the recommended methods differ slightly from manufacturer to manufacturer, but essentially the process is as follows:

- Before installation, and from the intended site of the wind vane (or as close to it as possible), use a sighting compass to determine true north (or another cardinal point if this is easier). If the mast or tower includes any ferrous components, the structure itself may affect the compass needle, and all bearings should be checked some distance away from the support if this is likely to be the case. Digital compasses which can be pre-adjusted to take magnetic declination into account in their readout, or some GPS units, which will display a bearing to 5° or even 1°, may be easier to use than a small magnetic compass.
- If this has not already been taken into account, adjust the alignment to take account of magnetic declination (see *Finding true north*).
- Take a sighting along the compass to a distant object or group of objects, preferably on the horizon, otherwise as far away as possible, and align the compass point to this reference. It may be easier to take a photograph and mark the alignment of the chosen compass point on the photograph for reference.
- Set up the wind vane in the desired location, and fix firmly into position, allowing for any minor adjustments that may still be needed. Ensure that there is a little slack in the wiring to permit final adjustments to be made without imposing strain on the cables and connectors, but not so much that cables will 'flap' in strong winds.
- Two people are needed for this next step – one next to the wind vane, the other by the display and able to call out the display reading. Align the wind vane to the designated cardinal point on the horizon (pointing the head as close TO that

point as possible), allow the display to settle for a few seconds then ensure the display reading coincides with the compass point chosen. (This step is easier with mechanical wind vanes but can be difficult to get right with ultrasonic sensors which usually have only a small marker to indicate north.) If not, loosen the wind vane fixings and adjust as necessary until good agreement is attained, then re-tighten securely.

- Without specialist equipment it is difficult to align a wind vane to better than 5° by eye – but in any case, only the most exacting applications will require the vane to be aligned to better than 5° accuracy.

Davis Instruments Vantage Pro2 anemometers make this step straightforward, in that the mounting arm of the anemometer needs merely to be accurately sighted to true north for the vane to be correctly aligned. (This can be changed if north is not a convenient alignment.)

If it is not possible to sight along and adjust the wind vane by eye (perhaps if the sensor package is to be fixed in place on a tall mast as one unit), then an alternative is to pre-fix the wind vane onto a cross-piece with north at 90 degrees to the long axis of the arm, then ensure the bearing of the cross-piece is exactly aligned (say) west-east by sighting along it from ground level using a digital compass, adjusting the angle of the cross-piece until exact alignment required has been achieved, then locking it in place.

Finally, firmly clamp all sensors into position (taking great care not to disturb the alignment of the wind vane when doing so) and then secure all cabling. Use ultra-violet-resistant plastic cable ties to secure cabling – do not use staples or a staple gun, because these may cut through the wiring. Any exterior connectors and building entry points for cables must be thoroughly weatherproofed to avoid water ingress.

If, after installation, it becomes apparent from comparison with neighbouring sites that the alignment is slightly incorrect, most weather station software will allow a fixed adjustment or offset to be made to the observed readings. (If the wind vane has been accidentally set up to indicate where the wind is going TO, rather than where it is coming FROM, simply set an offset of 180° into your software to correct this.) As with all adjustments, make a note in your site metadata of the correction applied, and the date it was introduced.

It is good practice, wherever possible, to compare recorded wind directions with other reliable local observations on a regular basis, as a gradually increasing or variable discrepancy may indicate that the wind vane fixings have worked loose. The alignment then needs to be reset and the mounting re-secured.

Finding true north

Because the north pole and the north magnetic pole are not in the same place on the Earth (at the time of writing the north magnetic pole lay near Ellesmere Island in northern Canada, at about 86.5°N, 164°E, and was moving towards Russia at 55–60 km per year), a magnetic compass points to magnetic north, not true north. Wind directions are referenced to true north, not magnetic north; therefore, for accurate alignment of a wind vane, the difference between the two must be corrected for when setting up a wind vane. The current value of the correction to magnetic north can be found from www.magnetic-declination.com – simply enter the latitude and longitude of the site.

Wind accessories and fixtures

Cup anemometer and wind vane sensors are often combined into a single unit in prebuilt AWSs, but if not a mounting frame or cross-frame to secure them to a mast will be helpful, as described earlier (see **Figure 9.14**). Aluminium tubular masts suitable for mounting lightweight wind sensors are readily available (as TV antenna supports) in 2 m or 3 m lengths from most DIY stockists; don't forget the fixing brackets too. Instrument suppliers (**Appendix 3**) can supply more substantial masts or lightweight towers, or local builders merchants can supply suitable scaffolding fixtures to construct a purpose-built unit.

Logging requirements

Chapter 2, *Choosing a weather station*, emphasised the importance of a rapid sampling time – no more than about 3 seconds – for an AWS if accurate depiction of gust speeds is required. Lower sampling intervals will not capture the detail of the high-frequency gusts. Where the logger can support this, WMO's guideline for wind sampling intervals is 0.25 seconds (4 Hz), although a non-overlapping interval up to 3 seconds is 'quite acceptable'. If the requirement is only to measure mean wind speeds rather than gusts, a site investigation for a wind turbine site for example, then such frequent sampling periods may not be necessary.

Where sampling and logging intervals can be separately specified, a sampling time of 0.25 to 1 second and a logging time from 1 minute to 1 hour will be suitable for almost all requirements.

Wind run

The term 'wind run' is still used in some older applications. 'Wind run' is a measure of mean wind speed, expressed in terms of 'distance covered over a given period': older 'cup counter' anemometers (often at 2 m above ground level) generated a 'wind run' distance display from the rotation of the anemometer cups via a geared mechanism in a manner similar to that of a vehicle odometer. A mean wind speed of 10 kn over one hour would generate a 'wind run' of 10 nautical miles; thus over 24 hours, 240 nautical miles would be recorded on such an instrument. A 10 m/s mean wind over 24 hours would result in a wind run of 10 m/s × 86,400 seconds = 864,000 m (864 km). Subtracting successive daily readings made at the morning observation provides a measure of mean wind speed for the intervening period. The reduction in size and cost of dataloggers has increasingly made such systems obsolete, as wind run is easily calculated if required. In addition, older cup counter anemometers suffered from high starting speeds, and are unlikely to be responsive at low wind speeds, leading to non-linear bias in their output.

Wind direction reporting

There are two methods of reporting wind directions over the logging interval – the 'modal' method, and the 'vector mean wind' approach.

The *modal method* counts the frequencies of observed wind directions within each sampling interval within pre-determined ranges or classes, then reports the class with the highest frequency as the logged wind direction. This method is used on all Davis Instruments AWSs: the classes are the 16 compass points (N, NNE, NE, ENE, etc.) and the wind direction is therefore available only to a precision of 22.5 degrees (360/16).

A more accurate method is the *vector mean wind* calculation, which is described in the text box *Vector mean winds*. The arithmetic required can be carried out by a programmable logger as it processes the wind samples, or it can be calculated retrospectively using a spreadsheet. The accuracy of the logged wind direction is limited only by the accuracy of vane alignment and the sensor specification, and ± 2 degrees is attainable, although ± 5 degrees is sufficient for most weather measurements. At all but class 1 or class 2 sites, some form of software damping on real-time wind direction values helps to reduce high-frequency fluctuations about the prevailing value; damping over a period of 5 to 10 seconds is normally adequate. Such software damping can be accomplished within programmable loggers, such as those sold by Campbell Scientific and others.

Vector mean winds

A vector mean wind calculation provides a means of generating a *resultant wind flow* from a series of varying wind velocities over time – wind velocity here denoting combined wind *direction* and scalar wind *speed*. The calculation resolves individual samples of wind velocity into east-west and north-south components using trigonometry, which can then be averaged numerically in the normal manner. The averaged value of the two components is then converted back into the resultant (think 'average') wind direction and speed.

This method of calculation is necessary because the use of polar co-ordinates (compass bearings) means they cannot be simply averaged numerically – the 'mean' of a north-westerly wind (315°) and a north-easterly wind (045°) is clearly not a southerly wind, as would be indicated by the numerical average of the two wind directions ((315+45)/2 = 180).

The expression is tedious to undertake by hand, but very easily automated in a spreadsheet: an Excel template and macro are listed in **Appendix 5** and available for download on www.measuringtheweather.net. The calculations can be performed over a minute, a day, a year or for any other time period.

Appendix 5 gives more details of the calculation method, but a simple example will suffice here.

Suppose that, in one hour, the wind blows from the south-west (225°) for the first 30 minutes at 20 knots, and for the next 30 minutes from the north-west (315°) at 10 knots (see **Figure 9.15**). To keep it simple, we assume that the wind direction and speed remain unchanged within these 30 minute periods.

The *scalar mean wind speed* for the hour is 15 knots *(the average of 10 and 20)*
The *vector mean wind* for the hour is 252°, 11.2 knots
It is perhaps easier to envisage this by drawing out the two vectors:

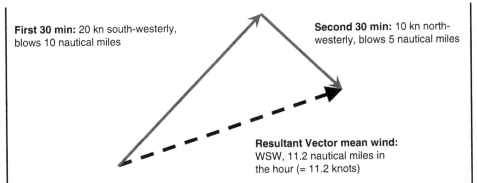

First 30 min: 20 kn south-westerly, blows 10 nautical miles

Second 30 min: 10 kn north-westerly, blows 5 nautical miles

Resultant Vector mean wind: WSW, 11.2 nautical miles in the hour (= 11.2 knots)

Figure 9.15 Vector mean wind visualisation; see text for details.

The output can be viewed as representing the end point of a parcel of air – say, a hot-air balloon at a constant height – at the close of the time period (1 hour, in this example) had the observed wind speeds and directions been replaced by a constant wind of that speed and direction. Note that of course the vector mean wind speed can only be the same as, or more often less than, the scalar mean speed: the difference between the two will increase with the variability of the wind. Also, the calculations are necessarily only two-dimensional in nature.

Advanced loggers include a vector mean wind option to summarise sampled wind speeds and directions in logged output. Where AWS output is given as compass points rather than as degrees, a simple Excel function (also listed in **Appendix 5** and downloadable from www.measuringtheweather.net) can be used within an Excel spreadsheet to convert compass bearings into equivalent degrees (so that, for example, all southerly winds will be converted to 180 degrees, SSW to 202.5° and so on). The vector mean wind calculation can then be run as earlier.

Wind data tabulations

Statistical summaries of wind speed and direction normally consist of hourly, daily, monthly and annual tabulations. For hourly summaries, the modal or vector mean direction (as set out earlier), scalar mean speed and the highest gust, usually with the time and direction of the highest gust, are usually quoted. Logging the 'lowest lull' (the minimum 3 second wind speed) is equally useful and enables the production of plots such as **Figures 9.1** and **9.2** which perhaps more visually portray the variation in wind speed over time than a line graph showing mean and gust speed alone.

For daily, monthly or annual summaries, the vector mean wind (direction and speed) along with the scalar mean speed, the highest hourly scalar mean wind speed and its direction with date and hour of occurrence, the highest gust and the date/time and direction of the highest gust are the norm, alongside statistical summaries of wind direction frequencies. A useful table is one of 12 × 30 degree sectors starting at north (350–010°, 020–040° and so on).

Wind roses present a more graphical means of expressing the frequency of various wind direction and speed combinations over time. Suitable software is suggested in Chapter 18, *Making sense of the data avalanche.*

Terminal hours

Daily wind speed and direction summaries normally refer to the civil day period midnight to midnight, disregarding any corrections for Summer or Daylight Savings time. Most AWS will generate the appropriate daily summaries for this period without adjustment, unless the logger is set to Summer Time. For more on time standards, see Chapter 12, *Observing hours and time standards.*

One-minute summary – *Measuring wind speed and direction*

- The wind is highly variable in both speed and direction, and obtaining good measurements of the wind poses particular challenges for instruments, logging equipment and site requirements.
- Wind is a *vector* quantity – it has both direction and speed. Wind *direction* refers to where the wind is coming from. A wind vane needs to be accurately aligned to true north, which is slightly different to magnetic north indicated by a magnetic compass.
- Mean wind *speeds* normally refer to 10 minute periods, gust speeds to 3 seconds. For accurate determination of gust speeds, a high sampling interval (no more than a few seconds) is essential, although the logging interval can be much longer than this.
- Wind direction and speed have traditionally been measured with separate instruments, typically a cup anemometer and a digital wind vane, although modern one-piece ultrasonic anemometers will output both direction and speed in digital form. Sonic anemometers have no moving parts (and thus are more reliable/less vulnerable to mechanical wear) and are much more sensitive: they are gradually replacing traditional instruments.
- WMO CIMO site guidelines for wind observations are set out. An ideal exposure for wind sensor(s) will be in an open location, well away from obstacles, at 10 m above ground level. However, such ideal sites are hard to come by, particularly in urban or suburban areas, and wind records are therefore necessarily more site-specific than most other weather measurements. Some corrections for the variation of mean wind speed with height are possible, and these are described in this chapter. Gust speeds should not be corrected.
- If a position 10 m above ground is not feasible, one as high as possible is preferred, commensurate with both safety and accessibility for installation and maintenance. An elevated exposure will increase the vulnerability of the instruments to extreme weather conditions, particularly snow or ice, lightning and of course high winds. Great care should be taken in installation and cabling to minimise the potential for subsequent weather-related reliability issues.
- The absolute accuracy of wind speed measurements is more likely to be limited by the exposure of the anemometer, rather than the accuracy of the sensor(s) themselves. The accuracy of wind direction measurements also depends critically upon careful alignment during installation.
- Planning permission or zoning approval is not normally required for domestic rooftop-mounted anemometers or wind vanes, and local authority case precedents exist within the UK. Specialist legal advice should be taken if in doubt.
- **Never take risks with personal safety when installing or servicing weather sensors at height.**

10 Measuring grass and earth temperatures

After air temperature, grass and earth temperatures are the most commonly observed surface climatological parameters. This chapter describes the methods used to measure temperatures other than those of the air and includes reference to the relevant World Meteorological Organization (WMO) guidelines [4].

The grass minimum temperature

Why measure grass temperature? Assuming there are no obstructions to outgoing radiation to space, the lowest temperatures on a clear night will be recorded at or close to ground level, where turbulent stirring and mixing of the cooling air is at a minimum. The nature of the surface, primarily whether it is a good or bad conductor of heat upwards from the earth, then determines its temperature under such conditions. Where the surface is covered by short grass, the lowest temperatures are attained just above the tips of the grass blades, because air trapped between the grass blades acts as a partial insulator to the upward heat transfer ('heat flux') from the warmer earth beneath. The minimum temperature attained in any given time interval is therefore known as the 'grass minimum temperature' (or simply 'grass min'), and it is measured using a thermometer (traditionally an alcohol-based minimum thermometer, now more commonly an electrical sensor) freely exposed at the tip of the grass blades (**Figure 10.1**).

Not all countries routinely measure grass minimum temperatures, sometimes referred to as the 'radiation minimum' or 'skin temperature', but in the UK they have been recorded for over 150 years in a few places. Some AWSs monitor ground surface temperatures using a downward-pointing infrared sensor mounted about 2 m above ground level; if the surface 'seen' by the sensor is short grass, overnight minimum temperatures measured in this fashion should be broadly similar to grass minimum temperatures observed with conventional thermometry.

The grass minimum, by virtue of its being exposed as a bare sensor, is more often than not covered in moisture, particularly at night, and thus is more accurately considered as a wet bulb. The errors introduced are not significant, however, as the air at grass-tip height quickly attains saturation on any clear night.

Grass minimum temperatures and 'ground frosts'

It is conventional in most countries making grass temperature measurements to term an occasion during which the grass minimum falls below 0 °C as a 'ground

Figure 10.1 Grass temperature sensor. (Photograph by the author)

frost' (as opposed to the minimum temperature in a thermometer screen falling below 0 °C, which is termed an 'air frost'): but see also 'When is a ground frost not a ground frost?'. Because minimum temperatures over grass are usually lower than air temperatures, the number of ground frosts in a year at any location is likely to be greater than the number of air frosts.

When is a ground frost not a ground frost?

The term 'ground frost' is so defined in the *Oxford Dictionary of Weather* (Oxford University Press, 2001), but in the UK until 1971 the threshold grass minimum temperature for a ground frost was –0.9 °C for some long-forgotten reason (see, for example, *Meteorological Glossary*, UK Met Office, 1961 edition). Older UK statistical summaries may therefore refer to the lower threshold. Where there is no danger of confusion with pre-1971 records, the term 'ground frost' is an acceptable term referring to 'grass minimum temperature below 0 °C'.

Exposing the grass minimum thermometer

At its simplest, the grass minimum sensor is laid on the surface of short grass, or (better) supported just above the tips of the grass blades by a pair of small pegs fashioned from wooden dowelling or similar. The supports should be gradually increased in height as the underlying grass grows. Although today an electrical temperature sensor connected to a datalogging system is more likely to be used than a traditional alcohol-based minimum thermometer, where alcohol-based thermometers are still used they should be placed with the bulb end slightly lower than the far end of the thermometer, as this will allow any alcohol evaporated during the heat of the day to find its way back down to the bulb without condensing or 'bubbling' in the thermometer stem. Where the sensor is liable to be disturbed or dislodged by animals or birds – foxes, rabbits, crows and magpies can be particularly troublesome – a loop or closed peg loosely enclosing the sensor in one or two places will generally solve the problem.

A robust 'bare' sensor is essential for a grass minimum sensor. Any extraneous mass, such as a bulky outer casing, an integral LCD display, battery housing or similar will inevitably increase thermal inertia and result in a more sluggish response, thus leading to artificially high readings. It is worth remembering that such sensors are exposed to the full force of the elements – including huge ranges in temperature on sunny summer days, and almost constant saturation in rainfall or by night – and these will quickly take their toll on battery life and find the slightest gaps in electrical connections or weatherproofing. The slight differences in radiative emission and response times between glass thermometers and metal probes can result in slightly different temperatures being recorded, although even identical sensors exposed close to each other in an identical exposure will usually differ slightly in any case. If close consistency with glass thermometers is sought, WMO guidelines ([4], section 2.2.3b) suggest enclosing the electrical sensor probe within a glass sheath. This may also provide additional resistance to water penetration into the sensor body, which will eventually cause sensor degradation.

The sensor should be placed in a position permitting maximum exposure to the sky (not underneath a tree or close to a wall, for example). It is essential that there be no obstructions – either natural or artificial, such as a protective cover – above the thermometer or sensor overnight once it is placed on the grass surface, because any obstruction will reduce outgoing terrestrial radiation and result in artificially high readings, perhaps by several degrees.

The grass surface should be kept mown fairly short, but not bowling-green short, as the radiative and insulating properties of very short grass are closer to those of bare earth than true short grass cover. Similarly, grass should not be allowed to grow over the sensor itself, for if the sensor becomes 'buried' within grass, observed temperatures will again be higher than the true 'grass minimum', owing to the obstruction of outgoing radiation by the grass, together with heat retention from the ground surface by insulating air pockets trapped within the grass. Under no circumstances should artificial surfaces, such as 'astroturf', be substituted for natural grass cover under a grass minimum sensor, as the physical properties of such materials differ substantially from those of natural vegetation.

It is advisable to mark out the location of the grass temperature sensor using white-painted pegs or a low barrier fence, no closer than at least 50 cm or so to the thermometer, to avoid accidental damage to sensors, particularly if fragile glass thermometers remain in use. The author can vouch for the fact that they are very easy (and very expensive) to step on by mistake, particularly in darkness or poor weather and especially when exposed in the middle of a large grass-covered area. A short post, 30–50 cm in height and painted in a colour other than white, will also make it easier to locate the instrument after a snowfall. In a domestic environment, children, pets, ball games and lawnmowers need to be kept well away from the sensor for the same reason. However, the fencing must be movable to allow easy access for cutting the grass (remember to move the thermometer/sensor to a place of safety FIRST!). If the grass minimum thermometer is a cabled thermistor or platinum resistance thermometer (PRT), ensure also that the cable will not be damaged by the lawnmower, strimmer or garden shears used to maintain the area where the instruments are located. Lawnmower blades, shears and strimmers will slice through AWS cables in an instant (even screened cable, where the screening otherwise provides some armouring). Minimise the risk of accidental damage from lawn tools by enclosing all but the half metre or so of the cable nearest the sensor itself in tough plastic conduit. This applies particularly on government or school sites where maintenance may be the responsibility of external contractors. Personnel may change on a regular basis, and not know to look out for the sensor until – oops, too late.

The 'grass minimum depression'

The difference between the air temperature and the grass temperature is known as the 'depression of the grass minimum'. This varies significantly during the hours of darkness (**Figure 10.2**), typically being greatest early in the night and least near dawn, because the ground surface cools more rapidly after dusk than the overlying air. Of course, **Figure 10.2** disregards readings from the sunshine-affected sensor during daylight.

A typical value of the grass minimum depression is 2–4 degC (4–7 degF) below the screen minimum following a clear night, but values vary significantly from minute to minute, day to day and location to location. Some sites regularly see very large grass minimum depressions (occasionally 10 degC / 18 degF or more), particularly over light or sandy soils, others the opposite. More sheltered sites tend to have smaller grass minimum depressions, because partial obstruction of the sky limits outgoing terrestrial radiation. Frost hollow sites also tend to have reduced grass minimum depressions, whereas exposed sites that shed cold air more easily can experience very large air–grass temperature differences. In thick and deep fog, low

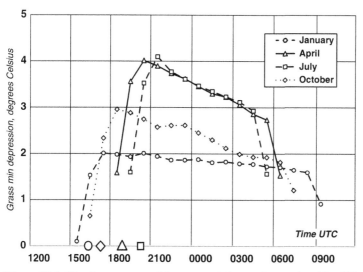

Figure 10.2 Hourly averages of the grass minimum depression (the difference between screen and grass temperatures), showing how the greatest difference from screen temperatures (at 1.25 m above ground) tend to occur within an hour or two of sunset. Sunset times (UTC) are indicated by filled symbols on the baseline – 15 January, 1624 UTC; 15 April, 1859 UTC; 15 July, 2015 UTC; 15 October, 1712 UTC. From a site in central southern England (51°N, 1 °W), averaged over 10 years.

cloud and after a period of warm weather the grass minimum temperature can occasionally be slightly higher than the screen minimum, but usually by no more than a few tenths of a degree Celsius. Anything more than this should prompt a calibration check of one or both sensors in use.

Snowfall and grass-tip thermometers

It takes only a couple of centimetres of snowfall (less than an inch) to bury a grass minimum sensor. Once the sensor is buried in snow, it will give a misleading reading, which should be discarded (but see the '*Calibration tip*' that follows). Once snow has stopped falling, the grass temperature sensor should be carefully removed from its covering of snow at the earliest opportunity and placed gently and level on or just above fresh and undisturbed snow cover. If snow is still falling, this process should be repeated at the next observation or when the snow stops falling. Fresh snow, particularly deep fresh snow, is a very effective insulator, as it contains so much trapped air, and the temperature on a snow surface can fall very rapidly indeed under clear skies. Exceptionally low surface temperatures can be attained under such circumstances.

In areas with frequent winter snowfalls, or in remote areas, some AWSs use downward-pointing infrared emission sensors to determine 'skin' or snow surface temperature. This minimises the risk of sensors becoming buried, and provided the sensor remains well above the snow surface, a variable depth of snow will not affect readings.

Calibration tip

Being buried in snow can provide a very useful natural 'ice-point' calibration check, because in these conditions the sensor will normally be held at exactly 0.0 °C for some time, owing to loss or gain of latent heat (see also Chapter 15, *Calibration*). Sometimes this condition will persist for several hours, until all the snow has either melted or frozen. If a logged electrical sensor is in use, the temperature observed during such long steady periods in snow can be easily determined, even if a fall or rise subsequently takes place. If, for example, during such an episode the thermometer remained steady at (say) –0.4 °C for an hour or so, more likely than not the sensor calibration is 0.4 degC too low. While this does provide a calibration check at just a single point, and calibration may differ at higher or lower temperatures, 0.0 °C is the single most useful calibration point for most electrical sensors.

Minimum temperatures above other types of ground surface

Surface temperatures just above concrete or tarmac are monitored and modelled to aid road surface temperature forecasting, particularly to assist winter gritting 'freeze/no freeze' decisions by highway authorities. For this reason, minimum temperatures above such surfaces, 'concrete minimum' or 'tarmac minimum', are made quite widely and in a similar fashion to grass minimum measurements as described earlier. The 'concrete [slab] minimum temperature' refers to the lowest reading of a sensor, mounted just above a concrete slab, which in the UK is of a standard size $100 \times 60 \times 5$ cm, and pale grey in colour. Many roadside AWSs report actual road surface (or just sub-surface) temperatures using sensors buried in the road surface.

A 'bare soil minimum' is measured in a similar fashion, the sensor exposed just above the surface of a patch of bare soil, typically about 1 m^2 in area, kept free of weed growth.

Terminal hours and daytime exposure considerations

WMO guidance ([4], section 2.3.2.2) is that the grass minimum temperature (and, by extension, other surface minimum temperature observations) should refer to the period from just before sunset to the normal morning observation hour (often 7 a.m. to 9 a.m., nominally 0900 UTC in the UK and Ireland). If an alcohol minimum thermometer is in use, it should be kept within the thermometer screen (if one is available) during the day to minimise the risk of the alcohol in the stem 'bubbling' in sunshine during daylight (see **Appendix 4** for guidance on 'debubbling' such thermometers). WMO terminal hour guidance specifying *overnight* surface minimum temperatures differs from the standard 'climatological day', where measurements normally refer to a full 24 hour period. However, this distinction is widely blurred at unmanned sites where there is no-one to stow the grass minimum sensor by day and replace it before sunset, or where temperatures are measured by a permanently exposed electrical sensor. This topic is covered more fully in Chapter 12, *Observing hours and time standards*.

Electrical temperature sensors can normally be left in position throughout the 24 hours. It is perfectly acceptable to remove or relocate a grass minimum sensor of any type from its normal exposure during the day, particularly if this minimises the risk of damage or theft, but it is important it be replaced before sunset, as grass temperatures fall more quickly than air temperatures during the early evening, and occasionally the night's grass minimum temperature will be reached at dusk. Note, however, that surface minimum temperatures, whether grass or other, will sometimes show considerable differences between sites where WMO's guidance on the 'sunset to next morning observation' record period is followed strictly, and those where a default 24 hour 'morning-to-morning' period is in use. Such differences are most frequent, and most pronounced, during the colder months when a single cold morning will usually leave its mark on two days records. To avoid such discrepancies, automated sites should ensure their datalogger is programmed to note the surface minimum temperature from the time of the earliest sunset on any date in the year to the standard morning observation hour next day – in the UK, this would be 1500–0900 or 1600–0900 UTC. A more elegant solution, where the AWS instrumentation set includes solar radiation sensors, is to programme the datalogger to commence logging the grass temperature sensor only once the global solar radiation falls below, say, 5 W/m^2, then ceasing logging at the requisite time of the morning observation and resetting the logged variable. Outside of this period, the 'grass minimum temperature' should be set to any arbitrary high value, such as 99 °C. At dusk, the first temperature observed as daylight begins to fade will then take the place of this value and so on as temperatures fall to their minimum level overnight. Only rarely would surface minimum temperatures be recorded during daylight hours, but even in temperate latitudes this occasionally happens during wintertime hail or snow showers. In polar latitudes, with longer hours of darkness in winter and longer hours of daylight in summer, the commencement hour should be adjusted seasonally to remain compliant with WMO guidance regarding 'overnight' intervals.

Logging intervals

For ease of comparison, it is best to sample and log grass temperature at the same intervals as screen temperatures. Ensure that the *minimum* temperature reached in each logging interval is logged, and not merely the spot value at that time. Grass temperatures in particular can fluctuate very rapidly, particularly on nights with variable cloud cover (indeed, overnight records from an electrical grass temperature sensor often provide good proxy evidence for the presence or absence of cloud cover). The true minimum would probably be missed even with 5 minute logging intervals if spot temperatures were logged instead of interval minima.

Earth or soil temperatures

Earth temperatures can be measured at a variety of depths, such measurements being particularly useful in agricultural applications as being more representative of the conditions experienced by growing plants. ('Soil temperature' is sometimes used to refer rather narrowly to shallow depths, and 'earth temperatures' to depths of 25–30 cm or greater, but to all intents and purposes the terms 'earth' or 'soil' are interchangeable in this context. For consistency, 'earth' is used throughout this book.) Such measurements are also useful in comparative climatology, where they can provide a clearer understanding of heat flow and retention within different soil types – sandy soils gain and lose heat more quickly than a heavy clay soil, for example – and for determining, amongst others, the depth at which the diurnal temperature change is reduced to zero, or how far frost penetrates into the ground in a cold spell.

Earth temperature depths and measurement methods

Standard depths

WMO guidelines for standard depths for earth temperature measurements are 5, 10, 20, 50 and 100 cm below the surface ([4], section 2.1.4.2.2), although few sites record temperatures at all of these depths: some locations maintain records at depths greater than 100 cm. In the UK and Ireland, many sites measure earth temperatures at 30 cm (1 foot) depth for continuity with historical records (earth temperatures at 30 cm have been measured since the mid-1870s in the United Kingdom – whereas records at 100 cm began only in 1971 when the previous standard depth of 4 feet / 122 cm was abandoned in favour of the metric 100 cm depth). The site selected for such measurements should be a *level* plot of ground representative of surrounding soil type and land use, fully exposed to sunshine, wind and rainfall (not unduly shaded), and usually within the station enclosure adjacent to other instruments in use. Earth temperature records made in very sheltered sites, in areas of heavy shade, or amongst growing plants, are likely to be representative of only the immediate surroundings. Where records of such environments are required, for agricultural crop trials and the like, the records may not be comparable with those made within a standard climatological enclosure.

Choice of sensors

Earth temperatures have traditionally been measured once or perhaps twice daily using mercury-in-glass thermometers. Most can be readily substituted by electronic sensors, usually PRTs or thermistors, connected to a suitable datalogger as described in what follows. For such readings, fast response (low time constant) sensors offer no advantage, but with reduced range in temperatures, particularly at depth, accurate calibration (and excellent waterproofing of connections) becomes more important. Typical entry-level and budget AWSs do not offer 'spare' sensor ports or 'trailing lead' thermistor options with lengths sufficient to measure earth temperatures. Some mid-range and advanced systems offer optional manufacturer-specific sensors (such as those offered by Davis Instruments as an option to their Vantage Pro2 AWS), although these tend to be more expensive and less flexible than standard thermistors or PRTs. Standalone single-channel loggers with trailing leads (such as Gemini Dataloggers' Tinytag models) can also be permanently dedicated to one earth temperature depth measurement. Time-stamped records are then easily consolidated with the other AWS outputs in a spreadsheet or database package.

Shallow depths (20 cm or less)

Earth temperatures at shallow depths (usually 20 cm or less) have traditionally been measured using mercury-in-glass thermometers with an elongated bent stem, the bulb being located at the desired depth with the stem of the thermometer lying flat within a small plot of bare earth, typically 1-2 m square, kept free of weeds and lying snow. (For obvious reasons, bent-stem thermometers cannot be sited on a grassed surface.) 'Bent-stem' glass thermometers using coloured alcohol rather than mercury can replace mercury thermometers, but shallow earth temperature measurements can usually be made more cheaply and conveniently by electronic sensors connected to a suitable datalogger. Logged records also enable a continuous 24 hour record to be obtained, rather than one or two manual readings per day.

Earth temperature measurements at shallow depths are also now made under grass more often than under bare soil, for continuity with other measurements at greater depth which have always been made under grass, but a change from 'under bare soil' to 'under grass' may introduce discontinuity into existing periods of record and should be noted in the site metadata.

Depths greater than 20 cm

Temperatures at depths of 30–50 cm and deeper were traditionally measured by mercury thermometers exposed in steel tubes of the appropriate depth, the thermometer being attached to a length of chain to allow it to be raised to the surface to be removed for reading. The thermometer bulb was heavily lagged with paraffin wax, to ensure its indication did not change appreciably while being briefly withdrawn for reading (**Figure 10.3**). The slow rate of change of temperature at such depths meant that the response time of the lagged thermometer was adequate for the intended purpose. However, significant errors associated with thermal conduction along the black-painted steel tubes have been identified [249], particularly during the warmer months when the exposed top of the tube can become very hot in sunshine. As a result, white or grey plastic tubes with much lower thermal conductivity than their

Figure 10.3 A traditional 30 cm depth earth thermometer (with wax-covered bulb), shown next to a steel earth thermometer tube. (Photograph by the author)

steel equivalents are preferred wherever possible, although steel tubes still tend to be preferred for depths greater than 30–40 cm owing to their greater strength and rigidity.

With the progressive withdrawal of mercury thermometers following the Minamata Convention, one or more of the following methods can be adopted to continue measurements of earth temperature at depths greater than 20 cm:

A. The simplest is to retain the existing thermometer tube(s) and replace the mercury thermometer with a suitable alternative thermometer. Lagged earth thermometers of similar construction to the traditional variety are available, using coloured alcohol as the thermometric liquid. Provided calibration checks are undertaken from time to time, such an approach should ensure continuity of measurements, albeit with known bias issues resulting from heat transfer along the length of the thermometer tube. If a steel tube is used, it is advisable to ensure any protruding surface is painted white to minimise thermal conduction bias. Manual records will normally be available for only a single hour once per day, usually the morning observation, which precludes analysis of diurnal temperature variations at depth.

B. An alternative approach is to replace the thermometer with an electronic sensor fitted within the existing tube, the sensor being connected to and logged by a suitable datalogger. Provided care is taken to ensure the sensor remains at the correct depth within the tube, then this method will provide more frequent records, typically hourly spot values, which will enable ground heat flux and diurnal temperature variations to be examined. An advantage of this method is that ready access to the sensor remains available for calibration checks; however, surface cable runs to the datalogger may become a problem, with consequent greater risk of damage/interruption of record from strimmers and lawnmowers, as well as a possible trip hazard. Of course, known bias issues resulting from heat transfer along the length of the thermometer tube remain.

C. In theory at least, the most accurate earth temperature readings will come from electrical sensors buried at the appropriate depth, although great care needs to be taken in ensuring minimum disruption to the existing soil profile when installing the sensors. An appropriate method is to prepare a trench to the depth required, then insert the sensors at the appropriate depth into the undis-turbed vertical face of the trench. The trench should then be carefully refilled with earth, taking care as far as possible to preserve the original strata and drainage characteristics. Once the ground has settled, buried sensors should provide a good measure of the earth temperature at that depth, without biases introduced by the presence of a foreign body (i.e., the metal thermometer tube, whose properties differ from the surrounding soil). However, this apparently ideal method suffers from a very significant drawback, in that once buried the sensors cannot easily be accessed for regular (at least annual) calibration checks and, eventually, replacement. Buried sensors will need replacing every few years, particularly those in the near-permanently wet soils typical of many temperate and polar latitudes. Access becomes impossible to achieve without considerable disturbance to the location where measurements are being made, which will itself disturb the earth temperature record being made. One possible way around this is to set up duplicate measurements at the site from two or even three locations a few metres apart and compare results. If differences of more than about a tenth of a degree Celsius become apparent over time, the sensor(s) in one site should be dug up, recalibrated and/or replaced and allowed to settle for a few weeks, before the exercise is repeated in the second location.

D. An alternative pragmatic approach is to place electrical sensors in narrow-bore plastic tubes whose (sealed) bases are located at the required depth but whose top (and watertight cable exit) terminates 10–15 cm under the surface. Here the sensor remains relatively easy to access for calibration or replacement as required (remember to mark the location so it can be found!), but as the plastic tube is of lower conductivity than steel and is not exposed to the greater diurnal temperature ranges at the surface arising from daily solar radiation input, con-duction and convection errors resulting from the presence of the tube are greatly reduced compared to those from a steel tube. In addition, the sensor cable (which should be safely buried in a tough plastic conduit) is out of danger from lawnmowers, etc. The author has employed this method for over a decade with excellent results, and annual calibration checks or sensor replacements are easily accomplished.

One inevitable consequence of any foreign body within the soil, whether that be a plastic or steel tube or the cable of an electronic sensor, is that rainwater tends to follow the path of least resistance. Heavy rain after a dry spell can produce remark-ably sudden apparent changes in earth temperatures at depths of 25 cm or more. Such rapid changes are at least partially attributable to accelerated rainwater drain-age to the depth of the temperature sensor along the length of the thermometer tube or sensor cable, the rainwater being at a higher or lower temperature than the soil.

'In-place' earth temperature measurements (methods C and D) tend to reduce the daily and annual temperature ranges at depth somewhat when compared with

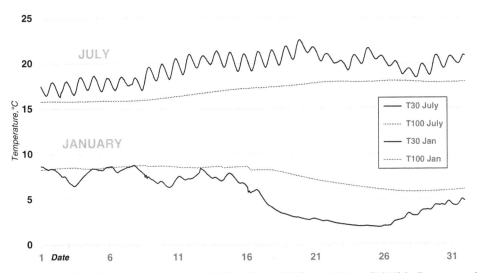

Figure 10.4 Hourly earth temperatures (°C) at 30 cm (T30) and 100 cm (T100) in January and July, showing the much greater diurnal variation in 30 cm temperatures during the summer months, and the negligible diurnal variation at 100 cm throughout the year. Observations from a site in central southern England (51°N, 1°W).

'traditional' steel tube records, because the heat flux bias arising from the conducting steel tube has been eliminated – winter temperatures slightly higher, summer temperatures lower [249]. Such measurements are more likely to be closer to the true earth temperature profile. Whichever method is adopted, or if an existing methodology changes, the station's metadata record (see Chapter 16, *Metadata, what is it and why is it important?*) should be annotated accordingly, in the event of any discontinuities in record subsequently becoming apparent.

Logging intervals

Traditional records of earth temperatures were made using mercury thermometers read manually once per day, usually at the morning observation. The use of logged electrical temperature sensors enables more frequent observations of earth temperatures to be made (**Figure 10.4**). While the shallow thermometers respond to the diurnal cycle of solar heating, the time lag from the screen maximum and minimum increases, and the daily range decreases, with depth. At 30 cm the daily maximum and minimum will be 6–12 hours behind their screen equivalents, while at 100 cm the daily variation is negligible. Exact relationships will vary from location to location and with soil type, soil moisture content and other factors. For shallow earth temperature records, 5 minute logging of spot values is ample; at greater depths, hourly spot values will normally suffice.

One-minute summary – *Measuring grass and earth temperatures*

- Grass and earth temperatures are the most commonly observed temperature measurements, after screen or air temperature.

- The lowest temperatures on a clear night will be recorded at or close to ground level. Where the surface is covered by short grass, the lowest temperatures are attained just above the tips of the grass blades. The so-called 'grass minimum temperature' is measured using a sensor freely exposed in this position. A 'ground frost' occurs when the grass minimum falls below 0 °C.
- Temperatures are also often measured above concrete or tarmac surfaces, or using sensors buried in road surfaces at roadside AWSs, to provide information on road surface temperatures for road forecasting models.
- Surface temperatures can easily be measured with a suitable trailing-lead electrical sensor (thermistor or platinum resistance thermometer, PRT) connected to a datalogger. Such sensors need to be small (for speed of response), weatherproof and robust as they will be exposed to all extremes of weather.
- WMO guidelines indicate that grass and surface minimum temperatures should relate to the period 'sunset to the morning observation on the following day', although the greater prevalence of unmanned sites is leading more locations to adopt the conventional 'morning to morning' 24 hour period. It is easy enough to program a datalogger to record the minimum temperature over a shorter period, for instance from near sunset to the next morning's observation.
- Earth temperatures are typically measured at depths of 5, 10, 20, 30, 50 and 100 cm below ground level, although few sites will include all depths. The location chosen for such measurements should remain fully exposed to sunshine, wind and rainfall.
- Earth temperatures have traditionally been measured by means of mercury thermometers to specific designs, bent-stem glass thermometers at depths of 20 cm or shallower or specially lagged thermometers hung on chains in steel tubes at greater depths. The use of steel tubes can cause bias in temperature measurements owing to enhanced heat flux along the tube itself. Existing liquid-in-glass earth thermometers in plastic or steel tubes are being progressively replaced by electrical sensors, although access for recalibration or replacement can be problematic unless carefully considered prior to installation. Waterproofed sensor connections are more important considerations for earth temp sensors than fast response times.
- Earth temperatures are normally quoted for a morning observation hour, although hourly values can easily be derived from logged electrical sensors. Hourly values provide useful insights into diurnal temperature variations below the earth's surface.
- Grass temperatures should be sampled and logged at the same interval as used for air or screen temperatures; for earth temperatures, particularly at depth, hourly logging intervals are normally sufficient.

Measuring sunshine and solar radiation

This chapter covers the instruments and methods used to measure both sunshine and solar radiation. Broadly, measurements of 'solar radiation' refer to the interception of radiant electromagnetic energy emanating from the Sun, measured by satellites at the top of the atmosphere or by instruments at the Earth's surface. Our eyes are sensitive to only a part of this stream of radiation, some of which is scattered, absorbed or reflected in its passage through the Earth's atmosphere. Measurements of 'sunshine' refer more specifically to the appearance or otherwise of the solar disk, and more particularly shadows cast by the Sun, when viewed from the surface of the Earth. 'Sunshine' can therefore be considered as a binary (yes/no) condition, occurring only when visible solar radiation exceeds a particular threshold. Solar radiation measurements are therefore the more complete and provide more useful and precise values for solar energy input, whether that be in relation to a domestic solar panel or a computer-based global climate model. However, presence or absence of sunshine is often the more significant in terms of human perception and health, and while few non-specialist members of the public would be immediately familiar with solar radiation records, almost everyone would identify with '12 hours of sunshine' as being a sunny day.

Records of the duration and intensity of sunlight are important aspects of the description of any climate, yet the number of sites which maintain records of these elements is fewer than those which measure temperature or rainfall. Perhaps that is because solar radiation is, after wind speed and direction, the most variable of all weather elements, and obtaining accurate measurements poses a number of unique challenges. The subject, and its instrumentation needs, can be daunting – there are more varieties, classes and sub-classes of specialised sensors in this field than in any other category of meteorological measurements. Some are hand-built research-grade precision sensors with world-class accuracy, found only in national observatories or international research institutions; others are rather more affordable instruments suitable for mid-range AWSs. This chapter describes the most common types of (surface-based) instruments in use, together with their advantages and disadvantages. Different instruments can give different results, sometimes significantly so.

Accurate measurements of solar radiation and sunshine require an open exposure: World Meteorological Organization (WMO) CIMO guidelines relating to sunshine and solar radiation site classes [4] are set out in this chapter. Very few 'perfect' unobstructed sites exist, however, and methods of estimating the potential losses resulting from nearby obstructions are also covered. Further reading recommendations are given for those who wish to explore this diverse, complex and fascinating field in more detail, followed finally by the 'One minute summary'.

The *measurand* – what is being measured?

Our nearest star, the body we call the Sun, is the source of (almost) all the energy redistributed around the globe by the global weather machine, and thus of all life on our planet. (A small amount of energy originates from the radioactive decay of elements in the Earth's crust, but the amount is insignificant compared to radiant solar energy receipts.) Solar radiation, the term covering the electromagnetic output from the Sun, covers a wide range of wavelengths. **Figure 11.1** summarises the distribution by wavelength of solar radiation received at the top of the atmosphere (bold line). The peak intensity lies in the visible wavelengths (i.e., the narrow part of the electromagnetic spectrum that we perceive as colour), with a wide spread into the (more energetic) ultraviolet part of the spectrum, and a longer and flatter tail in the (less energetic) infrared. A more detailed account of solar radiation is beyond the scope of this chapter, but there are numerous excellent reference works covering the subject in greater depth, including [99, 250, 251] amongst others.

Measurements of the radiant energy we receive from the solar disk are thus amongst the most important in climatology and climate change studies, if perhaps less so in day-to-day operational or synoptic meteorology. Particularly in mid-latitude climates, the presence or lack of sunshine has probably a greater impact on public perception of day-to-day weather than any other single element, while too much or too little solar radiation are known to have serious effects upon human health and well-being.

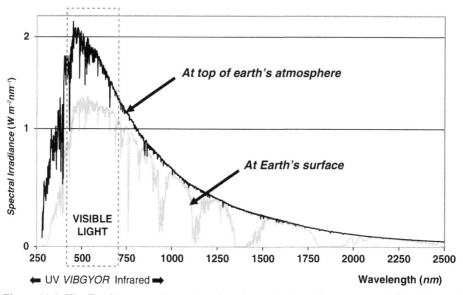

Figure 11.1 The Earth's atmosphere absorbs solar radiation. The upper dark line shows the spectral irradiance (W/m²/nm) by wavelength at the top of the atmosphere: the lower grey trace shows the typical irradiance by wavelength near sea-level at the Earth's surface, based on the ASTM G-173–03 standard (also known as ISO 9845–1). Visible region denoted by the colours VIBGYOR (Violet, Indigo, etc.). For sources, see [252].

All bodies above absolute zero emit thermal radiation, the peak wavelength of which depends upon the absolute temperature of the body (Wien's law): the higher the temperature of the body, the greater the emission of radiation (Stefan's law). The Earth and its atmosphere also emit radiation, but as both are much cooler than the Sun, the peak wavelength is much lower (in the infrared rather than the visible part of the spectrum), and the total emitted radiation is also much lower. **Table 11.1** summarises the wavelengths most relevant in meteorological measurements.

Table 11.1 *The main wavelengths of solar and terrestrial (or atmospheric) radiation used in meteorological measurements. Adapted from WMO CIMO [4] and Kipp & Zonen table at* www.kippzonen.com

Radiation category	Category name	Wavelength range	Sources
Short-wave solar radiation (ultraviolet)	UV-C	100–280 nm	Emitted from the Sun *Completely absorbed by the Earth's atmosphere before reaching the ground*
	UV-B	280–315 nm	Emitted from the Sun *Around 90 per cent absorbed by the Earth's atmosphere before reaching the ground, but what gets through is biologically very active – main cause of sunburn and skin cancer*
	UV-A	315–400 nm	Emitted from the Sun *Most reaches the ground, but less biologically active than higher-energy UV-B rays*
Visible light	Visible light spectrum	400–780 nm	Emitted from the Sun *The visible spectrum of colours from violet (400 nm) to red (780 nm)*
Long-wave radiation (infrared)	Near infrared	780–3000 nm (3 μm)	Heat radiation from the Sun
	Far infrared	3 μm – 50 μm	Emitted from the Earth and atmosphere *Heat radiation from the atmosphere, clouds, Earth and surroundings*

Units of measurement

The measurement unit of solar radiation intensity or *irradiance* is the watt per square metre (W/m^2), where 1 watt is 1 joule per second. When integrated over time, as is usual to express daily totals of solar radiation, the unit becomes the joule per square metre (J/m^2). This is a very small unit, and daily totals are more conveniently expressed as megajoules per square metre, or MJ/m^2 ($1 MJ = 10^6$ joules). Divide the values by 3.6 to convert from megajoules per square metre (MJ/m^2) to kilowatt-hours per square metre (kWh/m^2), if this measure is the more convenient (for example, in comparing solar panel output levels).

Sunshine measurements are usually stated as a total duration ('hours of sunshine'), whether for a day, a month or a year. This can also be expressed as a 'percentage of maximum possible' by expressing the actual sunshine duration as a percentage of the time the Sun is above the horizon (see Box, *Determining 'percentage of maximum possible sunshine'* later in this chapter).

Absorption by the atmosphere

Solar radiation is the only meteorological element where it makes sense directly to compare measurements made at the top of the atmosphere with those made at the Earth's surface. The average intensity of solar radiation just outside the Earth's atmosphere, perpendicular to the incoming solar beam, is 1361 W/m^2 (the so-called *solar constant*) [253]. This varies seasonally by about ± 3 per cent owing to the elliptical nature of the Earth's orbit around the Sun, being at a maximum in January when the Earth is closest to the sun, and minimum in July when the Earth is furthest away. Averaging across day and night, the seasons and all latitudes, about 340 W/m^2 arrives at the *top* of the Earth's atmosphere [254]. Of course, the amount of solar radiation received at the Earth's *surface* varies enormously: from day to night, between winter and summer, from poles to tropics, and under thick cloud or clear skies. Averaged over the year and across the globe, about 214 W/m^2 of the incoming solar radiation reaches the Earth's surface, the rest being absorbed by the atmosphere or scattered and reflected back to space from the air, clouds and particles of dust or other atmospheric aerosols [254]. Absorption by the atmosphere largely shields us from the most energetic sections of the solar spectrum, particularly the far ultraviolet, most of which is absorbed by the stratospheric ozone layer 25–45 km above our heads (**Table 11.1** and **Figure 11.1**). This is fortunate for all life on Earth, as ultraviolet radiation is known to result in genetic damage, including a clear causative link with skin cancer.

Instruments on satellites and at the Earth's surface routinely measure incoming solar radiation across the wavelength ranges shown in **Table 11.1**, but most surface-based instruments make their measurements only between the near ultraviolet and the near infrared, including the visible spectrum.

Standard methods of measuring sunshine and solar radiation

At the bottom of the Earth's atmosphere, after some of the solar beam has been absorbed, scattered and reflected on its passage through the depth of the atmosphere, there are two main components to solar radiation – *direct* and *diffuse*. (For simplicity, this ignores some smaller terms, such as reflected short-wave.) As the name implies, *direct* solar radiation is that received directly from the solar disk alone, while *diffuse* is that received from the rest of the sky as a result of atmospheric scattering and reflection. By definition:

Total (or global) solar radiation I = direct solar radiation D + diffuse solar radiation G

Note, however, that this summation applies strictly to equal-area measurements (all-sky or hemispheric values), and not to the readings from separate direct and global radiation sensors, which cannot simply be summed arithmetically. Direct radiation measurements apply only to a small area of the sky, while global radiation measurements require a cosine correction term to allow for solar angle.

On a very clear day, or at high altitudes, the direct component can account for 85 per cent or more of the global solar radiation; on a cloudy day, the direct contribution is zero. Measurements of diffuse radiation are useful in determining the scattering of inbound solar energy (essential for planetary radiation balance studies), and for monitoring the transparency (or otherwise) of the atmosphere and its constituents. Instruments are available to measure both direct and global solar radiation, although measurements of the latter are simpler and thus greatly outnumber the former.

Sunshine recorders can be considered as a subset of the wider category of solar radiation instruments. The principle of these instruments is simple enough – to provide an unambiguous binary (on/off) signal when the Sun is shining – although achieving this reliably and consistently is less simple. There are several instruments in this category.

What is the difference between radiometric terms and units (irradiance, W/m^2) and photometric terms (illuminance, lux)?

Radiometric quantities refer to measurements of radiation, of any wavelength, from any physical body – in this case, the Sun or the Earth. *Photometric* quantities describe how the human eye senses optical radiation, and therefore such measurements refer only to the visible part of the spectrum.

Irradiance is a radiometric term which refers to electromagnetic radiation incident upon a unit surface area; it is measured in watts per square metre (W/m^2).

Illuminance is a photometric term, referring to the incident flux of radiant energy emanating from a source within the visible spectrum and weighted by the response of the human eye to energy in visible wavelengths. It therefore simulates how bright the source appears to the human eye. A light-adapted eye generally has its maximum sensitivity at around 555 nm, which lies in the green region of the optical spectrum. Illuminance is measured in units of lux (lx).

Ultraviolet or infrared radiation from the Sun (or any other suitable source) will register on a suitable solar radiation detector, but will not register on a photometer (or lux meter) because it lies outside the visible spectrum, and therefore (by definition) has no illuminance.

Measuring direct solar radiation

Instruments to measure *direct* solar radiation are called *pyrheliometers*. A pyrheliometer consists essentially of a suitable electrical sensor – sensitive to a wide range of solar radiation, typically between 200 nm and 4000 nm – exposed at the end of a narrow internally blackened tube, pointed directly at the Sun to make a measurement (**Figure 11.2**). The detector has a field of view of about 5 degrees (approximately 10 times the apparent diameter of the solar disk) and thus excludes nearly all scattered radiation from the sky. The intensity of solar radiation from this narrow angle, perpendicular to the solar beam, is known formally as the *normal incidence direct irradiance*.

Measurements from manual pyrheliometers have been made for over a century, now largely replaced by automated versions kept pointing at the Sun by accurate tracking mechanisms and so providing continuous unattended records. Not surprisingly, such equipment is expensive and usually found at only a few well-equipped

Figure 11.2 Solar radiation sensors at Neumayer station, Antarctica. The platform tracks the sun, and the two black spheres continuously shade one pyranometer measuring diffuse solar radiation and one pyrgeometer measuring long-wave downward radiation. One direct-incidence pyrheliometer (an Eppley Normal Incidence Pyrheliometer) is mounted above a four-quadrant sensor to the right of the platform in the photograph, directly above the observer's glove. (Photograph courtesy of Gert König-Langlo of the World Radiation Monitoring Center at AWI)

research observatories within any given country. However, the principle of measuring direct solar radiation using a direct incidence pyrheliometer is important to grasp, because it forms the basis for the WMO definition of sunshine, as we shall see shortly.

Measuring global solar radiation

The most commonly used instrument to measure *global* solar radiation (i.e., the combined direct and indirect components) is the *pyranometer*, occasionally referred to as a *solarimeter* (**Figure 11.3**). A pyranometer is normally used to measure global solar radiation on a plane (flat and level) horizontal surface, as opposed to pyrheliometer measurements which are made perpendicular to the solar beam. The global solar radiation measurements obtained are more formally defined as *global solar radiation on a horizontal surface*. Measurements are sometimes made with pyranometers installed at an angle, or pointing only in one direction (such as 'facing east, angled at 45°'). These are typically undertaken for solar energy research purposes and the like, but for meteorological measurements global radiation sensors must be accurately horizontal.

Measurements of diffuse radiation can be made using a pyranometer fitted with a shadowing disk or shade ring which blocks radiation from the immediate area of the solar disk. A shade ring requires seasonal adjustments to follow the path of the Sun in the sky, or (as with the pyrheliometer) an accurate tracking mechanism can be used in which a small disk or sphere permanently eclipses the solar beam (as can be seen in **Figure 11.2**). Where both global and diffuse radiation measurements are made at the same site, two matched sensors are commonly mounted adjacent to each other, ensuring of course that neither instrument shadows or casts reflections upon the other. Where simultaneous measurements are made, the direct component can be inferred by subtracting the cosine-corrected diffuse component from the global. Alternatively, where both direct and global radiation are measured simultaneously,

Figure 11.3 Banks of pyranometers undergoing calibration testing at the UK Met Office test site in Exeter, England. (Photograph by the author)

the diffuse solar radiation can be determined by subtracting the direct component from the cosine-corrected global value.

As with many other meteorological instruments, there are a variety of instruments and suppliers to meet differing budgets and accuracy requirements (**Table 11.2**).

Budget solar radiation sensors

Many AWS systems at or above mid-range specification will offer one or more solar radiation sensors as standard or optional instruments, typically a silicon photodiode detector with a spectral response across the wavelength range 300 to 1100 nm (near ultraviolet, through the visible spectrum and into the near infrared). Such instruments can provide an interesting addition to the normal range of logged data, although their absolute accuracy can be difficult to ascertain. Typical is the Davis Instruments 6450 solar radiation sensor, included as standard on some AWS configurations (**Figure 11.4**). This sensor is based around a photodiode sensor, for which ±5 per cent accuracy at full scale (i.e. at 1000 W/m^2) is quoted, although it is difficult to verify either calibration accuracy or calibration drift over time for this or similar instruments in this class without side-by-side comparisons with a standard pyranometer of known calibration over a period of weeks to months. 'Near-neighbour' or regional comparisons against sites with calibrated instruments can provide useful information in this regard, although there are far fewer sites measuring solar radiation than for temperature and rainfall, for example, and reliable statistics can be more difficult to find. In addition, some photodiode sensors are fairly insensitive to rapid changes in solar radiation (the response time of the Davis 6450 unit is about 60 seconds), and as a result even 1 minute logging intervals will fail to follow the typically rapid swings in solar radiation on a day of broken cloud (see **Figure 11.6** later in this chapter).

Table 11.2 *Key specifications for various categories of pyranometer, comparing WMO guidelines for various classes of instrument with Kipp & Zonen's CMP3 and CMP10 thermopile pyranometers and the Davis Instruments model 6450 photodiode solar radiation sensor: guideline prices can be inferred as highest to lowest left to right across the table.*

Characteristic	WMO High quality	WMO Good quality	WMO Moderate quality	Kipp & Zonen CMP10 sensor	Kipp & Zonen CMP3 sensor	Davis Instruments 6450 sensor
Spectral range, nm	← 300 to 3000 →			270 to 3000	350 to 1500	400 to 1100
Response time to 95%, s	< 15	< 30	< 60	< 5	< 20	60
Resolution, W/m^2 *Smallest detectable change*	1	5	10	< 1	1	1
Temperature response, per cent *Maximum error due to 50 K change in ambient*	2	4	8	< 1	< 4	6
Calibration stability *Change per year, per cent full scale*	0.8	1.5	3.0	< 0.5	< 1	2
Non-linearity, per cent *Percentage deviation from sensor response at 500 W/m^2 due to a change of irradiance within the range 100 to 1000 W/m^2*	0.5	1	3	< 0.2	< 2	N/A
Directional response, W/m^2 *Directional error at 1000 W/m^2 assuming beam up to 80° offset*	10	20	30	< 10	< 20	N/A
Achievable uncertainty in daily totals, per cent	2	5	10	N/A	N/A	5

WMO specifications are from reference [4] (Table 7.4 therein), Kipp & Zonen and Davis Instruments from manufacturer literature/websites. In WMO terminology, 'High quality' refers to 'near state of the art', international or solar observatory reference instruments: 'Good quality' refers to national observatory network standards, 'Moderate quality' to low-cost monitoring networks. N/A indicates 'not available' or not stated by manufacturer.

Professional solar radiation sensors

Leading suppliers of solar radiation instruments include Kipp & Zonen (Delft, the Netherlands – part of the OTT Hydromet group), Hukseflux (also in Delft), EKO Instruments (Tokyo, Japan) and Eppley Labs (Newport, Rhode Island). All produce competitive ranges of high-quality solar radiation sensors which differ only in application type, accuracy and time constant (response time). Top-of-the-range instruments offer traceability to national and international calibration standards, at a price to match.

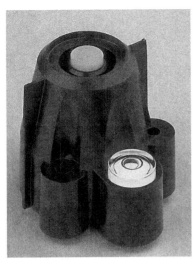

Figure 11.4 Davis Instruments' 6450 solar radiation sensor. (Photograph courtesy of Davis Instruments)

Most 'professional quality' pyranometers are thermopile instruments. A thermopile consists of an electrical circuit composed of two different metals, with one junction designated as 'hot' and the other as 'cold'. When there is a temperature difference between the hot and cold junctions, a small electrical potential is generated. In a pyranometer, the 'hot' junction (in reality, very many fine junctions) is painted matte black and exposed under a glass dome in the centre of the instrument, while the 'cold' junction is bonded to the instrument chassis (shaded from direct solar radiation) and thus at or close to the ambient air temperature. The dome is normally made of quartz glass, which is transparent to radiation over the wavelength range 250–2800 nm approximately, but opaque in the far infrared to reduce errors due to outgoing long-wave radiation. The glass dome (a double dome on higher-spec instruments) also protects delicate sensor elements from precipitation and dust, while minimising cooling effects resulting from ambient wind conditions.

When solar radiation falls on the instrument, the blackened hot junction becomes warmer than the ambient temperature, the difference in electrical potential being proportional to the intensity of the solar radiation. Thermopile instruments require no power (although some 'smart' sensors may do so) and are very responsive to rapid changes in solar radiation – a typical instrument in this class has a response time of only a few seconds to attain 63 per cent of a step change in solar radiation (see **Appendix 1** for more on instrument performance measures). Because the output signal is very small – typically only around 10 µV per W/m^2 – shielded cables and high-quality connections are essential for this class of instrument.

The Kipp & Zonen CMP3 pyranometer (**Figure 11.5**) is a small, light instrument ideal for use with AWSs and dataloggers, and covers the spectral range 350 to 1500 nm. Each instrument is supplied with an individual calibration certificate; year-to-year stability is quoted as within 1 per cent. Once installed (see below for exposure requirements), the instrument is maintenance-free apart from an occasional visual check and clean of the quartz glass dome and the shading disk covering the

Figure 11.5 Kipp & Zonen CMP3 pyranometer. (Photograph by the author)

instrument chassis. It is best to clean the glass dome of pyranometers daily wherever possible. During the morning is ideal, as deposits of dew, dust or pollen, and particularly frost and snowfall, can significantly impact output. At low solar angles, for example, a thin cap of frost may reflect solar radiation back onto the sensor and greatly increase the apparent output for an hour or more. During snowfall, anything more than a centimetre or so of snow remaining on the dome will result in near-zero solar radiation being recorded as long as the snow remains in place on the instrument, so the snow should be brushed off as soon as practicable to do so.

Pyranometer performance in darkness

During the hours of darkness, pyranometer output will normally fall slightly below zero. There are two reasons for these negative night-time values. Firstly, sudden changes in temperature can produce short-lasting negative or positive values, as the sensor element and the body of the pyranometer adjust to the change at different rates. Such sudden changes in temperature can be caused by weather conditions – a sudden increase in cloud, a shower of rain – or even by nocturnal birds perching on the sensor. The second, more significant and present on most nights, results from the normal situation whereby the Earth is warmer than the sky. On clear nights the sensor will 'see' the sky black-body radiation as a much lower temperature than the sensor body, which will be close to the ambient air temperature, and this temperature differential will result in a slight negative signal. Output can fall as low as -10 W/m^2 on very clear nights, particularly when the temperature is falling quickly, although it is rarely as much as half of this level with 'double-dome' pyranometers.

Reference-standard pyranometers can compensate for these effects to a certain extent, and units which include ventilation and heating fixtures almost eliminate them. Any persistent offsets under cloudy night-time skies at night may indicate a small zero error in the sensor, and adding the mean 'cloudy' offset can help compensate for a drifting zero point.

Although negative values can easily be suppressed by a simple line in the logger program, night-time records from a pyranometer can also be a useful, if rather qualitative, indicator of changes in overnight low and medium-level cloud cover. However, when calculating daily totals of solar radiation, all negative values should be treated as zero.

Measuring solar radiation from satellites

Some countries use geostationary satellite cloud data to provide estimates of surface solar radiation receipts. Cloud types, breaks and thicknesses at grid points are categorised from satellite imagery, and surface-level solar radiation levels estimated from these categories. In Australia, the Bureau of Meteorology uses this method to estimate daily solar radiation totals for thousands of locations across the country, the estimates being tied to ground truth from a much smaller network of surface solar radiation measurement sites [255]. Accuracies are better under clear-sky conditions (typically within 7 per cent of ground measurements) than under cloudy skies (± 20 per cent). Similar experimental approaches to estimating daily *sunshine* totals have been described for the UK [256] and for Poland [257], although it is fair to describe both results as 'mixed'.

Measuring ultraviolet (UV) radiation

The UV region (**Table 11.1**) spans the wavelength range from 100 to 400 nm. The spectral distribution of atmospheric UV irradiance is very variable, depending mainly on the elevation of the Sun and stratospheric ozone levels. Almost all of the higher-energy UV is absorbed by the Earth's atmosphere, but limited UV exposure has beneficial effects for human biology.

Measurements of UV radiation are important, not least because of the effects upon human health, but such measurements are difficult because the amount of energy reaching the Earth's surface is small and varies significantly with stratospheric ozone levels as well as tropospheric cloud cover. In addition, the rapid increase in UV levels with increasing wavelength above 280 nm demands precision in sensor sensitivity and range. Multiple scattering, particularly above a snow cover, may also make accurate observations more complex. Unlike most meteorological measurements, few standards of UV measurements yet exist, partly because in many countries measurements of UV radiation are made by health or environmental protection bodies rather than national meteorological services.

For all but the most specialised applications it is only necessary to monitor 'total UV' irradiance, which represents the combined UV-A and UV-B components (**Table 11.1**): UV-A radiation at the Earth's surface is normally 15–20 times greater than UV-B. (UV-C is completely absorbed by the Earth's atmosphere.) UV radiation measured with a similar response to the human skin is termed *Erythemally active UV irradiance* (UVE) and is often communicated as a Solar UV Index (UVI) on weather forecasts. Some budget and mid-range AWSs offer an optional 'ultraviolet sensor', providing an approximate indication of ultraviolet radiation in terms of the UVI index scale, although they should not be regarded as precision instruments in this regard.

More detailed information on measuring ultraviolet radiation is given by the WMO CIMO guide ([4], section 7.6). Where accurate measurements of UV radiation are required, application advice should be sought from specialist instrument manufacturers such as those cited earlier in this chapter and in **Appendix 3**.

Measuring sunshine

Sunshine recorders are a specialist subset of solar radiation sensors, in that most are designed to provide only a binary (yes/no) record at the times within a day when the intensity of the visible radiation from the solar disk at the surface of the Earth exceeds a particular threshold. To the human eye, this is more easily thought of as the appearance or disappearance of shadows behind illuminated objects in daylight, and for this reason 'sunshine' measurements relate to the visible spectrum only.

The duration of 'sunshine' is defined by WMO as ' ... the sum of the time for which the direct solar irradiance exceeds 120 W/m^2.' (Prior to 1981, the threshold was set at 210 W/m^2, and some older reference material still references this figure.) Note that this threshold refers to the measurement of *direct* solar radiation using a narrow-aperture pyrheliometer located perpendicular to the incoming beam as previously described, not to the *global* solar radiation on a horizontal surface measured from a pyranometer. The latter usually exceeds 120 W/m^2 at local noon in midlatitudes for most months of the year, even under cloudy skies. A simple tally of the duration of *global* solar radiation exceeding 120 W/m^2 will give a huge overestimate of the true duration of sunshine. As well as being normally expressed as a duration in hours, whether by day, month or year interval, sunshine totals can also be expressed as the 'percentage of maximum possible sunshine'; this is explained in more detail in the text box later in this chapter.

Figure 11.6 illustrates the highly variable relationships between global solar radiation and sunshine on three very different types of day.

In theory at least, the existence and duration of sunshine is best determined from minute-by-minute records made by a tracking pyrheliometer, because this is how 'sunshine' is defined by WMO. Even so, there remains some uncertainty relating to the definition of sunshine, for according to the WMO CIMO guide ([4], sections 7.2.1.3 and 8.2.1.2) direct pyrheliometer measurements can vary somewhat according to the field of view of the instrument: a tolerance of perhaps ± 10 per cent is suggested. The presence of rain or snow on the pyrheliometer viewing window can also produce erroneous results, whereas raindrops on a pyranometer dome usually have little significant effect. Unfortunately, the expense of such instruments renders widespread adoption impractical. Instead, less expensive and more practical instruments are used, their response and performance being carefully assessed against standard 'reference' instruments at regular WMO and World Radiation Center intercomparison events [258].

Over the years, many different instruments and operating principles have been devised to provide measurements of sunshine duration, but leaving aside direct radiation measurements made by a tracking pyrheliometer, most sunshine recorders in widespread use today fall into two main classes:

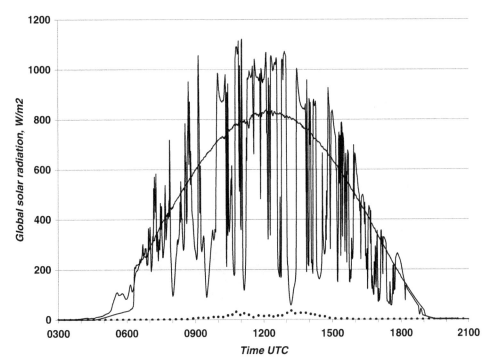

Figure 11.6 Global solar radiation measurements on three very different days. Thick line – an almost cloudless day in high summer (10 August 2022): the beginning and end of the record is affected by minor horizon obstructions. Thin line – a showery day close to the summer solstice (24 May 2022), showing briefly intense solar radiation partly reflected from convective cloud structures. Dashed line – a heavily overcast gloomy windwinter's day (4 December 2022). All records are from a Kipp & Zonen CMP10 pyranometer located in central southern England. The plots show 1 minute mean values, which themselves downplay the rapidity of changes in solar radiation on days with broken cloud. The global solar radiation receipts for these three days were, respectively, 25.1, 21.7 and 0.4 MJ/m^2.

Absolute versus relative measurements: solar radiation and sunshine duration

The total amount of solar radiation in any given period is an absolute measure of energy receipt at the site of the instrument. Daily solar radiation totals made at any one site throughout the year, or at different locations around the world, are directly comparable with each other.

In western European mid-latitudes, the mean daily solar radiation on a horizontal surface in June is typically 10 times that of the average day in December, whereas the mean daily duration of sunshine between the two months varies by a factor of just three or four. Daily sunshine durations can provide only an approximate relative measure of solar radiation inputs – a mid-latitude December day with 6 hours sunshine receives less than a fifth of that of a day in June with the same sunshine duration (**Figure 11.7**), because in midsummer the solar elevation is higher and the hours of daylight almost three times as long as in midwinter. These seasonal differences are greatest nearer the poles. It is for this reason that global solar radiation measurements are preferred to sunshine duration or 'percentage of possible sunshine' in climate, energy and agricultural modelling and similar applications.

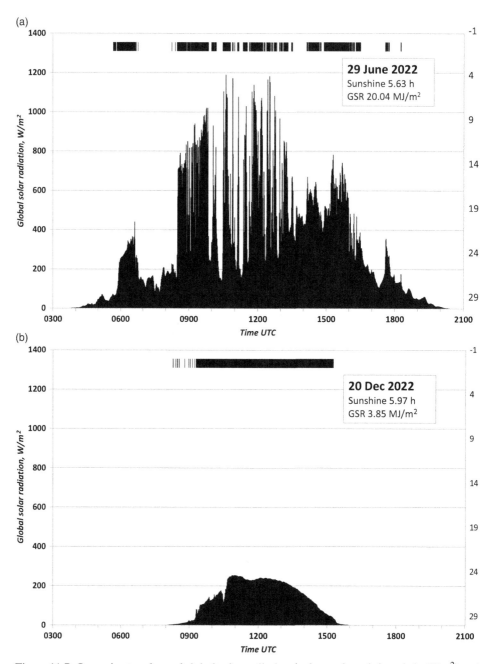

Figure 11.7 One-minute values of global solar radiation (column chart, left scale in W/m²) and minutes with recorded sunshine (vertical bars, at top of radiation plot). (a) A day near the midsummer solstice (29 June 2022) with almost 6 hours of sunshine, compared with (b, same scale) a day with a similar sunshine duration near the midwinter solstice (20 December 2022). Although the daily sunshine totals were almost identical, June's global solar radiation (GSR) receipt, 20.04 MJ/m², was more than five times that of December (3.85 MJ/m²). Logged using a Kipp & Zonen CMP10 pyranometer and CSD3 sunshine sensor, within the author's test site in central southern England, latitude 51°N.

- The first uses solar energy to change the properties of a recording medium. The earliest sunshine recorder, the Jordan pattern first described in the 1840s [6], used the action of solar rays to darken an insert of photographic paper, which was subsequently developed in a darkroom and the length of the trace measured. The iconic Campbell–Stokes sunshine recorder uses a spherical magnifying glass to focus the Sun's rays onto a pre-printed card, which chars in bright sunshine. This instrument is described in more detail subsequently.
- An alternative method uses electronics to infer the existence of shadows by comparing the difference in output between two photosensitive devices in the sensor body, one of which is exposed to direct solar radiation, the other kept in shade. When the difference between the two exceeds a given threshold, as it will in sunshine bright enough to cause hard-edged shadows, the instrument's electrical output changes in a step fashion, which can be detected by a suitable logger and/or associated timing circuit. The first type of electronic 'differential' or 'contrast' sunshine recorder was the Foster–Foskett sunshine switch, widely used in the United States between 1953 and 2009 [259, 260], but subsequently withdrawn from use. Similar principles are used in modern electronic sensors.

The difficulty with either method lies in setting a precise and consistent threshold between 'no sunshine' and 'sunshine', which is easier said than done. It is immediately obvious to the human eye that the Sun is *not* shining under a veil of thick cloud, and equally obvious that the Sun *is* shining in conditions of cloudless blue skies on a summer morning. The difficulty arises in the middle ground, for example in hazy skies under extensive high cloud cover, or as overnight fog begins to disperse, when the distinction is much less clear cut. The threshold intensity of solar radiation sufficient to trigger a sunshine recorder should be typically that which will just cast a *distinct*, hard-edged shadow, so very weak or hazy sunshine will normally not be recorded. In addition, the intensity of sunshine is rarely sufficient to register until the Sun is at least 2–3 degrees above the horizon (and of course this necessarily assumes a clear horizon, without obstructions from mountains, trees, buildings and the like). In mid-latitudes, this means that sunshine during the first and last 20–30 minutes or so of daylight will not normally be recorded, even if the sky is clear and (to the eye) the Sun looks to be shining. Very occasionally, reflections from cloud layers just above the horizon at sunrise or sunset can result in sunshine being registered by electronic sensors within a few minutes of astronomical sunrise or sunset, although such instances are due more to fortuitous atmospheric conditions than any particular performance characteristics of the sensor involved.

Estimating sunshine duration from pyranometer data

It would seem to be perfectly straightforward, in theory at least, to derive the duration of sunshine directly from logged global solar radiation records obtained from a pyranometer, using an appropriate threshold to determine the 'sunshine/no sunshine' cutoff point. The threshold varies with solar elevation (and thus season) and the elevation response of the pyranometer sensor (the cosine of the solar elevation). Various algorithms have been devised in an attempt to do this [260, 261, 262]; an outline of the approach is set out in the CIMO guide ([4], section 8.2.2). Most involve a comparison of the current or logged value of solar radiation with an estimate of the 'clear sky' maximum value for that date, time and place, using astronomical tables. When the current value of incident solar radiation exceeds a threshold, usually a fraction of that 'clear sky' maximum value, then that interval (minutes to an hour or more, depending upon method details) is counted as 'sunshine'. The arithmetic involved is rather laborious, although modern programmable loggers with powerful on-board processing capabilities can undertake the threshold calculations 'on the fly', comparing pyranometer readings minute-by-minute to produce a real-time binary (yes/no) sunshine output. These minute-by-minute totals are then summed by the logger to generate hourly and daily 'sunshine durations' or percentage sunshine values, as required. Alternatively, the logged output from a pyranometer, or a network of sites, can be retrospectively processed by computer. The main difficulty is undoubtedly in establishing a consistent and repeatable threshold from 'no sunshine' to 'sunshine' as this varies enormously with weather and cloud conditions, often on a minute-by-minute basis: for example, clearing fog can give rise to very high levels of diffuse radiation which can easily be interpreted as sunshine, even though it is unlikely that clear shadows would be cast under such circumstances. Days with thick high cloud cover, where the level of solar radiation may be very close to the 'sunshine/ no sunshine' threshold, or a day with broken cumulus, give very different results for only small changes in the assumptions made in the program. Variations in solar elevation, seasonal factors and site characteristics (coastal, inland or mountain sites may all react differently) further complicate the picture. A useful recent summary and comparison of several methods is given in reference [263].

Some countries have adopted such methods as their national standard, but it is fair to say that none have gained anything other than limited adoption. The Dutch state meteorological service, KNMI, pioneered algorithm-based estimates to derive estimates of sunshine duration from pyranometers in 1992 [264], replacing its network of Campbell–Stokes recorders entirely: KNMI's method is outlined in current WMO CIMO guidelines [4], although it is given as an example of the approach rather than as a recommendation (KNMI have since replaced it by an alternative algorithm [265]). But how do such estimates compare against measurements from dedicated sunshine recorders exposed at the same site? None of the various methods in use have yet been shown in independent evaluations to provide good and consistent agreement with the records from any particular type of sunshine recorder, particularly at daily and sub-daily timescales and in multiple geographies, although some methods claim a reasonable statistical agreement when comparing totals over monthly or annual timescales within a particular country or region. Lack of a common standard clearly poses numerous difficulties with regard to the continuity of existing historical records of sunshine (see Box, *Are today's sunshine measurements compatible with those made last year or last century?*). More work is urgently needed to verify and standardise the approaches taken. Should it become possible to develop a consistent and viable standard method with proven multi-region applicability, a recommendation for adoption through WMO CIMO could yet lead to a global standard measure of 'daily sunshine duration'. Until such a method can be devised, it is advisable to regard sunshine durations derived solely from pyranometers as no more than 'loose approximations' which are likely to differ – often by large amounts – from measurements made with dedicated sunshine recorders. For this reason, the source of any 'sunshine' measurements – instrumental or calculated, together with any changes – should be clearly identified in the station metadata (see Chapter 16, *Metadata, what is it and why is it important?*).

At the risk of stating the obvious, estimates of sunshine duration based purely on visual observation should not be regarded as reliable under any circumstances.

Determining 'Percentage of maximum possible sunshine'

The maximum possible period of daylight for any particular date is taken as the length of the period between sunrise and sunset. (For astronomical purposes precision to within a second or better is required, but for meteorological purposes tables accurate to a minute are perfectly adequate.) The 'percentage of maximum possible sunshine' is the daily duration of sunshine expressed as a percentage of the duration of daylight for that date.

A very useful calculator providing precise sunrise and sunset times and hours of daylight for any location (enter latitude, longitude and time zone, and year) for any specified year can be downloaded from www.esrl.noaa.gov/gmd/grad/solcalc/calcdetails.html – choose 'NOAA_Solar_Calculations_Year' in the spreadsheet format preferred.

Because sunshine recorders are insensitive to low solar angles, sunshine will generally not be registered until the Sun rises to about 3° above the horizon after dawn, or when it sets below 3° near sunset. The length of time taken for the Sun to reach 3° elevation after dawn, or to sink from 3° to the horizon at sunset, varies with latitude and season; on the equator it is as little as 13 minutes, but at 60°N it varies between 44 minutes near the summer solstice and 53 minutes at the winter solstice. (At higher latitudes the Sun does not reach 3° elevation in midwinter.) In midlatitudes, 20–30 minutes is typical. A 20 minute cut-off after sunrise and before sunset equates to roughly 7 per cent of the maximum possible daily duration averaged over the year. It is therefore unlikely that even the sunniest days at a site with a completely clear horizon will exceed about 94–95 per cent of the 'maximum possible duration' – the limits of current sensor sensitivity dictate that 100 per cent cannot *quite* be attained.

Weekly or monthly 'percentage of maximum possible sunshine' statistics are obtaining by summing the observed daily sunshine durations and expressing that as a percentage of the sum of the daily duration of daylight for the period. Such percentages are particularly useful for comparing relative amounts of sunshine between months of the year where the possible duration of daylight varies, such as in temperate or polar latitudes.

As there are only a small number of sunshine instruments in widespread use, each is described in a little more detail than is feasible for some of the other sensor types covered in this book. A recent comprehensive review of different sunshine sensors and their operating principles can be found in reference [263].

The Campbell–Stokes sunshine recorder

The idea of using the sun's rays to burn a record of sunshine duration was first described as far back as 1646 ([6], chapter VI), although it was not until John Campbell described a method of mounting a spherical glass lens in a wooden bowl

THE SUN SHINES THROUGH
THE GLASS SPHERE OF THIS
SUNSHINE RECORDER AND
BURNS A LINE ON A
MEASURING CARD

Figure 11.8 The Campbell–Stokes sunshine recorder at Chatsworth House, Derbyshire, England. (Photograph by the author)

in 1857 that the method became practical. Campbell's design was improved upon by the eminent Cambridge physicist and mathematician Professor Sir George Stokes in the late 1870s [266]: the instrument he described quickly became known as the Campbell–Stokes sunshine recorder, essentially a recording sundial. It has remained almost unchanged since (**Figure 11.8**) [267, 268]. With only slight modifications, it was adopted as the WMO Interim Reference Sunshine Recorder (IRSR) in 1962, until this designation was withdrawn in favour of more precise and repeatable direct pyrheliometer measurements in 1981. Although the Campbell–Stokes recorder forms the basis for most of the historical record of sunshine in many countries, including the United Kingdom (where it has been in continuous network use since 1880) and the Republic of Ireland, it was less widely used in some countries, including the United States, because of its lack of an electrical output to facilitate remote or automated monitoring.

This simple and iconic instrument consists of a precisely machined spherical glass lens mounted in a larger metal frame, shaped to reflect the path of the Sun through the sky throughout the year. Accurately cut grooves in the frame hold strips of blue-green ('Prussian Blue') coloured cardboard, graduated in hours. Because the path of the Sun in the sky varies during the year, the length and shape of the cards varies according to season, and models for both temperate and tropical latitudes are available. When the Sun is shining, solar rays are focused by the glass sphere onto the card, charring it (**Figure 11.9**). (The card is blue-green rather than black so that it chars rather than ignites: the burns are also easier to distinguish against the background.) The card is changed daily and the length of the burns on the card summed to determine the sunshine duration. Analysis of the width of the burn can also provide useful information on the intensity of the solar radiation [269].

The Campbell–Stokes sunshine recorder suffers from several disadvantages, as a result of which it is steadily being supplanted by newer electronic sensors (see below). In no particular order, these are:

- The cards from the instrument must be changed daily, requiring the physical attendance of the observer to do so. The changeover is best accomplished at or just after dusk, which is of course not always convenient: the change can be accomplished at the morning observation provided the changeover time does not vary by more than a few minutes from day-to-day. If the card is not changed daily, or is changed later one morning, each subsequent day's burn will simply 'over-burn' previous traces and the record will be lost, and records from individual days become indistinguishable one from another. The instrument is therefore unsuitable for use at sites where manual observations are not made 365 days per year, locations which are not manned outside normal working hours (including weekends), or at remote sites. The price of the sunshine cards is also a significant ongoing running expense and appears likely to become more so as the number of instruments in use declines. As the instrument does not provide an electrical output, it cannot easily be directly integrated into an automated datalogging system.

Figure 11.9 Campbell–Stokes sunshine recorder card (winter variant), showing the burn produced by a day's sunshine. The daily sunshine duration was 3.6 hours. (Photograph by the author)

- Short periods of strong sunshine – sometimes as little as a few seconds – produce an elongated burn, and in intermittent sunshine such individual short burns tend to overlap, leading to an exaggerated sunshine duration – an hour's burn might actually represent only 35–40 minutes sunshine under certain conditions. This 'overburn' effect has been known since the introduction of the instrument in the 1880s, and more recently has been quantified in comparison records with electronic sensors with the benefit of more precise timing of the start and end of spells of sunshine. Numerous attempts have been made to identify and apply simple corrections to current and historical records [270, 271, 272, 273], but as the relationship between the records of the different types of instrument considered depends also upon the intensity and character of the sunshine itself, such comparisons are problematic and can only be statistical approximations applied to monthly or annual totals, rather than hourly or daily records.

- The actual threshold for burning the card can vary considerably during the course of a day – less likely early and late when the Sun is low in the sky, and whether the card is wet or dry. For example, less sunshine might be recorded after a shower once the card has been wetted than before the shower, when the card was dry and easier to char. A burn is also less likely when the optical transparency of the glass sphere is reduced, for example by frost, dew or snowfall deposited on its surface. Various tests [270, 274] have shown that the 'sunshine burn threshold' for the instrument averages about 170 W/m^2, but that this can vary by a factor of four or more (between 70 and 301 W/m^2, depending on time of day and weather conditions). With dew or rain on the sphere, the burn threshold can exceed 400 W/m^2, and when covered in frost any burn is unlikely.

- In addition, the measurement of burn length itself is quite subjective, particularly on days with broken sunshine and near sunrise and sunset, and the measured duration of sunshine can differ by 10–20 per cent between different analysts. (It is not unknown for spas and coastal locations vying for the position of 'sunniest place in … ' to be 'generous' in their interpretation of Campbell–Stokes sunshine recorder cards.) Detailed guidelines on evaluating burn lengths are given by WMO ([4], section 8.2.3.2), although these are often only incompletely followed, if at all. It is best to have all cards independently analysed by two people if at all possible.

- Even the type and exact colour of the sunshine cards in use can make a significant difference. The WMO implementation of the Interim Reference Sunshine Recorder, IRSR (1962–81) specified the pattern of instrument as used in the UK Met Office, but fitted with cards to the specification of the French state weather service, Météo France. Tests at Kew Observatory in London in 1979/80 [270] showed that an IRSR instrument using the 'French' card specifications recorded 6 per cent less sunshine than one using the 'British' cards. The conclusion drawn was that the 'French' cards had a slightly higher burn threshold.

- Any instrument intended to record sunshine will, of course, require the site's horizon to be clear (or almost clear) of obstacles that could shadow the instrument, and of course the Campbell–Stokes instrument is no different to electronic sensors in this regard. Exposure can often be optimised by installation of the instrument at height, on the top of a building or a mast some way above ground level (instruments mounted well above ground level also tend to suffer less dewfall than those exposed at lower levels). However, suitable locations for a Campbell–Stokes instrument can be limited by the necessity for a safe daily access route to change the cards, whereas smaller and lighter electronic sensors requiring only a cable for remote reading tend to be easier to expose at height, improving the chances of a clearer horizon. Both types of instrument are less sensitive to low solar angles, a typical minimum being 2–3 degrees elevation, but records from a Campbell–Stokes instrument tend to be more affected by deposits of dew or frost than electronic instruments. As a result, all else being equal, sunshine is far more likely to be recorded near sunset, when the air tends to be warmer and drier, than just after dawn, when it is often at its coldest and near saturation. Based upon comparative trials of adjacent heated and unheated instruments, WMO ([4], section 8.2.3.3) assess the average error from this cause in northern European climates as ranging from 1 per cent of the monthly mean in summer to 5–10 per cent of the monthly mean in winter.

- At high latitudes, two instruments are required to cover the full range of solar azimuths at midsummer.

Even in geographies where the Campbell–Stokes has been the standard instrument for decades, it is steadily being replaced by remote-logging-friendly unattended electronic sensors. The UK Met Office dropped the instrument in favour of the Kipp & Zonen CSD sunshine sensor (described below) back in 2003, and the number of sunshine records made with Campbell–Stokes recorders declined after that to become a minority instrument today, retained mainly for overlap or continuity with historical records at long-period sites. (At the time of writing, the UK's sunshine recorder network of 139 sites comprises 54 Campbell–Stokes instruments and 85 Kipp & Zonen sensors, respectively 39 and 61 per cent of the total.) It is surprising, therefore, that at the time of writing the UK Met Office continues to reference its sunshine records to the Campbell–Stokes benchmark (see Box, *Are today's sunshine measurements compatible with those made last year or last century?*).

Electronic sunshine sensors

Sunshine is detected by electronic sunshine sensors as a difference in output between two photosensitive devices, one of which is exposed to direct solar radiation, the other being shaded. When the difference between the two exceeds a given threshold, closely equivalent to shadows being cast, the instrument output changes in a step fashion, an electrical signal which can easily be logged. The duration of sunshine is then simply the length of time the 'sunshine = yes' output is indicated (usually measured in minutes or hundredths of an hour, but easily enough to the nearest second). Such sensors provide a more objective and generally less weather-dependent output, but few detailed tests of relative performance have yet been published to quantify differences between different sensors, or even between same-model sensors on the same site. A few uncertainties remain with regard to the spectral response of different instruments, some of which may lie partly in the near infrared rather than in the visual range.

The Kipp & Zonen CSD sunshine recorder

This sensor (**Figure 11.10**) has been adopted as the 'standard sunshine sensor' within many countries. With increasing automation, and the continuing expense involved in sunshine recorder cards, the business case for this fairly expensive instrument is not too difficult to justify.

The instrument is mounted facing south (in the northern hemisphere) and angled at the site's latitude. It has no moving parts, and for operation uses three photodiodes with specially designed diffusers providing input to an analogue calculation comparing output from the solar beam with that from the sky. From this, internal electronics generate an approximate calculation of direct solar radiation, and the instrument is calibrated to switch between 'low' and 'high' output levels when this reaches the 120 W/m^2 threshold. (The calculated direct irradiance value is also available from the instrument's output channels, although it is not as accurate as that from a dedicated pyrheliometer). The instrument requires a 12 v DC power supply, and has optional internal heaters to evaporate dew, frost and light snowfall from instrument surfaces. The instrument's interior is kept dry by an internal desiccant cartridge: an external humidity indicator changes colour when the desiccant cartridge requires replacement. The requirement for safe access to inspect and occasionally change the desiccant cartridge should be allowed for when siting the instrument.

It is an impressive and reliable instrument, very fast in operation (response time less than a millisecond), capable of responding to short bursts of sunshine lasting only a second or two. This makes it easy to analyse the frequency and duration of spells of sunshine over any given period. The author's policy is to log sunshine duration in seconds exactly as logged, and convert this to hours and minutes in post-processing, but the UK Met Office use a slightly different approach, an algorithm which averages output each minute and counts sunshine only as the total of these minutes over the given threshold, and then adjust these records to emulate a Campbell–Stokes instrument benchmark (see Box, *Are today's sunshine measurements compatible with those made last year or last century?*).

A word of caution, however, as the threshold trigger point of the device does drift over time. Kipp & Zonen's specification is 2 per cent or less drift per year, but

Figure 11.10 Kipp & Zonen CSD electronic sunshine recorder. (Photograph by the author)

even over a short 5 year span a change in calibration point of 10 per cent in either direction will have a significant impact on logged sunshine durations. Recalibration by the manufacturers is expensive and will result in loss of record while the instrument is away unless a spare substitute unit is available. If there was a way in which Kipp & Zonen could offer rapid turnaround and less expensive single-point calibration checks in-country, or offer a swap-out scheme for a pre-calibrated sensor every 3 years or so, this would indeed be close to the 'perfect' automatic sensor.

The Instromet sunshine recorder

Instromet are a supplier of weather instruments based in Norfolk, England. Their simple, reliable, and above all reasonably priced sensor (**Figure 11.11**) has deservedly become the sunshine sensor of choice for most of the amateur and hobbyist weather observing community within the UK and Ireland. The operating principle is similar to the Kipp & Zonen unit described above, in this case comparing the output of eight miniature light-sensitive diodes surrounding a central shadow pillar. A shadow cast by this pillar (i.e. when the strength of the Sun exceeds a given threshold) will result in a reduced signal from one or more of the diodes, which in turn sets up an electrical imbalance within the unit's electronics module, switching output signal from low to high (response time less than 1 s), triggering the datalogger and/or the display unit to register sunshine. The display module accompanying the unit registers and displays cumulative daily sunshine totals, in 0.01 h increments (36 s), but when interfaced to a datalogger the sunshine duration in seconds, and hourly and daily totals thereof, are quickly and easily determined.

The sensor unit is small and light, easy to fix to a mast or rooftop, and relatively undemanding in alignment requirements, needing only to be aligned level and pointing approximately south (in the northern hemisphere). Once in place, it needs little or no maintenance, although if the unit is safely accessible it is advisable occasionally to check the glass dome for bird droppings and the like and give it a wipe from time to time. The sensor head does not seem to be unduly affected by dew or frost deposits, although even a thin covering of snow will result in loss of record.

Figure 11.11 The Instromet electronic sunshine recorder sensor. (Photograph by the author)

Cabling to the instrument consists of four wires – paired 12 v supply and signal, the output signal being processed in the electronics module which is not weather-resistant and so should be sited internally, close to the datalogger or display unit. The device is equally suited to standalone recording or interfacing to a suitable datalogger (both voltage and square-wave pulsed outputs are available): if a datalogger is used, the display unit becomes superfluous. Because the pulsed output is similar to the 'tip' from a tipping-bucket raingauge, for some AWS models it may be possible to substitute the output signal from an Instromet recorder for a tipping-bucket raingauge connection: setting the 'calibration' of the input from (say) 0.2 (mm, tipping-bucket raingauge) to 0.01 (h, sunshine recorder) will allow logging of sunshine duration in the field previously occupied by rainfall data. Note, however, that most prebuilt consumer AWSs do not include a 'spare' pulse counter input, and therefore a separate time- or event-based logger will be required to log output from one of these units. Given due care in installation and attention to connections, one of these instruments should provide many years of reliable records. One of the author's units has worked faultlessly for over two decades at the time of writing.

There are two main limitations to this instrument. As supplied, it is powered by a 12 v supply through a mains transformer. Unless this is backed up, perhaps using the datalogger power supply or an uninterruptible power supply, any loss of mains power will result in loss of record. Any such gap is not immediately obvious, for even when logged the gap is shown only as 'nil sunshine' rather than 'missing data'. Supplementary records (eye observations or logged solar radiation data) are there-fore required to assess whether the Sun was shining while the power was off and thus complete any break in the record.

But by far the most significant drawback regarding the Instromet sensor as supplied is its vague threshold calibration of 'sunshine'. The control unit includes a potentiometer which can quickly and easily be adjusted to vary this detection threshold. This requires only a minor adjustment with a screwdriver on a day when borderline hard-edged shadows are cast through thick cirrus or similar (best to set this once, check it a couple of times on suitable occasions, then leave it alone rather than constantly fiddle with it), but inevitably the threshold setting at different sites will be somewhat subjective. It is surprising that Instromet are unable to offer a laboratory-based calibrated threshold traceable to the WMO definition of sun-shine, perhaps for a small additional fee above the standard instrument pricing, which would ensure that the records from different locations were truly comparable with each other.

There are other types of electronic sunshine recorder available: a wider sum-mary of the field is given in [263].

Comparisons between different models of sunshine recorder

Various models of sunshine recorder each measure 'sunshine' slightly differ-ently. Experience from within the UK and the Netherlands has shown that electronic sunshine sensors tend to record slightly more sunshine than the traditional Campbell–Stokes model during the winter months, about the same on days with long spells of unbroken sunshine, and considerably less during spells of broken sunshine, particularly in summer. A short series of comparative trials of five sunshine sensors (including a Campbell–Stokes unit) were undertaken by the UK Met Office during 1998/99 [275], and it was as a result of these trials that the Kipp & Zonen CSD1 electronic sensor (now the CSD3 model) was eventually adopted by the UK Met Office as its standard.

Numerous side-by-side comparison trials between (loosely) 'traditional' and 'electronic' sensors have been carried out [258, 263, 270, 271, 272, 273, 274, 275, 276], but the results agree only on the conclusion that there is no simple 'conversion factor' to give an 'equivalent value' for one instrument based upon readings from the other, particularly as the 'conversion factors' themselves often vary seasonally owing to differences in the sunshine character at different times of the year. In particular, the greater prevalence of thermal-driven convective cloud in the warmer months, owing to increased levels of solar radiation associated with longer hours of daylight and greater maximum solar elevation, tends to result in more frequent breaks in sunshine than the clearer skies typical of some colder winter days, where short daylength and low solar elevation are less likely to result in convective cloud and the sunshine is less likely to be interrupted. **Table 11.3**, illustrated in **Figure 11.12**, shows monthly and annual average sunshine duration (in hours) as logged over a 3 year overlap period by co-located Kipp & Zonen CSD3 and Instromet sensors, based upon the author's own careful comparisons at a site in southern England. The seasonal pattern is clearly evident. Within this broad picture, hourly and daily relationships are often highly variable as can be seen in **Figure 11.13**. Personal observation has shown that the Kipp & Zonen sensor sometimes registers 'sunshine' on spring and summer days where thick high cloud permits a high level of hazy, diffuse radiation but few if any sharp shadows, whereas the Instromet unit tends to be better at distinguishing shadow contrasts. This performance difference accounts for many of the points plotted below the trendline in **Figure 11.13**, and on some of these days the lower Instromet duration certainly accorded more closely with

Table 11.3 *Side-by-side comparisons of average monthly sunshine duration from co-located Kipp & Zonen and Instromet models of sunshine recorder. Units are hours and percentages as indicated. Based on a careful 3 year overlap dataset by the author at a well-exposed site in southern England; see also Figure 11.12*

Month	Kipp & Zonen CSD3, hours	Instromet, hours	Instromet as per cent Kipp & Zonen
Jan	59.5	60.6	101.9
Feb	84.1	77.7	92.3
Mar	150.2	140.2	93.4
Apr	220.8	209.1	94.7
May	221.7	203.5	91.8
June	201.4	185.6	92.2
July	210.7	187.6	89.1
Aug	178.7	160.7	89.9
Sept	155.3	143.2	92.2
Oct	111.9	102.2	91.4
Nov	76.4	73.1	95.6
Dec	47.1	49.4	105.0
Year	1702.0	1576.4	92.6

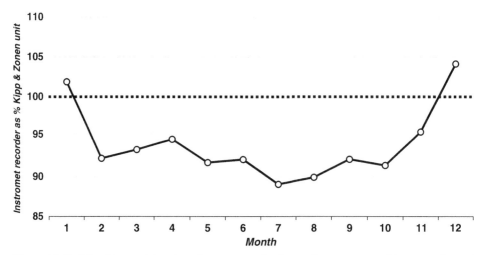

Figure 11.12 Monthly sunshine duration measured using an Instromet sunshine recorder unit exposed side-by-side with a Kipp & Zonen CSD3 unit, expressed as a percentage of the latter. Averaged over a three year period at a site in southern England. See also Table 11.3

a visual perception of 'sunshine'. In contrast, the winter months exhibited fewer outliers (**Figure 11.13**, a). Daily values for one instrument derived from another are often simply unreliable, although more consistent relationships may be possible when considering monthly or seasonal totals.

It is certainly possible to derive empirical monthly mean 'conversion factors' from such comparisons, but because different instruments operate on different principles, attempts to merge two or more sets of historical records in a seamless fashion are unlikely to be successful (see also Box, *Are today's sunshine measurements compatible with those made last year or last century?*). It does seem reasonable to expect that results obtained from different models of electronic sensor based upon similar principles, such as the Kipp & Zonen and the Instromet units described above, might be expected to be closer to each other than the output from the traditional Campbell–Stokes recorder, and that modern sensors are likely to provide more consistent and repeatable results than traditional devices. However, any station authority considering changing from one instrument to another should plan on a substantial overlap period – at least 2–3 years – to avoid irreparable damage to existing record homogeneity. Details of both instruments and their exposure characteristics should be carefully noted in site metadata.

Calibration of solar radiation and sunshine sensors

Pyrheliometers

Checking and calibration of national and international-standard reference pyrheliometers takes place at intercomparisons organised by WMO, which take place every five years at the World Radiation Center (WRC) at Davos in Switzerland [258].

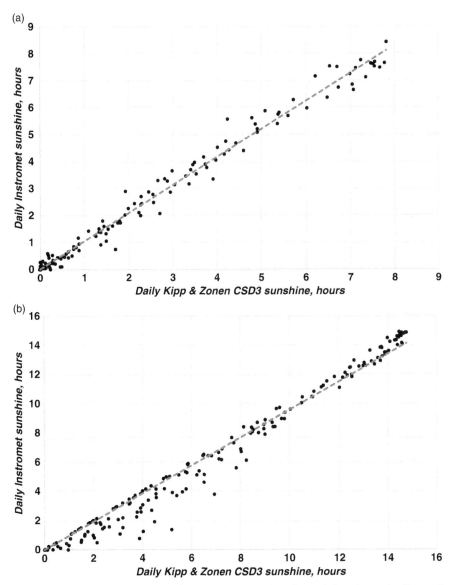

Figure 11.13 Scatterplots of daily sunshine duration (hours) in winter (a, December and January) and summer (b, June and July) over three years at a site in southern England as logged by adjacent Kipp & Zonen CSD3 (*x* axis) and Instromet sunshine sensors (*y* axis). Daily averages were as follows: winter, Kipp & Zonen 1.85 h/day, Instromet 1.93; summer, Kipp & Zonen 6.77 h/day, Instromet 6.27

The WRC ensures worldwide homogeneity of meteorological radiation measurements by maintaining reference instruments which are used to establish the World Radiometric Reference and transferable calibrations.

Are today's sunshine measurements compatible with those made last year – or last century?

The answer depends upon which country's records are being referred to, but the answer is almost certainly 'no'. As an example, let us look at the situation in three different countries – the Netherlands, the United States and the United Kingdom.

In **the Netherlands**, sunshine records were made using Campbell–Stokes recorders until 1992; records using this instrument commenced at De Bilt, near Utrecht, in 1901. In 1992 a new method of determining sunshine from records of global solar radiation was adopted throughout the country's climatological network [264]. Comparisons between old and new methods [277] from overlap observations made at De Bilt during 1993–2000 showed that the new method gave an average of 16 per cent greater sunshine duration than the Campbell–Stokes during the combined winter months (December, January and February) but 7 per cent less during the summer period (June, July and August combined). As summer months are sunnier than winter, average differences almost cancelled out over the year as a whole, although considerable variations remained from year to year depending upon the character of the weather in that year. It is clear, however, that the two periods of record (1901–1992, and since 1992) are not homogeneous. The new approach used by KNMI in Holland does produce an estimate of sunshine duration that is consistent from location to location across the country, together with a consistent dataset of 'sunshine' records from 1992, but at the cost of losing compatibility with those made in any other country. The KNMI method is included in the current WMO CIMO guidelines ([4], section 8.2.2.1), although as a suggested example approach rather than as recommended method, and to the best of my knowledge it has not been adopted elsewhere. Part of the reason may lie in the method ignoring all pyranometer readings below 5.7 degrees solar elevation. In Washington, D.C. (38.9°N) on 21 December, this would disqualify the first 40 minutes of record after sunrise, and the same period before sunset; in Amsterdam (52.4°N), the first and last hour; but in Reykjavik, Iceland (64.1°N), the Sun does not reach this elevation at any time between 23 November and 20 January. Although midwinter days are very short there, Iceland's capital *does* receive some winter sunshine!

In the **United States**, the historical sunshine record was based upon three 'standard' instruments in use at different periods [278] – the Jordan photographic sunshine recorder (in use between 1888 and 1907), the Maring–Marvin thermo-electric sunshine sensor (1893 to the mid 1960s), and the Foster–Foskett sunshine switch, progressively introduced from 1953, retired in 2009/10. (Curiously, very few American sites used the Campbell–Stokes recorder; one notable exception was the Blue Hill Observatory in Massachusetts (see Chapter 1), where sunshine records commenced in 1885, and continue today.) All three instruments differ in their method of recording and their sensitivity and responsiveness to solar radiation. For example, the 'sunshine' threshold of the Maring–Marvin thermoelectric sunshine recorder averaged 255 W/m², while its response time to the appearance or disappearance of the Sun was stated as '5 to 10 minutes' [279]. In contrast, the threshold of the Foster–Foskett sensor averaged just 87 W/m² – one-third of the previous device in use, and subject also to frequent and subjective manual adjustments. Although some analysts had previously asserted that the basic homogeneity of the US historical sunshine record remained intact [280], such optimism was surely misplaced given the enormous disparities in instrument performance. When considered alongside the long-standing documented policy to include manual 'guesstimations' of sunshine duration for low solar angles and missing data due to obstructions or defective record [259, 260, 278], which clearly introduce an additional variable subjective component into the records from each individual site, it is difficult to understand how the US historical sunshine record can be regarded as anything other than seriously flawed – doubts first raised in print as far back as 1985 [281] and 1990 [282]. The decision to drop the Foster–Foskett sunshine switch in 2009 came about mainly because of its continued reliance on manual fine-tuning (thus making integration into today's automated networks almost impossible), but also partly due to the realisation that records from this instrument simply lacked the consistent and repeatable performance necessary to be able to compare records between different sites and within long single-site records. With very few exceptions, 'sunshine duration' is no longer included in US climatological reports or averages.

In the **United Kingdom**, the earliest records made with a prototype Campbell–Stokes recorder date back to 1875/76. A handful of sites have sunshine records made at the same site with the same type of instrument – although almost certainly not the *same* instrument – extending back 100 years or more. Campbell–Stokes sunshine recorder records commenced at the Radcliffe Observatory in Oxford, England, in February 1880 [57] and continue unbroken to this day, the original instrument remaining in use until 1976. The introduction of electronic sunshine sensors was led by the UK Met Office following comparative trials in 1998/99 [275], following which the Kipp & Zonen CSD sunshine recorder was finally adopted as the standard or reference instrument in 2003. Overlap measurements were made at several sites, for a limited period, and the results published [272, 273]. Although no break in record homogeneity is welcome, it seems perverse not to embrace the benefits of the more accurate and consistent records available from modern electronic sensors, as described elsewhere in this chapter. The detrimental effects of any major change of instrument can be minimised by ensuring two or more representative series of overlapping records using both instruments are carried out prior to any change. (The overlap should be at least several years in length for sites with long records made using the older instrument.) Where records from the old and new instruments are unlikely to be truly seamless, as is the case here, it makes sense to adopt the new instrument across all other sites as quickly as possible, ensuring a degree of record overlap wherever feasible, to minimise the resulting period of ambiguity.

The policy of the UK Met Office in this regard is puzzling, however, in that sunshine durations reckoned from the newer instrument are adjusted to 'emulate' the *older* model, despite its known deficiencies, rather than vice versa. As automatic electronic sunshine sensors continue to supersede the traditional instruments, the point has already been reached where the *majority* of records are being adjusted simply to retain consistency with a fast-disappearing *minority*, with all sorts of downstream statistical fudges and confusion. In this case, surely it would be better to accept that the two types of instrument are essentially incompatible, and to move forward on the basis of overlap comparisons at as many sites as possible. The lack of clarity continues to impair the historical sunshine records of the United Kingdom.

Pyranometers

International Standard ISO 9847:2023, *Solar energy – Calibration of field pyranometers by comparison to a reference pyranometer* [283], specifies two preferred methods of calibrating pyranometers: an outdoor calibration method (with the pyranometer in a horizontal position, in a tilted position, or at normal incidence) or an indoor calibration (using an integrating sphere with shaded or unshaded lamp, or at normal incidence). Either method is valid for most types of field pyranometers regardless of the type of sensor, although the WMO CIMO guideline ([4], section 7.3.1) suggests that the indoor method is preferable for regular stability tests of field instruments.

The outdoor method can be briefly described as follows. Calibrations are undertaken using side-by-side comparisons between the records of an instrument of known calibration (pyrheliometer or pyranometer) and the sensor requiring calibration, by comparing logged records over a period of time and under various weather conditions and solar elevations. National meteorological services have instruments of known calibration, themselves calibrated against national or international standard instruments, which can be used as travelling standards for this purpose. Alternatively, the instrument manufacturer may be able to provide, or refer to, a calibration facility which can undertake the instrument calibration or recalibration. Calibration checks are advised at 2 year intervals.

Where no side-by-side calibration facilities or travelling standard are available, approximate calibrations can be obtained by comparing records with a nearby site using an instrument of known calibration, although the errors in doing so obviously increase with distance. Summarised records of solar radiation are published in most countries. Daily and monthly solar radiation totals between sites within a region of similar topography vary less than the equivalent statistics for sunshine duration, and interpolations between sites are often possible, even over considerable distances, provided there are no significant exposure or climate differences between the two regions (a valley site subject to persistent winter fogs would not be a good comparison against an upland location, for example).

Sunshine recorders

The WMO CIMO guide ([4], section 8.5) states unambiguously 'No standardised method to calibrate SD [sunshine duration] detectors is available ... Because of the differences between the design of the SD detectors and the reference instrument, as well as with regard to the natural variability of the measuring conditions, calibration results must be determined by [external] long-term comparisons (some months).'

Side-by-side comparisons with direct pyrheliometers are feasible only at well-equipped national or international observatories. Probably the best ongoing method of checking output is observing the threshold of detection, or loss of signal, at times when the sunshine is borderline – when shadows are becoming distinct or hard-edged. If the recorder is 'on' when no shadows are visible, or when indistinct shadows are evident, or 'off' when distinct shadows are present, the threshold may require adjustment. It is worth repeating this visual check several times over a period of a few weeks after installation, but leaving the setting alone thereafter. Frequent threshold adjustments introduce a subjective element into the record and should be avoided.

Site and exposure requirements

The preferred exposure for any sunshine or solar radiation sensor must obviously be a clear horizon below the level of the instrument throughout the azimuth range of the solar disk at any time of year: in northern temperate latitudes this will be broadly between north-east and south-east, and south-west to north-west. Shading owing to natural relief (hills, mountains) is not taken into account in shading assessments, although this will obviously reduce solar radiation receipts and recorded sunshine duration. Information on how to determine azimuth angles and elevations and making a site plan is given in Chapter 6, *Measuring precipitation*.

Obstructions will cast a shadow on the instrument and thus reduce sunshine duration or solar radiation receipts, although some obstructions may be seasonal in nature (such as deciduous trees) and their impact may vary over the course of a year. A flat roof will often provide a suitable location to locate solar radiation or sunshine sensors, provided safe and secure access is available, but ensure that reflecting surfaces such as windows (which may be below the instrument's horizon) cannot cause specular reflections onto the sensor during daylight at any time of year. Exposing the sensor or sensors on masts may also be acceptable, once again providing safe access can be ensured (see below). Where the horizon is largely clear, the instrument(s) may be mounted in the most open and unobstructed position available on a secure and rigid stand at a convenient height for access, such as a brick or concrete pillar. Where more than one solar instrument is co-located, care should be taken at installation to ensure that one instrument does not cast shadows or permit reflections to fall upon the other.

WMO site classes for sunshine and solar radiation instruments

Recognising that perfect sites are few and far between, the WMO CIMO guide sets out slightly different site classes for solar radiation instruments (**Table 11.4**) and sunshine sensors (**Table 11.5**), as follows. The section following explains how to assess the likely impact of shade on the solar radiation or sunshine duration records from any particular site.

Obstructions just above the instrument's horizon (up to 2–3 degrees elevation) will have negligible impact on the record, as discussed earlier in this chapter. Obstructions greater than about 3 degrees elevation (corresponding to an object 3 metres above the level of the instrument located 50 metres away) will result in some reduction in record, the effect varying with azimuth (compass bearing) and solar elevation (time of day and time of year). Close obstacles must be avoided. Narrow obstructions (less than about 2° in angular width, such as aerials and masts)

Table 11.4 *WMO class definitions for global and diffuse solar radiation sensors. Source: WMO CIMO guide ([4], section 8.3 and Annex 1.D)*

Class	No shade projected onto sensor at solar elevations …	Allowable shading or reflecting obstacles
1	> 5° (> 3° at latitudes ≥ 60°)	None above 5°, and total angular width < 10°
2	> 7° (> 5° at latitudes ≥ 60°)	None above 7°, and total angular width < 20°
3	> 10° (> 7° at latitudes ≥ 60°)	None above 15°, and total angular width < 45°
4	Shading < 30 per cent of daylight hours on any day in the year	
5	Shading > 30 per cent of daylight hours on any day in the year	

Table 11.5 *WMO class definitions for direct solar radiation sensors and sunshine recorders. Source: WMO CIMO guide ([4], section 8.3 and Annex 1.D)*

Class	No shade projected onto sensor at solar elevations …
1	$> 3°$
2	$> 5°$
3	$> 7°$
4	Shading < 30 per cent of daylight hours on any day in the year
5	Shading > 30 per cent of daylight hours on any day in the year

should be avoided where possible but will usually have little significant impact on sunshine or solar radiation measurements. Obstructions to the north of the instrument (in the northern hemisphere – vice versa for the southern hemisphere) will have much less effect than those from the south, providing they do not reflect solar radiation back onto the sensor(s). Any shading from deciduous trees will be reduced in the winter months.

In particular, a *sunshine recorder* requires (as far as is possible) an unobstructed horizon above 3° elevation for the range of azimuths where the Sun rises or sets during the course of the year – in mid-latitudes, roughly between north-east and south-east, and between south-west and north-west. Some obstruction to the south is permissible provided it does not extend above the elevation of the Sun at noon on the winter solstice (which at 50°N is about 17 degrees, and at 60°N 7 degrees). An obstruction in the path of the Sun around the sky will be recorded as 'no sunshine' for the period where the sensor is in shadow. (See *Assessing the impact of shade*, below.)

For *pyranometers*, obstructions will block most or all of the direct radiation but only some of the diffuse radiation from that part of the sky, and the effects of horizon obstructions on the record are therefore generally lessened. Provided that there is reasonable exposure between east and west through south, that there are few significant obstructions in or around the Sun's path in the sky within about 3 hours on either side of local noon, and assuming that the instrument is not badly over-sheltered by obstructions in other parts of the sky (or by other instruments on a mast, for example), daily solar radiation receipts should not be reduced by more than about 10 per cent compared with an unobstructed site – and this is the order of magnitude of instrumental uncertainty in any case. Sites subject to reflections from windows, light-coloured buildings or even a white thermometer screen poleward of the instrument should be avoided for obvious reasons.

Exposure at height

As the preferred exposure for solar radiation sensors or sunshine recorders is 'as open as possible', this often means 'as high as possible'. If the location chosen for the instruments cannot be accessed safely for installation or maintenance, with appropriate equipment such as a ladder or scaffolding tower, then choose another site. **Do not put yourself or others in danger when installing or maintaining meteorological instruments at height**. See also Box, *Important safety considerations for installing and maintaining weather instruments* in Chapter 4, *Site and exposure*.

Site security is a particular issue with solar radiation and sunshine instruments, particularly the glass sphere which is an integral part of the Campbell–Stokes recorder. Some sites within the British Isles suffer repeated thefts of these attractive objects.

Specialist contractors (TV and satellite aerial fitting companies, or local builders) will often be able to quote for installing meteorological sensors on roofs, disused chimneystacks and the like provided the requirements are made clear to them in advance. Alignment and level are critical on solar sensors, so check beforehand – perhaps by temporary installation at ground level – that the fittings to be used will hold the instruments accurately and securely in position, and that they are robust enough to survive strong winds, snowfall, UV exposure and other weather hazards, most of which will have greater impact at height. If arranging for a contractor to fit or maintain instruments which require accurate level or azimuth adjustments, ensure the operator is clear what is required *before* commencing operations – perhaps with a short prior demonstration at ground level. It is much easier to do this than when he or she is at the top of a ladder and cannot easily hear instructions! If the location chosen for the instrument(s) is not readily accessible, it is vital to ensure during installation that cabling is secured and all risk of chafing eliminated. All connections must be made secure and fully weatherproof, and all cables shielded as voltages are very small. Frequent contractor visits to fix minor issues (cables flapping in strong winds, battery replacements on wireless sensors or water ingress into connectors) will quickly become very expensive.

It is advisable to wipe over the dome of a pyranometer or sunshine sensor occasionally, but this may not be possible if the instruments are difficult to access, and in dusty locations dust accumulations on the sensor may affect readings. On the plus side, sensors located at height typically experience less condensation-related obstruction as a result of dewfall than sensors at ground level. It may be difficult to maintain level when instruments such as pyranometers are mounted on tall masts, particularly if the mast sways a little in the wind. Clearly, where the site is inaccessible some records may be lost owing to obstructions which cannot easily be cleared (particularly the accumulation of ice or snow), but **on no account should personal safety risks be taken to reach the instruments in difficult or dangerous weather conditions**.

Assessing the extent of shade, and its impact on the record

The following exercise is particularly valuable to identify in advance whether or not a particular site is suitable for solar radiation or sunshine instruments. If the site is unsuitable, the expense of the instruments can be avoided, or a suitable alternative site sought.

The likely extent of obstruction to a sunshine or solar radiation record can be accurately assessed by measuring local site obstructions and plotting them on to onto a *solar elevation diagram* (**Figure 11.14**). These are available as custom-drawn charts downloadable from solardat.uoregon.edu/SunChartProgram.html. Enter the site latitude, longitude and time zone to produce two site-specific diagrams, one for December to June, the other June to December. Each diagram shows the azimuth and elevation of the Sun at monthly intervals throughout the year; the curved lines crossing the date curves show the time in UTC or other chosen time zone. Thus, from

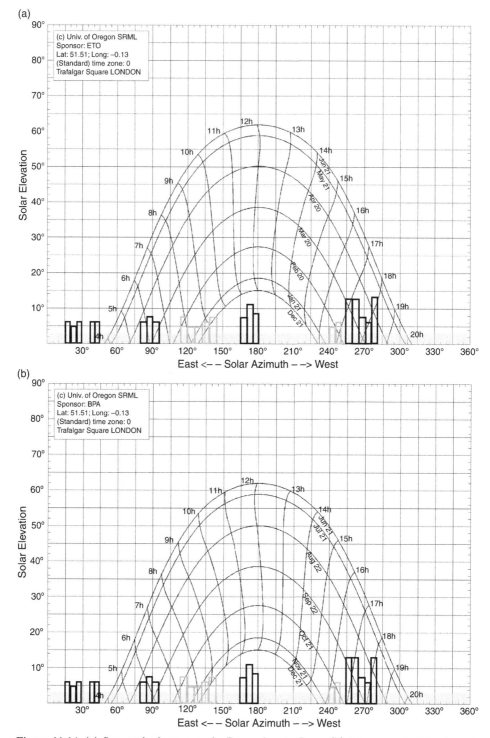

Figure 11.14 (a) Sun path chart, months December to June. (b) The same but for June to December. (Both plots courtesy of University of Oregon Solar Radiation Monitoring Laboratory)
Both plots include a suggestion of how measured elevations of horizon obstructions might be plotted, and from which obstructions to direct solar radiation can be quantified. Solar elevation

Figure 11.14a it can easily be determined that at this site at 1100 UTC on 20 April, the Sun will be 48 degrees above the horizon, and its azimuth 157° True.

Next, take a series of digital photographs from the instrument site (or planned site), and assemble these into two panoramas (there are a number of Internet freeware photographic 'stitch' utilities available which will do this). One should cover north-east to north-west through south, the other west to east through north. Use a good sighting compass to obtain accurate bearings, and mark azimuths at 5 degree intervals on to a printed version of the panorama photograph. Remember to allow for any variation between magnetic (compass) north and true north – www.magnetic-declination.com will provide this information. (Azimuth bearings should always be relative to true north.)

Then, with a clinometer (some compasses include a built-in clinometer), measure the elevation of obstructions above the horizon every 5 degrees around the compass, and mark the readings directly on the photograph. Remember to do this from as near as possible to the site where observations will be made, to ensure the elevations relate to the site and height above ground of the instrument, and not to a nearby ground level.

Caption for Figure 11.14 (cont.)

below 3° is shown lightly shaded: 'solid' obstructions such as buildings are shown edged in black, and 'porous' obstructions such as deciduous trees or other seasonal obstructions are edged in grey. As an example of using the plot to determine the approximate times of shading of the instrument and thus loss of record, consider the points on the plot for 20 March (top plot, a) and 22 August (bottom plot, b) as follows (hours shown as tenths):

		20 March (top plot, a)	*22 August (bottom plot, b)*
Morning	*Sunrise*	*06.1 h*	*05.0 h*
	3° threshold reached	*06.3 h*	*05.4 h*
	End of obstruction	*06.6 h*	*none*
Evening	*Start of obstruction*	*16.8 h*	*17.9–18.3 h only (0.4 h)*
	3° threshold reached	*17.8 h*	*18.8 h*
	Sunset	*18.1 h*	*19.1 h*
Daily values	*Duration sunrise to sunset*	*06.1 to 18.1 h*	*05.0 to 19.1 h*
		= 12.0 hours	*= 14.1 hours*
	Duration 3° to 3° (A)	*06.3 to 17.8 h*	*05.4 to 19.1 h*
		= 11.5 hours	*= 13.7 hours*
	Unobstructed duration (B)	*06.6 to 16.8 h*	*05.4 to 18.8 h less 0.4 h*
		= 10.2 hours	*= 13.0 hours*
	Obstruction (A-B)	*1.3 hours*	*0.4 hours*
	Percentage obstruction (A-B)/A	*1.3/11.5 = 11%*	*0.4/13.7 = 3%*

Repeating this process for each of the dated lines on the plot permits estimates to be made of the possible percentage obscuration throughout the year. The actual loss of record will be less than this, in the same ratio as the sunshine duration is to the maximum possible; for example, at a site where the average March sunshine represents about 25 per cent of possible, the actual loss due to obstruction would be typically 25 per cent of the 11% calculated above, namely 3% (0.25 × 11%); in August, if the average monthly sunshine represented 35 per cent of possible, a typical loss due to obstruction would be nearer 1% (0.35 × 3%). This necessarily assumes a similar ratio of actual to possible sunshine throughout daylight hours, which may not be realistic, but it does provide an objective indication of likely site losses due to horizon obstructions which are otherwise difficult to assess.

Next, mark in the 'skyline' of obstructions as a series of columns on one of the solar elevation diagrams using the elevations measured at each 5 degree point as suggested on **Figure 11.14b** (the azimuth on these plots is marked at 5 degree intervals, and elevation every 2 degrees). It can be helpful to mark 'solid' obstructions with dark shading and seasonal obstructions (such as deciduous trees) with lighter shading or cross-hatching, as this will simplify any required re-measurements when reviewing the assessment every two years.

When this is complete, copy the 'skyline' to the other solar diagram (the months are symmetrical about the solstices, only the hour curves differ between the first half and second half of the year). Using the two diagrams, it is easy to assess the extent of shading at the instrument location, and thence assign a site class using **Table 11.4** or **Table 11.5** as appropriate.

The potential effect of obstructions on the site record can also be quantified from these diagrams. First, carefully estimate the potential duration of obstruction for each date curve throughout the year, to the nearest 0.1 hour. Compare these to the 'realistic maximum possible' sunshine duration for these dates (in this context, the 'realistic maximum possible' duration can be regarded as the length of time the Sun is above 3 degrees elevation on that date curve, as obstructions below about 3 degrees have little effect). Then evaluate the potential loss of record on that date (obstructed hours / realistic maximum possible duration, as a percentage). The actual loss will be less than this because not every hour will be sunny – to a fair approximation, as previously stated it will be proportional to the climatological percentage of maximum possible sunshine in any month. In mid-latitudes, the actual reduction in sunshine records will be *very roughly* one-third of the possible obstruction percentage, and about half that figure for pyranometer records. In middle and high latitudes, or other regions where the monthly mean cloud cover or daylight hours vary significantly during the year, the average annual loss will be weighted towards the sunnier months, which normally experience lower obstruction losses owing to higher solar elevations.

The exercise should be repeated every two years to check on the growth of trees, or if a significant obstruction (a new building, perhaps) appears likely to affect the exposure. The slow growth of trees can insidiously wreck the exposure of solar instruments. Regular sets of panorama photographs taken at the same time in the year, two years apart, are useful to assess these slow skyline changes. Reductions in solar radiation or sunshine duration at the site due to tree growth may otherwise become apparent only after several years comparison with one or more unobstructed nearby sites – by which time the damage has been done, of course.

Should obstruction 'gaps in the record' be filled by estimates?

It is almost impossible to make accurate estimates of any sunshine 'lost' owing to obstructions unless they are very short, and first-hand observations or unobstructed solar radiation records are available. Doing so risks introducing a subjective variable element into the record, which should be avoided: sunshine or solar radiation measurements are best tabulated 'as recorded', with solar elevation/obstruction diagrams and WMO class included in site metadata.

Useful solar geometry sites
Sunrise/sunset times: www.esrl.noaa.gov/gmd/grad/solcalc/
Solar calculator: www.esrl.noaa.gov/gmd/grad/neubrew/SolarCalc.jsp
Solar azimuth/elevation plots:
http://solardat.uoregon.edu/SunChartProgram.html

Numerous smartphone apps can also provide sunrise, sunset and daylight duration data.

Levelling, azimuth and latitude adjustments

It is very important that solar radiation or sunshine instruments be mounted securely and accurately level. Particularly with solar radiation sensors, a tilt of even a few degrees towards or away from the Sun is likely to result in large errors in output measurements. Once installed, the level should be checked at least twice per year and adjusted as necessary. Most instruments have one or more small built-in spirit levels to facilitate regular level checks.

Almost all instruments require accurate azimuth alignments, usually to face due south. Some require very precise alignment, and for this an accurate compass is essential (local variations from magnetic north must also be taken into account). Pyranometers have a 360 degree field of view, and although in theory they are azimuth-independent, in some instruments the sensitive elements of the thermopile will have been calibrated assuming a particular alignment. If in doubt, or if no azimuth alignment is specified, orient the device so that the output cable emerges on the poleward side of the unit away from the noonday Sun.

Campbell–Stokes and Kipp & Zonen sunshine recorders also need to be adjusted for latitude at installation (this needs to be done once only). Ensure the latitude adjustment screw is firmly tightened after setup to avoid later slippage.

Logging requirements

Logging requirements for solar radiation and sunshine sensors vary with both instrument type and the application.

Solar radiation sensors: solar radiation intensity can and does vary very rapidly (**Figures 11.6** and **11.7**). If the requirement is to capture variations in daily solar radiation in fine detail, a high sampling and logging rate (1–5 seconds and 1 minute, or even less, respectively) will be required, assuming the response time of the instrument is fast enough to justify doing so, which may not be the case with budget sensors (**Table 11.2**). For many climatological purposes, hourly and daily means and extremes (perhaps the highest 1 minute or 5 minute mean within the hour) are sufficient for most purposes. The logging interval can be much less frequent than the sampling interval if only hourly or daily means are required.

Electronic sunshine recorders: for most such instruments an hourly logging interval (a sum of all counts or pulses within the hour) is sufficient for climatological purposes, although the unit count resolution (the minimum duration of a sunshine signal) should be 1 minute or less. Where the logger sampling time is 1 second (or less), and the instrument's response time is fast enough, it is easy enough to obtain period sunshine totals to a nominal precision of 1 second.

Calculating daily solar radiation measurements

The intensity of solar radiation (irradiance) is expressed in units of watts per square metre (W/m^2). These units are used for instantaneous values of solar radiation, or averages over short periods (up to about an hour). For periods longer than about an hour, total solar radiation inputs are integrated over time using the unit joules per square metre (or, more usually, megajoules per square metre) – one watt is one joule per second. The daily total solar radiation measure is the integral with time of the day's instantaneous values – easiest to envisage as the area under the curve of sampled solar radiation intensity shown in **Figures 11.6** and **11.7**.

The worked example below shows how to derive a daily total solar radiation from sub-daily records (best performed in a spreadsheet):

		Example
Derive the 24 hour mean solar radiation intensity in W/m2 from logged output	Ensure the full 24 hours are included, not just positive readings in daylight. Note that most	**224.6 W/m^2**
The number of readings does not affect the calculation, although obviously the sampling and logging frequency must be high enough to be representative of the day's conditions, especially on days where it changes rapidly: 1 minute is preferable, 15 minutes the maximum	pyranometers will show a slight negative value during darkness owing to outgoing terrestrial radiation, particularly under clear skies, and for accurate work it is best to set all negative readings to zero in the calculation of daily means	
Multiply by 86,400	The number of seconds in the day	**19,405,440**
Divide by 1,000,000	1 W = 1 J/s, so this factor scales from joules to megajoules to attain a more manageable number	**19.405 440**
	Total daily global solar radiation on a horizontal surface	**19.41 MJ/m^2**
	Alternatively if preferred: divide the values by 3.6 to convert from MJ/m^2 to kilowatt-hours per square metre, kWh/m^2:	
	Total daily global solar radiation on a horizontal surface	**5.39 kWh/m^2**

Time standards and terminal hours

Daily solar radiation totals and sunshine durations are normally quoted for the time zone's civil day, that is, midnight to midnight, excluding any summer time adjustments. More specialised solar radiation applications may require the use of Local Apparent Time (LAT), sometimes known as 'true solar time'. Noon LAT is, by definition, when the Sun reaches its highest elevation over the observing position (sundials and a Campbell–Stokes sunshine recorder show the time in LAT, not 'clock time'). Astronomical tables or Internet-based calculators can show LAT, which can vary by almost 20 minutes on 'local mean time'. A very useful downloadable calculator providing LAT, sunrise and sunset times and hours of daylight (useful for 'percentage of maximum possible' sunshine calculations) for any location (enter latitude, longitude and time zone, and year) can be downloaded from www.esrl.noaa.gov/gmd/grad/solcalc/calcdetails.html – choose 'NOAA_Solar_Calculations_Year' in the spreadsheet format preferred.

It is more difficult to set up a standard datalogger to log using LAT than standard clock hours, as the former varies continuously during the course of a year: it may be easier to log at 1 minute intervals and adjust hourly solar radiation means and extremes to LAT retrospectively by software if required.

More information on terminal hours is given in Chapter 12, *Observing hours and time standards*.

Accuracy versus precision

Excluding gross errors resulting from excess shadowing (class 4 or 5 sites, **Tables 11.4 and 11.5**), incorrect levelling of the instrument or poor calibration, global solar radiation measurements from sites in similar terrain which are reasonably close to each other are more generally comparable than their sunshine durations (see *Calibration of solar radiation and sunshine sensors* above). Daily global solar radiation totals using professional instruments, such as those from Kipp & Zonen and the like, can be expected to be within ±10 per cent of a reference standard (although a linearity drift of 1–2 per cent per year must also be allowed for): less expensive sensors are probably not too dissimilar when new, but performance is likely to degrade more quickly with age.

For reasons stated earlier, it will be apparent that measurements of sunshine duration are less consistent from instrument to instrument and site to site than for many other meteorological variables. Agreement to within 5 per cent (i.e., within 30 minutes on a summer's day with 10 hours sunshine) is probably about as good as can be expected. Although electronic sunshine recorders can generate daily totals to a precision of 1 second, in reality inter-instrument variation means that this level of precision is somewhat academic except in instrument comparison or calibration tests. Otherwise, WMO CIMO guidelines recommend daily sunshine duration totals are given to a precision of 0.1 h (6 minutes). Monthly totals are best quoted to 1 h.

One-minute summary – *Measuring sunshine and solar radiation*

- Radiation from the Sun consists of a wide range of wavelengths, from extreme ultraviolet to the far infrared, peaking in the visible region. Solar radiation is amongst the most variable of all weather elements and consists of two main components – *direct solar radiation* from the solar disk, and *diffuse solar radiation* from the rest of the sky, the latter as a result of the scattering and reflection of the direct beam in its passage through the atmosphere.
- The most common measurements made are of *sunshine duration*, using a sunshine recorder, and/or *global solar radiation on a horizontal surface*, using a pyranometer. 'Sunshine' is defined in terms of the intensity of a perpendicular beam of visible wavelength solar radiation from the solar disk.

The intensity of solar radiation is measured in watts per square metre (W/m^2), and daily totals in megajoules per square metre (MJ/m^2). Daily sunshine durations are measured in hours, to 0.1 h precision, or quoted as a percentage of the maximum possible duration.

- There are numerous models of sunshine recorder. The iconic Campbell–Stokes sunshine recorder has been in use since the late 1870s, although it is steadily being replaced by datalogger-friendly electronic sensors, which give slightly different readings – the Campbell–Stokes unit tending to over-record in broken sunshine. Estimates of sunshine can be derived from pyranometer data, although no method for doing this has yet been shown to provide consistent agreement with dedicated sunshine recorders. Changes in recorder types over time (for instance, the transition from the Campbell–Stokes device to modern electronic sensors using more precisely defined 'sunshine'/'no sunshine' thresholds) mean that today's measurements are not directly comparable with measurements made using different instruments in previous years.

- All solar radiation instruments require an open exposure, one with as clear a horizon as possible: a flat rooftop or a mast are often suitable locations. The WMO CIMO guidelines separately define five site classes for solar radiation and sunshine sensors. The effects of obstructions can be assessed using a solar elevation diagram in conjunction with a site survey, although obstructions within about 3 degrees of the horizon have little effect on the record. The instruments must also be accurately levelled, and most also require some form of azimuth alignment and/or latitude setting.

- Calibrations for solar radiation instruments tend to be based upon field comparisons with reference instruments. WMO organises instrument intercomparisons amongst national meteorological services every 5 years to ensure consistent and transferable measurement standards.

- A high sampling interval is advisable for electronic sensors, as solar radiation is amongst the most variable of all weather elements. The logging interval can be much less frequent than the sampling interval, and hourly totals or means will be sufficient for many applications.

- Sunshine and solar radiation instruments tend to be slightly more variable in their outputs than other meteorological sensors, and even adjacent instruments can be expected to vary somewhat in their readings. For this reason, all but the highest-specification sunshine and solar radiation measurements should be regarded as liable to errors of a few per cent in either direction.

12 Observing hours and time standards

For weather measurements to be comparable between different locations, the time (or times) at which observations are made, and the period covered by the measurements, should be common as far as possible. The World Meteorological Organization in Geneva (WMO) provides guidance on observation times for the main international synoptic observing networks, while 'climatological' observing practice tends to be defined at a country or regional level. It is outside the scope of this book to provide detailed guidance on all aspects of standard climatological observing practice for every country in the world, so this chapter outlines common observing routines, based around a typical once-daily morning observation. Examples based upon current practice within the United States, Australia, UK and Ireland are given where these illustrate generally applicable principles. The importance of common time standards and common time period(s) for once-daily values, such as maximum temperature or total rainfall, are stated, and the meaning, relevance and importance of 'terminal hours' is introduced.

Country-specific details on observing practices, including standard observing hours and 'terminal hours', can be found in the websites or publications of the world's state meteorological services listed on the WMO website [96].

Time standards – Local Time and Coordinated Universal Time (UTC)

By convention, operational weather measurements throughout the world are made to a common time standard – Coordinated Universal Time (UTC). For all meteorological purposes, and within this book, Greenwich Mean Time (GMT) and UTC are effectively equivalent.

'Local time' differs from UTC depending upon longitude, and whether or not clock adjustments for 'summer time' or 'daylight savings time' are in force – for example, Pacific Time in the western United States is UTC minus 8 hours. During the period of summer time – in the northern hemisphere, typically late March to late October – clocks are advanced 1 hour on local regional time. With few exceptions, standard observing hours are based on local regional time without 'summer time' adjustments, so that an observation made at 8 a.m. in the winter months would be made at 9 a.m. during the period of summer time. In the UK, this statement assumes that no changes are made to the long-established practice of adopting GMT (UTC) during the winter months, and adding an hour during the period of Summer Time / Daylight Savings Time (DST). In 2019 the European Parliament officially adopted a European Commission recommendation that DST be scrapped across the EU, but at the time of going to press no further action had been taken to progress the resolution.

Maintaining consistent observations is greatly simplified where one time standard is used throughout the year. For AWSs on automatic download, using clock time will result in the apparent loss of an hour's data when the clocks are put forward in spring, followed by the greater confusion of two sets of data apparently for the same hour when the clocks are put back in autumn. Most AWS system software can be set to 'ignore Daylight Savings Time' clock changes, regardless of whether or not the host computer's system clock is so updated, and thus maintain the observation database in a common time standard throughout. The station metadata (see Chapter 16, *Metadata, what is it and why is it important?*) should make clear which time standard is (or was) in use.

Observing hours

Many countries around the world implement a once-daily morning observation routine, the time of the morning observation being typically between 7 a.m. and 9 a.m., with some degree of flexibility and variation permissible. Many observers find it more convenient to make the observation at the same 'clock time' throughout the year regardless of summer time clock changes – some will be unable to make a 9 a.m. observation at 10 a.m. because of working hours, at least on weekdays, for instance. For many a regular once-daily manual morning observation time around 8 a.m. or a little earlier is the norm.

AWSs will of course perform observations throughout the 24 hours, and for instrumental data it is very easy to use AWS records to maintain an adjusted observation record conforming closely to the 'nominal' standard morning observation time (for example, 0900 UTC in the UK and Ireland), even if it is rarely possible to make manual observations at that time. Adopting the national or regional standard observing hour (or close to it) greatly simplifies comparisons of weather observations with other sites – particularly daily rainfall records. For this reason, a daily morning observation time within an hour or so of 8 a.m. to 9 a.m. wherever possible is greatly preferable to one made at other times of day.

Terminal hours

The once-daily morning observation hour has, in turn, defined the standard 24 hour period over which many 'daily' records, such as maximum temperature and total rainfall, are tabulated and reported. Some other elements, such as sunshine, fall more naturally within the 'civil day', midnight to midnight local regional time. The start and end time of these recording periods are known as the 'terminal hours' of that measurement. The term 'terminal hour' refers to the time of day at which one period of record ends and the next starts – whether this is when rainwater collected in the daily manual raingauge bottle is emptied and read, or when the datalogger clears its memory of the highest and lowest temperatures observed in the previous 24 hours and starts again at the time deemed to be the first minute of the new climatological day.

Elements whose terminal hours are based upon different time spans can (and do) cause occasional confusion and inconsistency, because the date upon which the value was recorded can differ from the one to which it is assigned by convention.

Table 12.1 *Typical terminal hours, by element*

	Terminal hours	Elements
These periods and terminal hours are normally defined by the state weather service for standard climatological observations	**Civil Day** 0000 to 0000 local regional time	Wind speed (means and extremes) Sunshine (daily totals) Global solar radiation (daily totals) Air pressure (daily means and extremes) Mean daily temperatures (derived from sub-daily data) Most 'days with …' element counts (see Chapter 14, *Non-instrumental weather observing*)
	Climatological Day Morning to morning, typically 9 a.m. to 9 a.m. local regional time	Maximum and minimum air temperatures Mean daily temperatures (derived from averaging maximum and minimum temperatures) Daily rainfall totals
	Sunset to 0900 regional	Grass minimum temperatures only
These periods and terminal hours are defined by WMO for the exchange of international observations	**Synoptic Day** Typically 0600 and 1800 UTC	Maximum and minimum air temperatures and 12 hour rainfall totals. The period of time covered by 24 hour maximum and minimum temperatures, and the synoptic hour at which these temperatures are reported, are determined by regional decision ([284], section 12.4.4) Within the UK and Ireland, maximum and minimum air temperatures over the periods 0900–2100 UTC and 2100–0900 UTC are also reported with the 0900 synoptic report

Historical convention has, rather confusingly, left us with four different terminal hour groups, as shown in **Table 12.1**. These vary slightly from country to country: each is covered in more detail below.

'Civil day' terminal hours

The civil day (midnight to midnight local regional time) appears to be the obvious and logical choice for terminal hours – simply because there is no possibility of the date of the occurrence being in doubt. It is also the default period for most AWS software. However, as long as many meteorological measurements (particularly rainfall) continue to be read manually by thousands of volunteer observers once daily, typically between 7 and 9 a.m., it is clearly unrealistic to mandate that these manual observations be made instead at midnight (1 a.m. in summer time).

Many elements are most conveniently assigned to a civil day – daily sunshine and solar radiation totals, for example (**Table 12.1**). For others, the choice of terminal hours makes no systematic difference to monthly means – wind speed and barometric pressure are examples here. For others, particularly temperature and rainfall, long historical convention and the need to retain homogeneity with existing records mandate that existing 'morning to morning' conventions be retained, at least for the foreseeable future. It does make good sense, however, to maintain parallel 'morning to morning' and civil day records for temperature and rainfall where AWS records permit doing so. The logic of the civil day is inescapable. Automated stations can provide civil day data as easily as any other 24 hour period, and as they increasingly assume a majority position in country networks, the civil day standard seems likely over time to replace the existing 'morning to morning' convention for all but rainfall observations.

'Climatological Day' terminal hours: air temperature

The standard period for 'once daily' extreme temperature records is normally 'morning to morning', as close to the national 'standard morning observation time' as is possible. In the UK and Ireland, this is 0900–0900 UTC, or as near as possible to 0900 UTC for sites that cannot observe at that hour; in the United States, typically 7 or 7.30 a.m. regional time.

By convention, the maximum temperature – which is most likely to have occurred the previous afternoon – is entered to the day *prior* to the observation, or 'thrown back', while the minimum temperature – most likely to have occurred on the morning of the observation – is entered to the day of reading. Unfortunately, the weather does not always co-operate with these tidy record-keeping conventions, and maximum and minimum temperatures can occur at any time of day. Particularly in temperate latitudes, this quite often results in maximum or minimum temperatures being credited to a different day to the one on which they actually occurred. The problem is most acute and frequent during the winter months.

The following example is not untypical (**Figure 12.1**):

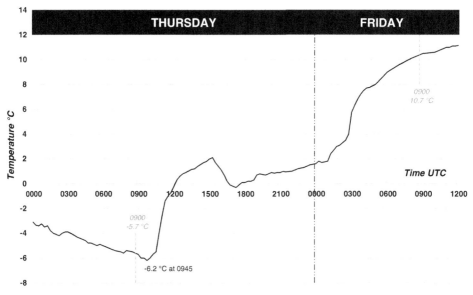

Figure 12.1 An example of the 'day minimum/night maximum' phenomena that occasionally results in bizarre climatological statistics – see text for details.

- Thursday morning was cold, with a heavy frost. The temperature at the 9 a.m. observation was –5.7 °C and still falling, and this was entered as the minimum for Thursday. The temperature continued falling slowly for another 45 minutes, reaching –6.2 °C at 9.45 a.m. before rising once more. Thursday was cold all day, the highest temperature attained during daylight hours being only +2.1 °C, but after an initial fall the temperature began to rise during the late evening and continued to rise all night. At the 9 a.m. observation on Friday the temperature had risen to 10.7 °C.
- The minimum entered to Friday, –6.2 °C, actually occurred at 9.45 a.m. on Thursday. Friday morning was mild, and the night was frost-free, despite a sharp frost being indicated by the 24 hour minimum temperature.
- The maximum entered to Thursday, 10.7 °C, actually occurred at 9 a.m. on Friday. Thursday was a cold day (barely 2 °C), yet the recorded maximum temperature would indicate the opposite.
- Another peculiarity of the method of assigning extremes using the 'morning to morning' convention can happen when the temperature is falling at the morning observation, and continues to fall thereafter. In this case, as on the Thursday example above, the minimum temperature for the date is the temperature at the 9 a.m. observation. However, if the temperature continues falling and does not subsequently reach the 9 a.m. value later in the day, the 9 a.m. temperature also becomes the maximum for the day. In this case the daily range (the difference between the maximum and minimum) will be zero, despite the records from electronic or mechanical recording instruments indicating clearly that this was not so.

During the winter months in temperate maritime latitudes, airmass and wind direction have far more effect on the air temperature than the Sun, and the 'morning to morning' convention can lead to bizarre results, as above. Particularly in unsettled conditions, maximum and minimum temperatures can be recorded at any time of the day or night.

Synoptic terminal hours

WMO operational guidance is for synoptic reporting stations to report maximum and minimum temperatures and rainfall totals covering the previous 12 hour period at the 0600 and 1800 UTC observations. Depending upon longitude and season, however, these may or may not coincide with 'day' or 'night' periods. In western Europe, the '0600 minimum' – the 1800–0600 UTC minimum reported as part of the 0600 UTC synoptic observation – is particularly prone to mislead, simply because for six or seven months of the year the air temperature is still falling at 0600 UTC and the minimum air temperature is not likely to be reached for 2–3 hours afterwards. The '0600 minimum' will therefore almost certainly not be the true minimum for the day. If the next reported minimum temperature is not until the 1800–0600 UTC period the following day and the temperature continues to fall after the 0600 UTC observation – as it very often does during the winter half-year – the true minimum for the day will not be reported at all.

The period of time covered by 24 hour maximum and minimum temperatures, and the synoptic hour at which these temperatures are reported, are determined by regional decision ([284], section 12.4.4).

Day maximum / night minimum

Within the UK and Ireland, synoptic sites usually report maximum and minimum temperatures at 0900 and 2100 UTC. Where sites report 'day' maxima and 'night' minima, the corresponding extremes in the 'other' 12 hour period (i.e., the 'day' minimum and the 'night' maximum) should also be reported, because only the full 24 hour span should be used when preparing climatological averages and extremes. The 0900–2100 maximum may appear to correspond more neatly to perceptions of the 'day maximum' near the Greenwich Meridian, and similarly the 2100–0900 minimum to 'night minimum'. (Although the time zones are specific, the problem is applicable to other time zones too.) It is tempting to regard these 12 hour periods as somehow more 'truthful' than the 24 hour 0900–0900 values, which (because the air minimum is often reached close to 0900 UTC in the winter months) can easily result in one cold morning showing up twice in the records.

This is flawed reasoning, however, because extremes reached in the other 12 hour period of the day are then lost to the record – and this may well include the highest or lowest temperature in any given month, especially in the winter half-year. Any climatological analysis that knowingly discards such events from the archive clearly cannot be regarded as presenting a complete and accurate picture. The phenomenon occurs so frequently in temperate latitudes that its effects are clearly seen in climatological averages and extremes. Observations from locations using 'day/night' terminal hours are not directly comparable with those using the full 24 hour span.

Table 12.2 shows that, over a recent 10 year period, at the author's site in central southern England, the day's maximum was reached outside the 0900–2100 UTC period an average of 31 days per year – 8 per cent of all observations. While these results are obviously specific to a single site, and variations can be expected between different locations and over different periods, they serve to illustrate this important issue.

Although occurrences were most frequent during the winter months (an average 8 days in December, almost twice a week), they did happen in every month of the year. On the occasions when they did occur, the average difference in *maximum temperature* between the 'daytime' 0900–2100 and '24 hour' 0900–0900 UTC periods was 1.2 degrees Celsius, and the largest single difference 7.6 degrees. The cumulative effect was to reduce the 0900–2100 UTC mean maximum slightly compared to the 0900–0900 UTC value, by around 0.1 degrees Celsius over the year as a whole, but by almost 0.3 degC in midwinter (December and January).

The differences are much greater for *minimum temperatures*. Over the same 20 year period, the 24 hour minimum occurred during the 'daytime' period 0900–2100 UTC on an average 55 nights per year (15 per cent of all observations). On these occasions the average difference between the 'night-time' 2100–0900 and '24 hour' 0900–0900 minimum temperature was 1.9 degrees Celsius, the largest single difference occurring one November when a very mild night followed a frosty morning, and the difference between the two amounted to 12.2 degrees. During midwinter (December and January), on typically one night in three the 24 hour minimum temperature did not occur 'overnight' (i.e. between 2100–0900 UTC). The cumulative effect was to increase the annual mean 2100–0900 UTC minimum temperature slightly compared to the 0900–0900 UTC value, by almost 0.3 degrees Celsius over the year as a whole, but by almost a degree in December and January.

Table 12.2 *Differences in daily maximum and minimum temperatures arising from various terminal hours. Data for the author's test site in central southern England, averaged over 10 years to 2022. In this table, + indicates warmer than standard 0900–0900 UTC period, and vice versa. Values in degrees Celsius.*

		J	F	M	A	M	J	J	A	S	O	N	D	Annual
Mean max	09–21 vs 09–09	−0.25	−0.15	−0.05	−0.06	−0.06	−0.04	−0.02	−0.02	−0.01	−0.07	−0.16	−0.29	−0.10
	00–00 vs 09–09	−0.04	−0.04	+0.01	−0.05	−0.05	−0.05	−0.02	−0.08	−0.01	−0.02	−0.05	−0.05	−0.04
Mean min	21–09 vs 09–09	+0.94	+0.50	+0.20	+0.03	+0.02	+0.01	+0.02	+0.02	+0.06	+0.25	+0.58	+0.80	+0.29
	00–00 vs 09–09	−0.06	−0.31	−0.45	−0.50	−0.57	−0.44	−0.44	−0.66	−0.68	−0.80	−0.35	−0.07	−0.45

Average number of days in each month when the temperature extremes differed from the 0900–0900 values:

	J	F	M	A	M	J	J	A	S	O	N	D	Annual
Max attained outside 09–21h	6.2	3.2	2.0	1.3	1.5	1.6	0.7	0.7	0.3	2.0	4.2	7.6	31.3
Min attained outside 21–09h	11.8	7.4	4.9	1.2	0.9	0.8	0.7	1.1	1.6	6.0	8.8	11.0	56.2
Max 00–00h differs from 09–09h	9.0	4.3	2.5	1.4	1.7	1.9	0.7	0.9	0.7	2.6	5.3	9.7	40.7
Min 00–00h differs from 09–09h	20.3	16.1	15.6	9.5	10.4	9.3	8.9	9.9	13.0	17.9	20.1	20.9	171.9

Although differences in monthly means of only a few tenths of a degree may sound insignificant, these are of course comparable to sensor calibration error, record biases resulting from sheltered exposure and urban heat island effects. The difference is stark – *air temperature records from sites using different terminal hours are not directly comparable.*

It is, however, very easy to adjust AWS data to tabulate daily extremes to a standard 'morning to morning' period, either by reprogramming the logger software where possible to do so, or by post-processing logged data using a spreadsheet template. This applies even where manual observations cannot conveniently be made at (or close to) the 'preferred' time of the morning terminal hour. See also the section in Chapter 14, *Non-instrumental weather observing* titled *Observing at set times.*

Comparisons between 0900–0900 UTC and 0000–0000 UTC terminal periods are even more difficult to generalise (**Table 12.2**). Over the same 20 year period at the same site, the 'civil day' 0000–0000 UTC maximum differed from the 'conventional climate day' 0900–0900 value on an average of 39 days annually (11 per cent of all observations), while the 0000–0000 UTC minimum differed from the 0900–0900 UTC value on very nearly one day in two (average 173 days per year). The effects on the mean maximum and minimum compared with 0900–0900 are much more variable, the largest effects on mean minimum occurring during the summer half of the year.

The logic for the adoption of the civil day as the reckoning period for extremes appears inescapable, and in doing so finally removing ambiguities regarding the dates of extremes and ensuring consistency with other elements already tabulated in this fashion. However, 'morning to morning' values will continue to be required for comparison with historical records for some time to come, particularly at sites where instruments continue to be read manually once daily at the nominal morning observation time.

The best advice for AWS records is to 'parallel log' daily maximum and minimum temperatures under both 'morning to morning' national standards and the 'midnight to midnight' civil day conventions wherever possible. That way, should national policies eventually change, both sets of records will already exist. Overlapping periods of record will ensure average differences between the two methods can be quickly and easily determined for any site (similar to **Table 12.2**) or site network. Any required adjustments to existing site-specific long-term records or averages can then be made.

Effects on mean temperature

By convention, the 'mean temperature' for any period (whether a day or a year) is defined as the mean of the daily (24 hour) maximum and minimum temperatures. Of course, the true daily mean temperature can be more reliably calculated from the average of the much greater number of hourly (or more frequent) observations where these are readily available from an AWS. Over time the 'civil day mean temperature' derived from higher-frequency automated observations is likely to replace the 'mean of the 24 hour maximum and minimum temperatures' as the preferred measure. However, where daily mean temperatures for any site are quoted or compared, the derivation should be clearly stated in the station metadata (Chapter 16, *Metadata, what is it and why is it important?*) to avoid possible confusion.

Grass minimum terminal hours

The only exception to the standard 'morning to morning' guideline for minimum temperatures concerns grass minimum temperatures. As covered in Chapter 10, *Measuring grass and earth temperatures*, WMO's CIMO guidelines ([4], section 2.3.2.2) specify that, where measured, the grass minimum temperature (and similarly other surface minimum temperature observations) should refer instead to the period from just before sunset to the following morning's observation terminal hour.

Precipitation

As with other elements, in most countries once-daily rainfall observations are made at a morning observation between 7 a.m. and 9 a.m. In the United Kingdom and Republic of Ireland the standard is 0900 UTC, the practice of reading the instruments at 9 a.m. first codified in the 1860s. There are very sound reasons here for retaining a morning observation time. Midnight is undoubtedly a more logical choice for AWS sites, avoiding any possible ambiguity about the date on which the rain actually fell, but the majority of the world's manually read daily raingauges are still read at a morning observation. A change in reading time to midnight (1 a.m. in summer time) would certainly meet with a less than enthusiastic response from thousands of rainfall observers, at least until the majority of the raingauge network is automated, or almost all manual gauges are paired alongside automated loggers.

The standard period for daily rainfall measurements in the United Kingdom and Republic of Ireland, 0900–0900 UTC, is known as the 'rainfall day' [147]. By convention, the rainfall measured at 0900 UTC is 'thrown back' (credited) to the *previous day* – the rationale being that 15 of the 24 hours lay in the previous day, compared with nine on the day of measurement. This applies even if one knows from personal observation that all the rainfall measured at 0900 fell in the 5 minutes preceding the observation. Similar conventions are widely observed in most other countries.

Daily rainfall observations between different sites become increasingly divergent the further apart the observing hours, and for comparison purposes a morning observation time is much to be preferred over, for example, a gauge read at 6 p.m. daily. Daily and perhaps monthly rainfall totals from two sites using different terminal hours are unlikely to show close agreement, even if the site and instruments are both standard in all other respects. An AWS which logs a TBR (or similar) data does enable this problem to be overcome very easily – simply use AWS period rainfall totals to adjust checkgauge readings made at other times to conform to the national standard morning observation time.

UK terminal hours survey – who uses what?

At the time of writing, 96 per cent of members in the largest community of 'amateur' observers in the UK and Ireland, the Climatological Observers Link (www.colweather.org.uk), adopted a 24 hour period for reporting daily climatological data, the majority a 'morning to morning' terminal hour policy [285]:

Morning terminal hours – 24 hour extremes 82 per cent
Midnight terminal hours – 24 hour extremes 14 per cent
Other terminal hours, or not 24 hour extremes 4 per cent

In the ten years since the first edition of this book, the number of 'midnight to midnight – 24 hour extreme' datasets has increased from 9 per cent to 14 per cent, reflecting no doubt the increase in use of AWSs, but the 'morning to morning – 24 hour extremes' has hardly changed.

For many amateur observers in the UK and Ireland, the 'morning observation' is often somewhat earlier than the UK standard 0900 UTC owing to employment commitments, and a typical morning observation time might be 8 a.m. clock time rather than 0900 UTC (9 a.m. in winter but 10 a.m. during summer time). Of course, using logged AWS data to adjust manual observations made at times other than 0900 UTC greatly simplifies adherence to the 0900 UTC standard, and thus improves comparability with other observations made to the same protocol.

One-minute summary – *Observing hours and time standards*

- By convention, weather measurements throughout the world are made to a common time standard – Coordinated Universal Time (UTC). For all practical meteorological purposes, UTC is identical to Greenwich Mean Time (GMT).
- For weather measurements to be comparable between different locations, the time(s) at which observations are made, and the period covered by the measurements, should be the same. WMO provides guidance on observation times for the main international synoptic observing networks, while observing practices for other station networks tend to be defined at a country or regional level.
- Many if not most countries around the world have adopted a once-daily morning observation as standard practice, the time typically between 7 a.m. and 9 a.m. Where AWS data are available, it is straightforward to adjust records to conform more closely to the 'nominal' standard morning observation time, even if it is perhaps inconvenient to make manual observations at that hour. Adopting the standard observing time (or close to it) greatly simplifies comparisons of weather observations with other sites – particularly daily rainfall records.
- The start and end times of these recording periods are known as the 'terminal hours' of that measurement. The term 'terminal hour' refers to the time of day at which one 'observation day' ends and the following one commences.
- The once-daily morning observation naturally establishes a standard 24 hour period over which many 'once-daily' values are tabulated, such as daily maximum and minimum air temperatures. However, WMO guidance is that the grass minimum temperature should refer instead to the period from just before sunset to the following morning observation terminal hour. Some other elements, such as sunshine, fall more naturally within the 'civil day' (midnight to midnight local regional time). Professional synoptic reporting sites will use observing and reporting times defined by WMO at regional levels.
- By convention, 24 hour minimum temperatures read or logged at the standard morning hour are entered to the day on which they were read, whereas the 24 hour maximum temperature and total rainfall are entered to the day *prior* to the observation (they are said to be 'thrown back'). Although this occasionally leads to some bizarre anomalies, a 'civil day' (midnight-to-midnight) record

period would be difficult to introduce at sites where instruments are read manually (particularly at rainfall-only locations).

- Terminal hours based around 'day maximum' and 'night minimum' temperatures (where the extremes span only 12 hour periods) will give results which are incompatible with '24 hour' sites to a greater or lesser extent, particularly in temperate latitudes in the winter months, and accordingly should be avoided.

13 AWS data flows, display and storage

A modern automatic weather station (AWS) is a sophisticated collection of various components, sensors and electronics modules tied together by software, together making up a data acquisition and processing system which greatly simplifies the collection and processing of almost any type of meteorological or environmental data. Such systems carry out a wide variety of tasks and activities, most of which remain transparent to the user until the final 'display output' stage, which may be on a local host computer or on a smartphone app from half a world away.

There are a very large number of AWS models and related capabilities available today, and of course manufacturers, systems, capabilities and prices can be expected to change over the lifetime of this book. Although detailed and model-specific technical descriptions or specifications quickly become out of date, many of today's and tomorrow's products follow a broadly similar set of basic processes, and a high-level understanding of what these are and why they do what they do can be helpful, regardless of price point. This chapter sets out to explain these basic processing steps within an AWS, keeping technical terminology to a minimum, and illustrating three different approaches to 'system architecture'. With the oversight provided by this chapter, the reader can become familiar with the key concepts, system approaches and application types, and from there review potential products and suppliers using Internet search facilities to ensure up-to-date product information. The chapter closes with the usual 'One minute summary'.

First things first ... power supply, and host computer

Before delving into the workings of the system itself, it is helpful to consider two important external factors – namely, the power supply and the computing platform.

Power supply

Most AWS systems are powered directly or indirectly from a mains/utility supply, typically a transformed low-voltage feed, most often 5–12 volts DC. To allow for outages to utility power, a sensible approach is to power the system from a battery or batteries wherever possible, batteries which are themselves kept charged from a mains/utility power supply or from renewable sources such as solar panels or a wind turbine. Where renewable sources are used, their battery backup facility should allow for continuous operation in the longest expected cloudy or calm spell, while retaining a healthy margin in reserve. (Solar panels will also be largely useless under snow cover, or riming.) Systems relying on mains power alone will suffer loss of record for the duration of any power failure; voltage spikes from nearby electrical storms can also result in corruption or loss of record, and a backup power supply is a must. An uninterruptible power supply (UPS) is a good investment, not only keeping any host computer or display console operating for a time, but suitably configured it will maintain other related infrastructure including powered sensors and components, the host computer or AWS console, and wireless and Internet connections. Power supply interruptions are themselves most likely during spells of adverse weather, and battery backup should be capable of providing at least 48 hours cover. There are few things more frustrating than to lose the AWS record during a period of severe weather, but provided there is both adequate memory capacity and backup batteries to continue to operate the main system components (including the Internet connection where data storage is devolved to 'the cloud'), records from the event should be safe until they can be uploaded to permanent storage once power is restored. Ensure backup batteries are recharged or replaced immediately following any prolonged power outage, as they may not otherwise be sufficiently charged if a second outage follows in quick succession. Such cautions apply particularly to remote sites, where access may become difficult or impossible after a spell of severe weather, such as a heavy snowfall or damaging flood.

If the AWS or datalogger is permanently connected to a host computer via USB or the like, check beforehand whether a power-saving 'sleep' mode on the host will suspend USB port connections or disk access, and if so, disable sleep mode for those components to avoid frequent and prolonged 'dropouts'.

Computing platform

The choice of 'host' computing platform is less relevant than it once was, and for consumer-focused products all three common platforms (namely Microsoft Windows, Apple MacOS or Linux) are widely supported, albeit with a few functionality differences between operating systems. For many low-end and budget systems the main data access/display route is increasingly through a standard web browser in any case (modern smartphones or tablets, whichever operating system is used, tend to be used as display devices rather than as the primary data handling/management computer). Most professional systems run on Microsoft Windows PCs or laptops, as does much legacy AWS software, including Davis Instruments' original (and now obsolete, but still widely used) Weatherlink for Windows datalogger product.

Modern single board computers, such as a Raspberry Pi typically running some flavour of Linux, are capable computers in their own right, possessing ample processing power for most weather software, especially when used as a dedicated weather computer. Some AWS software is available as 'open source' or 'build your own' for those comfortable with building a bespoke system from scratch, and there are numerous websites offering tips and support to get started. There is an important distinction here between devices with a full microprocessor (CPU) and local storage offering much more powerful built-in computing facilities, such as Windows, Apple or Raspberry Pi computers and later Davis Instruments consoles, and most display-only consoles which are typically driven by microcontrollers and dedicated firmware.

AWS processes

The main processes carried out by any AWS system are summarised in **Table 13.1**. This table can be considered as a top-to-bottom recursive flowchart, although the actual implementation varies by model and manufacturer, as subsequent examples make clear. Exactly where each of these processes occurs within a given multi-process station can vary considerably depending on the data architecture of the unit.

Table 13.1 *The key processes carried out by any AWS system*

System	Setup, configuration and installation on host computer system or network System resource monitoring, memory use, sampling and logging timing, etc.
Sensor management and sampling	Interface and communicate with sensors – via cabled connections or wireless links Poll current conditions at defined intervalsConvert raw sensor signals (see Data types, Table 13.2) into appropriate measurement units, perform range quality control checks, and time stamp data
Data handling architecture	Process sensor data (sums, averages, apply calibrations, extract highest/lowest values, and the like)
	Package current timestep data into a data package
	Store (log) timestamped measurements data package in working memory at predefined intervals
	Store timestamped logged data in working memory internal database
	Apply any additional processing, such as evaluation of period averages, totals or extremes and the like
	Store sampled and processed data in system memory ready for uploading to permanent storage (the 'datalogger' function)
Communications	Communicate timestamped data to storage location for subsequent display, upload, etc.
	Convert internal/console data interface into standard computer interfaceprotocol such as USB, Ethernet or Wi-Fi
Display *(if required)*	Display current and recent data to a console or computer monitor in tabular and/ or graphical format, whether from datafiles uploaded to host computer or via web browser from remote storage option (some third-party software options also available)
Permanent storage	Upload collected and processed timestamped data files to permanent storage (whether local or remote/cloud storage) at predefined intervals, or on request, for permanent archiving
Access and export	Access stored current and archive data for viewing on a computer screen or via a web browser
	Export facilities for additional processing, storage or display in other data handling applications, such as databases or spreadsheets

Data types

Most meteorological sensors generate one of the types of output signal shown in **Table 13.2**; a few can provide two or more alternative options. There are other more specialised data formats or communication protocols for particular sensors or applications, covered later in this chapter. Pre-built systems, such as entry-level to mid-range systems, will already be built and programmed to suit the sensors included, but if

Table 13.2 *Most common datalogger input types*

Sensor input type	Method of operation	Typical measurements
Analogue	The datalogger reads a voltage generated by the sensor, or the capacitance/resistance/voltage drop across a sensor, one whose properties vary with the element being measured. This is then converted into a digital value using a analogue-to-digital (A/D) converter, which may be an integral part of the sensor or a separate component. This value is itself then subsequently converted into appropriate meteorological units using an internal conversion/calibration algorithm.	Most continuous output variables – temperature, humidity, barometric pressure, wind direction and solar radiation sensors are analogue.
Analogue-to-digital (A/D) conversion	A/D precision varies; on older or budget-level systems it may be 8 bit, on modern high-spec dataloggers 24 bit A/D is used.	An 8 bit A/D conversion can distinguish 2^8 or 256 discrete values in an analogue signal; 13 bit 2^{13} or 8192 values; 16 bit 2^{16} or 65,536 values; and 24 bit 2^{24} or 16,777,216 values.
Switch or pulse	The sensor generates a pulse output, which is detected and counted by the datalogger. The count is then converted into meteorological units using appropriate internal algorithms or conversion or calibration factors – such as every tip of a 0.2 mm capacity tipping-bucket raingauge accumulating 0.2 mm precipitation.	Wind speed, sunshine and TBR sensors are usually pulsed outputs.
Binary code	The position of a sensor, such as a wind vane, is defined by a binary code (the 'Gray code') driven by a contactless optical coding mechanism.	Some wind vanes.

planning an AWS with a range of sensors built around a dedicated datalogger (as described in the following section), it is important to ensure that the required number of available inputs, and their types or required communications protocol if necessary, can be accommodated on the intended device.

Two important but often overlooked 'data type' considerations relate to date formats and time zone. Date conventions differ between North American (month-day-year, m-d-y) and the rest of the world (typically day-month-year, d-m-y, but sometimes year-month-day, y-m-d). A date such as 2 November 2028, for example, would be written as 11.2.28 in the United States and Canada (only), but as 2.11.28 within Europe, Australia and New Zealand. Within AWS software it is important to check which date convention is in use, and whether display and output modes can be altered to suit regional preferences. Without checking, 'American' 2 November will be transposed to 11 February in 'Rest of world', with entirely predictable downstream consequences!

Equally important are the related settings for time zone and particularly the handling of Daylight Savings Time/Summer Time and its various iterations world-wide, and most importantly whether the changeover is automated or not. There is more on Summer Time in Chapter 12, *Observing hours and time standards*.

Alternative system architectures

It is always difficult to set out hard and fast distinctions between various AWS categories, but at the time of writing three main 'system architecture' approaches to packaging and delivering the processes outlined in **Table 13.1** are evident, and these are outlined in **Figure 13.1**. Of course, some products blur these distinctions, and no doubt further variations of system design will appear during the lifetime of this book. Whichever approach is preferred, careful evaluation of system architecture is important in matching any particular AWS brand or model capabilities to a particular requirement, now or in the future.

The 'black box'

The first, **Figure 13.1a**, amounts to an AWS 'black box' connected to one or more sensors, perhaps via an intermediate external electronics controller such as the Davis Instruments Integrated Sensor Suite (ISS). The functions and processes in **Table 13.1** are carried out in firmware within the AWS console 'black box', including a 'virtual'

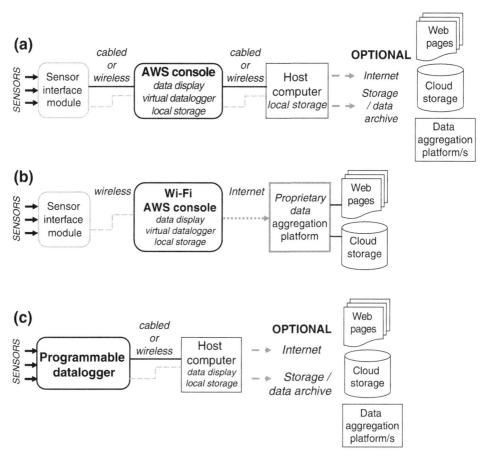

Figure 13.1 Illustrations of the three main types of AWS system architecture; see text for details.

datalogger which may be little more than a dedicated memory chip. A host computer regularly uploads data for post-processing and display (if required) and local storage, using manufacturer or third-party software running on the host. Records can then be stored on the host drive(s), exported to other software (databases, spreadsheets), or copied to individual websites, cloud storage or data aggregation service (for remote access via a web browser) as required. This approach is most typical of entry-level and budget systems (see Chapter 2, *Choosing a weather station* for more details), where sensor range, specifications and sampling/logging programming are usually pre-determined by the manufacturer, although some offer limited expansion options. This type of approach also works best for standalone products such as the Nielsen-Kellerman Kestrel portable AWSs, covered in Chapter 3, *Buying a weather station*, and was for many years the traditional approach taken by Davis Instruments with their Weatherlink for Windows software for mid-range AWS products: current Davis Instruments now use a completely different architecture, as discussed below.

The advantages of this approach are simplicity and ease of installation – 'plug and play', no programming involved. The disadvantages are limited (or no) customisation in sampling and logging intervals and limited choice of sensors, and possibly limited onboard memory. Third-party software, such as CumulusMX, Weather Display and weewx, is available for most popular models in this category as an alternative display option to the original equipment manufacturer's bundled option.

The 'black box' combined with Internet and cloud storage

The second option is essentially development and extension of the 'black box' AWS model coupled with access to an always-on Internet connection (**Figure 13.1b**). In such systems, sensor interfacing and basic data handling/datalogging processes are managed within a wireless unit (which may double as the display console), and this device uploads processed data packets to a remote data aggregation service run by the manufacturer. Such wireless units or data consoles can, in many cases, handle more than one set of sensors, provided of course they remain within wireless range. Post-processing and storage are managed 'in the cloud'. There is therefore much less need for the traditional host computer, because data can be viewed locally if there is a data console, and perhaps more usefully from the data aggregation site using a web browser, although of course the choice is dictated more by personal preference than mandated by technical necessity.

This type of system design is used by current Davis Instruments products, together with NetAtmo and a few other companies. The advantages of this approach are similar to the 'black box' method (simplicity and ease of installation) combined with fuss-free near-instant access to current data from the manufacturer's webpages via a web browser on smartphone, tablet or laptop from anywhere in the world with an Internet connection. Probably the best example here is Davis Instruments' weatherlink.com platform, which displays data from 'hundreds of thousands' of sites worldwide (see also Chapter 19, *Sharing your observations*).

The disadvantages are also similar – limited (or no) customisation in sampling and logging intervals and limited choice of sensors: third-party software is also less easily integrated with this approach. In addition, most such options involve subscription charges, while the legal ownership of the uploaded data is something of a grey area. If you decide to upgrade to a Brand X AWS in the future, will it be possible to download/export the last few years of Brand Y data from Brand Y's site to continue your record?

The programmable datalogger

The final systems approach, **Figure 13.1c**, is one built around a (user-)programmable datalogger.[*] Here, sensors are connected directly to the datalogger (or sometimes via an intermediate device, such as a multiplexer to expand the number of sensors that can be handled), and the datalogger runs all sensor interface and data handling processes, together with communications with the host computer as and when required. The latter may be permanently connected (with a real-time monitor display option, for example), or connected only occasionally for data upload – such as via a laptop or mobile phone/satellite communications network at remote sites. Records can then be stored on the host drive(s) or network, exported to other software (typically databases or spreadsheets with delimited text or CSV comma separated variable file import facilities), or copied to individual websites, cloud storage or data aggregation services allowing remote access to permitted users via a web browser as required. Sophisticated cloud storage options and data aggregation facilities are also available, usually on a subscription basis.

This design approach is most often used in professional systems from the likes of Campbell Scientific and Lambrecht meteo. It offers the advantage of almost unlimited flexibility – almost any electrical sensor can be accommodated in dataloggers, with similarly flexible programming options. There are a large number of datalogger products on the market, and while most have the capability to log standard sensors, there are fewer offerings with 'off the shelf' code for meteorological purposes, such as calculating relative humidity and dew point from the readings of a dry and wet bulb PRT psychrometer, or determining vector mean winds over a period of time. Logger programming flexibility extends to sampling and logging intervals, such that (for example) air temperature and relative humidity could be sampled every few seconds and averages logged every minute, while elements which change more slowly, such as earth temperatures, can instead be sampled and logged alongside other data elements only once per hour or less, thereby reducing the size of datafiles. Advanced capabilities such as event-based logging (see Box, *Event-based logging*) may also be supported.

There are some disadvantages to the programmable datalogger approach, mainly in respect of programming requirements. Bespoke logger code is straightforward enough for anyone with a little coding experience, or can be supplied within a contract by the logger manufacturer (and is in any case a one-off task unless additional sensors are added or requirements change). Logger code can easily be replicated across multiple sites for networked rollouts. A little more care is also needed to set up physical sensor connections, to observe maximum permitted cable lengths and current ratings, to ensure secure terminal block connections, observe correct polarity or overvoltage limits to or from the sensor and suchlike, all of which will be set out within the datalogger operating manual. Some sensors need only a simple two-wire cable connection to the datalogger, others may require more complicated wiring arrangements. Some dataloggers support communication protocols permitting more than one instrument to be connected to a particular port; you may come across references to such protocols in product datasheets, so some of the most common terms are explained in **Table 13.2**. The use of such protocols can greatly simplify wiring requirements while at the same time reducing processor and memory loads.

---

[*] As distinct from pre-built and pre-programmed datalogging components which cannot be altered from factory setup.

High-end dataloggers and their operating software (such as Campbell Scientific's LoggerNet suite) are expensive, although second-user components do appear occasionally on eBay and similar sites. However, both hardware and software components are extremely capable and robust, with tens of thousands of systems installed worldwide in every climate imaginable, and a 20 year working life is not unrealistic.

Event-based datalogging

There are many types of datalogger on the market, but for meteorological purposes there are two main categories, namely *time-based* and *event-based* loggers. The first is by far the most commonly used in meteorological monitoring systems, but the second type opens an entirely new class of measurement possibilities, particularly for some sensor types – rainfall is the usual example, but it can usefully apply to other elements too.

Time-based loggers, as the name implies, sample or 'poll' one or more sensors at a specified sampling interval, then log data at a particular logging interval (the 'archive interval' on Davis Instruments AWSs: see Chapter 2, *Choosing a weather station* for more on sampling and logging intervals). Most loggers allow user choice of the *logging* interval (typical values are between 1 minute and 1 hour), although generally only the more advanced programmable loggers will allow user selection of *sampling* intervals, which may vary from sub-second to minutes, or multiple logging intervals (such as 1 minute, *and* hourly *and* daily).

In contrast, *event-based loggers* initiate a log sequence only when a predetermined trigger event takes place. A typical use is logging output from a tipping-bucket raingauge (TBR), discussed with an example in Chapter 6, *Measuring precipitation*. The logger device itself may be a small, dedicated battery-powered unit where logged data are periodically uploaded to a host computer or laptop, or a programmable logger capable of combining both time-based and event-based methods within one device. Either can work alongside other logging infrastructure, but if the same device is logged by two independent devices, care should be taken to ensure timekeeping remains consistent on both items of equipment to avoid any subsequent ambiguity.

As described in Chapter 6, precipitation measurements are undoubtedly the main application for event-based logging, but there are many others, including (for example) the onset and cessation of particular events, perhaps the duration of air temperatures above or below a particular level, or flagging system-relevant episodes such as critically low solar panel power. A real example is given in **Table 13.3**, showing a small sample of wind speed output during a severe gale in southern England. On this occasion the logger was programmed to log a line of data every time the 3 s mean wind speed (gust) reached or exceeded 20 m/s (39 knots), together with the instantaneous wind direction and 10 minute mean wind speed at that point. To minimise duplication, only the highest entry in each 3 second gust period has been shown in the sample output shown. Using such an approach ensures critical data, in this case high wind gusts, are logged only when they occur – which may be at long recurrence intervals. At the same time, important details of gust speed, wind direction and exact occurrence time to the nearest second, are easily and precisely determined.

Table 13.3 *Sample event-based logging of highest gust wind speeds (≥ 20 m/s)*
during a severe gale in central southern England: for details see source notes [286].

Date and time of gust, UTC	Gust speed (m/s)	Wind direction (deg True)	Spot 10 min mean wind speed (m/s)
18 Feb 10:52:37	24.5	219	10.76
18 Feb 10:53:02	22.1	216	11.13
18 Feb 10:54:54	24.5	216	11.34
18 Feb 10:55:05	23.3	219	11.63
18 Feb 10:58:15	21.9	215	12.51
18 Feb 10:59:07	21.3	218	12.73
18 Feb 11:00:02	21.3	212	13.11
18 Feb 11:01:00	28.1	217	14.20
18 Feb 11:02:36	27.1	217	14.67
18 Feb 11:03:16	22.2	217	14.88
18 Feb 11:05:02	23.0	222	14.74
18 Feb 11:07:44	21.5	220	14.48
18 Feb 11:08:00	21.6	214	14.54
18 Feb 11:09:18	20.9	217	14.32

Table 13.4 *Common datalogger communication protocols*

SDI–12	SDI–12 (Serial Digital Interface at 1200 baud) is an asynchronous serial communications protocol for intelligent sensors that monitor environment data. These instruments are typically low-power (12 v), are used at remote locations, and usually communicate with a datalogger or similar data acquisition device. The protocol follows a client-server configuration whereby a data logger (SDI–12 recorder) requests data from the intelligent sensors (SDI–12 sensors), each identified with a unique address.
SDM	The SDM port and communication protocol allows the connection to a datalogger of individual modules or sensors through an addressing scheme; multiple SDMs (in any combination) can be connected to one datalogger. Communications are serial using RS–232, RS–485 or RS–422 signals.
SPI	Serial Peripheral Interface (SPI) is a *de facto* standard synchronous serial communication interface specification used for short-distance communication, sometimes referred to as a four-wire serial bus. SPI devices communicate in full duplex mode using a client-server architecture, and multiple client devices may be supported.
Modbus	Modbus is an older *de facto* standard communication protocol, commonly used to connect electronic devices; it is openly published and royalty-free, relatively easy to deploy and maintain compared to other standards, and places few restrictions on the format of the data to be transmitted. It uses character serial communication lines, Ethernet, or the Internet protocol suite as a transport layer, and supports communication to and from multiple devices connected to the same cable or Ethernet network. This may include (for example) a temperature sensor and a humidity sensor connected to the same cable, both communicating measurements to the same computer or datalogger, using the Modbus protocol.
PakBus®	PakBus® describes a proprietary family of protocols created by Campbell Scientific for communication between connected devices. Similar in many ways to TCP/IP Internet protocols, PakBus is a packet-switched network protocol with routing capabilities.

One-minute summary – *AWS data flows, display and storage*

- Today's AWS products are sophisticated data acquisition and processing systems which enable straightforward collection and processing of almost any type of meteorological or environmental data. Host computer platform and power source are important early considerations: if a mains/utility power supply is used, an uninterruptible power supply with a surge protector is a wise investment.

- Most products follow a broadly similar set of basic processes, which can be briefly summarised as system maintenance, sensor management and sampling, data handling architecture, communications, display (where required), permanent data storage, and access and export facilities. These steps are explained and illustrated using three different approaches to 'system architecture'. Which to choose is very much dependent upon requirement, budget and expected or planned future expansion: all three have their advantages and disadvantages.

- 'Black box' systems usually carry out the required functions and processes within firmware, perhaps via an external communications controller module. A host computer regularly uploads data for post-processing and display (if required) and local storage, using manufacturer or third-party software running on the host, and from here records can be archived, exported, written to websites and so on. This approach is most typical of entry-level and budget systems and has the advantage of simplicity and ease of use.

- The 'black box combined with Internet and cloud storage' system type is an extension of the above, coupled with access to an always-on Internet connection. Sensor interfacing and basic data handling/datalogging processes are managed within a wireless console, and data are regularly uploaded to a remote proprietary data aggregation service. Post-processing and storage are managed 'in the cloud', and data viewed from anywhere using a web browser. Such systems are also straightforward in installation and use, but may incur subscription costs.

- The 'programmable datalogger' approach uses a dedicated datalogger for all sensor interface and data handling processes, and data are uploaded through a host computer and communications network as and when required for onward processing, storage, display and cloud storage or data aggregation services, including web browser access. This approach offers almost unlimited flexibility, including advanced capabilities such as high-resolution event-based logging. Such devices are more expensive than the other types, however, and may require specialist programming.

- Finally, it is important to check and test all sensor / system / upload communication links thoroughly, over a period of at least a few days, before permanent hardware installation or embarking on any long-term data collection. This is particularly important if access to some or all of the sensors is more difficult – wind sensors at height, for instance, or remote or relatively inaccessible sites.

14 Non-instrumental weather observing

Instrumental readings are of course vital when making weather observations, but non-instrumental 'eye observations' (such as cloud amounts and types) and brief notes (such as short weather diary entries) help to help build a more complete picture. This chapter sets out how to include these types of record, along with documentation regarding the occurrence of fog, snowfall, thunderstorms and other elements, in a practical and useful series.

Observations at set times

The conventions regarding observing hours were set out in Chapter 12, *Observing hours and time standards*. The convention and convenience of a daily 'morning observation' daily, typically between 7 and 9 a.m. (nominally 0900 UTC in the UK and Ireland), is followed in most countries worldwide. At this 'morning ob', instruments are read and reset for the coming 24 hours, and various eye observations made, as detailed below.

As described in more detail in Chapter 12, the importance of a once-daily observation at a set time has declined, as AWSs have made it easier to adjust instrumental observations made at other times, or variable times, to the national or regional standard 'climatological day' and in so doing have greatly simplified comparison of observations between sites. Even with an AWS, it is preferable wherever possible to make at least one 'manual' observation every day at approximately the same time, as home and work schedules permit.

I have an AWS – why do I need to do a manual observation?

Many new weather observers, particularly those whose first experience is with a domestic AWS rather than with 'traditional' instruments, ask this very logical question. The answer is, of course, *you don't* have to, but there are many good reasons for doing so nonetheless. As well as reading and resetting any manual instruments, such as the standard raingauge (see Chapter 6, *Measuring precipitation*) or legacy liquid-in-glass screen, earth or grass thermometers (see Chapters 5 and 10, and **Appendix 4**), a manual observation provides the opportunity to note various important non-instrumental elements such as cloud amounts and types, visibility, the depth and extent of any snow cover, and so on, many of which are difficult to perform with an AWS. When observations of this nature are made in a consistent manner, they quickly build into a valuable complementary weather record.

A daily manual observation is also the ideal time to make a quick visual check that all instruments are functioning correctly, that the TBR funnel is clear of debris, and that no temporary obstacles, obstructions, damaged cabling or instrument defects are affecting readings, that the grass is not too long and so on. When combined with inspection of the logged/displayed AWS record, regular visual checks reduce the risk of a long period of lost or defective record owing to undiagnosed instrument or signal failure. The 'morning ob' need take only a few minutes to complete.

I work irregular hours and can't observe at a fixed time ...

An AWS will, of course, provide 24 × 7 observational cover, regardless of whether regular manual observations are or can be made. As outlined in Chapter 12, *Observing hours and time standards*, there are many advantages to maintaining an observational routine built around standard 'terminal hours', whether or not these actually coincide with regular manual observations. Even those with irregular working patterns may also find noting the occurrence of snowfall, thunderstorms, hail and the like to be useful in building a more complete picture of local weather conditions than can be provided solely with the digital output from an AWS. Occasional visual checks on the integrity of both instruments and data are probably even more important for those with irregular routines, as a minor problem may take longer to become apparent and be put right.

The daily observation

As previously suggested, the best time to do a once-daily observation is in the morning, usually between 7 and 9 a.m. (in the UK and Ireland 0900 UTC, 9 a.m. clock time in winter and 10 a.m. clock time during summer time). If this is not possible, make the observation at a more convenient time for you, and use AWS records to adjust the readings to the country standard 'daily climatological day', as described more fully in Chapter 12, *Observing hours and time standards*. No observation time is perfect, but a consistent slot is usually easier to fit into a daily routine.

Observational routine

After reading and resetting any manual instruments as required, note non-instrumental observations as detailed below using a small notepad or similar. You may find it advisable to use a notepad and pencil for the outside observation and copy up the observation into a 'fair copy' logbook immediately afterwards to avoid the pages and previous manuscript observations being smudged should rain be falling at the time. Rule up a blank sheet with columns for a month's observations and photocopy it, then file or scan each monthly sheet once completed. It is assumed here that observations of temperature, humidity, barometric pressure and perhaps other elements will be logged by the AWS, although it is good practice to make manual assessments of wind direction and speed even if you have AWS wind instruments, and thereby check the accuracy of your estimates. Best practice is to include a quick visual check of all instruments during 'the ob'.

Cloud amount and type(s)

Cloud amount is estimated by eye, in eighths of the sky (or *oktas*: see Box, *Why assess cloud amounts in eighths?*) The easiest way to do this is to mentally divide the sky into quarters and assess whether each quarter is clear of cloud, partially covered by cloud, or completely covered by cloud. A clear sky will obviously be 0 oktas (0/8), and a complete cloud cover 8 oktas (8/8). A trace of cloud will be 1, and a chink of blue sky in an otherwise cloudy sky is coded as 7. It is important to assess cloud cover without regard to the thickness or density of the cloud – it is quite possible to have unbroken sunshine with 8/8 cloud cover of cirrus or cirrostratus, for example. When the sky is obscured (usually by fog, occasionally by falling snow), code it as × (not '9'). Aircraft contrails should be counted as cloud cover, unless they last for only a short period before evaporating completely – less than about 30 seconds, say, but be consistent in your approach. Cloud amounts can be quickly and accurately estimated by eye with only a little practice, and after a time it becomes automatic.

Why assess cloud amounts in eighths?

The term *okta* is attributed to Colonel Ernest Gold, CB, DSO, OBE, FRS (1881–1976), one-time Deputy Director of the UK Met Office, and came into general use when meteorological reporting codes were revised in 1948 [287]. Why eighths in a decimal world? Because when observations are coded for international distribution on the meteorological communications networks, reporting cloud cover N in eighths requires only a single character. Prior to 1948, cloud amounts were reported in tenths, which required two characters. $N = 9$ in coded synoptic reports is used to indicate 'sky obscured' (due to fog, dust or sand or snow, for example), although for statistical purposes $N = 9$ should be reckoned as 'missing' to avoid distorting average cloud cover calculations. Synoptic AWSs cannot report cloud extent, and therefore always report N as /.

A full tutorial on *cloud types* is beyond the scope of this book (and its illustration budget ...). The interested reader is referred to two excellent and well-illustrated books on this subject – Richard Hamblyn's *The Cloud Book* [288] and Gavin Pretor-Pinney's *The Cloudspotter's Guide* [289], together with the World Meteorological Organization's online *International Cloud Atlas*, updated in 2017, at cloudatlas.wmo.int. With a little practice, and a good 'spotter's guide' to start from, the major cloud types can be identified quickly and confidently. There are also many excellent and regularly updated cloud photographs on the Cloud Appreciation Society website at cloudappreciationsociety.org their 'cloud selector information wheel' is also a cleverly thought-out little device which works well across all age groups and knowledge levels.

Visibility

Visibility is defined by WMO as ' ... the greatest distance at which a black object of suitable dimensions (located on the ground) can be seen and recognised when observed against the horizon sky during daylight, or could be seen and recognised during the night if the general illumination were raised to the normal daylight level' ([4], section 9.1.1). Visibility is an important element in operational meteorology, especially in aviation applications: poor visibility in bad weather is a major cause of aircraft accidents. In climatology, the main purpose of visibility observations is to define the frequency of fog. In the meteorological context, fog is said to occur when the horizontal visibility falls below 1000 m (1100 yards), and thick fog when the visibility is below 200 m (220 yards). Poor visibility – between about 1000 and 4000 m visibility – is referred to as 'mist' when the relative humidity is above about 95 per cent, and 'haze' when below 95 per cent. 'Shallow fog' – fog in a layer no more than 2 m deep – may not reduce visibility at eye level, in which case the visibility at both 'eye level' and closer to the ground should be estimated. If the visibility is markedly worse in different directions at or near the same elevation of the observing site, perhaps due to fog patches, then the lowest visibility should be reported and 'shallow fog' noted.

Eye observations of visibility, and thus the determination of fog frequency, are easily made with a little preparation. Look around the horizon from your observing site and note clearly identifiable objects or distinct landmarks at various distances. Nearby, these could be buildings – a church steeple, for example: at greater distances, a prominent hilltop or skyline. Note their azimuth (compass bearing) and then, using maps of the area – Google Earth is particularly useful – determine their straight-line distance from your observation site. To determine the occurrence of fog and thick fog accurately, ideally objects both a little below and a little above 200 m and 1000 m distances, respectively, are preferable. If there is no suitable object or landmark close to those distances, careful interpolation from other objects will be necessary. Visibility in fog is very rarely less than about 20 m, but you should define a selection of 'visibility objects' (as they are known) from 10 m to the most distant points visible on a clear day – ideally 30 km (20 miles) or more. Where possible, it is best to choose objects close to the boundary of the category – for example, the visibility or otherwise of an object close to 4000 m distant would establish clearly whether the meteorological visibility was 'poor' or 'moderate'. Visibility objects do not all have to be in the same direction of view (see *Visibility objects*).

Visibility objects

A typical list of visibility objects might look something like the following:

Visibility objects at <station name>. *Objects are viewed from* <e.g. the thermometer screen>.

Visibility category	Object (Examples)	Bearing	Distance
Very dense fog (< 20 m)	Sunshine recorder pillar	ESE	9 m
Dense fog (< 50 m)	School clock tower	NW	45 m
Thick fog (< 200 m)	Poplar tree	SSW	185 m
	White-fronted building	E	220 m
Fog limit (< 1000 m)	Pylon	SW	1050 m
Poor visibility (1000–4000 m)	Church spire	N	2.1 km
	TV mast	ESE	3.7 km
Moderate visibility (4–10 km)	Trees on skyline	SE	6.5 km
Good visibility (10 km or more)	Cluster of tall buildings	NE	9.7 km
	Gap in scarp slope	ENE	28 km
Excellent visibility (> 40 km)	Distant hills	WNW	42 km

At the observation, note which is the furthest object to be clearly visible. If the visibility varies by direction, patchy fog for example, the lowest visibility should be noted, together with the direction. Note 'fog at observation' if the visibility is below the fog limit (1000 m) for whatever reason; it might be owing to heavy snow, or even a dust storm, rather than fog.

Present weather

Note the occurrence of any significant weather (rainfall, fog, snowfall, etc.) at the time of the observation, together with its intensity and persistence – using unambiguous terms such as 'continuous light rainfall' or 'heavy snowfall ceased within the past hour'. A short written description is perfectly adequate for most purposes, but if abbreviations are used, ensure they are set out within the logbook or site metadata to avoid confusion if the records are reviewed by someone else at some future date. Operational reporting observing sites use the 'present weather code', a two-digit code (ww, code 4677) [284] which provides a convenient structured reporting protocol to cover almost all weather types.

Note any other observations as required, for example the extent and depth of snow cover at the time of the observation (see also *Days with* section, below).

Away from home?

An AWS is ideal for providing unbroken instrumental records during observer absences – whether due to family holidays, lack of weekend observational cover at, for example, local authority observing sites, or gaps due to school breaks, but non-instrumental observations can be more difficult to cover. Short periods of absence can often be filled in by asking a family member, neighbour or another local observer either to make observations in your absence, or at least 'keep a weather eye open' so that occurrences of thunderstorms and the like are not omitted from the observational record. Ensure the AWS will record for the intended period of absence, if necessary by setting up an automatic download in good time, and by checking batteries are in good condition. If using a Windows host computer to upload logged data at regular intervals by remaining on throughout the period of absence, pause automated system updates as these will reboot the computer; unless the AWS software is included in the startup routine, it may not otherwise restart and records will be lost if system memory becomes full.

Photographic records

Weather observation sites with webcams can provide a permanent record of cloud and weather conditions. There are also many advantages to keeping a camera handy at all times, to record notable weather events photographically – significant flooding, large hail, tree damage after severe gales … Over time, such images build into a valuable visual record of noteworthy events, and can help document environmental changes as a result of individual events, or a prolonged sequence of unusual weather.

Occurrence frequencies or 'Days with ... '

Climatological statistics often include the frequency of various meteorological phenomena at a particular location. Some of these (such as counts of air or ground frosts) are derived from thresholds applied to instrumental observations, but many are derived from eye observations of particular weather types, such as the occurrence of snow or thunderstorms. For accurate reporting, a 24 × 7 × 365 weather watch is required (at least in theory), something that very few operational meteorological sites manage these days but one which many amateur observers pride themselves on covering as fully as possible. An amateur observer with a home-based weather station can often provide a more complete 'weather watch', particularly at weekends, public holidays or in severe weather, than the professional meteorological networks (and at much lower cost). For this reason, eye observations are of enormous value, provided standard definitions and terms are used. Many meteorological services actively encourage 'citizen science' reporting of such conditions on social media networks in particular weather conditions, such as borderline rain/sleet/snow situations, or in tornado warning areas, to obtain broader coverage in space and time than conventional meteorological reporting sites permit.

'Days with' frequencies usually refer to the civil day (midnight to midnight local regional time, excluding summer time adjustments), although others may refer to a specific time, such as 'Fog at the morning observation'. Definitions of the most common 'days with' elements are listed in **Table 14.1**: those derived from instrumental observations are also included for convenient reference.

Table 14.1 *Definitions of the most commonly used 'days with' climatological elements. Most feature the current definitions in use within the UK and Ireland, with US differences shown where appropriate. Definitions and terms do vary slightly from country to country – check these by referring to your state weather service publications or website [96].*

The definitions in the following section refer to occurrences within the civil day, midnight to midnight local regional time, excluding any summer time adjustments	
Thunder heard	*Definition:* The audible sound produced by a lightning discharge, arising from the intense heating and expansion of the air by the electrical current [290].
	A day with 'thunder heard' is logged whenever thunder is heard. There is no minimum duration or intensity, even a single weak rumble of thunder counts as a 'day with thunder heard', although great care should be taken to distinguish genuine thunder from other noises such as aircraft. 'Thunder heard' does not have to be accompanied by visible lightning, although it is helpful to note whether lightning was also seen.
	A thunderstorm which starts at 2350 h and finishes at 0005 h counts as two 'days with thunder'. 'Lightning seen' (only) does not count as a 'day of thunder' as it can sometimes be seen from a great distance: thunder is not normally audible beyond about 15–20 km from the parent storm, whereas lightning at night can sometime be visible more than 100 km distant.
Gale	*Definition:* A surface wind attaining a mean speed 34 knots (17 m/s) or more, averaged over a period of at least 10 minutes [291].
	Note that the surface wind refers to that measured at 10 m above ground, and thus the mean speed threshold will be lower nearer ground level (see Chapter 9, *Measuring wind speed and direction*).
	Where instrumental wind speed records are available, these should be used to assess whether the threshold has been attained. If instrumental records are not available, estimates using the Beaufort Scale (see **Table 9.1**, page 227) should be made. The source should be stated in the site metadata.

Table 14.1 (*cont.*)

The definitions in the following section refer to occurrences within the civil day, midnight to midnight local regional time, excluding any summer time adjustments

Snow or sleet observed to fall	*Definition of snow:* Solid precipitation in the form of individual tiny ice crystals when temperatures are low, or larger snowflakes when the air temperature is near 0 °C [290].

Definition of sleet in UK/Ireland: A mixture of rain or drizzle and melting snow.

Definition of sleet in the United States: Frozen precipitation that results when raindrops freeze while falling, before hitting the ground.

A day with 'snow or sleet observed to fall' is logged whenever snow or sleet is seen to fall. There is no minimum duration or intensity – even a single flake of snow counts, at least in theory. It can be difficult to decide whether to include or exclude an event following slight snow showers, overnight snowfall or blowing snow after a snowfall, and considerable alertness is needed in marginal situations. Snow showers from shallow convective clouds can occasionally reach the ground after the parent cloud has evaporated, or be blown some way from the parent cloud, leading to the apparent fall of snow from a largely clear sky. Particular care should be taken to distinguish genuine sleet from cold rain at temperatures between about 1 and 3 °C.

Snow (or sleet) which starts at 2350 h and finishes at 0005 h counts as two 'days with snow (or sleet)'. For statistical purposes, it is helpful to distinguish 'days with snow' from 'days with sleet only'. A day when both occur is counted as one with 'snow' only.

Within the UK and Ireland, the following types of wintry precipitation also count as 'snow' for statistical purposes:

Snow – definition above

Snow pellets – opaque, white ice particles, which can be spherical or conical in shape (*graupel*), with a diameter generally 2–3 mm or less

Snow grains – very small, opaque white particles which appear flat or elongated, and normally 1 mm or less in diameter; the wintry equivalent of drizzle

Ice pellets – transparent ice particles, spherical or irregular in shape, and typically 1–5 mm in diameter, rarely more. This type of precipitation is known as 'sleet' in the United States – see above.

At low temperatures, small **ice crystals, ice prisms** or **ice needles** may fall from freezing fog or occasionally, below about −10 °C, from a clear or almost clear sky (sometimes very appropriately referred to as 'diamond dust'). These should not be counted as 'snow' in statistical summaries.

Hail observed to fall	*Definition*: Solid precipitation in the form of balls or pellets of ice [290].

For statistical purposes, falls of hail are usually classified as 'small hail' (where the ice particles are less than 5 mm in diameter), and 'hail' (5 mm or more in diameter). When large hailstones occur (10 mm or more in diameter), it is useful to note measurements of the mean and maximum stone size, and (where possible), photographs should be taken and samples of the stones collected promptly for preservation in a freezer.

A simple hailgauge can assist considerably in the accurate reporting of hail, as hail often occurs for very short periods and/or mixed with other forms of precipitation and can easily be missed (see Box, *A simple hailgauge*).

Table 14.1 (*cont.*)

The definitions in the following section refer to occurrences within the standard 'climatological day' (morning to morning) – in the UK and Ireland, 0900 to 0900 UTC

Air frost	*Definition*: A minimum air temperature below 0.0 °C.
	The temperature refers to measurements made in a standard thermometer screen or equivalent (see Chapter 5, *Measuring the temperature of the air*), unless clearly stated otherwise: 'Aspirated thermometer air frost', for example.
Ground frost	*Definition*: A minimum temperature below 0.0 °C observed by a thermometer exposed above short grass.
	WMO guidance on the observation period is 'sunset to the subsequent morning observation', but at unmanned sites where the grass minimum thermometer cannot be exposed shortly before sunset the 'morning to morning' period necessarily becomes the default. See the notes on this topic in Chapter 10, *Measuring grass and earth temperatures*.
	Grass temperatures on a clear night are typically 2–5 degrees Celsius below air temperature, although this varies with exposure, soil type, intermittent cloud cover and wind speed.
Rain day	*Definition*: A day on which 0.2 mm or more of precipitation is recorded during the 'rainfall day', the 24 hour period commencing at the time of the morning observation (within the UK and Ireland, conventionally 0900 to 0900 UTC). First set out by George Symons, founder of the British Rainfall Organization, in *British Rainfall 1865* [168]. Where the inch is the unit of measurement, the rain day is usually 0.01 inches (0.25 mm) or more within the rainfall day.
	Occasionally, thick wet fog or a heavy dewfall can deposit 0.2 mm or more in the raingauge. This should be measured and recorded as (for example) 'Fog 0.2' in the register and counted as a 'rain day'.
Wet day	*Definition*: A day on which 1.0 mm or more of precipitation is recorded during the 'rainfall day', the 24 hour period commencing at the time of the morning observation (within the UK and Ireland, conventionally 0900 to 0900 UTC).
	This definition first appeared in *British Rainfall 1920* and was the result of a UK Met Office metrication push at the time. However, it was not until 1971 that all UK rainfall records were finally made in millimetres [168].

The following definitions refer to observations made at the morning observation

---

Snow lying at morning observation	*Definition*: Snow lying occurs when snow covers one half or more of the ground of an open area representative of the site at the morning climatological observation [248].
	Within the UK and Ireland there is no minimum depth for 'snow lying'– sometimes just a fine dusting can count as a 'snow cover', as it is the coverage rather than the depth that is the criterion for the event. In the United States, the minimum depth for 'snow lying' is 'a measurable depth'– which usually means 0.5 inch (1.3 cm) or more. 'Representative of the site' excludes rivers and lakes, cleared paths or roads, areas of tarmac, house roofs, cars and the like as well as areas with significantly different altitude, such as nearby hills or mountains. The remains of snow drifts, whether natural or artificial (such as those from snow clearing operations) should not be included in the assessment. If the observation is not made at the normal time, the time should be included – for example, 'snow lying at 0630 h, depth 2 cm, melted by 0900 h'.
	On occasions when snow is lying, the snow depth at the observation should also be carefully measured and noted – see Chapter 6, *Measuring precipitation* for details on how to do this.
Fog at morning observation	*Definition*: A visible suspension of water droplets in the atmosphere near the surface, and defined by international agreement as reducing visibility to less than 1000 metres / 1100 yards [290].

Table 14.1 (*cont.*)

The following definitions refer to observations made at the morning observation	

	When the surface visibility is below 1000 metres at the morning observation, a 'day with fog' should be noted – even if the obscuration is not due to fog itself (smoke, blowing dust or sand, heavy snow or torrential rain can also reduce visibility below 1000 m). If the time of observation is significantly different from the standard morning observation time, the time of the observation should be stated. In most areas fog frequencies are much higher at dawn than later in the morning, and where the morning observation is made earlier than at other sites (say at 7 a.m. rather than 9 a.m.) the observed frequency of fog will be greater. To avoid possible confusion, the normal time of observation should be clearly stated in the site metadata (Chapter 16, *Metadata, what is it and why is it important?*).
Thick fog at morning observation	As above, but when the visibility is below 200 m (220 yards).

A simple hailgauge

The determination of the number of days on which hail falls is an important climatological statistic, but the fleeting nature of most such events is such that all but the heaviest falls are easy to miss for even the most conscientious observer. The reported frequency is therefore more dependent upon the alertness of the observer than any other single factor.

One way to improve matters is to use a hailpad or hailgauge [292]. This simple instrument is easy to make and is underused. In its simplest form, a hailgauge can be made from a small sheet of thin metal foil (aluminium/ aluminum foil as sold in supermarkets is ideal). The foil is simply stretched across a small open frame such that an area of perhaps 200 cm^2 is left unsupported, or alternatively the foil can be lightly supported on a backing material such as a block of polystyrene foam (**Figure 14.1**). Fix the instrument firmly in a position where it is well-exposed to precipitation (and where it will not blow away in strong winds), and where it is also easy and convenient to examine the surface of the foil on a daily basis.

When hail falls, the surface of the foil becomes pitted, the extent depending upon the intensity of the hail. Minor hailfalls may be marked by only a few small dents, whereas heavy hail or large stones may shred the foil. Hailgauges with an area of unsupported foil tend to be a little more sensitive but are more easily punctured by a few stones: supported variants are more robust but less sensitive. The raw materials are, of course, easily obtainable and inexpensive, and a combination of both types gives good results. Different shapes and sizes can be used to augment results (for instance, an upright cylinder or cone can be used to estimate the direction of hailfall and precipitation angle).

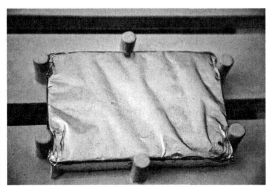

Figure 14.1 A simple hailgauge, constructed from a small sheet of aluminium foil stretched over a block of polystyrene and freely exposed to the weather. Dimensions approximately 95 × 135 mm. (Photograph by the author)

Keeping a weather diary

The 'narrative element' of weather observation is frequently overlooked, and yet it can often serve as a more meaningful and easily recalled description of a day's weather than a raw table of instrumental statistics. Keeping a simple daily weather diary takes very little time, typically just a couple of minutes per day, and it quickly builds into an important and invaluable reference.

The exact form of the document is largely a matter of personal choice, whether that be a daily manuscript entry in a bound annual desk diary, or a daily paragraph added to a document on a laptop computer or smartphone (but don't forget to back it up from time to time – see Chapter 17, *Collecting and storing data*). A voice recorder is another alternative, perhaps a smartphone app, although subsequent searches for any particular event is more difficult with audio files. The advantage of computer-based records is that they can be easily searched ('How many times have I noted rainbows in the last 5 years?'), copies or extracts can be easily provided to others (for example, times and details regarding a severe thunderstorm), and if necessary photographs or tables/graphical output of AWS data can easily be included alongside descriptive text. They are also easy to store alongside other computer records – bound volumes of desk diaries can take up a lot of bookshelf space over time, and manuscript records are of course irreplaceable in event of theft or fire or water damage.

The amount of detail included is also largely a personal decision. Obviously, the more detail included, the longer the entries will take to complete each day. My personal preference is to write around 100–150 words per day's weather – although that varies from only a line or two on a day with unbroken overcast and no significant weather, to a longer entry for a major weather event such as prolonged and heavy snowfall. It is best to write down the detail of such events as soon as possible, preferably on the day itself, while the details are still fresh in the memory. I find it next to impossible to recall the detail of most days' weather even a few days afterwards without a 'prompt' from instrumental data. Having a written description prepared immediately after the event does aid recall of any event, even months or years later.

One-minute summary – *Non-instrumental weather observing*

- Instrumental readings are of course vital in making observations of the weather, but for a complete picture non-instrumental and 'narrative' weather observations are equally important, especially so for the analysis of severe weather events.
- A once-daily 'morning observation' is the best time to read/reset any manual instruments in use, as well as perform visual checks on the continuing operation of all instruments, including the AWS sensors: raingauge funnels are especially likely to become blocked, and the obstruction may not become obvious until the next time rain falls. A manual observation also provides a convenient opportunity to note current weather details such as the amount and types of cloud, the surface visibility, present weather, the occurrence of lying snow and so on. Weather observing need not be restricted to viewing graphical or tabular output on a computer screen!
- With a little practice, maintaining a near 24 hour weather watch becomes second nature, and with some assistance from friends and family, colleagues or neighbours a 365 day, 24 hour coverage of significant weather is not difficult. When combined with the instrumental observations from an AWS and a brief daily descriptive weather diary, a high-quality combined weather record quickly builds up.

> *'The person who has only one watch knows what time it is, but the person who has two is ... not sure.'*

A favourite saying of professional metrologists, sometimes referred to as 'Segal's Law'

Instrument calibrations are both one of the most important, and yet sometimes one of the most neglected, areas of weather measurement. Calibration can be defined as 'the comparison of measurement values delivered by a device under test with those of a calibration standard of known accuracy', where the reference 'calibration standard' will normally be traceable to a national or international standard held by a national metrology body, such as the National Physical Laboratory in the UK or the National Institute for Science and Technology in the US, and similar bodies in other countries. The term is also applied to a second step, in which the derived calibration comparison is applied to establish relationships between the indication of a given instrument and the quantity being measured, together with some measure of related measurement uncertainty. In this manner, an instrument's calibration can be expressed by a statement, calibration function, calibration diagram, calibration curve, or calibration table, often in the form of an additive or multiplicative correction to the instrument's indication.

We have already seen in Chapter 2, *Choosing a weather station* that *precision* is not the same as *accuracy*. To make *accurate* weather measurements, the instruments themselves need to be accurately calibrated, or at least regularly compared against instruments of known calibration to quantify and adjust for any differences, or error. Calibrations can and do drift over time, and therefore instrumental calibrations should be checked regularly, and adjusted if necessary. An error of 1 degree Celsius in temperature, or 20 per cent in rainfall, may not seem very significant on a day-to-day basis, but if monthly or annual values are adrift by even half this amount, the readings obtained will not be comparable with other locations, or with historical records. A 1 degree Celsius difference in mean air temperature corresponds on average to about 150 metres (500 ft) difference in altitude, or to the difference in annual mean temperature between London and Paris, or between Boston and New York.

One practical difficulty that applies to calibration checking on operational weather instruments is that, without a duplicate set of instruments, removing the sensor (and sometimes the AWS or datalogger too) for offsite calibration will involve the record from that device being lost while it is away, perhaps for several weeks. The expense may also be considerable, perhaps more than the original purchase price of the sensor itself. Therefore, methods which allow *in situ* calibration checks on instruments in use are sometimes preferable, if perhaps less rigorous than results obtained under laboratory conditions. Such 'local' calibration methods can often be easily accomplished using 'absolute' or 'fixed point' methods, or by comparing readings over a period with a portable reference instrument whose calibration is accurately known. Regular checks in this manner cannot be a complete substitute for formal traceable calibration checks by an accredited laboratory, but carefully done they are certainly better than no tests at all.

This chapter describes straightforward methods to check and adjust calibrations for the most common meteorological instruments – precipitation (rainfall), temperature, humidity, and air pressure sensors.

Calibrating a recording raingauge

For the reasons outlined in Chapter 6, *Measuring precipitation*, it is always advisable to ensure that the 'reference' precipitation measurement is made using a standard 'manual' raingauge (a five-inch gauge in the UK and Ireland, eight-inch pattern in the United States). Recording gauges (such as tipping-bucket or weighing raingauges) will almost always read a little lower than the standard gauge, owing to both instrumental and evaporative losses and differences in exposure. By definition, a standard manual raingauge, when correctly exposed, gives the 'reference' rainfall total. Minor differences between the standard gauge and a recording raingauge are therefore to be expected: rarely will two raingauges record exactly the same amount of rainfall. An automated raingauge should *not* be adjusted merely to attempt exact agreement or near-agreement with the standard gauge – instead, carry out the method below to derive an absolute calibration for the unit by passing a known volume of water through the unit and comparing its measured output.

The method below assumes a tipping-bucket raingauge, but the principle is the same for almost any type of recording gauge.

For the test, the recording raingauge should be connected either to its normal display or logging system, or to a pulse counter, whichever is easier. If a tipping-bucket raingauge is in use, the calibration check can be performed *in situ* on a dry day (remember to delete the calibration test tips from the record afterwards). Ensure the gauge is absolutely level before and during the test.

First, carefully measure out 500 ml of water at room temperature (preferably distilled or de-ionised, pure water). This volume will be sufficient for most raingauges with funnels of diameter 100–200 mm (4 to 8 inches) or so; larger or smaller funnels may need more or less in proportion to their funnel diameter. The exact amount is not critical (although of course it must be accurately known), but it must be sufficient to generate at least 100 tips (of a 0.2 mm tipping-bucket unit) to minimise random counting errors. The actual quantity of water should be measured as accurately as possible, preferably with a laboratory balance, but with digital scales if not. At room temperature, 1 ml (= 1 cm^3 or 1 c.c.) of pure water weighs 1 g, so measure 500 g of water, netting off the weight of the container of course. Strictly, this relationship applies at a water temperature of 4 °C, but at 20 °C the difference in specific gravity is less than 0.2 per cent (1.000 g/cm^3 at 4 °C, 0.9982 g/cm^3 at 20 °C). Uncertainty in the weighing device is likely to be larger than this.

The water is then carefully poured through the tipping-bucket raingauge. Pouring it in too rapidly will simply overload the buckets (they will stick in the 'tipped' position and the resulting calibration will therefore be inaccurate), so the rate of inflow needs to be reduced to a steady trickle. A large plastic funnel with sufficient capacity to hold at least 500 ml water, obtainable from hardware stores, can be adapted to do this. Push a blob of Blu-Tack, putty or similar material well down into the spout of the funnel so that it blocks it. Using a small screwdriver, carefully make a small hole in the Blu-Tack. Fix the plastic funnel securely in place above the raingauge funnel and tipping-bucket unit (make sure the gauge is perfectly level, and in a position where the water from the emptying buckets can safely drain away). Pour a cupful of water (not the measured 500 ml sample yet) into the funnel and allow it to drip into the raingauge funnel, at a rate to ensure the buckets tip no more often than once per minute or so – simple arithmetic will show that this corresponds to a rainfall rate of 12 mm/h for a 0.2 mm tipping-bucket unit, a convenient and typical calibration of rainfall intensity. Adjust the hole size as necessary. (This also serves to pre-wet the surfaces of the funnel and the tipping buckets.) Too rapid a rate of flow risks the buckets overflowing – too slow a rate will simply mean that the test takes hours to complete.

Once the 'test' water has completely drained through, remove the raingauge funnel and empty (tip) any partially filled buckets by hand. Replace the raingauge funnel. Note the rainfall reading or pulse count at this point; this is the zero point of the calibration test.

Re-fix the plastic funnel in position and *very carefully* pour the measured 500 ml into it, ensuring that none is spilt and that as little as possible remains in the original vessel. Allow it to flow slowly through the partially obstructed funnel into the tipping-bucket raingauge – this will take an hour or so.

After all the water has passed through – check both the plastic funnel and the raingauge funnel to ensure none remains – note the logged rainfall reading or pulse counter value.

The volume of water v collected by a cylindrical raingauge funnel is given by $v = \pi r^2 h$, where r is the *radius* of the funnel (half the *diameter*) and h the height of the cylinder (= the measured depth of rainfall).

$$\text{Rearranging in terms of } h \text{ gives } h = \frac{v}{\pi r^2}$$

Measure the radius r of the raingauge funnel opening, in millimetres, as accurately as possible. Use this measurement to calculate the depth of water (= amount of rainfall) that passing through 500 ml of water – or any other amount – should cause the gauge to indicate. This method is also useful to check the calibration of a measuring cylinder for a standard raingauge, for example. For a UK-standard five-inch gauge, 1 mm of rainfall corresponds to 12.7 ml of water; for a US-standard eight-inch gauge, 0.1 inches of rainfall is 82.4 ml.

Worked example: using a Davis Instruments Vantage Pro2 TBR (funnel diameter 165 mm, radius 82.5 mm = 8.25 cm), and working in *centimetres* throughout: the depth of 'rainfall' h corresponding to exactly 500 ml input water is

$$h = \frac{500}{3.14 \times 8.25 \times 8.25} = 2.34 \text{ cm} = 23.4 \text{ mm}$$

Calculate the calibration from the comparison with the measured amount during the test. For example, if the indicated amount shown by the display after 500 ml input was 19.8 mm, then the calibration is 19.8 / 23.4 = 85 per cent and the tipping-bucket unit reads 15 per cent low (a not atypical value 'out-of-the box'). The nominal 0.2 mm tip capacity in this case is therefore actually 0.17 mm (85 per cent of 0.2 mm). If using a pulse counter, multiply the number of tips by the nominal bucket capacity: so, for example, 108 tips of a 0.2 mm unit would give 108 × 0.2 = 21.6 mm, which would mean this gauge read almost 8 per cent low (21.6/23.4) and so on in proportion.

The same method can be used for *inch measurements* if the bucket is calibrated assuming 0.01 inch capacity (although the calculations are much simpler using metric units!). Using the same values as in the previous example, 500 ml water poured carefully through the gauge should result in 23.4 mm rainfall equivalent or 0.92 inches, which assuming a 0.01 inch bucket capacity (0.25 mm) should result in 92 tips (compared with the 108 for the slightly smaller 0.2 mm bucket). The absolute bucket capacity varies with the number of tips – thus, for example, if only 88 tips were logged, then the true bucket capacity (assuming equal bucket sizes) would be 92/88 × 0.01 = 0.0105 inches, or 5 per cent too high, and so on in proportion.

It is advisable to repeat the test at least once more and compare results. If the two derived calibrations differ by more than 5 per cent, repeat a third time and average the two closest results.

If the derived bucket calibration is more than about 5 per cent different from the nominal value, 0.2 mm in this case (i.e., outside the range 0.19 to 0.21 mm), the tipping capacity of the buckets themselves should be adjusted. The manufacturer's manual should be checked for the recommended way to do this, but usually this is achieved by adjusting the base fitting upon which the buckets rest in the empty position, by means of an adjusting screw or nut (**Figure 6.12**, page 161). Lowering the height above the base plate increases the bucket capacity (more water required to tip the bucket) and vice versa; the objective should be to adjust the tip capacity to as close to 0.2 mm (or the nominal bucket capacity otherwise) as possible. It is very important that both buckets are adjusted evenly, and it may be helpful to mark the screw heads or nuts to ensure the same amount of adjustment is made to both sides. Particularly on new units, the initial setting of the buckets may be unbalanced. If this is the case, the calibration of the gauge will vary according to which bucket is in use. This can be checked by carefully timing the intervals for 10 or so tip times on each side as the water drips through. If the average tip time for one side is noticeably different from the other, then the bucket tipping capacities differ. If so, both buckets should be adjusted to one end or other of their adjustment and then 'wound back' evenly so that they are at the same adjustment position. The calibration test, and the tip timing measures, should then be performed again until the discrepancy between the two has been eliminated and both tip at the equivalent of 0.2 mm ± 5 per cent.

Once any adjustments have been made, repeat the calibration process and check results. Calibration within 5 per cent of 0.20 mm is satisfactory: with care, 2 per cent may be achievable on some units.

If time and resources permit, it is worthwhile to repeat similar calibration 'runs' at different flow rates to assess the variation of calibration with rainfall intensity. The resulting matrix of calibration factors versus rainfall intensity becomes a useful aid in the accurate analysis of intense rainstorms. For a 0.2 mm capacity bucket, a simulated 5 mm/h rainfall rate will generate one tip every

2.4 minutes; at 60 mm/h the tip time is 12 sec; at 200 mm/h 3.6 sec; at 500 mm/h 1.4 sec. (Note that the tip rate will slow over time as the hydrostatic pressure of the head of water is progressively reduced.) Even assuming the inflow pipe diameter can handle such intensities, above this level splashing, 'continuous tipping' or multiple bounce-tips become increasingly significant and repeated calibration runs may generate different results. Where high-intensity rainfall is a regular occurrence, higher capacity tipping buckets matched with wider inflow pipes will yield more reliable intensity profiles, at the cost of decreased resolution for low-intensity rainfall events. Alternatively, field or laboratory calibration testing kits are available to assess this in more detail (for example, EML's Rain Gauge Field Verification Kit www.emltd.net/rfvk.html).

The calibration test should be repeated at least once every 12 months. The derived calibration may show seasonal variations, particularly with tipping-bucket raingauges using small buckets (the Davis Instruments Vantage Pro2 0.2 mm capacity bucket holds only 4.3 cm^3, for example) and therefore it is best to perform the calibration test at a temperature close to the annual mean. (The density of water varies little over normal air temperature ranges, but its viscosity, and thus surface tension, increases significantly at lower temperatures, and this may lead to incomplete emptying of small buckets at lower temperatures.) Most AWS software will permit the actual bucket calibration, where known, to be substituted for the nominal (and default) 0.2 mm capacity.

Calibrating temperature sensors

One way to obtain accurate temperature calibrations is to send off the sensors (for electronic sensors, probably the logger too), to a professional calibration facility. As well as being expensive, it is also quite impractical because (except in the case of legacy mercury thermometers) the sensors, wiring and logger will have to be de-installed then re-installed on their return. Unless duplicate equipment is available as a backup, this may mean the loss of several weeks' records.

For temperature sensors, there are two calibration methods which are easy enough to perform *in situ*: the *absolute* method, using the melting point of ice as a fixed point, and the *comparative* method, comparing results over time against a sensor of known calibration. Of the two, the absolute method is to be preferred, but as it involves immersing the sensor or sensors to be calibrated in an ice/water mixture, the method is not suitable for some devices. In addition, the design of many budget AWS models may make it difficult or impossible to access or remove the temperature sensor(s) for an immersion calibration test, in which case the comparative method may be the only available option.

Absolute temperature calibration using an ice/water mixture

This method uses a fixed point, namely the melting point of ice at 0.01 °C, to establish an accurate calibration point. Using a similar approach, estimates of uncertainty from −5 °C to +40 °C can be found.

This method is easiest to undertake with electrical sensors in steel probes, but is unsuited to many budget or mid-range consumer AWSs, simply because it requires that the temperature sensors be immersed in the ice/water mixture (as described below): for many of these models the sensor is either inaccessible, surrounded by circuitry which would be damaged by exposure to water, or both. It should also **not** be undertaken with combined temperature/humidity sensors, because immersing the humidity element in water will irreparably damage it. For these sensor types, the comparative method described subsequently is a better option.

The method is very straightforward. It requires a small, insulated container (a 500 ml Thermos-type flask or 'Chilly bottle' is perfect) and a supply of ice – ice cubes from the freezer are fine (preferably made with distilled water). Partially crush the ice cubes to fit them into the flask and fill it almost full with crushed ice. Add a little cold water, just sufficient to allow the ice to 'float' almost to the brim of the flask, and shake vigorously for a minute or two before carefully inserting the sensors. Gently pour off any excess water.

Electrical temperature sensors should be connected to a datalogger (preferably the logger that will be used with the sensor when operational) and logged during the test. Be careful not to damage the sensor probe or connected wiring by excessive force (the ice-water mixture should not be so 'stiff' or heavily granular that the sensor cannot be easily inserted such that it is fully surrounded by the semi-liquid combination, or to make subsequent stirring or shaking difficult). If space permits, more than one device can be checked at the same time – simply secure the probes together with an elastic band so that they can be inserted and removed from the flask as one unit. The temperature-sensitive areas (probe ends on electrical sensors) should be as close to each other as possible.

Carefully insert the sensor or sensors into the flask. Ensure the probe sensors are fully immersed into the ice/water mixture and sit roughly in the middle of the flask volume, with part of the trailing cable(s) also immersed.

Gently and continuously shake or stir the flask gently for several minutes, to ensure an even temperature distribution within the ice/water mixture. Allow the sensors a few minutes to adjust to the flask temperature and settle to a steady reading.

Over a period of at least 10 minutes, take several readings of each sensor, stirring or gently shaking throughout. The reading from electrical probes should be logged every 30–60 seconds or read off the displayed output as frequently. Note and average the 'steady state' temperatures, ignoring the highest and lowest values. If the sensor is correctly calibrated, the average should be 0.0 °C. An average of, say, −0.3 °C would indicate that the sensor was reading 0.3 degC too low, and thus the correction to be applied at this temperature would be +0.3 degC.

Ice point calibration for platinum resistance thermometers (PRTs)

This method is particularly suitable for checking the calibration of PRTs. If the sensor is an ISO standard unit, its change of resistance with temperature is accurately defined (the world-wide standard for 'Pt100' platinum RTDs, IEC 60751:2022 [293], requires the unit to have an electrical resistance of 100.00 Ω at 0 °C and a temperature coefficient of resistance of 0.00385 Ω/degC between 0 and 100 °C). Once the sensor error (if any) at 0.0 °C has been determined, and provided of course the correct temperature coefficient of resistance is used in the logger, then this simple offset correction should be applicable to all other temperatures measured by the sensor.

This method is not applicable to thermistors, whose change in resistance with temperature may not be linear, and therefore an offset at any one temperature is unlikely to be applicable at another temperature.

Establishing other calibration points using this method

The method can easily be extended to assess calibrations at other temperature points. A mixture of crushed ice and salt in the flask can be used to obtain temperatures down to −5 °C, or a little lower. Removing the ice and salt and adding warmer water allows flask temperatures to be obtained at various points up to about +40 °C. An insulated flask and continuous gentle stirring is essential to maintain a steady temperature, particularly where the flask temperature differs considerably from the ambient air temperature.

The method is the same as for the ice-point test, but for all points above 0 °C, an accurate temperature reference is required. If one of the sensors is a PRT, applying the offset determined from the ice point test should give a temperature accurate within 0.1 degrees Celsius. A pre-calibrated temperature logger with a flying lead (see Box, *Using a pre-calibrated temperature logger*) is ideal for this purpose. Once the ice point and extended point tests have been completed, prepare a calibration table for the sensor similar to that shown in **Table 15.1**. Points between calibration points can be obtained by interpolation. If an electrical sensor is being used, a calibration table such as **Table 15.1** can be manually applied, although it is better to include the calibration algorithm into the logger programming or setup configuration wherever possible.

Note the test date, results and calibrations applied in the site metadata, particularly whether or not corrections have already been included in observations from the site (to avoid mistakenly applying them once more when the observations are archived). Calibrations on electrical sensors should be checked at least every 12 months, or immediately if any sudden change in calibration is suspected.

Using a pre-calibrated temperature logger

Temperature sensor calibrations can be performed quite easily and accurately (with care, to 0.1 degC) using a calibrated portable reference device which itself possesses a recent calibration certificate or correction table from a reputable source. The Tinytag loggers made by Gemini Dataloggers (www.geminidataloggers.com; see **Appendix 3** for supplier details) are perfect for this purpose; similar units are available from other suppliers. These small, rugged, accurate units can be calibrated to a traceable national standard by the manufacturer, and then exposed alongside existing equipment for an extended comparison period (days to weeks). The most suitable device is a Tinytag Plus2 unit fitted with a PRT (Pt100) probe sensor on a short flying lead (the loggers with built-in sensors are slower to respond owing to thermal inertia and should not be used for calibration purposes – see **Appendix 1**). Portable loggers such as these are ideally suited to being left in a thermometer screen or close to a small radiation screen, or alongside exterior sensors such as grass or earth temperature units (see Chapter 10). The probe can usually be wiggled inside the radiation shield of small AWS modules, such as budget and mid-range units, the logger itself being left securely attached outside by cable ties or similar. The calibration process itself is described within this chapter.

Within the UK, the Climatological Observers Link (COL) operates a Tinytag logger loan scheme for members. For a nominal fee plus shipping costs, a Tinytag Plus2 logger can be borrowed for up to a month to conduct cross-calibration tests on your own equipment. Contact details for COL are given in **Appendix 3**.

Table 15.1 *Example: Simple calibration and correction table, derived*
from the fixed-point calibration method described in this chapter

Pt100 sensor Serial no. 12345/67
Corrections to be applied at various temperatures
Based on fixed-point calibration tests [date].
Calibration introduced from [date]
Recalibration due [date]

At observed °C	Add correction to observed reading, degrees C
−15.0	+0.3
−10.0	+0.3
−5.0	+0.2
0.0	+0.2
5.0	+0.1
10.0	+0.1
15.0	0
20.0	0
25.0	−0.1
30.0	−0.1

Comparative temperature calibration using a sensor of known calibration

The second method is cross-calibration alongside a sensor of known accuracy. For temperature sensor calibrations, this can be done quite easily and accurately (with care, to 0.1 degC) with a calibrated portable reference device which itself possesses a recent calibration certificate or correction table from a reputable source (see Box, *Using a pre-calibrated temperature logger*).

1. Obtain a calibrated datalogger

A suitable temperature logger with a sensitive PRT on a flying lead is required, together with a recent calibration table from the supplier (specify three calibration points at −10 °C, +10 °C and +30 °C when ordering the logger), logger software and USB cable to connect the logger unit to a host computer. The calibration process is undertaken within spreadsheets (it can be undertaken by hand but is much quicker using a spreadsheet). Sample spreadsheets are available from www.measuringtheweather.net. In the examples below, Microsoft Excel has been used, but most spreadsheet packages should be able to duplicate the functions described easily enough.

2. Set up the logger

Connect the PRT to the logger, install the software if not already installed, then launch the datalogger. Check that the battery is fully charged, and that the logger is working satisfactorily by leaving it to log for an hour or two with a short logging interval (say, 1 minute). After this period, check that logged data has been uploaded satisfactorily to the host computer, and that both logged timestamps and temperature records appear realistic.

Once everything has been tested and is working satisfactorily, reset the logging interval to be the same as the logging interval on the AWS for the element being monitored (which may be 5 minutes for air temperatures, for example, or perhaps hourly for earth temperatures – see Chapter 3, *Buying a weather station* for more on logging intervals). Choose to log either temperature only at the set interval, or maximum and minimum temperatures attained during the logging interval – the latter provides a closer calibration against maximum and minimum observed temperatures logged by the sensor under test, where these are observed. Relaunch the logger. Make a diary note of the date when the logger memory will become full and require downloading.

Choosing more parameters and shorter logging intervals will of course use memory more quickly and so shorten the interval between logger downloads. Selecting 5 minute resolution with spot temperature, maximum and minimum recording permits at least 4 weeks record before the memory becomes full and starts over-writing once more. A few minutes' data will inevitably be lost when the logger is temporarily removed for downloading, so try not to change the logger near a time of maximum or minimum temperature – the morning observation is often a suitable time to do this.

3. Expose the temperature sensor adjacent to the equipment to be checked

Expose the flying-lead PRT adjacent to the sensor whose calibration is being checked. Assuming the sensor has a minimum 500 mm or so of cable, it should be easy enough to locate it exactly where it is required. In a Stevenson screen (or similar), expose the PRT close to (but not touching) the main air temperature sensor. Things are a little more complicated with smaller plastic AWS radiation shields, as it is more difficult to see where the sensors are, but try to fix the sensor in place as close as possible to the AWS temperature sensing element without actually touching it. Check that it is not exposed to direct or indirect solar radiation through the 'saucers' of the radiation screen where one is in use (try shining a small torch through them at dusk). Whether in an AWS screen or conventional thermometer screen, secure the PRT and its lead with cable ties or weatherproof tape to ensure it cannot work loose. (When removing the unit, ensure you do not accidentally snip through the sensor lead as well as the cable ties.)

When checking other sensors, for example grass or earth temperature sensors, locate the unit as close as possible to the sensor to be checked. For a grass temperature sensor, ensure the calibration PRT is located alongside rather than on top of the unit being checked, as otherwise outgoing radiation and thus the indicated temperatures may be affected.

Finally, connect the calibration PRT to the calibrated logger itself. Minimise any thermal inertia effects from the body of the logger itself by locating it some distance from the sensor(s) in use. In small radiation screens, there is unlikely to be sufficient internal space to house the logger, so trail the lead carefully outside the screen and secure the logger to a convenient external mounting point. Ensure the cable connecting logger to PRT is not snagged or stretched, as it is easily damaged: the cable connector plug must also be screwed tight into the logger port to avoid moisture ingress. The logger should not be located where it will itself affect the temperature record within the radiation screen (by warming up in sunshine, or blocking ventilation, for example). The logger itself is weatherproof and can safely be left exposed to the elements, provided the risk of theft or vandalism is small.

Allow the sensor and logger to settle to the outside temperature before commencing logging (a delay-start option is ideal for this purpose), or ignore the first 30 minutes or so of readings to allow for settling. If the logger has been in a centrally heated room and is then taken outside in winter, it may be 20 degrees or more warmer than the ambient air temperature, and while cooling down it will affect the readings obtained to a decreasing extent.

4.   Log comparison data

Leave the calibration logger to record alongside your existing sensors for as long as possible (at least 2–3 weeks). The logger itself will require removal for downloading to the host computer at regular intervals as its memory becomes full, but this need involve only a few minutes loss of record (if possible, disconnect the sensor from the logger, leaving the sensor in place so that its temperature is unaffected). The larger the range of temperature covered during the period, the better, because this provides a better estimate of the calibration curve (see below).

5.   Download logger data and apply calibration to logged temperatures

At the end of the logging period, remove the logger, connect to the host computer and download the data using the logger software. Export it into a suitable spreadsheet.

The manufacturer's calibration certificate provided with the instrument will give the logger calibration: this is normally linear across the range of calibration temperatures. Plot these on a graph and determine the slope of the calibration curve (a few mouse clicks in Excel will do this). For example, if the calibration values were as follows:

At −10 °C Subtract 0.25 degC,
At +10 °C Subtract 0.15 degC,
At +30 °C ... Subtract 0.05 degC,

... then in this example the calibration offset at any logged value would be given by

$$\text{Calibration offset} = (\text{Observed temperature } ^\circ\text{C} - 40) \times 0.005.$$

The calibrated temperature is then simply observed temperature + calibration offset.

Lay out a spreadsheet something like this (sample calibration spreadsheets in Excel format, including one to determine the slope of the calibration curve with temperature, are available for free download at www.measuringtheweather.net):

Logger record	Date/time	Logger observed temperature	Calibration offset	Calibrated logger temperature
1	23 Feb 18:15	10.00 °C	−0.15	9.85 °C
2	23 Feb 18:20	9.62 °C	−0.15	9.47 °C
3	23 Feb 18:25			
4				

6. To check logged temperature sensors

Ensure the logged values from the calibration logger and the sensor being checked are at the same time interval (5 minutes is ideal) and that both observations are made simultaneously, for example, at 0900, 0905, 0910 and so on: logged values should be coincident within 60 seconds.

Check the system documentation as to whether the temperature sensor being cross-calibrated outputs a 'spot' or 'sample' value at the logging interval, or whether all samples are averaged over the logging interval (some AWSs allow toggling between these two options).

If the values are 'spot' values, then these can be compared directly with the 'spot' calibrated logger values as described in section 5 above, although bear in mind both are subject to their respective sensor time constant, which may differ.

If they are averaged over the logging interval, it is best to compare them with a pseudo-average derived from the calibrated logger data. For short data intervals, the average of (spot value at beginning of logging period + spot value at end of logging period + observed maximum during logging period + observed minimum during logging period) will be very close to the sampled average – and this is very easy to calculate in the Excel table.

Paste into the existing logger spreadsheet the appropriate data from the sensor being checked, taking care to ensure that all data values are coincident in time. (During summer time, ensure both loggers are operating to the same time standard – UTC, local regional time or summer time. If the transition from summer to winter time, or vice versa, happens during the comparison period, check both loggers have handled the transition correctly. It is much simpler to use one time standard throughout the year, of course.) Once completed, the comparison table will now look like this:

Logger record	*Date/time*	*Logger observed temperature*	*Calibration to be applied*	*Calibrated logger temperature °C (from Step 5)*	*Temperature of sensor being checked °C*	*Difference*
1	23 Feb 18:15	10.00 °C	−0.15	**9.85**	**9.82**	**+0.03**
2	23 Feb 18:20	9.62 °C	−0.15	**9.47**	**9.45**	**+0.02**
3	23 Feb 18:25					
4						

The 'difference' column is (calibrated logger value minus sensor to be checked) and this is the correction to be applied to the sensor being checked to indicate the same temperature shown by the (calibrated) logger.

With logged data at frequent intervals over a period of several weeks, many more observations are available to provide a good comparison and the optimum times to check cross-calibrations can be extracted from the record. These are cloudy, windy, dry conditions at night (no solar radiation), when the temperature is stable (from the author's experience, a rate of temperature change less than 1 degC per hour is preferable). Occasions to avoid include times when the temperature is changing rapidly, because relatively large transient differences may arise due to response time/ lag effects rather than genuine calibration differences (see **Appendix 1**). Excluding these occasions enables the construction of a more consistent and thus accurate calibration curve. Here is how to extend the spreadsheet to filter out these specifics:

Temperature change In a new column, define a variable that is 1 when the temperature has changed less than 0.25 degC (in either direction) in the previous 15 minutes (i.e., a rate of 1 degC/hour). This value is not critical, and it may be increased a little if too many cells are being excluded from the analysis to obtain representative results.

Day/Night Set out a new column that has Day = 1 and Night = 0. Defining day/ night periods is easy if solar radiation data are available from a pyranometer (day = pyranometer output positive: make the threshold slightly above zero to allow for the slight zero offset of these instruments). If no pyranometer record is available, a table of sunrise/sunset times will provide these (see Chapter 11, *Measuring sunshine and solar radiation for sources*); enter 0 or 1 in the column for each observation time.

Wind speed If wind speed data are available, set out a third column to indicate 1 if the wind speed at the observation time is above a pre-set value, say 10 knots or 5 m/s to start with, else leave it zero. Remember that if the anemometer is located well above screen height, the anemometer-indicated wind speed may not be representative of screen-level wind speeds and a higher threshold may need to be chosen, and in any case airflow within the screen or radiation shield is unlikely to be any more than about one tenth of the external wind speed [112]. Again, adjust the value if there remain too few cells in the analysis when the filter is included.

If wind speed data are not available, use only the temperature change and day/ night splits. Rainfall is another factor that can make a difference when comparing between screen types (louvred screens tend to stay wetter for longer than the smaller plastic AWS radiation screens and can therefore appear cooler for a time owing to evaporative cooling effects), but it is difficult to define from recording raingauge data alone how long a surface will remain wet once the rain has stopped.

Next draw a scatter plot of the observed sensor value (horizontal *x* axis) versus the difference from the calibrated logger values (vertical *y* axis), as in the previous example. Using Excel, evaluate the equation of this line, which may be linear (varies in a straight line with the thermometer reading) or polynomial (a curve which includes more than one term). The better the spread of data points, the better the calibration result. Using the Excel Filter function, evaluate the curves for (a) all observations and (b) cloudy, windy nights only with steady temperature – the latter will have far fewer observations (and a reduced temperature range) but a smaller range of variance and thus perhaps a more accurate derived calibration.

The calibration curve obtained should then be applied to all future logged values – some systems will allow programmed calibrations to be applied as the values are logged, with others they must be adjusted in a post-download spreadsheet. Make a note in the station metadata (see Chapter 16, *Metadata, what is it and why is it important?*) of the calibrations applied, and the date they were introduced. Retain the calibration comparison test results, as these will be useful to refine calibrations if a wider range of temperature data becomes available (if perhaps the initial calibration run in a winter month can be followed up some months later by a summer comparison), or when the exercise is repeated every 12 months when checking for calibration drift.

Strictly speaking, the derived calibration curve is valid only over the observed range of temperatures examined (this is the reason why it is a good idea to undertake one cross-calibration run in winter and another in summer and pool the results), but in practice the results can normally be extrapolated using the derived calibration equation for at least a few degrees Celsius above and below the upper and lower observed experimental values.

7. Check regularly for calibration drift or sensor malfunction

The calibration of any temperature sensors can change over time: electrical sensors occasionally go awry for no obvious reason. Whichever type of instrument is in use, it is therefore advisable to repeat this calibration test every 12 months, or as soon as practicable if the sensor is suspected of malfunctioning.

Calibrating humidity sensors

Humidity sensors can be calibrated in a laboratory environment using a variety of chemicals which will produce a known relative humidity within an enclosed environment. However, this approach is rarely practical for *in situ* calibration, and while a cross-calibration process is less rigorous than a laboratory calibration, it may be better suited for operational sensors. The process is essentially identical to that for cross-calibrating temperature sensors, with the following provisos:

- No two humidity sensors will agree exactly for very long; agreement to within 2–3 per cent is perfectly satisfactory.
- Avoid using observations where one sensor remains close to saturated while the other begins to dry out. These circumstances can give rise to large transient differences owing to time constant/lag effects and hysteresis (see Chapter 8, *Measuring humidity* and **Appendix 1**) rather than true differences in calibration. Including them in the calibration curve will bias the results obtained.
- The sensor 'ceiling' (maximum indicated humidity) in saturated air may be as low as 94–95% on some sensors. Calibration comparisons at high humidities should be treated with care.
- Best results will be obtained for readings in the range of 50 to 85 per cent humidity, with reasonable sensor ventilation (= screen-level breeze), and when the humidity is not changing too quickly. Afternoon humidity values can vary rapidly in turbulent boundary layer conditions, by several per cent in a few minutes depending upon airflow, and it may be best to smooth both compared and calibrated values by averaging over, say, 15 minute periods, and comparing these results, rather than 5 minute 'spot' values.
- Humidity sensors tend to have a shorter 'calibration lifetime' than temperature sensors, and calibration checks should be carried out every 6–12 months, or if readings become erratic.
- The raw readings should be adjusted in line with the revised calibration as appropriate.

As with other sensors, note any calibrations derived and applied in the site metadata, with the date they were applied.

Checking calibration drift on pressure sensors, including barographs

Chapter 7, *Measuring atmospheric pressure* gave details of setting or correcting
barometric pressure sensors to mean sea level. Entry-level AWS models often
output barometric pressure readings to a precision of only 1 hPa (1 hectopascal, or
hPa = 1 millibar, or mbar), despite their poor accuracy, which may be ± 5–10 hPa. For
accurate meteorological and climatological purposes, a precision of 0.1 hPa is
required. However, sensor *precision* to 0.1 hPa does not imply *accuracy* to 0.1 hPa,
and the calibration should be checked – and regularly rechecked, at least every six
months – to guard against calibration drift. Drift is inevitable, even in the best
sensors: a good electronic sensor should drift by 0.1 hPa per year or less, but
a household aneroid barometer or small-scale barograph may drift by much more,
perhaps several hPa over a year. Differences in range or scale may also become
apparent across a wide range of pressures – a barometer or barograph that is correct
at, say, 1030 hPa may read 5 hPa high or low at 970 hPa, owing to insufficient
magnification of the sensor movement.

Calibration of pressure sensors in a pressure chamber is expensive, but sensor
accuracy (after correction to mean sea level, or MSL) can be quite easily bench-
marked against the synoptic pressure field using essentially a more detailed version
of the method given in Chapter 7 (section *MSL pressure corrections* – method 1, page
195). This more accurate method uses more stations and requires original readings
to 0.1 hPa. Unfortunately, many state weather service websites display pressure
observations only to 1 hPa (1 mbar) precision, which is not precise enough for
accurately determining calibration drift.

Unless your site is very close to a main reporting synoptic station (within about
10–15 km / 10 miles or so, and at a similar height above sea level – in which case
a single station is sufficient), select at least four synoptic stations, ideally at
similar distances to the north, east, south and west. Locations (latitude and longitude) and maps of observing sites, or plotted maps of station pressures, may be
available on state weather service websites or data aggregation sites (see Chapter 19,
Sharing your observations). Alternatively, MSL pressures from synoptic reporting
stations can be decoded from transmitted observations to the required 0.1 hPa
accuracy from several locations on the web – such as ogimet.com. The pressure
observation is contained within the third (surface pressure) and fourth (MSL pressure) coded groups within the synoptic observation. (Details of the synoptic codes
and how to decode the coded pressure value can be obtained from various meteorological reference sites on the web including Wikipedia – search WMO surface
SYNOP code FM-12). Plot the locations of your nearest reporting stations on
a sketch map with your observing location at the centre. Plot their reported MSL
pressures on an overlay or photocopy of this map, with the date and time of the
observations, then draw isobars (lines of equal pressure) at 0.5 hPa intervals.
Estimate the pressure at your location from the isobars to 0.1 hPa and compare
this value with your own observations made at the same time. (Remember that the
synoptic station observations will always be in UTC – see Chapter 12, *Observing
hours and time standards* for more on time standards – so if your observations are in
local clock time, or summer time, remember to correct for the difference.) Repeat
this at different times of day over a couple of weeks and keep track of the results in
a small spreadsheet. If possible, do the exercise in a period with significant pressure
changes as the calibration or magnification error may vary with pressure, and at

different times of the year as results will probably vary with external air temperatures too.

Include as many observations as possible to minimise outlier errors – occasional large stray differences may result from showery activity, rapid pressure changes at frontal passages, slight timing differences or even observational error. Check every data point and discard any that are obviously outside the normal range to avoid biasing the results obtained. Analyse the results to ascertain how close your barometer readings are to the background field, then adjust future observations/offset your sensor output reading accordingly. On barographs, there is an adjustment wheel to enable the position of the pen arm to be reset in this manner.

When undertaken carefully, this method can identify calibration drift errors down to 0.1 hPa per year. Once set up in a spreadsheet, it becomes easy to repeat every few months as required.

Calibrating other sensors

It is possible to cross-calibrate other sensors *in situ* using similar methods to those for temperature, but the relatively high cost of additional calibrated sensors for other elements (such as solar radiation) makes this an expensive exercise unless a spare calibrated unit can be borrowed for the duration of the test. Of course, unless the calibration of the 'reference' unit is reliably known, using another instrument to adjust calibrations on existing sensors may make matters worse.

For anemometers, the exposure of the instrument is likely to have a greater effect on the readings obtained than any relatively small uncertainties in calibration. Accurate calibration of wind instruments is important, but less important than getting the best possible exposure – see Chapter 9, *Measuring wind speed and direction* for details.

One-minute summary – *Calibration*

- Instrument calibrations are one of the most important, yet sometimes one of the most neglected, areas of weather measurement. Making accurate weather measurements requires accurately calibrated instruments.
- Recording raingauges can be easily and accurately calibrated by passing a known volume of water through the gauge and comparing with the indicated measurement. 'Out of the box' errors for some AWS tipping-bucket raingauges of this type can exceed 20 per cent, so this is a vital test for all new instruments at first installation. Recording raingauges should not be adjusted merely to attempt exact agreement, or near-agreement, with a standard raingauge, because instrumental and exposure differences will always lead to slight variations in the amount of rainfall recorded.
- Two calibration methods are described for temperature sensors. The first is a quick and easy method based on the fixed point of melting ice at 0.0 °C. An extension of the approach can extend the range of calibration points from −5 °C to +40 °C when used with an accurately calibrated reference thermometer. However, this method is not suitable for certain types of sensor, and on some AWS models the temperature elements may not be accessible to allow this test to be undertaken.

- The second temperature calibration method involves careful comparison over a period with a portable reference unit of known calibration. Both sensors (calibrated reference and test) are exposed in identical adjacent surroundings exposures for a period (days to weeks). Plotting differences between the test instrument and the calibrated standard can then be used to prepare a calibration table or datalogger adjustment algorithm, which is then used to apply the corrections obtained to the sensor readings going forward.
- Calibration checks, and checks for calibration drift or scale magnification errors, on pressure sensors can be made using plotted pressure reports from nearby synoptic station reports over a period of a few days or weeks.
- Make a note in the site metadata of all calibrations applied, and the date. Keep a copy of the calibration table or algorithms used in the metadata file. Retain the calibration test results.
- Calibrations can drift over time, so calibrations should be checked (and adjusted if necessary) regularly – at least once every 6 months for pressure sensors and humidity probes, and every 12 months for temperature sensors.

16   Metadata: What is it, and why is it important?

Metadata is literally 'data about data'. In the context of weather records, it is a description of the site and its surroundings, the instruments in use and their units together with any significant changes over time, information about observational databases, where the site's records are archived, and any other details about the measurements that may be relevant.

Why is it important? Because it provides the essential information for any other user of the records to understand more about the location and characteristics of the data, and therefore enables more informed use of the dataset. For example, metadata could make it clear that the anemometer in use at the site was at a different height for the first few years of the record, and that records before and after the change are not homogeneous. Such details may have been known to the observer or station authority at the time, but may not be immediately obvious from the records themselves at some later date. Metadata are especially important for elements which are particularly sensitive to exposure, such as precipitation, wind and temperature. A comprehensive site and instruments description also allows you and other observers to compare records with a degree of confidence, and to be sure that you are 'comparing apples with apples'.

For professional sites, a good account of the site, instruments and their calibration (and any changes) together with details of observing practices, is particularly important for long-period records, or where the records themselves may be required for legal evidence – at a local authority pollution monitoring site, for example. Many 'amateur observer' weather station owners will have this information in their head rather than written down, but it is good practice to write it down and keep it with the station records (and on your website, if you have one), updating it occasionally as things change. Why? Because memory can fade, observers do not live for ever, and good site metadata may enable others to make use of your carefully collected data in the future, perhaps long after you have passed away.

This chapter deals with what should be recorded about records made at the site, including reference to WMO guidelines and template forms ([4], Annex 1. F, Station exposure description), and brief example documentation [294].

More detail on collecting and storing the measurements themselves, particularly digital files, is given in the following chapter.

Metadata – what should it include?

Site metadata should include whatever is relevant, and a short written or tabular description is normally sufficient. The following headings are suggested:

- Station name, location, geographical co-ordinates (latitude, longitude) and height above mean sea level
- Station authority
- Brief geographical context of site and locality, including local topographical map
- Site description and scaled sketch map of site and instruments
- Date records began (and ended, if the site is no longer current)
- Observing hours
- Instruments in use, calibration details, exposure information, record length and changes over time
- Site and instrument photographs
- Station records – location, format, units and so on
- Any other relevant information.

Two examples of metadata files are given at the end of this section, but any format giving the same information will suffice. A sample form in Microsoft Word format is given on the www.measuringtheweather.net website, or alternatively the basic WMO form can be used ([4], Annex 1.F). Use either, and the examples presented here, as a starting point and adapt as required to document your own site information.

Metadata information can be in manuscript or computer file format. It is suggested that a hard copy be retained with printed station records, and soft copy (word processor file or PDF document) be kept with other digital station observational records, including links on websites. (For privacy reasons, and to minimise the risks of vandalism or theft, you may wish to redact some details, such as private postal or e-mail addresses, postal codes, phone numbers and the like, and 'round' positional information slightly, before placing it online or on social media.) The computer file should be included in the same directory as the site data files, and clearly identifiable with the word 'Metadata' in the file name. It is good practice to review the file annually and update it as necessary.

Station name, location and geographical co-ordinates

Include the city, town or village name in the station name, together with an identifier to distinguish it from other sites in the area (perhaps other observing sites you have previously maintained in the same locality). The station name 'White House' is not too specific; better would be 'Washington, D.C. – White House, 1600 Pennsylvania Avenue'.

The location statement should provide brief detail on the area around the observing site, while the geographical co-ordinates (latitude and longitude) define the site precisely. The co-ordinates should include the station altitude above mean sea level.

Latitude and longitude can be obtained most easily from Google Earth (many online mapping sites can also provide reasonably accurate latitude and longitude from an address or postal code), or from a handheld GPS device. When stating latitude and longitude, be clear whether decimal notation (e.g. 41.564°N) or degree notation (41° 33′ 50″N) are being used. (Showing latitude and longitude to three decimal places is accurate to about 108 m in latitude and 76 m in longitude at 45°N.) Altitude is best obtained from detailed local topographic maps, such as the USGS 1:50 000 or 1:24 000 maps or UK Ordnance Survey 1:25 000 series. Altitudes can be read off from Google Earth and some GPS units, but are much less reliable than from local topographical maps (an accurate altitude is essential for reliable barometric pressure corrections to mean sea level, as described in Chapter 7, *Measuring atmospheric pressure*).

Station authority

Specify who runs or manages the station, which may be a private observer, a network authority, a company or some other body.

Geographical context

Give a brief pen picture of the area surrounding the observing site – is it a 'city canyon', a high-density suburb, an open university campus, or rural moorland? Is it a hilltop or valley site? If the former, is the site open and windswept, or does it have local shelter? If the latter, does it tend to collect or dam cold air? What about soil types – is the site in an area of light, sandy soils (which would tend to amplify the observed range in air and grass temperatures), or on heavy clay (which tends to suppress temperature ranges)? How close is the site to the coast, inland lakes and rivers or other significant bodies of water? Proximity to open water may affect many observed weather elements, particularly air temperatures, humidity, and wind speed and direction, while the daily temperature regime at coastal sites may be heavily influenced by the state of the tide.

Include an extract from a local topographical map showing the surroundings for about 3 km / 2 miles around the site, marking the location of the instruments. If the instruments are split across two or more sites, or the site has moved from a nearby location, indicate all relevant locations.

Site description

Narrow down the description to the observing location. Is it a suburban back garden site, a city park, or an exposed hilltop? Is it well exposed to sunshine, wind and rainfall from all directions, or is it partially sheltered by hills, houses or forest? If so, from which direction(s)? Are there buildings or trees nearby – how far away are they? For airfield sites, indicate direction and distance to large areas of artificial surfaces such as runways or aircraft parking areas. Are any of the items noted likely to affect the readings?

Include a sketch plan, to scale, of the observing location, the instruments and the immediate surroundings, identifying the nature and height of all obstructions within 5–10 times their height from the screen or raingauge. For hedges and trees, give their heights (with a date) and indicate whether they are deciduous or evergreen. Suggested methods for surveying sites are given in Chapter 6, *Measuring precipitation*.

Date records began

Include the date records commenced – if this is different for various elements, include the start dates for each. If the site has since closed, include the last date of records too, and whether the record recommenced elsewhere. If the instruments and records at this site were moved from a nearby site, give details, together with any overlapping period of records.

Observing hours

Include details on the observing hour(s) and the terminal hours used for each element. Examples might include:

> 'The observing hour is 8 a.m. clock time throughout the year: maximum and minimum temperatures and rainfall totals for the previous 24 hours are read and reset at this time.'
> 'No manual observations are made at this site: max, min and rainfall are logged by the AWS and refer to the period 00–00 h Pacific Time.'

If maximum and rainfall observations read at the morning observation are 'thrown back' to the previous day in the station record (see Chapter 12, *Observing hours and time standards*), then that should be clearly stated.

Instruments in use, calibration details, exposure information, record length and changes over time

This is necessarily the most detailed section of the metadata summary. List here sufficient detail of the instruments in use to give any future user of the data a clear idea of the exposure, length, quality and reliability of the records obtained, and any significant changes in instruments or exposure over time (for example, the introduction of an AWS replacing thermometers exposed in a Stevenson screen, with the date of the changeover). Use the WMO site classes by instrument type as set out in Chapters 5 to 11. It is particularly important to note details of the thermometer screen, if one is in use, and any changes over time.

Where more than one instrument is used to record the same element (for example, where both a manual and a recording raingauge are in use), state which measurement is regarded as the site or legal standard. Include brief details of the instrument manufacturer and any instrumental calibrations applied, where known, and the date(s) when any changes in calibrations were introduced (see Chapter 15, *Calibration*). Serial numbers and details of calibrations can be referenced or hyperlinked as necessary. For anemometers, note particularly the height of the anemometer and any changes over time, and surrounding obstructions with distances and heights. In the case of AWS, note the logging interval (and, if known, the sampling interval), by element.

For ease of preparation and subsequent reference, it is best to divide the description by element – air temperature, grass and earth temperature(s), rainfall, sunshine, wind speed and direction and so on: see the examples at the end of this section. Where no measurements are made of any particular element, the section can be annotated 'no records of [sunshine] made at this location', or simply omitted altogether.

Site photographs

Include a link in the metadata document (and on any related weather website) to a selection of site photographs, preferably one taken showing the site and instruments from each cardinal compass point (facing north, east, south and west). Take a set in both winter and summer, as the presence or absence of growing vegetation such as hedges and trees may indicate significant seasonal variations in site shelter. Take a series of photographs every year or two, from the same viewpoints if possible, to document any changes in exposure caused by growing trees or the cutting-back of vegetation. Take a hard copy (and your own soft copy) of anything created outside your control, as Google Earth imagery (for example) changes from time to time and may not show how your site looked before that new business park or housing development was built just upwind. If an aerial photograph of the area is available, include that too.

For posterity, why not include a photograph of the observer(s) as well?

Station records – location, format, units and so on

Summarise in this section information about what records are available (elements measured), date of start and perhaps end of records, the time frequency of observations (one record daily, or 1 minute resolution AWS data?) and where the station records can be found (for example, in the local library, or county or state archives). State also what format the records are in – hard copy and/or computer files (more details on storing computerised records are given in Chapter 17, *Collecting and storing data*). For computer files, state the application that was used to create them (such as Microsoft Excel), and include duplicate records in standard, portable formats such as text, CSV or Portable Document Format (PDF).

State the units used for each measured element – the difference between temperature records in Fahrenheit and Celsius will be obvious on inspection, but are wind speeds in miles per hour, metres per second or knots? Have any units changed during the period of record? It is good practice also to include units details in the metadata header for each computerised record dataset – more on archiving the site records themselves in Chapter 17.

Any other relevant information

If previous or local sites have been used to extend the record length or to estimate long-period averages and extremes, for example, it would be appropriate here to include relevant details, periods of record, data sources and suchlike.

Examples of metadata files

Tables 16.1 and **16.2** are examples of metadata statements from two fictional sites. Both files are available on www.measuringtheweather.net and can easily be downloaded and adapted to suit. Something is always better than nothing when it comes to metadata!

The metadata record should be only as long, or as short, as required. Where appropriate, additional details (instrument serial numbers, photographs, site plans and the like) should be referenced in this document, or linked on a website, rather than included in the text, to simplify preparation and maintenance. A document that is easy to maintain is more likely to be updated than one which is not.

Table 16.1 *Example site metadata*

Site metadata for Slapton-on-the-Hill, Devon, England
Compiled on 1 January 2029

Station name, location and geographical co-ordinates	Slapton-on-the-Hill This site is located near Hilltop Farm, Slapton-in-the-Slush Latitude 50.727°N, longitude 3.474°W National Grid Reference SX (20) 960 930 Altitude 225 metres above MSL Latitude and Longitude have been obtained from GPS, NGR and altitude from Ordnance Survey 1:10 000 map
Site authority	University of Slapton, Department of Agriculture
Geographical context	Open and well-drained moorland site, near the summit of a low hill; nearest buildings are 2 km to the north
Site description	Exposed site. Raingauge exposed within a turf wall to minimise overexposure. Site enclosure approx 12 m × 12 m, protected from sheep and passers-by with open-link fencing, no closer than 5 m from gauge. Plan of site is given at (website URL, or location of hard copy) and photographs from cardinal points at (location of images). The exposed hill climate results in some over-exposure of the gauge, and snow falling in strong winds is probably under-represented in the records.
Date records began at this site	Rainfall – 1 May 2025 No other measurements are made at this site
Date records ended	Records continue at the date of writing
Observing hours	Raingauge is read manually once weekly, normally on Mondays, although the date and time are somewhat weather-dependent. Details of the observation date and time are noted with each monthly total.
Instruments in use: Precipitation	Standard five-inch monthly raingauge, measured with standard 10 mm measuring cylinder. Daily rainfall totals are not available for this site.
Site photographs	A selection of site photographs taken on 2 September 2028 are available at [location]
Station records	Records from this site are held at the University of Slapton Geography Department; copies are sent monthly via website to the Environment Agency and the Met Office.

Table 16.2 *Example site metadata*

Site metadata for Elk City, Wyoming, USA
Compiled on 24 January 2028

Station name, location and geographical co-ordinates	Elk City (White River) The site is located on the White River Farmstead, Elk City, Wyoming 82342 Latitude 43.57°N, longitude 110.83°W Altitude 1937 m (6,355 ft) above MSL Latitude, longitude and altitude have been obtained from USGS 1:24 000 map
Site authority	This is a privately run site: Observer – John Doe
Geographical context	Low-density suburban site. Valley side, gentle well-drained south-east-facing slope, sandy loam soil, drains freely to the McKay river, approximately one mile east at closest point.

Table 16.2 (*cont.*)

Site metadata for **Elk City, Wyoming, USA**
Compiled on 24 January 2028

Site description	Suburban backyard, approx 90 × 125 ft, longer axis running roughly ENE-WSW. Reasonably well exposed between south-east and west, rather sheltered by buildings and trees especially to north and north-east. Nearest buildings: house, 38 ft tall at apex, 85 ft N of raingauge. Deciduous tree 30 ft tall 50 ft NE of raingauge. Plot bordered by open fence 6 ft tall to south and on all other sides by conifer hedging 4–5 ft high.
	Plan of site centred on the standard eight-inch rain gauge is given at [location or hard copy] and photographs from cardinal points also centred on the rain gauge at [location of photographs].
Date records began at this site	Precipitation – 1 October 2012
	Air temperature – 1 March 2014
	Wind speed and direction – 1 May 2016
	Solar radiation – 1 March 2020
	Instrumental details given below
Date records ended	Records continue at the date of writing.
Observing hours	Manual observation (manual raingauge) normally at 8 a.m. clock time throughout the year.
	Continuous 5 minute AWS record since 1 May 2016: air temperature, relative humidity, wind speed (mean and gust) and wind direction, rainfall and barometric pressure. AWS time is Mountain Time all year round (not adjusted for Daylight Savings Time).
Instruments in use	Wireless tipping-bucket raingauge (TBR) capacity 0.05 inch installed 1 October 2012: hourly records continued until May 2016, when 0.01 inch TBR record available from AWS
Precipitation	
	Standard eight-inch NWS raingauge installed on 15 April 2015, mounted above short grass, rim approximately 4 ft 6 in above ground. Read once daily using standard measuring tube at 8 a.m. clock time. Units: inches. Daily rainfall totals taken from this instrument since installation date (a 12 month overlap with the wireless unit showed the latter to read approximately 18 per cent low). Resolution: 0.01 in
	Davis Instruments Vantage Pro2 AWS installed 1 May 2016, located 3 m east of the raingauge: 5 minute rainfall totals are available from AWS TBR. TBR calibration checked and adjusted annually (last checked on 11 March 2027)
Air temperature	Daily maximum and minimum temperature record from liquid-in-glass thermometers in double-louvred Cotton Region Shelter commenced 1 March 2014. Thermometers 5 ft above short grass. Record discontinued August 2017 after 12 month overlap with AWS record.
	AWS temperature records from Davis Instruments Vantage Pro2 AWS commenced 1 May 2016, mounted in Davis AWS passive radiation shield (not aspirated) at 5 ft above short grass, 12 ft east of Stevenson screen. Maximum and minimum temperatures from AWS refer to civil day 00–00 h Mountain Time. Calibration of temperature sensor checked and adjusted over 6 week period March–April 2022 and again June 2027 using a portable calibrated reference logger (see [location of calibration test results]). Calibration adjustments included in real time by means of software offset.
	Units: °F. Resolution: 0.1 degF.
Relative humidity and dew point	Humidity measurements made at 5 min resolution using the Davis AWS since May 2016.

Table 16.2 (*cont.*)

	Manufacturer's sensor calibration used without adjustment (not checked). Humidity units: % RH. Resolution: 1%.
	Dew point calculated by AWS software from observed temperature and relative humidity. Units: °F. Resolution: 1 degF.
Barometric pressure	Barometric pressure (5 min data) from Davis Instruments Vantage Pro2 AWS, since May 2016. Pressure sensor mounted within AWS display unit located in a second-storey room of observer's residence, logged as station level, not mean sea level, owing to location altitude. Calibration checked twice annually against hourly pressure readings from Brighton airport (14 miles west) and Flag City USAF (7 miles south of the site), allowing for altitude differences. Units: millibars. Resolution: 0.1 mbar.
Wind speed and direction	Wind speed and direction records commenced with installation of the Davis AWS in May 2016.
	From May 2016 to June 2025, anemometer and wind vane were mounted on 10 ft mast (the AWS system unit is also mounted on this mast), with wind speed logged in miles per hour.
	On 20 June 2025 the Davis anemometer and wind vane were relocated to a mast affixed to a chimneystack located 6 ft above the building roofline, at 44 ft above ground level, to improve exposure. From this date, wind speeds are logged in knots.
	Owing to the instrument move, the pre- and post-June 2025 records are not homogeneous.
	Both wind speed (mean speeds and highest gust) and wind direction (16 point compass) logged at 5 minute resolution throughout. Manufacturer's calibration of anemometer sensor has been used. Wind vane referenced to True North, not compass north.
Sunshine and solar radiation	Solar radiation records are made using a Davis Instruments solar radiation sensor, mounted on a 10 ft mast at 10 ft above ground level. Records commenced on 1 May 2016. Exposure is partially obstructed to south-east by deciduous trees (exposure diagram is available at [location]). Instrument is a silicon photodiode, logged at 5 min intervals by Davis Instruments Vantage Pro2 AWS. Units: W/m^2.
	No sunshine records are available for this site.
Site photographs	Site photographs from each of the four cardinal points are taken annually and are available at [location]. The most recent set of photographs were taken on 20 June 2027. A Google Earth aerial photograph of the site and immediate area (half mile radius) is also available at [location].
Station records	Records from this site are held by the observer, John Doe, at White River Farmstead, Elk City, Wyoming 82342. Monthly files are forwarded to the Wyoming State Climate Centre and annual files to the Wyoming Archives Office in Cheyenne.
Averages and extremes	Daily rainfall records exist from September 1974 to date at Grand Rapids Farm, 3.6 miles south-east and at a slightly lower altitude (6195 ft). Over the 10 year period 2016–2025 this site recorded on average 2.6 per cent less rainfall than the current location. The estimated 1991–2020 average annual rainfall for this site, 20.71 in (526 mm), has been derived by applying this factor to the monthly averages for 1991–2020 from the Grand Rapids Farm rainfall record.

One-minute summary – *Metadata – what is it, and why is it important?*

- Metadata is literally 'data about data'. In the context of weather records, it is a description of the site and its surroundings, the instruments in use and any changes over time, information about observational databases and units used, and any other details about the measurements that may be relevant.
- Metadata statements are important because they provide the essential information for any other user of the records to understand more about the location and characteristics of weather records made at any site, thereby enabling more informed use of the data to be made. The 'other user' may be weeks or decades in the future, quite possibly long after all observer recollection of the circumstances of the site and its record have faded.
- A metadata statement is best prepared as a short, structured text document, and retained alongside data files in soft copy or hard copy. A copy or link should also be included on the site weather website, if there is one. Links should also be provided to site photographs, instrument calibration certificates and other related documents.
- Review the metadata statement whenever instrument or site details change, and at least annually. Update as required. Retain previous site descriptions and photographs, which will assist in documenting site, instrument and exposure changes over the years.

PART III

MAKING THE MOST OF YOUR OBSERVATIONS

Having described in turn each of the elements in a typical weather observation, the remaining chapters focus on managing and using the collected records.

17 Collecting and storing data

Making the best use of collected weather observations involves a little thought being given to record management and storage, for after all simply collecting more and more observations is usually a means to an end, rather than an end in itself. Traditional once-daily observations of just a few elements pose few data storage concerns, as they can easily be written up in manuscript in an observations logbook and/or typed up into a small spreadsheet for archiving and analysis purposes. With an AWS in place, however, careful consideration should be given to managing and storing the avalanche of digital data, which otherwise can quickly become unmanageable and difficult to use. Depending upon your needs, the solution may be as simple as a logical file-naming structure (for example – year month data content observation location – such as *2028 04 5 minute AWS data Newtown Observatory*), or as complex as a multifunction spreadsheet or database package. Even a few months of observations can provide useful insights about the climate of a particular location, or high-resolution records of particular occasions such as a heavy thunderstorm rainfall event. The longer the record, the more useful it becomes, and more thought given at the outset to implementing an effective record-keeping strategy will greatly simplify data collection, management and analysis options as the record length grows. The more effectively records are stored, the quicker and easier it becomes to analyse and use them productively – the subject of the following chapter, Chapter 18, *Making the data avalanche work for you*. This statement applies equally to both professional and amateur observers.

This chapter provides tried and tested suggestions on methods for collecting, storing and archiving data from both manual observations and AWSs. The next chapter outlines techniques for analysing data, building upon the foundations set out in this section. Together they should assist in making best use of collected observations, now and in the future.

Familiarity with the use of spreadsheet software is assumed, as detailed instructions are beyond the scope of this book. Many good 'teach yourself' guides are available for the major packages in both hardcopy and online formats.

Sampling and logging intervals

The concepts of sampling and logging intervals were introduced in Chapters 2 and 3. The sampling interval is how often the instrument or sensor is read ('polled' in the case of dataloggers), while the logging interval (the 'archive interval' in Davis Instruments terminology), as its name suggests, refers to how often the element is logged – this may consist of one or many individual samples or averages over a period of time. For efficiency, consistency and simplification in record-keeping, it is best to define a logging interval that meets your particular requirements, and then stick to it. Management and analysis of an AWS record over several years becomes more complicated if the dataset comprises a mixture of records made at several different logging intervals. More 'advanced' AWS software packages may allow more than one logging interval to be set, providing a more flexible approach to data capture – for example, a system could be configured to log a few elements at 1 minute and 5 minute resolution, others at hourly intervals, and daily extremes just once daily at a predefined time, say 9 a.m. It is useful, for example, to log 1 minute records of elements which can and do change rapidly, such as wind speed and direction, but there is little benefit in cluttering up the database with other elements which change much more slowly, such as earth temperatures. The records from such multiple logging interval systems capture both high-resolution data, which is more useful for studying particular events, while simultaneously generating hourly and daily averages and extremes, which go on to build useful climatological datasets.

Collecting data

Storing manual observations

Traditional once-daily manual observations can easily be written up in manuscript in an observations logbook: it is easy enough to rule up blank sheets to create your own bespoke forms, then have them photocopied for record keeping, thereafter filing them in ring binders or having them bound into annual volumes. Over the years the completed logbook volumes will build into a useful hardcopy record of the weather, although of course records in this format do not permit easy computer analysis. Better still, type the observations up into a small spreadsheet for archiving and analysis purposes (more on spreadsheet formats below). Storing records on computer also facilitates making backup copies of the observations in the event of loss or damage to the manuscript records. Unfortunately, all too often observation records which exist only in manuscript registers are lost forever in house clearances after the death of the observer. Maintaining digital records as well as hardcopy reduces the risk of total loss, particularly where the records are regularly copied to a dedicated backup/archive site for such records, as provided in the UK by the Climatological Observers Link Digital Archive for their members [295].

Storing AWS observations

There are three good reasons for archiving AWS records in spreadsheet format – usually in addition to the 'native' or proprietary AWS file format used within the AWS software itself. Exporting the data into a spreadsheet allows for the creation of larger and mixed datasets (longer record lengths, data from different sources), much improved analysis, graphing and presentation capabilities, and (where required) the inclusion of calibration adjustments and automated error checking/quality control methods. Each of these is considered in turn below.

Fortunately, it is normally very easy to generate spreadsheet files from AWS output. Every AWS software package, whether run on a local computer or to a data aggregation site (Chapter 13, *AWS data flows, display and storage*), includes an '*Export* ... ' function, whereby records for a specified date or time period can be exported. Different export file formats are available, the most common (and near-universal) being text-only (.txt files) and/or Comma Separated Variable (.csv) files: these file formats can be read and opened by all popular spreadsheet packages on both Windows and Apple computers. Once imported, the file is then saved in the normal spreadsheet format ('*Save as* ... '). Subsequent records can simply be added onto the end of existing files at a later stage using standard cut-and-paste methods.

Data volumes

Without some thought given to record management, data volumes can rapidly become overwhelming, as **Table 17.1** illustrates. As an indication, logging 30 parameters at 5 minute resolution for 12 months will accumulate in excess of 3 million data points, and the resulting Excel file will be a little over 100 MB in size. If 30 parameters seems a lot, bear in mind that the dataset will need to hold both date (date-month-year format as one number) together with date, month and year as separate fields; that air temperatures need to be stored as 'spot' hourly values, maximum and minimum readings, quite possibly with times of occurrence of the extremes; that wind speeds need to include both mean and gusts, with time and direction of the maximum gust; and so on. These items alone account for 13 parameters over just three data categories (date/time, air temperature, wind speed). For a multi-element AWS, the parameter count can quickly increase to tens or hundreds of individual datapoints for every logging interval, and file sizes and complexity in proportion.

Table 17.1 *Number of unique data points, and consequent approximate file sizes, for various combinations of logging interval and element count*

Logging interval	Records per element per day	Records per element per year	Data length to fill Excel's row limit (1 M rows)	Number of datapoints, 10 years × 30 elements	Approximate file size for 10 years data
24 h	1	365	} > 100 years	109,500	1 MB
2 h	12	4,380		1,314,000	10 MB
1 h	24	8,760		2,628,000	20 MB
15 min	96	35,040	~ 30 years	10,512,000	82 MB
5 min	288	105,120	~ 10 years	31,536,000	245 MB
1 min	1440	525,600	23 months	~ 5 × 10^8	*Not supported*

Note: File size depends upon element type and is given as an approximation only (Microsoft Excel 64 bit, xlsx format)

Most AWS software is intended to poll, process, log and display real-time data, rather than store and process multi-year files; even those that do quickly become unwieldy as their database file sizes become ever larger (Davis Instruments' original WeatherLink software holds a maximum 25 years data, for example, and simply fails once this limit is reached). Once output is held in an external file, however, given sufficient computing resources, there is no upper limit on the size of the resulting database, although access and processing times will increase non-linearly as the file size grows. With spreadsheets, however, there are upper limits. At the time of writing, a single 64 bit Microsoft Excel worksheet (.xlsx format) can contain up to 1,048,576 rows and 16,384 columns of data, and an Excel workbook can contain up to three such sheets (earlier .xls file formats could contain only 65,536 × 256 cells). However, files of this size (hundreds of megabytes) require multiple fast processing cores and plentiful system memory to process efficiently. Keeping spreadsheet files considerably smaller than the maximum permitted makes them more manageable, particularly on laptops and tablets. **Table 17.1** illustrates how long various logging intervals will take to 'fill' Excel's row limit, although it is certainly not recommended to approach this limit. A year of *1 minute* data will require 525,600 rows, plus a few others for means and extremes and so on, and is probably about the realistic maximum convenient working file size; but an Excel file of *hourly* records, with 365 × 24 = 8,760 rows in a non-leap year, will comfortably hold decades of data.

File backups

There is a further important point here regarding backup of working/operational files. The best backup systems are those which run automatically, to a backup location (separate local disk, network attached storage device or cloud storage) and at intervals which you decide. Most host computer-based AWS software runs quietly in the background, using minimum system resources except when (for example) a sensor poll is running. This can create difficulty for some types of backup software, which may simply bypass any working files that are in use when the backup runs to avoid the risk of file corruption. As the AWS software is always running, this may mean that AWS files in use never get backed up. To avoid doubt, check that the date and timestamp of your AWS file backups are current. If they are being bypassed by your backup software, pause the AWS software for a few minutes once a week or so to allow the backup to complete. If all else fails, a simple manual copy of the AWS software output file(s) to another location is better than nothing, but hardly robust.

Where AWS data are written directly to web-based cloud storage rather than a local host computer (see Chapter 13, *AWS data flows, display and storage*), it is probably reasonable to assume that these are regularly backed up, but it does no harm to check that this is indeed the case.

Analysis and presentation

The second main reason to use a spreadsheet to store long-term records is that most AWS software lacks sophisticated data analysis and graphics capabilities, written as it is primarily to poll sensors, to copy short-term data to a logger database and to display fairly basic graphical representations of current real-time or stored data on request. Most spreadsheet packages include a much richer set of data analysis and graphical output functions, greatly simplifying statistical analysis and presentation of stored historical data. (Almost all of the examples in this book have been created using standard spreadsheet software; there are more details in the next chapter.)

Inclusion of calibration and error checking

Some AWS software will permit limited manual editing of the stored records (for example, removing spurious tips from a tipping-bucket raingauge on occasions when the instrument was moved to mow the grass around it), although a line-by-line editing process can be very time-intensive and thus suitable for only very limited corrections to be incorporated in the record. Storing records in a spreadsheet allows for more extensive edits to be made more quickly and easily – perhaps to delete several days of logged TBR records if the funnel became blocked, for example, or to amend observed air temperatures while sensor calibrations were being checked (see Chapter 15, *Calibration*). Simple routines can also be set up to provide basic quality control (to check for sensor readings falling outside normal limits owing to intermittent faulty connections, for instance), flagging appropriate corrections or deletions as a result.

More sophisticated AWS systems and some dataloggers will allow adjustments to calibration corrections to be applied across a range directly to the sensor reading as it is logged, although most budget AWS software includes only 'fixed offset' capabilities at best. Where required, calibration corrections or offsets can be quickly and easily applied to 'raw' AWS data in a post-processing step by using a spreadsheet. Both 'raw/as logged' and 'corrected' values should be archived, so that any changes can be undone or amended at a later stage if necessary.

Missing data

It is usual practice to indicate missing data in climatological datasheets by an entry such as –999 or similar, but in Excel spreadsheets missing data cells are best left blank to avoid corrupting statistical analyses of means and extremes. Blank cells are simply ignored in calculations. However, if records are to be exported from Excel for use in other software or to be included in other datasets, it is best to check whether there is any preferred import specification to mark such as 'missing data' cells, to avoid exporting incorrect or misaligned entries.

Legality of altering records

Where amendments are made to an AWS record, whether they be amendments, estimated data or deletions, the change should be recorded in a metadata entry detailing both 'as logged' and 'corrected' entry or entries, with a brief explanation for the deletions or amendments. No changes should be made to the logged record without good reason, but of course it is valid to make corrections or deletions to maintain record accuracy or completeness. (One example, already referred to, would be deleting or estimating air temperatures while the various sensors were undergoing calibration checks.) Note that if a corrected AWS record is produced in support of a legal case then both 'as read' and 'corrected' records (if any) should be presented, along with the explanation for the amendment, to avoid raising any possible doubts regarding data integrity.

Managing the data avalanche

Climatological observations tend to fall naturally into four types of dataset, namely sub-hourly, hourly, daily and monthly observations. Where data are available for each timescale, a good way of managing the record is to set up a separate spreadsheet for each – because the elements covered will be different. For example, the hourly and sub-hourly datasheets might include 'spot' temperatures at each hour, the daily dataset 24 hour maximum and minimum temperatures, and the monthly dataset monthly and annual means and extremes of daily maximum and minimum temperatures. It is perhaps easiest to think of each spreadsheet table as a logbook page, albeit a page of almost infinite length.

Using Microsoft Excel's 'tabbed worksheets' feature, data for each timescale can be held in separate tabs within a single spreadsheet file, although as the tabbed worksheets grow over time the file size will become rather unwieldy and separate files, rather than separate tabs within one file, become easier to manage. Once-daily manual observations can be easily accommodated by a single spreadsheet containing two tabbed worksheets, one for daily values and the other summarising monthly totals or means.

Content, format and layout suggestions for each of the four types of spreadsheet are given below.

Sub-hourly spreadsheet

Sub-hourly data are most useful for detailed analysis of particular events or series of events, such as examining changes in wind speed and direction, air temperature and dew point during the passage of a frontal system, or a minute-by-minute analysis of rainfall intensity during a heavy thunderstorm. The data quantity (half a million rows of data per year for 1 minute logging intervals) simply rules out concatenating such files into multi-year sub-hourly records for combined analysis unless more sophisticated data processing facilities are available. Long experience with high-resolution AWS data suggests that the optimum archive strategy is to write sub-hourly output files into annual folders using simple but consistent file structures and file naming conventions (for example: year month – data content – observation location – example *2028 04 5 minute AWS data Newtown Observatory*), then backing those files up in the normal manner. Doing so ensures the files remain readily accessible should sub-hourly data be required for any particular date or series of events in the future.

Hourly spreadsheet

Data for the hourly worksheet comes from an AWS – either directly, where the logging interval is 1 hour, or by 'distillation' from higher-frequency logged data (such as 5 minute observations), most easily pre-processed using another small template spreadsheet and then cut-and-pasted into the hourly table.

Distilling observations

Where only a single logging interval option is available (typical of many AWS below professional systems), then elements logged at sub-hourly intervals (such as 5 or 10 minutes) can always be 'distilled' to derive hourly records, although of course the reverse is not the case. 'Distilling' sub-hourly records into hourly, hourly into daily and then daily into monthly records greatly simplifies data handling and is easily accomplished using Microsoft Excel's Pivot Table function (see Chapter 18, *Making sense of the data avalanche*) or with template spreadsheets. For the latter (sample templates for 5 minute data to hourly, hourly to daily and daily to monthly are provided on www.measuringtheweather.net and these can be easily adapted to suit) simply cut-and-paste blocks of exported AWS data into the 'data' tab, then copy and paste the required distillation from the 'output' tab. The arithmetic is straightforward (for example, the 'hourly to daily' spreadsheet simply sums all the hourly rainfall values to give a 24 hour total covering the relevant period).

Note that the sample templates assume identical numbers of observations per interval (namely 12 × 5 minute observations in every hour), and that if one or more rows of observations are missing, duplicated or replaced by observations at a different logging interval then the pre-programmed functions will produce erroneous results. In such cases, use Excel's Pivot Table feature in preference.

Content obviously depends upon instrumentation and archiving requirements, but typical elements in a basic dataset might include the elements listed in **Table 17.2**. If possible, ensure the order of elements in the spreadsheet is the same as in the exported AWS file, as this will simplify the cut-and-paste operation.

It is a good idea to include a tabbed metadata worksheet detailing the origin of each of the measurements (sensors, brief exposure details, height of anemometer, etc.), and the units used. It may be obvious from inspection whether air temperatures are in degrees Celsius or degrees Fahrenheit, but not so obvious whether wind speeds are in knots, miles per hour, kilometres per hour or metres per second. Note also that some elements will be 'spot' values (such as air temperature 'on the hour') while others will be period means or extremes (such as hourly mean wind speed, and maximum gust speed) – the exact derivation should be stated in the metadata sheet as again it may not be obvious, or it may change over time.

This fairly basic hourly spreadsheet layout (**Table 17.2**) includes 15 elements and would be suitable for preparing sub-hourly output from a Davis Instruments Vantage Pro2 AWS or similar, with little amendment. (Sample Excel templates, with metadata tab included, can be found on www.measuringtheweather.net.)

Table 17.2 *Example of hourly dataset layout and format*

Column	Content of column	Notes
Date	Include date in standard numerical form (d.mm.yyyy – example 20.06.2099). Within most spreadsheet software the actual date can be entered in one way and output formatted another – so the dates might come from the AWS logger as '20.12.99', but they can easily be output as '20 December 2099' (or almost any other combination) if required. This is useful when European (d m y) and American (m d y) date conventions differ. To avoid transposition errors between date and month, check which date convention is in use within the AWS/logger software, and whether it can be altered to suit	As will be seen in the next chapter, there are analysis advantages to holding the date in separate dd, mm and yyyy entries in addition to the date in standard form. Note that if dates are held in yyyymmdd format, they will be automatically sorted into date order in file structures held on computer – this is especially useful when creating file names (for example '2099–05 observations' and '2099–06 observations' will by default be filed in the correct order, whereas 'May 2099 observations' and 'June 2099 observations' will not (they will appear alphabetically, adjacent to the May and June observations for other years)

Table 17.2 (*cont.*)

Column	Content of column	Notes
dd	Numerical day number from date	In Microsoft Excel, this is the =DAY(*) function, where * is the cell containing the full date
mm	Numerical month number from date	In Microsoft Excel, this is the =MONTH (*) function
yyyy	Numerical year number from date	In Microsoft Excel, this is the =YEAR(*) function
HH	Hour – use one time standard (UTC, or local regional time) throughout	Specify in metadata whether the averages and extremes refer to the hour *ending* at this time (most AWS default to this), or the hour *commencing* at this time The spot values should refer to the logging interval ending 'on the hour'
TT Tx Tn	Air temperature – hourly spot value 'on the hour', maximum Tx and minimum Tn temperatures within the hour (three variables)	State in metadata whether the sensor is in a screen, with type and height, etc., units of measurement, and any calibration applied. Best practice is to include units in column names if space permits – for example, 'Tx_C' for maximum temperatures in °C, or 'Tx_F' in Fahrenheit
RH_%	Relative Humidity – normally a spot value 'on the hour'	Give instrument details in metadata. Units can be assumed to be %.
Td	Dew point temperature 'on the hour'	Normally an AWS derivation from air temperature and RH readings rather than a measurement. State units
ff mean ff gust	Wind speed	Normally the (scalar) mean wind speed and the highest wind gust in the preceding hour. State in metadata anemometer height and type, any calibration applied and especially units of measurement
dd2	Wind direction	Normally the value 'on the hour', although a macro can be set up to calculate the more accurate vector mean wind (speed and direction) over the hour from higher-frequency measurements (see Chapter 9, *Measuring wind speed and direction* and **Appendix 5**). Note however that Davis Instruments systems log 'modal wind direction': using spreadsheet macros, compass points can be easily converted into bearings (i.e., SW becomes 225 deg) which simplifies numerical analyses. An Excel macro to do this is available from www.measuringtheweather.net. State in metadata wind vane orientation (True north or compass north), height (if different from anemometer) and derivation of measurement (vector mean wind or modal wind direction)

Table 17.2 *(cont.)*

Column	Content of column	Notes
Pressure	Barometric pressure 'on the hour'	State in metadata whether the pressure is station level, or corrected to mean sea level, and if so, what method is used: also units of measurement. Include both station level (P_stn) and msl pressure (P_msl) if possible.
Rain	Hourly precipitation total	State in metadata raingauge type, height, tipping-bucket capacity and units of measurement.
Optional parameters		For analysis and summary purposes, it can be useful to derive certain parameters by reference to the logged value – for instance, if a parameter 'RainHour' is set = 1 when 'rainfall in the hour > 0' and 0 otherwise, then hours with rainfall can be flagged for analysis using Excel's 'filter' functions (see next chapter). The template spreadsheet includes a few examples; others can easily be added over time as required.

Depending upon sensor availability, hourly sunshine, hourly mean and maximum solar radiation, earth temperatures and other elements can easily be included. If it is planned to add other sensors at a later date, setting out a spreadsheet format which accommodates current observations yet is flexible enough to expand as required will avoid duplication of effort at some future date.

Daily spreadsheet

Data for this worksheet may consist of a mix of AWS and manual observations. AWS data can be distilled down to daily totals/averages/extremes within the terminal hours from hourly records as required. Specific content is again dependent upon instrumentation and archiving requirements, but typical elements in a basic daily dataset might include the items in **Table 17.3**. As there is likely to be a mixture of terminal hours (some will be morning to morning, some will be midnight to midnight, others will be spot observations – see Chapter 12, *Observing hours and time standards*), brief metadata descriptions are essential to describe content (some of the metadata will be identical to the hourly spreadsheet and can be cut-and-pasted).

Many observers also use the daily dataset to record the elements from one or more manual observations during the day (see Chapter 14, *Non-instrumental weather observing* for details). It would be easy enough to include columns to include, say, cloud cover and types, wind direction and speed, barometric pressure,

Table 17.3 *Example of daily dataset layout and format*

Column	Content	Notes
Date	Include date in standard numerical form (d.mm.yyyy – example 20.06.2099) or yyyymmdd, according to preference	As will be seen in the next chapter, there are analysis advantages to holding the date in separate dd, mm and yyyy parameters in addition to the date in standard form.
dd	Numerical day number from date	In Microsoft Excel, this is the =DAY(*) function, where * is the cell containing the full date.
mm	Numerical month number from date	In Microsoft Excel, this is the =MONTH(*) function.
yyyy	Numerical year number from date	In Microsoft Excel, this is the =YEAR(*) function. Hold as four digits rather than two, to ensure dates on either side of year 2000 are sorted into correct order, and to ensure correct calculation of period lengths.
TT	Mean air temperature	Derive the midnight to midnight mean temperature from the mean of the 24 hourly observations.
		State in metadata whether the sensor is in a screen, with type and height, etc., units of measurement, terminal hour, and any calibrations that have been applied.
TTmax	Maximum air temperature during the day	State period covered in terminal hours metadata – is it morning to morning or midnight to midnight? If the former, is the maximum temperature 'thrown back' to the day preceding the date of morning observation?
TTmin	Minimum air temperature during the day	State period covered in terminal hours metadata – is it morning to morning or midnight to midnight?
Rain	Total precipitation during the 24 hours	State period covered in terminal hours metadata – is it morning to morning or midnight to midnight? If the former, is the rainfall 'thrown back' to the day preceding the date of morning observation?
		Is this from a manual checkgauge, or the sum of hourly TBR data? (Both can be held, of course.)
		State in metadata raingauge type/s, height, tipping-bucket capacity and units of measurement.
ff	Mean daily wind speed	Normally a daily mean (scalar) wind speed, midnight to midnight.
		State in metadata anemometer height and type, units of measurement, and any calibration applied.
ff-gust	Highest wind gust	Normally refers to the same period as the mean daily wind speed. Instrument and units can be assumed same as for wind speed.
dd	Wind direction	Use either the AWS 'daily' output (Davis Instruments AWSs output a daily 'modal wind direction') or vector mean wind from sub-daily observations – see **Appendix 5** for calculation details.
		State in metadata wind vane orientation (True north or compass north), height (if different from anemometer) and derivation of measurement (vector mean wind or modal wind direction).
MSLP	Daily mean barometric pressure	Normally a daily mean midnight to midnight. State in metadata whether the pressure is corrected to mean seal level, and if so, what method is used: also units of measurement. You may also wish to extract daily (00-00h) maximum and minimum pressures too.

air temperature and humidity from a daily 8 a.m. morning observation if required, alongside AWS-generated fields. Obviously, the manual observation records will need to be entered manually using a keyboard rather than imported from the AWS.

This basic daily spreadsheet example in **Table 17.3** includes 12 elements: a template can be downloaded from www.measuringtheweather.net for editing to individual requirements.

As well as details from non-instrumental 'eye' observations, the daily spreadsheet can also include records of 'days with' elements – include columns for snow or sleet observed to fall, thunder heard and suchlike. (Prepopulate them all with 0, and when one of these events occurs within the relevant time period, amend the 0 to 1. This will enable both automatic summing and filtering/analysis of such events.) Many observers, including myself, make a coding distinction between some elements, such as differentiating between different sizes of hail by means of a second column. In such cases, the first column is a 'binary' entry (0 or 1 only), the second column in this case could be used to note the maximum size of hailstones (state units in metadata), remaining as 0 when no hail is observed. To avoid double-entry, the first column can easily be set automatically to 1 whenever there is an entry greater than zero in the second column. Similar coding methods could be used to note thunderstorm intensity on a scale from 1 to 5, intensity of snow events and so on. The coding convention is up to you, but note it in your metadata (and be consistent over time)

It is also good practice to include a free text column to note 'significant weather' such as the time of thunderstorms, snow depths and any other 'significant weather' if such information is available.

You may also wish to include also 'derived' or calculated binary flags (0 = no, 1 = yes) for the following parameters, as these will simplify later analyses:

- Air frost (minimum temperature < 0 °C)
- Ground frost (grass minimum temperature < 0 °C)
- Ice days (maximum temperature < 0 °C)
- Hot days (maximum temperature ≥ 25 °C) – or any other applicable threshold
- Rain days (daily rainfall 0.2 mm/0.01 in or more), wet days (daily rainfall 1.0 mm or more), or other rainfall thresholds as required
- Nil sunshine (daily sunshine = 0)
- ... or any other specific thresholds that may be of interest.

The sample spreadsheet includes some of these as examples, but others can easily be added as required, or included at a later date.

Monthly spreadsheet

The monthly spreadsheet is normally derived from the daily datasheet and will usually include summarised monthly values such as totals, means and/or extremes, with dates, of the elements in the daily observational record. An example spreadsheet is available at www.measuringtheweather.net.

Preserving your observations

Weather measurements are more interesting and useful when analysed and shared, either in real-time on weather websites or forums, or perhaps via monthly observation-exchange agreements with other sites in the area, or by joining enthusiast organisations such as CoCoRaHS in the United States and Canada, the UK's Climatological Observers Link or one of the other national amateur observer organisations (more in Chapter 19, *Sharing your observations*, and details of meteorological societies in **Appendix 3**). Professional weather observers can probably assume that their observations are transmitted and stored on reliable computer systems, regularly backed up, and that the data will eventually be securely archived in a purpose-built data storage facility. Unfortunately, amateur observers cannot rely on this happening by default unless they take those steps themselves: the rest of this section is intended primarily for such observers.

You may wish to ensure that your observations are preserved for future researchers, perhaps many years after your death. But what would happen to your records if you were to drop dead tomorrow? The sad fact is that many amateur observers' records (and often their instruments, too) are simply thrown away in post-funeral house clearances – and this applies just as much to home computer or laptop records as well as manuscript logbooks. It is tempting to think that computerised records are 'future proof', but in fact the opposite is true – frequent changes in hardware, software, storage media and operating systems make computerised records the *least likely to survive* for any length of time without careful record management. If you feel this is unnecessarily pessimistic, consider whether your records stored in VisiCalc, written under MS-DOS on an 8088-based PC and stored since 1984 on an eight-inch floppy disk, would now be readable by anyone else? At the time of writing, I can still read archived observations written to Excel 97–2003 worksheets over 20 years ago, but only by overriding a security default in Microsoft 365. Will such files, even USB memory sticks, still be readable in a decade or two?

Most people have a hugely over-optimistic opinion of the lifetime of various computer media. Consider the following examples:

- Modern solid-state disks, widely used on desktop and laptop computers, have a finite 'read/write' limit and will eventually expire. This limit is unlikely to be reached in normal usage, but an 'always on' computer serving an AWS writing data to and from a disk every few seconds may be at greater risk of premature failure;
- Similarly, the magnetic media from which memory sticks are made has a limited number of read/write cycles, typically only a few thousand, and cannot be relied upon to retain data for more than a few years, sometimes less;
- The expected lifetime of an external USB 'spinning disk' hard drive is typically only 3 years, less if permanently 'on' or subject to knocks and vibration, and a catastrophic failure (all data lost) becomes more likely than not after that time.
- At one time, optical media such as CD-ROMs were seen to be the answer to long-term archival storage, but even here the media can be expected to decay sufficiently to introduce data errors within 10–15 years. In any case, very few modern computers are now fitted with optical drives, although add-on USB units are still available; the media are now relatively expensive in comparison with memory sticks or solid state drives.
- Beware of 'archive-quality' media options, typically at higher prices. It is entirely academic whether a 'ruggedised' memory stick or gold-coated CD or DVD 'guaranteed' to last 50 years (will you still be around to claim if it doesn't?)

is of benefit if, firstly, there are no longer any CD readers to read the disk and, secondly, whether the file formats themselves will still be readable more than a decade or two in the future.

The history of the personal computer industry over the last 30 years or so does not give grounds for optimism regarding future compatibility!

Even hard copy (printed) output is not future-proof. Laser-printed hard copy output may deteriorate beyond readability, or facing pages adhere to each other, in 10 years or less, 20 at the most. And if you keep all your observations in a manuscript book, what happens if your house is struck by lightning, flooded, burgled or has a disastrous fire? The answer is, of course, you will probably lose the lot.

However, taking some basic record management/archiving decisions now can significantly increase the chances that your records will survive more than a decade or two. Here are a few suggestions:

- Holding multiple copies of your records, in different places and in different formats (hard copy, computer records), will significantly increase the chances of your records surviving a catastrophe. Keep multiple copies – computer files, hard copy printed material of the key documents and cloud storage/memory stick/external hard disk – and ensure digital files are written and rewritten regularly into the current version of a popular format, such as Adobe PDF or Microsoft Office (Word, Excel) for documents, or JPEG/TIFF for image files. Open old files (including those written using previous versions of the software currently in use) and periodically 'Save as' into current version file formats.
- Make sure the file naming convention and backup management process avoids any risk of overwriting newer files with older ones of the same name (by including the date or a version number in the file name, for example), and by always backing up only from one specific location to another backup volume – never the other way around, unless of course a damaged file is being restored from the backup medium.
- Sharing observations with other observers or groups which publish their records, such as one of the various national amateur observer organisations, will increase the chances of copies of your observations surviving.
- Back up all key files (sub-hourly, hourly, daily and monthly spreadsheets, AWS files, weather diary files, metadata, station photographs, etc.) to a separate and preferably offsite location – daily. Backup software is not expensive and will automate the process – if your computer remains permanently on, the software can be set up to run at off-peak times, perhaps during the early hours of the morning. Keeping copies of key files away from your main or host computer (on a networked storage drive or on an external USB disk) reduces the chance of losing everything if your main system disk crashes. 'Cloud storage' is ubiquitous and inexpensive, and ideal for digital records such as these, but do check you can retrieve files quickly if you need to! That way, if anything happens to your house or computer (fire, flooding, storm, burglary . . .) your records are safely offsite, at least up to the most recent backup.

- When backing up AWS native or proprietary file formats separately from the AWS software that generated them, if possible keep a duplicate copy of the AWS software itself on the remote media, as this will increase the chances that the files can be read in the future.
- Back up your key files weekly to another external USB hard drive and keep it in a separate location, perhaps at the office or with a neighbour or relative. Such drives are inexpensive, and with multi-terabyte capacity they are more than sufficient to hold a lifetime's records. Ensure more recent files are regularly added, perhaps by swapping drives over on a 'one local, one remote' basis.
- For manageability, keep a minimum number of files – one Excel file containing 20 years records is much easier to maintain and use than 20 separate annual data files. Use simple and logical file names as suggested earlier; by including *year month* in your file name you reduce the risk of overwriting files with otherwise near-identical names.
- At least once per year, write all important data, metadata and site photograph files to an external USB disk or new memory stick and keep it offsite – perhaps at the office, or give a copy to the local library or public records office. Replace or update the media at least every 2–3 years to minimise failure risks.
- Laser-printed material may begin to deteriorate within a decade or two (unless printed on archive-quality paper and stored in archive conditions) and may need reprinting. If your files are in a current format, that should not be a problem, but if the original software to read/write/print the files becomes obsolete, then the file content may not be recoverable.
- If your records are kept in manuscript form in a hard copy logbook, photocopy recent pages every so often and keep copies safe in a second location.

Keep all your records organised, so that if you should drop dead tomorrow someone knows where your records are and what to do with them. Include a codicil to your will specifying what should happen to your instruments and records; you may wish to donate them to your local County or State Museum, Archive or Records Office, for example, or their equivalent in other countries. Clear and comprehensive accompanying metadata will increase their chances of acceptance and retention. If you are a member of a regional or national amateur observer organisation, you may find they can help seek a home for station records and instruments if these are not required by surviving family members.

One-minute summary – *Collecting and storing data*

- Making weather measurements, particularly using an AWS, can quickly generate vast amounts of data and these can become unmanageable without some thought being given to how records are to be kept and used.
- Spreadsheets are ideal for archiving weather records and provide more comprehensive analysis tools than the AWS software used to log the sensors. Holding and archiving data in separate sub-hourly, hourly, daily and monthly spreadsheets is easy to do, simplifies record-keeping and makes subsequent analysis much more straightforward.

- Each spreadsheet should include an integral metadata sheet or 'tab' detailing the instruments used, their exposure, units of measurement, record length and any other essential information.
- Months or years of data can be lost in an instant if held in a single file on a single computer disk. An entire lifetime's manuscript record could just as easily be lost forever in a house fire or burglary. Taking simple steps in advance, including putting in place a multiple backup strategy, will greatly improve the chances that records (and instruments) will survive to be available for future users.

18 Making the data avalanche work for you

For the majority of observers, collecting weather records is a means to an end, rather than an end in itself, because making use of the records obtained is ultimately more rewarding still. Records build rapidly into useful datasets, particularly from AWSs, and even a few months' worth of data can provide fascinating and sometimes unexpected insights into local weather patterns. Current records can be combined or compared with historical and local digital climate information from national records, which are freely available online in many countries. Modern spreadsheet software provides a wide variety of sophisticated, yet easy-to use, graphical and statistical analysis tools which allow the potential of the records to be much more quickly and easily realised than when held in manuscript form.

This chapter builds upon the suggested ways of collecting and storing data outlined in the previous chapter, to show how quickly and easily presentation-quality graphics and sophisticated statistical analyses of meteorological records can be generated using everyday software. (A reasonable working knowledge of spreadsheets has been assumed, as it is beyond the scope of this book to attempt to provide detailed software tuition.) Numerous examples are presented to aid understanding and provide a starting point for those with more limited spreadsheet literacy: there are also many excellent book and video-based tutorials available for all software literacy levels. Readers are encouraged to use these ideas and concepts to develop their own projects using their own observational records. Practice quickly builds into expertise.

Most examples use Microsoft Excel 64 bit, the industry standard spreadsheet, and the examples given here relate to the Windows version; Apple MacOS commands are very similar. Nothing in this chapter precludes the use of other spreadsheet packages, or code-driven methods such as Python or MATLAB, and indeed the concepts and approaches suggested in the following pages remain broadly similar whichever software is preferred.

Managing and analysing the data avalanche

The previous chapter suggested ways to store weather records, both manual observations and those exported directly from an AWS. Four separate dataset formats, namely a spreadsheet each for sub-hourly, hourly, daily and monthly observations, were suggested, although the first of these is not considered in this analysis chapter simply because the file sizes quickly become unwieldy. You may not need all three, and of course you can define your own format to meet your specific requirements (or start off with one of the template examples on www.measuringtheweather.net and edit to suit), then grow your dataset(s) over time. Don't forget to include a 'metadata' worksheet tab or 'readme' file to provide essential details regarding site, instruments, observing practices, data content and units (see Chapter 16, *Metadata, what is it and why is it important?*).

Separating hourly, daily and monthly record types makes good sense because of the different timescales and nature of the elements observed. For instance, the amount of rain that falls by wind direction can be quickly and easily analysed within an hourly dataset, because normally the wind direction does not vary much within an hour, but clearly to do this for a day, or even a month, with just one figure for rainfall amount and another single data point for a prevailing wind direction would not permit the granularity essential for a comprehensive analysis. Similarly, analysing 30 years of hourly rainfall data to derive long-period monthly and annual average rainfall amounts would involve a huge number of records – much simpler to use just the 30 monthly and annual totals.

Updating the datasets with recent observations is best done regularly, every month or two. Once hourly, daily and monthly records have been 'distilled' from the high-frequency AWS record, augmented if you wish by manually entered observations such as cloud amounts, snow depths and the like, updates can be reduced to a series of regular and straightforward cut-and-paste operations.

Where to start? Ask a question!

Examining, selecting, graphing, analysing and sorting weather records is made much easier with records held on computer, whether these be manually entered by keyboard from manuscript records or logbooks, higher-frequency digital records exported from an AWS, or local long-period records downloaded from a national climate archive. Basic graphing, averages and extremes can be performed in a few clicks: beyond that almost anything is possible, when allied with a keen curiosity. The best way to get started is to think of questions waiting to be asked:

- Which is wetter (or sunnier, or warmer ...) – a westerly wind, or an easterly?
- When is the snowiest time of the year? And the most thundery?
- Which is the wettest day of the week?
- Is it windier when it is raining? Is it warmer when the Sun is shining, or not?

Some of these will be presented as examples in this chapter, but there are an infinite number of other possibilities. The real interest is in asking, and answering, other questions of your own collected data. You may sometimes be very surprised by the answers!

Microsoft Excel basics

Microsoft Excel offers an excellent – even bewildering – range of presentation quality graphical output facilities. At the basic level, it takes only a few clicks of the mouse to generate simple line or bar graphs (**Figure 18.1** shows graphed daily

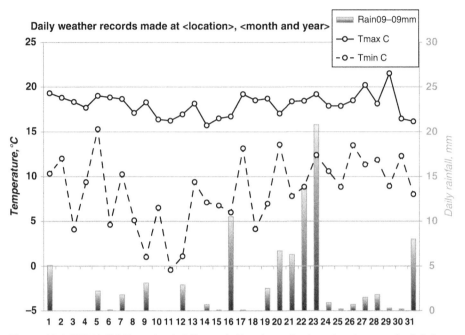

Figure 18.1 EXAMPLE – a basic Microsoft Excel graph plot, showing graphed daily max-imum and minimum temperatures (left scale, °C) and daily rainfall totals (right scale, mm) for 1 month. This file, and the data from which it was prepared, are available for download and editing on www.measuringtheweather.net

maximum and minimum temperatures and daily rainfall for one month), scatter plots (**Figure 18.2** – daily sunshine totals versus daily solar radiation amounts for 5 years in June), and many other combinations. Copies of the illustrations, example worksheets and templates are given on www.measuringtheweather.net, including those in **Figures 18.1** and **18.2**. The illustrations presented in this book are necessarily in monochrome only, and can only give a hint of what is possible – colour illustrations can be found on www.measuringtheweather.net, listed by Figure number. Correctly applied, colour can make even the largest or most complex table or diagram instantly understandable.

Try out the basic graphical capabilities to gain experience, then experiment with some of the more sophisticated analysis and plotting techniques.

Averages and extremes

Spreadsheets are ideal for handling and analysing rows and columns of numerical data, and generating statistical outputs from a set of meteorological records is very straight-forward. In most spreadsheets, the command takes the form of a **function name** (such as *average*) and the spreadsheet *cell range* over which that function is to be applied. In Excel, for example, averaging a month's daily maximum temperatures contained in columns B2 to B32 in the spreadsheet illustrated in **Table 18.1** (the file can also be downloaded from www.measuringtheweather.net) would take the form:

= average(b2:b32)

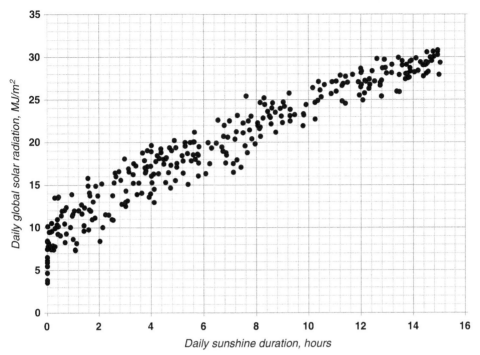

Figure 18.2 EXAMPLE – Microsoft Excel scatterplot showing the relationship between daily sunshine duration and daily global solar radiation totals at a site in southern England in June, over a 10 year period (300 observations) [296]

for the mean daily maximum, the calculation result being placed in the cell which contains the function. (It does not matter whether the function and cell references are in upper-case or lower-case.) Note that the last cell reference will need to be changed if there are less than 31 days in the month (or simply leave those cells blank).

If the month's daily minimum temperatures are in column C, as here, copying and pasting the cell that calculated the average daily maximum in column B *to the same relative position* in column C will repeat the calculation for the data in column C, without needing to retype the formula. Doing so makes it very quick and easy to duplicate the function, the mean of a month's records in this case, anywhere else on the sheet where the range (not necessarily the cells involved) remains the same.

Similar syntax goes for the highest and lowest in the range of cells, and here the Excel functions are **=*max*** and **=*min***. So to place the highest and lowest daily maximum temperatures of the month's data contained in columns B2 to B32 in columns B37 and B38, enter the instructions:

(In cell B37) **=max(b2:b32)** – for the highest daily maximum, and

(In cell B38) **=min(b2:b32)** – for the lowest daily maximum

Similarly, copying and pasting these two cells to the equivalent position in the 'Tmin_C' column C will repeat the result for that element.

Daily rainfall totals are given in column D. Copying the average, max and min functions from columns B and C into the equivalent cells in column D will calculate the same functions for the daily rainfall totals. However, neither the mean daily rainfall or the minimum daily rainfall within a month are particularly useful statistics, so better to replace the **=average** function with **=total (d2:d32)** to sum total rainfall, a more useful figure, while also deleting the cell deriving the minimum daily rainfall value.

This simple example illustrates basic cell selection and handling of statistical functions. The process is identical for, say, 10 years of monthly rainfall observations used to derive a decadal site average together with the highest and lowest monthly totals during that period (**Table 18.2**). This table is also given on www.measuringtheweather.net.

The most useful Excel functions are listed in **Table 18.3**.

Using Microsoft Excel to graph a month's observations

It is very easy to produce a graph of several elements for one month (**Figure 18.1**).

1. Create a simple two-dimensional table of daily values arranged by date in a vertical column. The dates should be included as the left-most column, followed by one or more sets of daily values in successive columns to the right of the date (**Table 18.1**). The first row should give the column headings and units, such as Date, 'Tmax_C', 'Rain09–09_mm' and so on as appropriate.
2. Using a mouse, select the Tmax_C, Tmin_C and Rain09–09_mm columns of data (including the column headings). When all the columns are highlighted, click on the 'Insert' toolbar (immediately to the right of the 'Home' toolbar at the top of the page), then 'Line' (for line chart). Excel will then create a simple line chart similar to **Figure 18.1**.
3. Click on the maximum and minimum chart lines, then right-click and select 'Format Data Series'. The options on this menu will allow setting of line colours and styles, widths, markers and various other formatting tools – try it and see.
4. Click on the rainfall graph generated. Right-click on the line, select 'Format Data Series', then 'Series Options', then 'Plot on Secondary Axis'. The rainfall graph will now have its own axis (on the right). To change it from a line graph to a histogram, select it again, go to 'Chart tools', 'Change Chart Type', and select one of the column options. The columns can then be formatted to suit. The graph is now a histogram of rainfall amounts by day of the month, plotted together with the daily maximum and minimum temps, but with its own scale of values.
5. Formatting, layout, colours and titling can all be changed as required. Experiment with the various styles and types of graphs and graphical presentations to gain familiarity with the capability of the software.

Sorting

It is very easy to sort any selection of data into a defined order – for numerical data such as the observations in **Table 18.1**, an ascending or descending rank order is very useful.

Sorting and its related function, filtering, are extremely powerful tools when working with large volumes of data, such as 10 years of hourly records or 100 years of monthly rainfall totals.

Table 18.1 *A very basic data table shown as it would appear within Microsoft Excel on-screen; data used to generate Figure 18.1*

	A	B	C	D
1	Date	Tmax C	Tmin C	Rain09–09 mm
2	1 Oct 2028	19.3	10.3	5.1
3	2 Oct 2028	18.8	12.0	trace
4	3 Oct 2028	18.3	4.1	trace
5	4 Oct 2028	17.7	9.4	0
6	5 Oct 2028	19.0	15.3	2.2
7	6 Oct 2028	18.8	4.6	0.1
8	7 Oct 2028	18.7	10.3	1.8
9	8 Oct 2028	17.1	5.1	0
10	9 Oct 2028	18.3	1.0	3.1
11	10 Oct 2028	16.4	6.5	0
12	11 Oct 2028	16.2	-0.4	0
13	12 Oct 2028	17.0	1.1	2.9
14	13 Oct 2028	18.2	9.4	0
15	14 Oct 2028	15.7	7.1	0.7
16	15 Oct 2028	16.5	6.8	0.1
17	16 Oct 2028	16.7	6.0	10.5
18	17 Oct 2028	19.2	13.1	0.1
19	18 Oct 2028	18.5	4.1	0
20	19 Oct 2028	18.7	7.0	2.5
21	20 Oct 2028	17.0	13.6	6.7
22	21 Oct 2028	18.4	7.8	6.3
23	22 Oct 2028	18.4	8.9	13.9
24	23 Oct 2028	19.2	12.4	20.8
25	24 Oct 2028	17.9	10.6	0.9
26	25 Oct 2028	17.9	8.8	0.2
27	26 Oct 2028	18.5	13.5	0.7
28	27 Oct 2028	20.2	11.3	1.5
29	28 Oct 2028	18.1	11.9	1.8
30	29 Oct 2028	21.5	8.9	0.3
31	30 Oct 2028	16.5	12.3	0.2
32	31 Oct 2028	16.2	8.0	8

Table 18.2 *EXAMPLE – monthly and annual rainfall totals at the Radcliffe Meteorological Station, Oxford, England (51.76°N, 1.26°W, 63 m AMSL), for the ten years 1801–10, in millimetres (Source: [57], Table A5.1) The annual total is the sum of the individual monthly values, using the* **=sum** *function*

	Jan	Feb	Mar	Apr	May	June	July	Aug	Sept	Oct	Nov	Dec	Year
1801	34.4	10.4	32.2	5.1	70.7	39.2	54.9	51.8	42.4	46.5	71.4	49.6	508.6
1802	8.7	49.6	6.5	6.5	42.8	43.6	81.8	6.8	6.3	39.2	40.9	56.6	389.5
1803	50.1	39.2	5.8	16.7	19.4	68.7	42.6	18.4	21.8	6.8	107.5	105.3	502.3
1804	70.7	16.2	77.5	50.6	52.8	14.5	86.2	50.8	8.0	73.1	65.1	20.6	586.0
1805	51.1	25.2	15.2	40.7	37.5	43.6	36.1	50.6	41.9	44.3	21.3	48.2	455.6
1806	82.1	33.9	42.8	22.0	24.9	17.7	87.4	45.0	37.3	24.7	69.0	86.4	573.2
1807	30.0	29.8	11.4	9.2	81.6	30.3	43.3	38.5	59.8	45.7	98.0	32.9	510.5
1808	24.0	22.8	8.5	69.5	48.7	31.7	74.3	51.3	61.2	85.0	55.9	31.7	564.5
1809	104.8	63.2	30.5	85.0	31.7	35.8	77.5	77.9	78.2	4.1	36.1	67.1	691.8
1810	15.7	36.8	55.4	54.5	44.3	26.9	84.5	62.2	28.8	51.6	131.0	82.8	674.4

Useful Excel tips

Wrong totals?

If Excel does not seem to be calculating results properly, firstly check the cell range included in your calculation is correct, then check within the range for any errone-ous values. Typical glitches include accidental omission of the decimal point (7.9 mm with entered as 79 mm, for example), missing values entered as –999 or similar instead of being left blank, and values that lie outside their typical climatological range. Text entries within otherwise numerical entry columns, such as 'missing' or 'trace', are best avoided as results can be unpredictable.

Backup copies, Autosave and Undo

Make a backup of all data before you commence any spreadsheet operations – particularly if the data table you are working with is fresh from logger memory and no other copy exists. Excel can do this for you automatically using Autosave if you save to the local OneDrive option, or set up your own regular save option.

Data copied from a paper observation sheet or logbook will not be lost should the software crash, but it is possible to lose or corrupt data in the spreadsheet by entering a wrong command. The 'undo' function is very useful here, but if the error is not spotted for several steps, it may not be possible to 'undo' all the intervening steps without causing additional damage. Autosave creates and works on a backup copy of the file, which is only saved when the 'Save' or 'Save as' button is pressed. Once a save has been made, however, Autosave and Undo start from scratch once more, so if you have any doubts about the accuracy or reliability of the entries you have just made, 'Save as' into a new but related file name – perhaps 'Averages table v2 (date)' rather than the original file 'Averages table'. To avoid creating confusion with many multiple versions of similar files, however, it is important to clear out all 'version' variants as soon as they are no longer needed. If all important files are backed up daily by automated backup software (see Chapter 17, *Collecting and storing data for details*) then even a major crash should not lose more than a few hours' work.

Table 18.3 *Common Excel functions useful in climatological analyses*

Function	Purpose	Excel command
Maximum and minimum	Evaluating the highest and lowest values within a selection of *numerical* data; for example, the highest and lowest maximum temperatures within a month. Text entries (such as 'trace' in rainfall data) are usually ignored, but best omitted altogether	**=max(cell1:cell2)** **=min(cell1:cell2)**

Table 18.3 (*cont.*)

Function	Purpose	Excel command
Average	Taking the mean of a selection of *numerical* values; for example, calculating average monthly rainfall over a number of years	**=average(cell1:cell2)**
Summation	Summing a selection of *numerical* values, usually of one element; for instance, the number of days with snow falling in March during the previous 10 years	**=sum(cell1:cell2)**
Round	Useful for rounding data values to a lower precision, perhaps to decrease the number of class sizes. Note that 'number' formatting will change the precision of displayed numbers, but **=round** will **permanently** change the precision of the stored values and should therefore be used with extreme caution	**=round(cell reference, number of decimal places required)**
Sort	Ranking one or more elements from largest to smallest, or vice versa: useful for finding the extremes of any element, and for performing quality control (for example, checking for spurious barometric pressures outside 950–1050 hPa/ mbar range)	On Data tab: sort button. Select the cells to be ranked. Excel will ask whether to include other columns. Respond 'yes' to ensure that the values remain linked to the date and the other values on that date. Ensure the row is kept as one unit to avoid mis-references to other elements on the row such as the date of occurrence, etc.
Match	Use to find a specified item in a range of cells, then return the relative position of that item in the range. For example, if the range A1:A3 contains the values 15, 62, and 9, then the formula **=match(62,A1:A3,0)** returns the number 2, because 62 is the second item in the range. Use Match instead of one of the LOOKUP functions when the position of an item in a range instead of the item itself is required – such as the date in a month when the highest or lowest value occurred	**=match(lookup_value, lookup_array, [match_type])** The value in match_type controls whether the value MATCHed is higher, lower or the same as the value being checked – in the example on the left, 0 indicates the value being looked for must be an exact match

Table 18.3 (*cont.*)

Function	Purpose	Excel command
Filter	Picking out all records satisfying one or more criteria; for example, all air temperatures above 30 °C in any month or series of years, from hourly, daily or monthly datasets	On Data tab: filter button. Select the cells to be filtered, and enter the filter criteria (values less than x, or more than y, as required)
Conditional formatting	Applies highlighting (variable colour, highlighted borders, icons and the like) to cells meeting certain criteria: for example, highlighting all monthly rainfall totals below 5 mm in a long-period dataset. Can be used in conjunction with other functions.	On Home tab: Conditional formatting button. Select the cells to be highlighted, and enter the criteria for display, or simply rank the cells and apply a colour range from highest to lowest
IF	The **=if** function permits another operation to take place only when the value is within the range specified by the IF statement; for example, a column could be set to mark all days with wind gusts over 10 m/s by checking the highest gust and setting the column entry from 0 to 1 if this value was exceeded.	Examples are given under 'dependent variables' below. Another variant, useful for monthly or annual frequency counts, is the **=countif** function, which for a given cell range will count all values above, below or equal to a specified value, and return that number in the cell: for example, to count the number of days in a month with air frost. If that is the only requirement, COUNTIF avoids the requirement for a separate column of 0s and 1s on a spreadsheet.
Undo	Most functions can be 'undone'	If in doubt, save regularly, or work from a copy of the 'master' spreadsheet.

Filtering

As datasets become larger, it is very useful to be able to pick out observations that meet only certain criteria, and then either display those records or perform statistical analyses on that subset alone. For example, in a 20 year record it would take only a few mouse clicks to pick out the last time the minimum temperature fell below −10 °C, or to calculate the average July maximum temperature on all days with (say) more than 10 hours sunshine duration.

Filtering can also be undertaken with two or more parameters: the second example above could be extended to filter for days with more than 10 hours sunshine duration AND a mean daily wind speed of less than 2 metres per second. It is very easy to sort and/or filter selected events or combinations of events.

It is perfectly acceptable to derive averages for subset conditions (for example, the mean daily maximum temperature on all days in July with 10 hours or more sunshine duration), provided that sample sizes remain large enough to be meaningful – 30 or more items is preferable.

After filtering, the table can be restored to its original layout by removing the Filter commands – de-select by reversing the way in which the filter was set up. If using multiple filters, be sure to de-select all the filters applied.

Within Excel, there is a much easier and more powerful way of deriving averages, extremes and frequency counts of data tables using sorting and filtering – namely, the **pivot table** function, which is outlined later in this chapter.

Dataset sizes

Datasets obviously expand as more records are added. The hourly dataset will be the largest – after about 10 years data, a typical hourly dataset, containing 30 measurements each hour, will be around 20 MB (see **Table 17.1**). Almost 90,000 rows of hourly records will be held in this file, far too many to check or analyse by eye. With modern computers it is very easy to analyse, average, and display entire datasets tens of megabytes in size, to sort and/or filter selected events or combinations of events, or to produce sophisticated statistical tables, all in just a few mouse clicks.

Dependent variables

A useful way to filter and analyse meteorological datasets is to assign *dependent variables*. As the name implies, these are tabular elements whose value is dependent upon another element. These can be assigned automatically using an Excel function, or they can be entered manually. Examples are given below, and are included in **Table 18.1** on the website www.measuringtheweather.net:

- *(Automatic)* Set a cell value to 1 if the daily rainfall is greater than a certain value, otherwise assign 0
- *(Manual)* Insert a cell value of 1 if thunder is heard on that day, else leave it as default 0
- *(Automatic)* Use an Excel function to specify a numerical value corresponding to the day of the week
- *(Automatic)* For any particular day's data, define an arithmetical expression to calculate the week number within the year

Example – to mark rain days and wet days:

In **Table 18.1**, create two new columns headed 'RainDays' (rainfall 0.2 mm or more) and 'WetDays' (rainfall 1.0 mm or more). Other values can be used, as you wish.
In cell E2 (rain days) enter

$$\text{=if(d2>0.19,1,0)}$$

... and then copy-paste to all the other entries for the month
In cell F2 (wet days) enter

$$\text{=if(d2>0.99,1,0)}$$

... and then copy-paste to all the other entries for the month
Excel will assign a value of 1 to these cells if the daily rainfall is greater than a certain value, here 0.19 mm or 0.99 mm, otherwise 0 will be entered into the cell. Note that the values 0.19 and 0.99 have been used as the function is strictly 'greater than' rather than 'equal to or greater than': setting the threshold to 0.20 and 1.00 mm would not count records *equal to* these values. This function will count any text entries, such as 'trace', as fulfilling the inequality, so to avoid spurious counts of rain days, for example, ensure the column being checked contains only numerical data. For period counts, **=countif** may be more useful.

The count of rain days or wet days is then simply the sum of the monthly column of 0s and 1s. Similar working goes for air and ground frosts, days with/without sunshine or other daily data elements as required.

To differentiate between data imported or manually entered and that calculated automatically in this fashion, it can be useful to show calculated cells in a different colour on the spreadsheet – perhaps dark blue rather than black. This also helps prevent accidental over-writing of cells containing formulae.

Lookup tables

Sometimes it can be helpful to reduce the number of class sizes to avoid generating an unwieldy number of classes, each containing low sample counts. One way of doing this is to use the **=round** command, which will (permanently) round any given value or values to a given number of decimal places. For elements where the class size may be variable, the **=lookup** function is the more useful. When analysing wind speed records, for example, it can be useful to sift observed hourly or daily mean speeds into the equivalent Beaufort Force: here the class sizes increase with wind speed (see 'The Beaufort Scale', **Table 9.1**, page 227). This can be quickly and easily achieved using an Excel 'lookup table' with its associated **=vlookup** function. Essentially a lookup table provides upper limits for classes, so the lookup table to convert mean wind speeds in knots into Beaufort Force would be as shown in **Table 18.4**, cells A2:B13.

In this table, an hourly mean wind speed of 6.3 kn would be categorised as '2' (i.e., Beaufort Force 2), while one of 6.6 kn would be '3'. Similar lookup tables can be used to convert wind directions in degrees into compass points, or other similar functions as required. Full syntax details are given in the Excel online help function.

Table 18.4 *VLOOKUP table entries*
for sorting wind speed into Beaufort
Force classes values in knots.

	A	B
1	ff_kn	Beaufort Force
2	0	0
3	1	1
4	3.5	2
5	6.5	3
6	10.5	4
7	16.5	5
8	21.5	6
9	27.5	7
10	33.5	8
11	40.5	9
12	47.5	10
13	100	100

Conditional formatting

Conditional formatting provides a quick and easy way to highlight cells which meet specified selection criteria. It differs from the filter command in that the highlighted cells remain visible within the complete table of data, rather than displayed as a reduced subset. It is very useful to pick out (for example) the highest or lowest values in a set of data, to colour-code values to provide easier visual assessment of the displayed information, or to apply rule-based quality control to a set of records, perhaps to identify erroneous or missing values.

Table 18.5 is based upon Table 18.2. Here the basic table of 10 years of monthly rainfall totals for Oxford from 200 years ago has had conditional formatting applied, so that all monthly totals in excess of 100 mm have been highlighted in **bold**, and all monthly totals below 20 mm have been shown in *italic* font. On a computer monitor, colour shaded cells, or other distinguishing formatting, could have been used just as easily. (Note, of course, that the conditional formatting applied to the monthly cells has *not* been applied to the annual totals, otherwise all would show > 100 mm; separate conditional formatting rules reflecting more appropriate threshold(s) could be applied here in a similar fashion to the monthly totals.)

Conditional formatting is quick and easy to apply and is useful for rapidly picking out the key features in complex numerical tables such as this. To apply conditional formatting: using the mouse, select the cell range to which the conditional formatting is to be applied, then select 'Conditional formatting' on the 'Home' toolbar; then select the formatting and threshold(s) required. Try it and see for yourself.

Table 18.5 *EXAMPLE – monthly and annual rainfall totals at the Radcliffe Meteorological Station, Oxford, England (51.76°N, 1.26°W, 63 m AMSL), for the ten years 1801–10, in millimetres, with conditional formatting applied for* dry *and* **wet** *months (Source: [57], Table A5.1)*

	Jan	Feb	Mar	Apr	May	June	July	Aug	Sept	Oct	Nov	Dec	Year
1801	34.4	*10.4*	32.2	*5.1*	70.7	39.2	54.9	51.8	42.4	46.5	71.4	49.6	508.6
1802	*8.7*	49.6	*6.5*	*6.5*	42.8	43.6	81.8	*6.8*	*6.3*	39.2	40.9	56.6	389.5
1803	50.1	39.2	*5.8*	16.7	*19.4*	68.7	42.6	18.4	21.8	*6.8*	**107.5**	**105.3**	502.3
1804	70.7	*16.2*	77.5	50.6	52.8	14.5	86.2	50.8	8.0	73.1	65.1	20.6	586.0
1805	51.1	25.2	15.2	40.7	37.5	43.6	36.1	50.6	41.9	44.3	21.3	48.2	455.6
1806	82.1	33.9	42.8	22.0	24.9	17.7	87.4	45.0	37.3	24.7	69.0	86.4	573.2
1807	30.0	29.8	*11.4*	*9.2*	81.6	30.3	43.3	38.5	59.8	45.7	98.0	32.9	510.5
1808	24.0	22.8	*8.5*	69.5	48.7	31.7	74.3	51.3	61.2	85.0	55.9	31.7	564.5
1809	**104.8**	63.2	30.5	85.0	31.7	35.8	77.5	77.9	78.2	4.1	36.1	67.1	691.8
1810	*15.7*	36.8	55.4	54.5	44.3	26.9	84.5	62.2	28.8	51.6	**131.0**	82.8	674.4
Avg	47.2	32.7	28.6	36.0	45.4	35.2	66.9	45.3	38.6	42.1	69.6	58.1	545.6
Max	104.8	63.2	77.5	85.0	81.6	68.7	87.4	77.9	78.2	85.0	131.0	105.3	691.8
Min	8.7	10.4	5.8	5.1	19.4	14.5	36.1	6.8	6.3	4.1	21.3	20.6	389.5

Excel analysis and processing tips

Storing dates. There are many analysis advantages to holding the date in separate date, month and year parameters (i.e., dd, mm and yyyy), as well as the complete date as dd-mmm-yyyy. To create these additional numerical date parameters within Excel, the **=day (*date cell reference*)** function will split out just the date (dd, 01–31) from a cell containing day-month-year such as 19.Sep.2028 (although the date must be later than 1 Jan 1900). Similarly, **=month (*date cell reference*)** will select only the month (mm, 01–12), while and **=year (*date cell reference*)** will do the same for year. Example: 4-Sep-28 would be held as dd = 04, mm = 09, yyyy = 2028. These can then be formatted as required in Excel (for instance, the month could be formatted on output as *9, 09, Sep* or *September*, even in different languages, depending upon requirements and preference). Some of the subsequent analyses in this chapter show the benefits of splitting dates in this fashion.

Continuous data. Separate continuous data into discrete classes for ease of analysis. For example, to analyse rainfall totals by wind direction, the range of hourly vector mean wind directions could be from 0.0 to 360.0 degrees, at 0.1 degree intervals, giving 3600 × 0.1 degree classes, almost all of which will contain only a few observations. Defining a smaller number of classes, perhaps grouping into 10 or 20 degree sectors using the **=lookup** function, gives a more manageable number of classes and thus a larger number of points in each sample, reducing table sizes and improving both legibility and statistical reliability.

Defined events. Include 'binaries' (1 or 0) for 'defined events', for example 'hours with rain' or 'air frost'. For the latter the Excel function would be

---

=if(cell containing temperature record< 0,1,0)

... assuming temperature values are in °C (else the inequality would read <32 for °F)

Note that in this case a blank cell (missing data) will give the same result as a negative entry, so to avoid errors check beforehand that no blanks are included in the selected cell range(s).

Beware of small class sizes. Too many classes or too small a sample will result in some classes having very few observations: averages or extremes from these small cell sizes will therefore be unreliable. This can be a problem with infrequent wind directions (in the south of England, for example, winds from the east-south-east occur on only a handful of days every year), and of course some observations will in any case fall into the statistical 'tail' of the distribution. For obvious reasons, not many entries from events with a '1 in 100 year' recurrence can be expected in a 5 year climate record. For such analyses it may be better to increase the class size to generate more reliable averages. Ideally, each cell should have 30 or more data points. This is not always possible, of course, and provided the small class size is appropriately qualified smaller class sizes may be acceptable.

Pivot tables

The concept of Excel's pivot table function is easily understood, although at first glance it can appear somewhat daunting: once again, starting with simple analyses will quickly build familiarity. Pivot tables have an easy 'drag and drop' metaphor but take a little practice to become proficient. They are well worth taking the trouble to get to know as they are enormously powerful analysis tools, simplifying even highly complex statistical analyses, as the following examples will demonstrate.

A pivot table essentially provides the means in just a few clicks to summarise a larger body of data, perhaps decades of weather observations, into one or more subset tables. A wide range of statistical functions are available, of which the most useful in meteorology are totals and averages, maximum, minimum and frequency counts. Providing the original dataset permits splitting the data into a manageable number of classes, pivot table analyses are quick and easy to generate. They are ideal for both quick 'what if ... ' or 'how often ... ' questions, as well as for more formal structured analyses such as constructing tables of long-period averages.

All of the following examples used pivot tables to generate the analysis dataset, which was then graphed in Excel. The pivot tables took typically less than a minute to set up, provide output, and graph. The results illustrated here all use hourly, daily and monthly observation spreadsheets from the author's own records in southern England covering 10 years or more. All use real data throughout.

Setting up a Pivot table

Click on any cell within the spreadsheet. The data must be contiguous – individual cells can be blank, but there must be no blank rows or columns, and all columns must include a heading in the first row with a suitable abbreviated title such as 'Tmax_C', 'RH_%' and so on, as in **Table 18.1**. These will be the field names used to set out the pivot table grid, so it is best to make them both meaningful and unique.

- Within spreadsheet data cells, click on one cell within the table range then click on 'Insert' within the top ribbon, then click on 'Pivot Table' (on far left)
- Choose default data range (contiguous cells, no gaps), or reduce range if entire data table is not required, then click OK
- A blank pivot table will then be displayed, with pivot table fields shown in a column on the right of the page. Drag and drop one or more fields into 'columns' box and similarly one or more fields into the 'rows' box. Keep it simple at first to get the hang of it; the column heading might be 'hours' and the row headings could be 'months' for example
- Then drag and drop one or more elements into the 'values' box. The pivot table should refresh with entries in each cell, usually 'sums' by default
- Then click within the element field chosen in the 'Values' box to select the type of calculation required from the drop-down menu that appears – select sums, averages, maximum values, whatever is required
- The data table will then refresh to show (for example) hourly mean wind speeds by hour and month over the table range selected
- Experiment with drag/drop layouts for best results
- To filter (for example, by year any other value within the dataset), drag field (such as year) into 'filter' box, then use drop-down tick boxes at top left of table to select.
- Format to suit; coloured formatting using Conditional Formatting is particularly effective at making sense of what may otherwise just be a large table of numbers.

A final word of warning: Pivot tables increase the size of your data files if they are saved as a tab within the original worksheet. Large datasheets and complicated pivot tables can quickly become unwieldy; keep it simple, and if necessary save pivot table sheets separately to keep file sizes manageable.

Pivot table Example 1: when is the sunniest time of year?

This example uses 20 years of hourly sunshine totals from the author's test site in southern England, over 169,000 observations in all [297]. The table is very simple – hour of day (column) by month (row), with the mean sunshine (in hours) for each cell calculated using the pivot table function. All available records are included (it would be just another couple of mouse clicks to include the number of observations in each cell, if required). On-screen conditional formatting could instead colour-code the entries from grey to orange ('heatmapping'), but for the sake of example in monochrome it is easy enough to pick out only the Top 5 values automatically (bold font with borders).

From opening the dataset to producing the fully formatted table (**Figure 18.3**) took just 79 seconds. (The colour example is included on www.measuringtheweather.net.) It can be very quickly seen that the sunniest times of the year at this observing location (over the 20 years considered) are mornings in April and July, each averaging a little over 50 per cent of the possible duration.

This layout is also suitable for hourly mean temperatures, pressures, humidity, wind speed – most elements. Try it (or something similar) based upon your own records – your conclusions may be very different.

One other important point regarding pivot tables is that, once set up, they are easily expanded or amended. **Figure 18.3** was based upon 20 years data: to add another (say) 5 or 10 years data in due course would require only adding the additional observations to the existing file, extending the table cell range to include the new data, then clicking 'refresh' within the pivot table to update the analysis. To include (say) only the most recent 10 years data would merely require selecting the years required using the 'Filter' option, then clicking 'Refresh' on the pivot table.

YYYY	(Multiple Items)												
Average of SUNSHINE Hourly duration, h	**Column Labels**												
Row Labels	**1**	**2**	**3**	**4**	**5**	**6**	**7**	**8**	**9**	**10**	**11**	**12**	**Grand Total**
0000	0	0	0	0	0	0	0	0	0	0	0	0	0
0100	0	0	0	0	0	0	0	0	0	0	0	0	0
0200	0	0	0	0	0	0	0	0	0	0	0	0	0
0300	0	0	0	0	0	0	0	0	0	0	0	0	0
0400	0	0	0	0	0	0	0	0	0	0	0	0	0
0500	0	0	0	0.00	0.11	0.17	0.11	0.01	0	0	0	0	0.03
0600	0	0	0.00	0.18	0.38	0.39	0.40	0.25	0.02	0	0	0	0.14
0700	0	0.00	0.11	0.44	0.45	0.44	0.45	0.40	0.30	0.05	0	0	0.22
0800	0.00	0.10	0.32	0.51	0.49	0.50	**0.52**	0.47	0.43	0.28	0.08	0	0.31
0900	0.15	0.27	0.39	**0.54**	0.50	0.51	**0.52**	0.50	0.48	0.38	0.29	0.15	0.39
1000	0.30	0.35	0.44	**0.54**	0.49	0.49	0.51	0.49	0.51	0.42	0.35	0.27	0.43
1100	0.32	0.37	0.44	**0.53**	0.46	0.48	0.46	0.46	0.50	0.41	0.36	0.31	0.43
1200	0.34	0.35	0.41	0.49	0.42	0.44	0.43	0.42	0.46	0.40	0.35	0.32	0.40
1300	0.34	0.34	0.39	0.46	0.40	0.40	0.40	0.40	0.44	0.38	0.34	0.31	0.38
1400	0.33	0.32	0.41	0.47	0.40	0.40	0.41	0.42	0.45	0.38	0.34	0.29	0.38
1500	0.29	0.33	0.43	0.47	0.42	0.42	0.45	0.42	0.44	0.39	0.32	0.26	0.39
1600	0.22	0.33	0.42	0.47	0.43	0.43	0.46	0.42	0.43	0.34	0.22	0.13	0.36
1700	0.03	0.24	0.39	0.45	0.44	0.44	0.45	0.44	0.43	0.21	0.01	0	0.30
1800	0	0.02	0.23	0.42	0.44	0.44	0.46	0.42	0.28	0.00	0	0	0.23
1900	0	0	0.00	0.17	0.40	0.44	0.42	0.28	0.01	0	0	0	0.14
2000	0	0	0	0	0.11	0.23	0.22	0.03	0	0	0	0	0.05
2100	0	0	0	0	0	0	0	0	0	0	0	0	0
2200	0	0	0	0	0	0	0	0	0	0	0	0	0
2300	0	0	0	0	0	0	0	0	0	0	0	0	0
Grand Total	**0.10**	**0.13**	**0.18**	**0.26**	**0.26**	**0.28**	**0.28**	**0.24**	**0.22**	**0.15**	**0.11**	**0.08**	**0.19**

Figure 18.3 Microsoft Excel pivot table analysis showing hourly mean sunshine (in hours) for every hour of the year at a site in central southern England, averaged over 20 years. Automatic (conditional) formatting has been used to highlight the top five values using bold font within bordered cells. The columns are month (January = 1, February = 2, etc.) and the rows the hour ending, UTC. A colour version of this plot using colour conditional formatting is available on www.measuringtheweather.net. Source notes: [297]

Example 2: the variation of rainfall amount by wind direction

It is also very easy to produce analyses based upon cell totals, rather than means, and to analyse one element in terms of another. This two-dimensional table (**Figure 18.4**) shows the total rainfall by wind direction at the same location over the same 20 year period used in the previous example [298]. Depending on the format of the observational records, this can be done by compass point (SW, WSW, W, etc.), or by azimuth degrees as here.

If analysed by compass point, Excel will by default lay the column headings out in alphabetical order, so to obtain the correct order around the compass some cut-and-pasting into a separate table will be required.

| YYYY | (Multiple Items) |
| WIND mean speed, Bft force | (Multiple Items) |

Row Labels	Sum of RAIN Hourly rainfall (0.2 mm TBR), mm
10	240
20	251
30	161
40	161
50	153
60	148
70	158
80	87
90	80
100	141
110	111
120	144
130	147
140	189
150	148
160	191
170	247
180	565
190	665
200	901
210	1001
220	1083
230	559
240	346
250	190
260	205
270	93
280	105
290	87
300	60
310	39
320	40
330	30
340	82
350	100
360	307
Grand Total	9215

Figure 18.4 Pivot table analysis in 'data bar' style showing rainfall totals (mm) by 10 degree wind direction classes in central southern England over the same 20 year period as in Figure 18.5 – tabular presentation using Excel's conditional formatting tools. Source notes: [298]

If analysed by degrees, to avoid an unmanageable number of column headings (and the problem of small cell counts), it is best to aggregate the original observations into 10 degree segments. In Excel this is best done using a dependent variable, rounding to the nearest 10 degrees (using Excel's **=round** command). Some care is needed to allow for wind directions going through north; here it is best to assign wind directions between 355.1 deg and 004.9 degrees to '360' rather than '000' to avoid confusion with 'calm'.

Here the pivot table is constructed from hourly observations using 10 degree wind direction classes: the total rainfall within each class is summed. To demonstrate exclusion filtering (drag/drop field name to 'Filter' box at head of resulting pivot table and select entries required), hours with light winds (hourly mean wind speed less than Beaufort Force 2) were omitted: this exclusion could just as easily have been 'hours with MSL barometric pressure below 1000 hPa', or 'daylight hours only in June, July or August', or almost any other database element, with just a couple of mouse clicks.

This table took only 46 seconds to generate from 20 years of hourly observations (over 169,000 records). **Figure 18.4** was then formatted using the 'data bars' variant from the conditional formatting drop-down box, which took a further 53 seconds. This columnar layout gives a clear visual indication of the dominance of rainfall from between 180 and 230 degrees (south to south-west) in central southern England – more than 50 per cent of all rain falls with winds in this sector. Creating such a table by hand would take weeks of work.

A more visually striking display method is to re-plot the data within Excel using polar co-ordinates (a 'radar plot' in Excel terminology). **Figure 18.5** visually

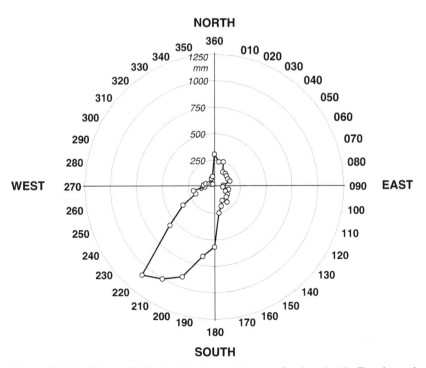

Figure 18.5 As Figure 18.4 but plotted on polar axes ('radar plot' in Excel terminology).

illustrates both the dominance of rain-bearing winds from between south and south-west in central southern England, and the relative lack of rainfall on north-westerly winds.

The shape of this plot will vary enormously with geography – New Orleans or New South Wales would look very different, for example. Try this with your own hourly observations – even a year or two of data should be sufficient to provide a good indication. Is it what you expected? Is it different from the appearance of the data presented here? Why?

Example 3: the variation of rainfall intensity by wind direction

If we simply repeat the pivot table above, but instead select *average* hourly rainfall by wind direction rather than *sum*, then a very different picture emerges (**Figures 18.6** and **18.7**). Note the plotted values are now in millimetres per hour.

Contrasting analyses such as these are amongst the most interesting uses of accumulated weather records and can quickly lead to real insights. In this case, we can quickly see that while winds from between south and south-west produce the *majority* of the rainfall at this site, the *heaviest* precipitation tends to occur with winds between east-south-east (100°) and south or south-south-west (180–190°). These directions are often associated with cyclonic and warm frontal rainfall events in temperate latitudes. Note also the above-average intensities of rainfall from between west-north-west and north (290–360°) at this site – although the quantity of rainfall is small from these directions, when it does rain the rain can be quite heavy. These are often showers at or immediately after a cold front, or sometimes showers originating from the Irish Sea through the 'Cheshire Gap' and passing across the English Midlands. Again, repeat this with your own observations. What conclusions do you draw?

Similar analyses are quickly and easily generated for temperature, sunshine or almost any other element in the database (which can include daylight yes/no, days of the week, windchill . . .). The benefit of such analyses is not merely providing quantitative confirmation of what we may know or suspect already, but sometimes in throwing out very surprising results – the reader may care to consider that a similar analysis for sunshine showed that winds from east-south-east are also the sunniest, for example. Why should one wind direction experience both the heaviest rainfall and the sunniest conditions? Compare and contrast with your own observations.

Example 4: when is the snowiest/most thundery period of the year?

There are many advantages in weekly analyses – they provide improved granularity over monthly statistics, and reduce the statistical variability associated with low class sizes when using daily data, particularly with relatively short periods of record. This example examines the annual variation of the frequency of snowfall ('snow or sleet observed to fall 00–24 h') and 'thunder heard' by week. Definitions of these elements were given in Chapter 14, *Non-instrumental weather observing*. Note that in this example we are now using the *daily* dataset, not the hourly values.

YYYY	(Multiple Items)
WIND mean speed, Bft force	(Multiple Items)

Row Labels	Average of RAIN Hourly rainfall (0.2 mm TBR), mm	
10		0.09
20		0.08
30		0.05
40		0.06
50		0.07
60		0.09
70		0.11
80		0.12
90		0.14
100		0.22
110		0.18
120		0.25
130		0.25
140		0.33
150		0.27
160		0.30
170		0.22
180		0.25
190		0.22
200		0.17
210		0.11
220		0.07
230		0.07
240		0.07
250		0.04
260		0.06
270		0.04
280		0.06
290		0.08
300		0.09
310		0.05
320		0.06
330		0.06
340		0.11
350		0.06
360		0.11
Grand Total		0.10

Figure 18.6 As Figure 18.4 but for rainfall intensity (mm/h).

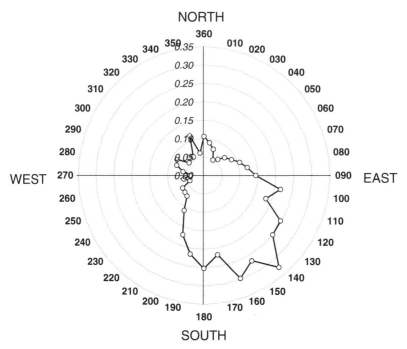

Figure 18.7 As Figure 18.5 but for rainfall intensity (mm/h).

To do this, a new variable needs to be assigned in the data table, to provide the week number. There are two ways to calculate this:

Excel provides two options for the **=weeknum** function. The first option ('system 1') defines the week containing the date 1 January to be the first week of the year, which is then numbered week 1. The second option, or 'system 2', considers instead the week containing the first Thursday of the year as the first week of the year, which is then numbered as week 1 (this system is the method-ology specified in ISO 8601, which is commonly known as the European week numbering system). (For more details and syntax see Excel's help function.) The second option is preferable to the first, because when 1 January falls on a Saturday, this becomes week 1, thereafter Sunday and the rest of that week will be week 2 and so on. Week 1 therefore consists of a single day's entry in some years, and seven days in others. This renders any statistical analysis of week 1 data highly unsatisfactory, as the number of days in the sample will vary from year to year.

A statistically preferable method is not to use the Excel function, but instead to ensure that, every year, week 1 consists of the observations from 1 to 7 January, week 2 is 8–14 January and so on, regardless of the weekday upon which the week starts. To do this we need to calculate the *yearday*, which simply increments from 1 January = yearday 1, 2 January = yearday 2 and so on. This is very easy to do in Excel simply by calculating a yearday value in a separate column: 1 January is 1, then every subsequent day increments this value by 1 (thus yearday for 3 January = yearday for 2 January +1). The week number is then simply (yearday + 3) / 52, rounded to an integer. (Remember to start again with 1 next 1 January.) The Excel function is as follows:

$$=round((yearday\ cell+3)/7,0)$$

Of course, as the number of weeks in a year is not an exact multiple of 365 (365 × 52 = 364), this leaves week 53 with only one or two days; but this is easier to allow for than a variable number of days in week 1. Including this function in the daily data spreadsheet permits easy analysis of variables by week in the year, due allowance being made of course for week 53 always being 'short'. This also has the advantage of including 29 February in its rightful place within the year in leap years.

Generating a pivot table analysis to examine 30 years records of days with snowfall and days with thunder data (manual spreadsheet 'binary' data entry: 0 = none, 1 = observed) by week of the year takes only a few mouse clicks to derive **Figure 18.8**. Note that in the pivot table *sum* should be used to give the total number of days, not *count*, which counts all days with data ('0' being a valid data point). Applying conditional formatting in the form of 'data bars' enables a rapid visual analysis of the data. (This table took less than 90 seconds to specify, output and format as shown here.)

It can be quickly seen that during these 30 years, the annual frequency of snow or sleet and thunder heard was about the same (339 days with snow or sleet, 317 with thunder), but the frequency of each within the year was very different. The snowiest week of the year was week 9, starting on 26 February. In the 30 years analysed it snowed on 30 days, so on average about one day in that week could be expected to see a snowfall.

Thunder has almost the opposite distribution throughout the year in southern England: only at the beginning of April is there both a reasonable and approximately even chance of either occurring. The majority of thundery activity occurs between weeks 17 and 38 (23 April to 17 September). Between them these 22 weeks accounted for 242 days of the 317 days with thunder in 30 years, or 76 per cent of all occurrences. The most thundery week of all was the week commencing 2 July, which saw 21 occurrences in 30 years – on average, a little less than one day in that week.

The risk of thunder, or snow, or the variations of any other element by week can be very quickly analysed this way. This form of analysis can be very useful for shortlisting possible dates for weddings, holidays, sporting events, school fêtes and the like. It is very important to remember, however, that the outcome is a statistical probability based on previous observations, and *not* a weather forecast as such!

Example 5: which is the sunniest / driest day of the week?

Excel offers a simple function to derive a numerical value for the day of the week, making statistical analyses by weekday very easy to set up.

Example – days of the week:

- In a daily data table, create a new column headed 'DayOfWeek'. In the first row, enter in the cell **=weekday(*cell containing full date dd-mmm-yyyy*)**

  . . . and then simply copy-paste to all the other entries by date.

YYYY   (Multiple Items)

Row Labels	Min of Date	Sum of Days with Snow or sleet (BINARY)	Sum of Days with Thunder heard 00–24h
1	1 Jan	27	2
2	8 Jan	17	3
3	15 Jan	9	2
4	22 Jan	12	2
5	29 Jan	23	1
6	5 Feb	28	2
7	12 Feb	16	1
8	19 Feb	26	1
9	26 Feb	30	3
10	5 Mar	12	1
11	12 Mar	11	2
12	19 Mar	11	3
13	26 Mar	10	6
14	2 Apr	8	6
15	9 Apr	11	7
16	16 Apr	5	6
17	23 Apr	4	10
18	30 Apr	0	4
19	7 May	0	16
20	14 May	0	8
21	21 May	0	8
22	28 May	0	9
23	4 Jun	0	10
24	11 Jun	0	10
25	18 Jun	0	9
26	25 Jun	0	10
27	2 Jul	0	21
28	9 Jul	0	9
29	16 Jul	0	13
30	23 Jul	0	15
31	30 Jul	0	18
32	6 Aug	0	20
33	13 Aug	0	13
34	20 Aug	0	8
35	27 Aug	0	7
36	3 Sep	0	5
37	10 Sep	0	10
38	17 Sep	0	9
39	24 Sep	0	2
40	1 Oct	0	4
41	8 Oct	0	5
42	15 Oct	0	6
43	22 Oct	0	1
44	29 Oct	2	0
45	5 Nov	1	1
46	12 Nov	1	0
47	19 Nov	9	4
48	26 Nov	5	1
49	3 Dec	9	1
50	10 Dec	8	1
51	17 Dec	25	1
52	24 Dec	17	0
53	31 Dec	2	0
Grand Total	**1 Jan**	**339**	**317**

Figure 18.8 Total number of days in 30 years with snow or sleet observed to fall (left column plot) and thunder heard (right column plot) by week number in the year at one site in central southern England (51°N, 1°W). Source notes: [299]

Excel will assign a weekday numerical value to these cells based on the date: Sunday = 1, Monday = 2 and so on to Saturday = 7 (the numerical values and start day can be changed if required, using an optional second parameter: for more details and syntax, see Excel's help function).

Table 18.6 *Mean daily values of 0900–0900 UTC rainfall (mm) and daily sunshine duration (hours) by day of the week at the author's test site in central southern England, over the 30 year WMO standard averaging period 1991–2020, derived using a simple Excel pivot table analysis*

	Sun	Mon	Tue	Wed	Thu	Fri	Sat	Week
Mean daily rainfall (mm)	1.88	1.96	2.03	1.89	1.83	2.02	1.95	1.94
Mean daily sunshine (h)	4.61	4.46	4.45	4.39	4.43	4.29	4.68	4.47

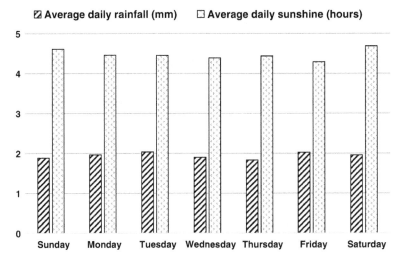

Figure 18.9 Mean values of daily rainfall (mm) and daily sunshine duration (hours) by weekday at the author's observing location in central southern England over the 30 years 1991–2020 (see also Table 18.6)

Applying a pivot table analysis and conditional formatting to 30 years daily records of rainfall (mm) and sunshine (hours), we can derive **Figure 18.9** and **Table 18.6** in less than a minute. It would require only another couple of clicks to evaluate mean temperatures, frequency of snowfall and so on within the same table.

(An obligatory statistical health warning at this point: while many of these differences may look real, it is more likely that the slight differences in the means are merely coincidental and without statistical significance – different periods of data may give different results. Such analyses are always popular for non-specialist audience presentations, however, and here provide a good example of pivot tables using a dependent variable, and the possibilities of more creative output formatting options.)

It is interesting (if perhaps not very statistically significant) to know that at this site during the 30 years analysed:

- Tuesday was the *wettest* day of the week, and Thursday the *driest*: the difference in mean daily rainfall between the wettest and driest days of the week was 17 per cent.
- Saturday was the *sunniest* day of the week – and Friday the least sunny: the difference in means is 9 per cent. Contrary to popular opinion, the weekends are

the sunniest days, and slightly drier, than the days of the working week … at least in southern England.

The reader is left to draw his or her own conclusions.

Example 6: is it windier when it's raining?

To answer this question requires a slightly more complicated pivot table analysis of hourly data once more over a number of years, as follows.

Firstly, to categorise all occasions with 'rain', define a parameter which is 1 when the hourly rainfall total (usually from a tipping-bucket raingauge or similar) is not 0, and 0 otherwise (it could be a higher threshold if required, say 0.5 mm). The cell entry will be:

=if(cell containing hourly rainfall total> 0,1,0)

Then, using the pivot table function, produce means of hourly wind speeds for all occasions with rain and those without rain. Graph and compare (**Figure 18.10**).

It can be seen that, in central southern England at least, mean wind speeds *are* higher when it is raining than when it is dry. This is true throughout the year, although the difference is less marked during the summer months.

Of course, this analysis can quickly be expanded by the interested reader. Do the findings hold true for all wind directions? Are winds stronger when rain falls on winds from the main rain-bearing wind directions (see Example 2, above)? What about gust speeds? Gusts (and thus the gust ratio, see page 232) might be expected to be higher in showery rain rather than frontal rain situations. Is this borne out by the observations?

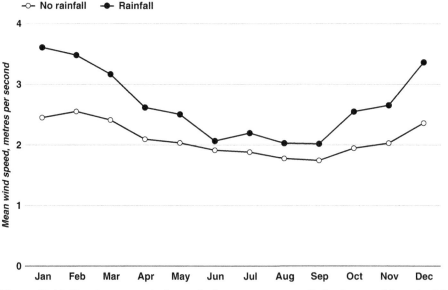

Figure 18.10 Hourly mean wind speeds (metres per second) for hours with and without recorded rainfall at the author's test site in central southern England. Source notes: [300]

Be careful of introducing unintentional statistical bias. Why would repeating the above analysis for sunshine instead of rainfall introduce a very significant bias? Finally, could this analysis be replicated for snowfall?

Example 7: different ways of looking at observational records

There are many different ways of looking at, analysing and presenting weather data. Some are familiar, while the analysis and graphical facilities within Excel make it easy to experiment with different, more creative methods of displaying information. One example is given in **Figure 18.11**, which shows 5 minute sunshine data throughout a whole year. A similar diagram appeared in Edward Tufte's inspiring book *The visual display of quantitative information* [301] many years ago (page 165). Preparing this manually would take days. I puzzled how best this could be generated quickly within Excel. Using conditional formatting, it took only a few minutes, as follows.

The starting point is an Excel table of 5 minute logged AWS sunshine data for the selected year across the full 24 hours × 365 days. A pivot table was prepared with date horizontally along the top (columns), hours vertically along the left side (rows), with the actual value of the sunshine duration in seconds output in each cell. In any 5 minute period, the maximum amount of sunshine is 300 seconds. Using conditional formatting, all values were assigned different graduated shades, from black (nil sunshine) through light grey (100 seconds sunshine in the 5 minute period) to white (200 seconds or more). The visual metaphor thus corresponds clearly to the data, namely the lightest areas being the times when the Sun shone for longest.

There is no easy way to suppress the actual cell values in the pivot table, so the font size was set to the minimum possible and positive values were output in a white

Figure 18.11 A pivot table plot showing 5 minute sunshine data for an entire year, summarising 105,000 datapoints and over 6 million seconds of sunshine. Source notes: [302]

Figure 18.12 The position and intensity of sunlight through a pinhole camera for the six months from the winter solstice 2018 to the summer solstice 2019 left this impression on photographic paper. This exposure was made at the same location as Figure 18.11 (but covering a different period), using a Solarcan pinhole camera, available from solarcan.co.uk

font, zero in black, so that they are too small to show. The height and width of the grid were reduced so that individual cells were represented by tiny squares. Individual dates/times are not then visible, of course, but if desired, hour markers at 6 hour intervals can be added to the plot.

The final result contains a little over 105,000 separate data points and summarises 6,583,797 seconds of sunshine in the year plotted. The graphical output emulates how the burns from a year's worth of Campbell–Stokes sunshine recorder cards would look if they could all be neatly assembled into one annual array. It illustrates the variation of sunshine both throughout the day (longer days in the summer months) and throughout the year, in a diagram that is both striking and quite easy to produce. It also bears a pleasing resonance with the output from a pinhole camera containing photographic paper left exposed to sunshine for 6 months (**Figure 18.12**).

Tufte's extraordinary books [301, 303, 304], which pre-date modern data analysis and graphics software, will suggest many other inspiring examples of striking data presentation formats to the interested reader. Tufte's legacy lives on in David McCandless's website InformationIsBeautiful.net which has many beautiful and interactive examples, although it should be appreciated most of these are well beyond Excel's current capabilities ... But with a little imagination, the presentation of meteorological statistics need be anything but dull!

Wind roses

A wind rose is a polar area plot, used to show the relative frequency of combinations of wind speed and direction. Microsoft Excel cannot prepare wind roses directly, although it is helpful to use it to prepare the input files for software packages that can do so. There are a few Windows-based wind rose packages: the examples shown here have been prepared using WRPLOT View software from Lakes Environmental

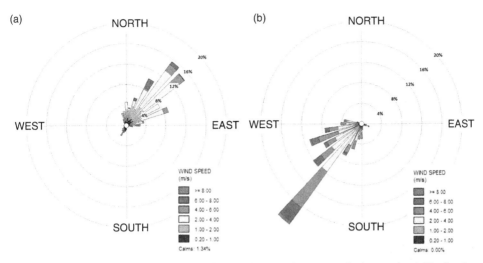

Figure 18.13 Wind roses for two strikingly different winter months in southern England – (a) cold, wet and cloudy March 2013 and (b) mild and windy February 2022. Wind speed classes are in metres per second; the frequency of calms (< 0.2 m/s) is given in the accompanying table. Prepared using WRPLOT View from Lakes Environmental Software [305].

Software [305]. The diagrams here are in black and white, although the colour originals are available from www.measuringtheweather.net.

To prepare the wind rose, the software reads a pre-prepared datafile of wind direction and speed, typically hourly data from an Excel file or from an AWS. The software works best with an unfiltered dataset (all observations), but filtered subsets (for example, all hours with more than 5 mm of rainfall, or all sunny hours) can be prepared and the software run on those if required – provided class sizes do not become too small. The options (class intervals, colours, titling and other presentation aspects) can be varied as required, and the analysis can be as detailed as ten-degree segments as shown here. The software produces standard-format graphics files which can be pasted into other applications.

Figure 18.13 compares wind roses from southern England, between two winter months, one cold and the other very mild. The scales on the two plots are identical. The difference is striking – and instantly obvious using a wind rose.

Putting current records into perspective: averages and extremes

It is often useful to compare or extend records with long-period climate data from local sites, whether this is raw hourly/daily/monthly data or summarised in 30 year averages. (The WMO standard period for climatological averages is 30 years, the current averaging period at the time of writing being 1991–2020.)

In many countries, weather records obtained and archived at public expense are free or attract only nominal data preparation charges. Many provide for free access, inspection and download over the Internet. A few examples are given below: if your country is not listed here, check the WMO list of national weather services' websites, and search for 'climate records'.

Australia

The Australian Bureau of Meteorology provides an exemplary climate website. Access to the nation's weather records, many extending back over 100 years, is mostly free and very easy to find, with a well laid-out search facility. For most sites, even rainfall-only locations, online records are updated daily. There is an impressive wealth of information on all aspects of Australia's weather and climate readily available in tabular, graphed or mapped form. I have used the site extensively in preparing this book.

Australian climate data: www.bom.gov.au/climate/data

Canada

Environment Canada's online climate data facility provides a similar impressive level of data accessibility and functionality to Australia's site. Easy drop-down menus permit direct access to a vast range of historical hourly, daily and monthly climate data by province and location, average and extreme temperature and precipitation values for particular sites, and whole-country summaries of averages and extremes for particular months and years. The site integrates detailed climate change extrapolations of current climatology to 2100 CE. More specialised records such as upper-air information, precipitation radar, short-period rainfall intensity-duration-frequency statistics and many others are also available for inspection or download as required, together with a wide range of Historical Environment Canada publications available in digital form. Full information on dataset layout, content and definitions of measurement units is also available online. Data not available on the website can be obtained on application to Environment Canada for a basic charge.

Canada climate information: https://climatedata.ca

The Netherlands

The KNMI (Koninklijk Nederlands Meteorologisch Instituut, the Royal Netherlands Meteorological Institute) climate website provides free online access to daily records of temperature, sunshine, cloud cover and visibility, air pressure, wind and precipitation from 36 weather stations across the Netherlands. Some have more than 100 years of records. Information is updated daily for all current sites. Climate change scenario options are also available through this Climate Explorer site.

The Netherlands: https://climexp.knmi.nl/start.cgi

New Zealand

Access to New Zealand's climate database has been free since April 2007, when data charging was scrapped to permit and promote more widespread use of New Zealand's national climate database. The database holds data from about 6,500 locations (including rainfall-only sites) which have been operating for various periods since the earliest observations were made in 1850. The database continues to receive data from more than 600 currently operating stations.

The operation is run by NIWA, the National Institute of Water and Atmospheric Research (a Crown Research Institute established in 1992), and web access is through the CliFlo system (link below). CliFlo returns raw data and statistical summaries. Raw data include 10 minute, hourly and daily records: statistical data include about 80 different types of monthly and annual statistics and six types of 30 year normals. CliFlo Data is free: access is via an online, biennial registration. Some restrictions to climate data apply, such as Pacific Island sites.

New Zealand national climate database: https://cliflo.niwa.co.nz

United States

NOAA's National Centers for Environmental Information (NCEI) offers a huge range of online climate data, averages, reports and publications, for both the United States and many other countries. All raw physical climate data available from NOAA's various climate observing systems as well as the output data from state-of-the-science climate models are openly available in as timely a manner as possible. Much is free and downloadable online with just a few clicks of the mouse, although charges may be levied for large datasets or long runs of station records. Access and selection are easy enough, and the site below is a good place to start:

NOAA's National Centers for Environmental Information pages: www.ncei .noaa.gov/access

A comprehensive set of latest climate averages (for the current 1991–2020 period) are also available at www.ncei.noaa.gov/products/land-based-station/ us-climate-normals.

United Kingdom

Part of the UK Met Office public service remit is to 'provide public access to historic weather information via our Library and Archive and climatological records'. The earliest instrumental data for the UK extend back to the late seventeenth century; today there are about 260 'climate' sites and some 2700 'rainfall-only' stations (see Chapter 4, *Site and exposure* for more details). The Met Office began digitising the UK's climatological and rainfall records in the early 1960s; today most information arrives in digital form rather than on paper. The Met Office Library and Archive site at www.metoffice.gov.uk/research/library-and-archive has a huge range of material available online, including the entire library catalogue database and scanned copies of most Met Office periodical publications such as *British Rainfall* (published 1860 to 1993), the *Daily Weather Report* (first published September 1860, with a digital version of it still published today), and the *Monthly Weather Report* (published 1884 to 1993). There are also useful monthly and annual summaries of recent weather, including extremes and map data of both observed values and differences from normal, published online within a few days of the end of the month or season to which they refer at www.metoffice.gov.uk/research/climate/maps-and-data/ summaries/index.

Since the first edition of this book, public access to UK climate data has improved considerably, although it still lags far behind the exemplary online facilities of other countries, particularly Australia and Canada. The data are collected, quality-controlled and archived entirely at the public expense, and following quality-control procedures are then archived afterwards in the Centre for Environmental Data Analysis (CEDA) Archive, part of the Natural Environment Research Council (NERC)'s Environmental Data Service (EDS). CEDA holds data from atmospheric and earth observation research, petabytes of data in all from climate models, satellites, aircraft, climate observations, and other sources, including the Met Office Integrated Data Archive System (MIDAS) [306]. Access is (normally) free of charge, but prior registration is required. The size of the collection, the poor user interface and very non-user-friendly documentation make navigation and use far from easy, but it is possible to find and download runs of station data. A good place to start is the MIDAS station catalogue at https://archive.ceda.ac.uk/tools/midas_ stations (no pre-registration is required to access this). There are many 'undocumented features' in the dataset which are traps for the unwary: one particularly unhelpful example is that maximum temperature and daily precipitation data are entered to the 'day of reading' rather than being 'thrown back' per normal protocols (see Chapters 5 and 6, and 'thrownback' in the index for more on this point). This poorly thought-through archiving decision ensures that assigning correct event dates, and correctly tabulating monthly and annual totals and means, becomes very much more complicated than it needs to be. The matter is apparently 'in hand', but has been for a decade or more with unfortunately little improvement evident at the time of writing.

Two other useful sources of long-period climatological data deserve mention. The British Rainfall project saw the scanning of the entire monthly paper archives of the British Rainfall Organization between 1677 and 1960, some 66,000 individual sheets listing monthly and annual rainfall totals at thousands of locations (there are very few places within UK that have not had at least one rainfall record within 5 km within the last 150 years). A citizen science project in 2020 followed up by digitising the records of every site [307]. As a direct result of this work, the UK's national rainfall record was extended back to 1836, while neatly listed and downloadable monthly and annual rainfall totals at literally thousands of locations throughout the UK were made publicly available online at 10.5281/zenodo.5770389, along with PDFs of the scanned records themselves containing details of the site, the observers and much other fascinating historical material besides.

The UK Climatological Observers Link (www.colweather.org.uk) holds a database of monthly records from several hundred member stations, some extending back to the 1940s, all updated monthly. Members can request copies of station records from sites within their area at no charge. Averages and extremes for several hundred stations have been produced directly from these records, covering both the 1981–2010 and 1991–2020 normals [308, 309].

One-minute summary – *Making the data avalanche work for you*

- There are many advantages in storing data in computer-accessible formats such as spreadsheets. Holding and archiving data in hourly, daily and monthly spreadsheets is easy to do, simplifies record-keeping and makes subsequent analysis much more straightforward.
- Spreadsheets are ideal for archiving weather records, and provide more comprehensive analysis and presentation tools than the AWS software more typically employed to read, display and store current and recent sensor output.
- Develop a format and structure that works for you – and stick with it. The files will build rapidly into useful datasets, and even a few months' observations can

reveal interesting local weather patterns and peculiarities. Don't forget a 'meta-data' sheet giving details of the records in the spreadsheet contents.

- Current local records can often be augmented and compared with historical records from national climate archives. In many countries, online access and downloads are free or available at a nominal charge.
- The examples in this chapter and on www.measuringtheweather.net suggest a few ideas for analysis and how to perform them using Microsoft Excel spread-sheet software. Excel's Pivot Table function is particularly useful for analysing weather records.
- Other specialist graphics plotting software is also relevant for certain types of analysis, such as the preparation of wind roses.
- As with any software, practice builds experience. Experiment with simple graphing and analysis to become familiar with the spreadsheet functions, then experiment with question-based analysis along the lines of the examples given in this chapter. Potential topics and questions are infinite.

19 Sharing your observations

Up to this penultimate chapter, all previous topics have been about a single weather station with one or more sets of instruments at one site. Of course, weather knows no boundaries. Interest in 'measuring the weather' at any particular location improves with the exchange and comparison of observations with others – locally/regionally, nationally or internationally. This chapter suggests ways to exchange information with other sites and other observers, under three main headings – online or real-time sharing using the Internet, online or offline reporting to informal or voluntary networks, and co-operation with national weather services and other official bodies.

Real-time information exchange

There are two main ways to share real-time (or near real-time) weather information via the web: using a site-specific website, one or more data consolidation/aggregation sites or newsgroups which accept data feeds from a number of locations, or both. With a relatively dense network of reporting locations in populated areas, together with a fast update/refresh rate, highly detailed mesoscale displays of current weather conditions can be made almost instantly available on the web, whether on desktop, laptop or smartphone display.

Site-specific websites

Almost all the popular consumer-level AWS packages include built-in Internet connectivity options, sufficient to enable users with minimal computer expertise to post their own data online, and details were covered in Chapter 13, *AWS data flows, display and storage*. Once configured, the data feed runs in the background and regularly refreshes current weather conditions such as air temperature, humidity, wind speed and direction, barometric pressure and tendency and so on. Assuming the AWS, host computer or console and Internet connection remain permanently connected, some AWS software can update conditions every minute or so, although a 5–15 minute update interval is normally ample. The various model options available offer a wide choice of communication/data storage/web display options, and choosing between them becomes a matter of personal choice. Users possessing sufficient technical or web coding skills may even wish to write their own applications.

Include essential details of your site on your webpage

It is good practice to include basic site/exposure, instrument details (metadata) and the length of record, together with site photographs (from different directions) and contact details if necessary, on all 'weather websites' fed by real-time or near real-time AWS data. Without this information it is impossible to judge how representative the station records are of the locality. Are temperature records taken from a sensor protected from sunshine? Does the raingauge sit at ground level or on a roof? Is the anemometer sited at 1 m or 10 m above ground? See Chapter 16, *Metadata, what is it and why is it important?* for a more comprehensive site information metadata template.

Weather station software suppliers – why not include a metadata section on your web page design to encourage users to provide site, exposure, and instrument information?

Unwanted visitors

Sometimes, weather stations can attract unwelcome attention from certain members of the community. If this happens, the user may wish to consider whether the site location details or photographs are sufficiently accurate on, say, a Google Earth map such that a potential vandal or thief could find the site with little difficulty. A location reference accurate to 100 m or so is good enough to locate and correctly plot the observations on all but the most detailed of mapping scales. Consider whether more accurate site location references, including the site address, should perhaps be made available only to genuine enquirers upon request.

Data aggregation websites

By definition, data aggregation sites accept inputs from many sources (which may include, for example, satellite and radar data in addition to AWS feeds) to display current and recent data, usually on scalable map backgrounds such as Google Earth. Most also include options to 'drill down' into stored data from individual sites using a variety of graphics and tabular formats. Larger sites may include data no more than an hour or so old from many thousands of locations worldwide, including updated reports from 'official' observation sites such as airports as well as 'personal' weather stations. The combination of many different data feeds can provide high-density observation coverage in some areas, particularly useful when monitoring fast-moving frontal systems or severe local storms, and national meteorological services are beginning to appreciate the potential value of such dense observational networks (see, for example, [310]). Some sites can display recent historical data, usually in map form, for a specific area and time, which enables retrospective analysis of specific past weather events – a map of 'temperatures at 0700 this morning' when examining a sharp spring frost, for example, or a snapshot of conditions around the time of a heavy thunderstorm event.

Some data aggregation sites are vendor-specific, such as Davis Instruments' weatherlink.com (**Figure 19.1**) which restricts information to data feeds from Davis Instruments products only (no synoptic or other 'official' sites are included),

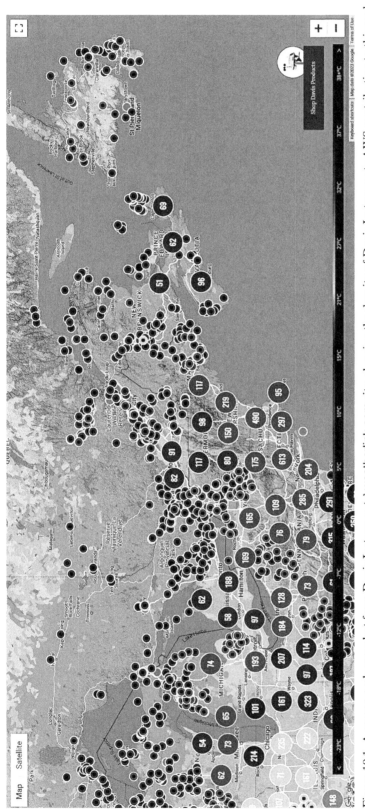

Figure 19.1 A screen grab snapshot from Davis Instruments' weatherlink.com site, showing the density of Davis Instruments AWS contributing to this real-time display.

while others such as Weatherobs.com use global data feeds from the synoptic communications channels together with the Citizen Weather Observer Program (CWOP). CWOP is described in more detail later in this chapter.

Lack of information on site and exposure, and thus data quality, at individual observing locations remains the biggest limitation with most data aggregation sites, other than those such as NOAA's or the US mesonet networks (details and links in Chapter 4, *Site and exposure*). In most cases, there is no easy way to assess instrument accuracy, reliability or exposure. For most 'electronic' sensors, apart perhaps from barometric pressure, variations in instrument exposure are likely to dwarf instrumental calibration errors. Knowing whether thermometers are sited in an open location and exposed within a radiation shelter, or mounted on a sheltered south-facing patio, or whether anemometers are in a sheltered urban garden or in a standard exposure on a windy hilltop, is clearly essential for almost any purpose.

Should you decide to submit your weather station data to a data aggregation website, whether directly or via a third party, be aware of the site's Terms and Conditions, which generally permit site owners to use your data without restriction, or notification, for any purpose, and often without one's knowledge.

The following notes and pointers to the leading data aggregation sites are deliberately brief, as the detail of all such sites changes frequently. All information and Internet links were correct at the time of going to press.

The Citizen Weather Observer Program – CWOP: http://wxqa.com

CWOP (sometimes known by its original name of APRSWXNET) is the oldest and largest online data aggregation site, having commenced US operations in the 1990s. Up-to-date information on the number of contributing sites is difficult to find (the website pages are many years out of date), but at the time of going to press in excess of 10,000 observation sites in North America report hourly through CWOP, and a similar number for the 'rest of world' combined. CWOP collects surface observations from a wide variety of AWSs, most privately owned. Data are sent from either a computer/Internet link or via ham radio protocols and follow simple setup steps within the display/logging software. Following temporal and spatial consistency quality control checks, meteorological records are fed into NOAA's Meteorological Assimilation Data Ingest System (MADIS: see also https://madis.noaa.gov). From there, they are used in NOAA's short-term forecast models and in data aggregation sites worldwide. One significant drawback is that input values in metric units are converted to imperial units (Fahrenheit, inches, miles per hour, etc.), and then rounded (to the nearest degree Fahrenheit for air temperatures and dew points, for example) before being added to the dataset, thus introducing rounding errors to the original observational record. Unfortunately, this limits the precision with which the observations can subsequently be viewed and analysed on the numerous aggregation sites ingesting CWOP input via MADIS.

The CWOP home page contains details on how to register and submit observations. There are a range of options for viewing observations, either as data plots on a map background (only a few available outside North America) or as tabular observations from individual sites. Sites submitting observations can request automated quality control reports on their data, using a near-neighbour comparison method to correct systematic sensor errors, in barometric pressure for example. For US sites, units default to Fahrenheit for temperature and inches for rainfall. Few pages include a 'metric units' option, which limits usefulness outside North America.

Weather Underground: www.wunderground.com

Weather Underground, originally developed within the University of Michigan in 1993, is now a part of The Weather Company, an IBM business. They describe themselves as 'a global community of people, connecting data from environmental sensors like weather stations and air quality monitors so we can provide the rich, hyperlocal data you need to power your passions'. Weather Underground claims upwards of 250,000 connected sites worldwide, this figure including not only individual AWSs but inputs from CWOP or MADIS feeds, WMO synoptic observations and METAR reports (aviation weather observations from airfields). After basic quality-control procedures, data are assimilated into proprietary forecast models run by The Weather Company. Weather Underground's focus is on North America, although with good coverage of amateur AWSs in many other countries including the UK and Ireland. The default 'start-up page' can be set to a location and scale of the user's preference. Adding your own AWS data is easy enough for any Internet-connected computer, as a wide range of AWS software (both current and legacy) is supported and a helpful step-by-step guide to doing so is included on the site.

The site display uses a Google Earth background map and displays current observations using a 'station circle' model, with a limited set of elements (**Figure 19.2**). Most plots can be displayed on a variety of output devices, including tablets or smartphone displays, allowing access to local weather information on handheld devices while travelling, assuming of course Internet access is available. Multivariate plots outside the rather narrow menu options are not supported (for instance, both air and dew point temperatures cannot be displayed simultaneously). The default units are Fahrenheit for temperature, miles per hour for wind speed, and inches for rainfall and barometric pressure, although metric units can be selected from a setting menu. Archived data from individual stations can be viewed, plotted and even downloaded (useful for looking back at past weather events), although the time standard in use is often unclear, particularly during 'summer time' operation, and rounding errors from imperial to metric units are sometimes evident. A few stations continually appear in the wrong locations, which also limits the usefulness of the site, although in fairness to Weather Underground positional errors may be due to the 'feed' from other networks. Unfortunately, almost no site or equipment metadata information is shown, making it next to impossible to assess the reliability of the data. This would be a useful addition to the site.

Weather Observations Website WOW: https://wow.metoffice.gov.uk

The UK Met Office's Weather Observations Website, WOW, was launched in 2011, with support from the Royal Meteorological Society and the Department for Education, initially for weather observers across the UK. It exists to provide a platform for the sharing of current weather observations from all around the globe, regardless of where they come from, what level of detail or the frequency of reports. Data sources include voluntary climate observers, personal weather stations (whether input manually, or added automatically after completing a setup routine), together with short-period rainfall data from England's Environment Agency. The additional coverage offered by private weather stations benefits the resolution of the 'official' (but thinly spread) Met Office observing network, particularly in urban and suburban areas. In turn, this improves the detection and monitoring of small-scale

Figure 19.2 A screen grab snapshot from Weather Underground (original in colour). Temperatures are shown in °C, with wind direction and speed shown with standard synoptic notation; the grey shading shows radar rainfall coverage.

weather phenomena such as severe local storms, which can otherwise slip through the relatively coarse network of reporting sites. The Met Office uses WOW data in short-period forecasting and has prepared case history analyses of some events where the additional high-resolution detail WOW data has made a significant difference to data assimilation in particular storms [310, 311]. At the time of writing, the site ingests upwards of 1.3 million observations per day, and probably as a result the site is often very slow to load. Observation coverage is normally excellent almost everywhere within the UK and Republic of Ireland, with good coverage in many other European countries, the United States and Canada, Australia and New Zealand. As with Weather Underground, almost no site or equipment metadata information is available, making it next to impossible to assess the reliability of the data.

Met Éireann Weather Observations Website WOW-IE: https://wow.met.ie

Met Éireann's Weather Observations Website WOW-IE provides similar functionality to the UK Met Office site. 'Personal' weather stations are shown in addition to the records from the national observing networks.

Weatherobs.com

Weatherobs.com is a global data aggregation site, the observations originating from the international meteorological data exchange networks (observation types include METAR, SYNOP, SHIP, BUOY and TAF reports, together with CWOP data via MADIS). The site is managed by MetDesk Limited, a private weather services company based in Buckinghamshire in southern England. Most elements (air temperature, wind speed, precipitation amounts) can be displayed on a variety of map scales, and data from individual locations (including extremes over the preceding seven days) can be 'drilled down' to graphical plots and tabular displays using a 'dashboard' menu.

MADIS display: https://madis-data.ncep.noaa.gov/MadisSurface

For North America, NOAA's MADIS display pages provide details of all MADIS data on scalable maps, down to individual cities and counties. Around 10,000 sites are available hourly, many with data from multiple instruments (temperature, humidity, wind speed and direction and so on) from state-wide networks such as mesonets, but to avoid slow browser load times a maximum of 600 locations are plotted. Clicking on the site marker brings up a drop-down panel giving latest observational details and basic graphing facilities of the main parameters. The quality of MADIS site records varies considerably, from high-quality reference sites with professional instrumentation in pristine locations to budget all-in-one AWSs in less than perfect locations, and users should be aware of this limitation.

AWEKAS: www.awekas.at

An Austrian site, AWEKAS (the acronym stands for Automatisches WEtterKArten System, or automatic weather map system) provides similar functionality to Weather Underground as it displays overview maps of weather data from participating private weather stations. There are various scales of maps covering regions of Europe, the United States and Canada, and other regions of the world. Metric or imperial units are available. Data displays are of the 'coloured symbol' variety, rather than more useful numerics, but individual sites can be selected either for current observations, or an observer's own website, with just a couple of clicks. The site, which operates on a member subscription basis, also includes some details of instruments used, although no details are shown regarding the site or exposure of the instruments. At the time of going to press AWEKAS listed over 16,000 listed sites, making it one of the largest European data aggregation sites.

Other weather station data aggregation sites: country-specific weather networks

Other sites exist, some for only a year or two: they blink in and out of existence, and the coverage and quality of the information are often highly variable. Since the first edition of this book was published, a few more professionally run and maintained supersites have expanded and increasingly replaced some of these smaller operations with improved organisation, layout and navigation. There remains a place for local or regional sites, such as the following examples of country-specific weather websites, and no doubt many more exist:

Scotland – www.scottishweather.net
Ireland – www.irelandweather.eu
Other regional or country websites are also linked on these pages.

Weather newsgroups and blogs

There are numerous online newsgroups, blogs and Facebook pages covering weather-related topics, and these can be a useful source of observations, information and comment, particularly at times of interesting weather. Some offer generalised weather 'chat', others (such as the US storm-chasing sites) are more specific in their topics. For the UK and Ireland, the Climatological Observers Link (COL) moderated Facebook page *Weather Group – UK & Ireland* offers regular comment and updates on weather events. Unfortunately, most non-moderated interactive Internet newsgroups suffer from the aggressive activities of a few social misfits posting under pseudonyms to spoil the experience of the majority: some sites are best avoided altogether. As with the data aggregation sites, some of these blink in and out of existence quite frequently, and any substantial listing would quickly become out-of-date.

Information exchange via like-minded groups or organisations

Get to know other weather observers in your area: most are happy to exchange observations, upon request, whether in real-time via websites or on a monthly basis by e-mail PDFs. Virtual meetings via Zoom video links or similar can quickly and easily connect local or regional observing communities to discuss weather-related matters of interest. Local comparisons are interesting, and highlight the variability of rainfall patterns, urban and elevation influences, the sites most prone to fog or frost, and the like. Making contact with local observers may also lead you to a source of long-period records for sites in your area, assistance with observational techniques or site guidance for new observers, a calibration check on your instruments, and numerous other benefits. A small local network consisting of a few local observers can prove useful to help fill gaps in observational records, particularly non-instrument observations such as the occurrence of snowfall, thunderstorms and so on, when a primary observer cannot make observations due to a business trip, family holiday, or other absence from home.

The Climatological Observers Link (COL): www.colweather.org.uk

One of the largest networks of co-operating voluntary or amateur observers in the world is the UK's Climatological Observers Link (COL), now more than 50 years old [312]. COL publishes a comprehensive monthly weather review towards the end of the following month, including monthly summary listings of observations made at member sites (the majority within the UK and Ireland) together with observers' notes on the month, a synoptic summary, and an active letters section. The monthly COL bulletin is distributed via email PDF, and back copies are available for members to download. Membership is open to everyone, and includes a surprisingly wide range of occupations. In addition to the monthly bulletin, the strengths of COL membership include access to other like-minded members' help and experience, access to local and regular virtual meetings via Zoom, and an annual members meeting with an interesting range of presentations. An online member directory, containing details of several hundred past and present COL sites and contributing members, with station photographs, is regularly updated. For those looking to make contact with like-minded weather observers within the UK or Ireland, COL is an excellent place to start. Contact details are given in **Appendix 3**.

Similar organisations exist in some other countries, and some have the active support of their national weather services; in the United States, the Community Collaborative Rain, Hail, and Snow Network (CoCoRaHS) is a notable example.

The Community Collaborative Rain, Hail, and Snow Network (CoCoRaHS)

CoCoRaHS is a community network of volunteers of all ages and backgrounds working together to measure and map precipitation across the United States, Canada, Puerto Rico, the US Virgin Islands, the Bahamas and Guam. Using low-cost plastic raingauges (**Figure 6.9**) and a high-quality, high-content interactive website with online training and education resources, CoCoRaHS provides precipitation data to a high standard for natural resource, education and research applications throughout the United States and other countries. A prime objective of the program is increasing the density of precipitation data available by encouraging volunteer weather observing. CoCoRaHS also encourages citizens to have fun participating in meteorological science while heightening their weather awareness.

From a modest beginning within Colorado State University's Climate Center in 1998, today the network includes over 25,000 active volunteers and is now the largest provider of daily precipitation measurements in the United States (**Figure 19.3**). Its major sponsors are the National Oceanic and Atmospheric Administration (NOAA), the PRISM Climate Group and donations from the observer community. CoCoRaHS welcomes anyone within their coverage regions with an enthusiasm for watching and reporting weather conditions and encourages observers to enter their daily precipitation reports on the www.cocorahs.org website. Users range from

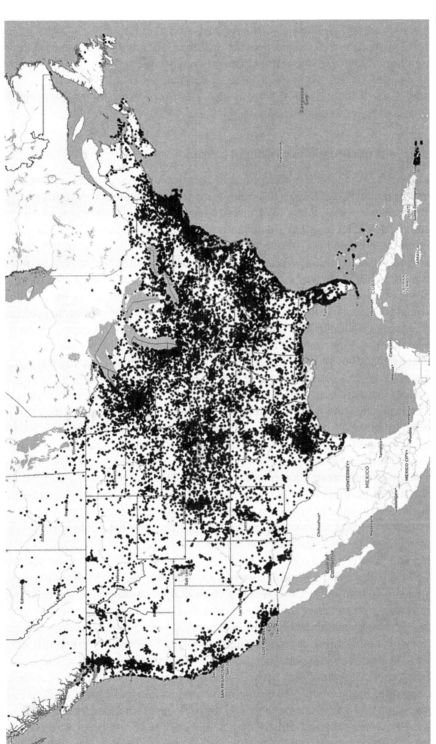

Figure 19.3 Operational CoCoRaHS rainfall sites within the continental United States, as at March 2023 (courtesy Henry Reges, CoCoRaHS)

national official bodies such as the National Weather Service and the National Drought Mediation Center to local farmers, municipalities, teachers, and students, and CoCoRaHS data are regularly used in event case histories and climate analyses [149]. The large daily sample size of CoCoRaHS and COOP networks forms a basis for monitoring, mapping, and categorising daily precipitation extremes, and other aspects of extreme precipitation, with enhanced spatial detail.

The CoCoRaHS web pages provide information on the necessary type of instrumentation and how to join the network. Quality, easy-to-read training materials and animations are available as free downloads from the site, with links to YouTube, X and Facebook, and there is also *The Catch*, a chatty and informative monthly e-mail letter to all participants.

Providing your observations to official bodies

Most national weather services welcome and encourage the contribution of weather observations made by individuals or bodies outside the professional synoptic or real-time observing networks, many of which are geared towards aviation reporting or forecasting requirements rather than weather and climate monitoring. Such observers generally provide less frequent observations (typically once daily) but at a higher spatial density than the sparse official networks. Some 'co-operating observers' go on to provide years or even decades of observations at little or no cost other than minor investments in training, encouragement, and equipment by the state meteorological agencies. Information from such 'second tier' climatological and rainfall networks is vital to fill in the gaps in mapping weather and climate between the government agency-run 'tier one' sites, at all timescales and perspectives from sub-daily to decades in length. Detailed mapping of rainfall patterns, the tracks and impact of severe local storms, and research into city climates including the spatial variation and intensity of urban heat islands, are just a few of the research areas that benefit enormously from relatively dense networks of weather observations run by volunteers. In Australia, for example, the Bureau of Meteorology in Melbourne has maintained a voluntary observer network since 1908, which at the time of writing comprised a national network of 300 cooperative observers, 6,000 daily-reading rainfall observers and over 1,000 river level gauges [313].

Volunteering information to the state meteorological service is an important benefit to society, and in doing so both ensures your observations are made to agreed standards and that they will also become a permanent part of the national weather record. To ensure compliance to common observing practices, site exposure and instruments at co-operating sites must conform to a set of standards set out by the relevant state meteorological authority. In some cases, the necessary equipment may be provided free of charge, or on a long-term loan basis. Observers are given training in observing and reporting. Most sites can expect an inspection visit every 2 to 5 years to ensure equipment operates properly, verify instrument calibrations, and generally ensure the exposure remains satisfactory (sometimes with a recommendation to cut back trees, trim shrubs, etc.). Such visits, together with occasional contact via newsletters or training events, also build a mutually beneficial relationship between the observer and the state meteorological service, contributing enormously to both observer motivation and data quality.

Some voluntary observers make weather observations for many years. Australia recognises the consistent and outstanding achievements of volunteer observers by awarding excellence awards to individual volunteers who reach 50 years records, or to family volunteers who attain 100 years. Similarly, the US cooperating network (see below) provides a comprehensive range of length-of-service awards for observers who contribute observations for 10 years or more [314]. The scheme includes the Richard Hendrickson Award for completing 80 (!) years of observations, established in honour of Richard Hendrickson who completed a remarkable 80 years of continuous observations at Bridgehampton, New York, in 2010 (**Figure 19.4**). He continued his work as a primary observer until he retired in 2015 at age 103. Other prestigious awards in this scheme include the John Campanius Holm Award (named for John Campanius Holm, a Lutheran minister who was the first person known to have taken systematic weather observations in the American Colonies in 1644–45: see also Chapter 1). Each year, up to 25 cooperative observers are honoured with this award for outstanding public service in the provision of daily observations in support of the climate and weather programs of the National Weather Service. Finally, the Thomas Jefferson Award is the most prestigious award a cooperative observer can receive: named after the third President of the United States (b. 1743 – d. 1826), who kept an almost unbroken series of weather records from 1776 to 1816. Only five cooperative observers are honoured each year with the Jefferson Award for outstanding and distinctive achievements.

The UK Met Office had a similar programme of awards to voluntary observers, including presentation barographs and nominations for national honours such as MBE and OBE for those contributing 30 or 40 years of records, or more. Two UK voluntary observers are known to have contributed 75 years observations. John Walker from Ruddington, near Nottingham, maintained a rainfall record from 1873 to 1947 [315]. In January 2024, Tom Bown MBE, from Llwydiarth-Esgob, on Anglesey, was presented with an award for 75 years rainfall records. Tom began daily rainfall observations at age 10 in 1948, following on from his grandfather, whose records on the family farm commenced in 1890: he was awarded MBE after 50 years contributions [168]. Unfortunately, today's UK Met Office has dismantled almost all measures of formal appreciation of long-term contributions to the nation's weather records rendered by such volunteers, in sharp contrast with successful and cost-effective co-operating weather observer programmes run within the United States, Australia and other countries.

The US cooperating observers network

The National Weather Service (NWS) Cooperative Observer Program (COOP) represents the nation's volunteer weather and climate observing network: more than 8,700 volunteers take observations on farms, in urban and suburban areas, National Parks, seashores, and mountaintops, ensuring the records are truly representative of where people live, work and play [316].

The COOP program was formally created in 1890, and its mission is twofold:

- To provide observational meteorological data, usually consisting of daily maximum and minimum temperatures, snowfall, and 24 hour precipitation totals, required to define the climate of the United States and to help measure long-term climate changes; and
- To provide observational meteorological data in near real-time to support forecast, warning and other public service programs of the NWS.

COOP observational data supports the NWS climate program and field operations. Amongst others, the program responsibilities include selecting data sites,

Figure 19.4 The US National Weather Service Richard Hendrickson Award for completing 80 years weather observations. (Courtesy NOAA)

recruiting, appointing and training observers, installing and maintaining equipment and station documentation, collecting data and delivering it to users, maintaining data quality control, and of course managing fiscal and human resources required to accomplish program objectives.

A cooperative station is a site where observations are taken or other services rendered by volunteers or contractors; today, some are augmented by automated observing equipment, and some are co-located with other types of observing stations, such as standard observation sites, flight service stations and so on. Equipment used at NWS cooperative stations may be owned by the NWS, the observer, or by a company or other government agency, as long as it meets NWS equipment standards. COOP observers generally record temperature and precipitation daily and electronically send those reports daily to the NWS and the National Centers for Environmental Information (NCEI). Many cooperative observers provide additional hydrological or meteorological data, such as evaporation or soil temperatures.

Because of its many decades of relatively stable operation, high station density, and high proportion of rural locations, the Cooperative Network has been recognised as the most definitive source of information on US climate trends for temperature and precipitation. Cooperative stations form the core of the US Historical Network (HCN) and the US Reference Climate Network. More information on the US programme, and how to contribute, is available from www.weather.gov/coop.

The UK co-operating observers network

In 1978, the UK climatological network was described by the UK Met Office as ' ... a remarkable institution . . . ' [317], and the contribution of the co-operating observer community was acknowledged as a vital and highly cost-effective part of the networks maintained by the state meteorological service. At that time, the network consisted of 624 climatological observing sites within the UK (only 20 per cent of which were directly administered or manned by Met Office personnel), and a little more than 6,000 rainfall sites, most of them reporting daily observations monthly in arrears. By mid-2023, both networks had fallen to little more than half of their numbers 45 years previously, the majority of sites had been automated, and the proportion of voluntary co-operating observers fallen to low levels. In this respect the approach of the UK national weather service is very different from the United States and Australia, for example, for it has actively discouraged the registration of 'private' voluntary co-operating climatological sites for several decades. Aside from its Weather Observations Website (WOW [310]), all references to 'making voluntary weather observations' have been withdrawn from its public website, despite advances in flexible and affordable technology and a higher than ever public interest in taking weather observations.

In 2001, the day-to-day administration of the remaining rainfall networks devolved to the Environment Agency in England, Natural Resources Wales/ Cyfoeth Naturiol Cymru in Wales, and the Scottish Environment Protection Agency (SEPA) in Scotland. These three regional agencies now maintain the majority (upwards of 90 per cent) of the voluntary rainfall observer network across the British Isles. Within these agency networks voluntary rainfall observers observe to set standards, after a short introductory training session, using agency-supplied equipment consisting of a standard five-inch raingauge (see Chapter 6, *Measuring precipitation*) and/or a tipping-bucket automated gauge, some with data-logger or telemetering capability. Site inspections normally take place every 3 years.

New observers are sought where significant gaps exist in the rainfall monitoring network, or when existing observers withdraw from service. Those interested in providing rainfall records, or offering a site for suitable equipment, should contact their regional office of the respective agency as above, who can advise whether additional local records are sought; those offering to run a voluntary climate site should contact www.metoffice.gov.uk/about-us/contact for further information.

The co-operating observers network in Ireland

Met Éireann's observation network gathers weather data across the country for use in weather forecasts, aviation, marine, agriculture, climate analysis, forecast verification and meteorological research. There are five staffed airport weather stations, supplemented by 20 fully automatic sites, around 75 climatological stations (also mostly automated) and over 500 rainfall stations. All records are quality-controlled and archived in Met Éireann's HQ in Dublin, and many can be viewed in real time at Met Éireann's Weather Observations Website WOW-IE, https://wow.met.ie.

Approximately 80 per cent of the rainfall and climatological sites are voluntary, the balance made up from bodies such as the Irish Agriculture and Food Development Authority (Teagasc), Ireland's Electricity Supply Board (ESB) or various local authorities.

Offers of additional observation sites are welcome, subject to site requirements, although offers may be declined if a particular area has good coverage. Efforts are made wherever possible to fill in any gaps occurring within a network, and from time to time Met Éireann actively seek new observers, usually by inviting existing observers to pass on requirements by word of mouth. Occasionally, a large gap might be filled following local visits to recruit a suitable voluntary observer in the area. Standard equipment and training is provided by Met Éireann in order to maintain WMO standards. To enquire, contact Met Éireann at met.eireann@met.ie.

One-minute summary – *Sharing your observations*

- Weather knows no boundaries. The inherent interest and benefit of making weather observations is greatly enhanced by exchanging and comparing observations with others locally, nationally or internationally. There are three main methods of doing so: online or real-time sharing using the Internet, online or offline reporting to informal or voluntary networks, and more formal co-operation with national weather services and other official bodies.
- Sharing real-time weather information from a digital weather station over the Internet via a site-specific website, or submitting the output automatically to one or more data aggregation sites, the largest of which store and display observations from many thousands of locations across the world, can help build a clearer picture of weather conditions within a town, city or country, help pin down the tracks of showers or thunderstorms, or map an urban heat island.
- With a relatively dense network of reporting locations in populated areas, together with a fast update/refresh rate, highly detailed mesoscale displays of current weather conditions are almost instantly available on the web and on smartphones.
- National forums and publications to assist the exchange of data between those with an interest in 'measuring the weather' are available in several countries: the Climatological Observers Link (COL) in the UK and the Community Collaborative Rain, Hail, and Snow Network (CoCoRaHS) network in the United States are good examples.
- Most national weather services welcome and encourage the contribution of weather observations made by private individuals or organisations, as these provide a richer network of observing points to supplement the wider spacing of professional observing networks. For more than 120 years in the United States, the Cooperative Observer Program has proven itself as a cost-effective method in weather data collection, and currently administers about 8,700 observing sites. The Australian Bureau of Meteorology oversees in excess of 6,000 daily-read rainfall stations across the continent, the majority of which are operated through voluntary co-operating observer arrangements.
- Agreeing to provide observations to a state meteorological service requires minimum standards of site, exposure and instrumentation: the controlling agency may provide the instruments on a free loan basis where the observing site fills a gap in the network. For observers collecting data for a state meteorological agency, they also have the benefit of contributing to their community, and knowing their observations become a part of the nation's permanent weather archive.

- Voluntary observers provide the backbone of most countries observing networks and tend to do so for many years. There are examples within the UK and the United States of a few individuals completing 70 years or more of high-quality weather records. Without doubt, the longer the record, the more interesting it becomes to look back upon notable events.

20 Summary and getting started

This final chapter summarises all of the '*One minute summary*' sections of previous chapters in one location. Page numbers for the relevant chapters are also included, and together with the subject index these facilitate easy reference to any of the information contained within this book.

Chapter 2 One-minute summary – *Choosing a weather station*

- There are many different varieties of automatic weather stations (AWSs) available, and a huge range of different applications for them. To ensure any specific system satisfies any particular requirement, consider carefully, in advance of purchase, what are the main purposes for which it will be used, then consider and prioritise the features and benefits of suitable systems to choose the best solution from those available.
- The choices can be complex, and a number of important factors may not be immediately obvious to the first-time purchaser. Deciding a few months down the line that the unit purchased is unsuitable and difficult to use (or simply does not do what you want it to) is likely to prove expensive, as very few entry-level and budget systems can be upgraded or expanded.
- Decide firstly what the AWS will mainly be used for: some potential uses may not be immediately obvious. Once that is clear, review the relevant decision-making factors as outlined in this chapter, then prioritise them against your requirements.
- An AWS does *not* have to be the first rung on the weather measurement ladder. Short of funds? Not sure whether you'll keep the records going and don't want to spend a lot until you have given it a few months? Not sure where to start? Different options are covered in this and subsequent chapters.
- Consider firstly whether the site where the instruments will be used is suitable. There is little value in spending hundreds or thousands of dollars on a sophisticated and flexible AWS if the location where it will be used is poorly exposed to the weather it seeks to measure. In general, a budget AWS exposed in a good location will give more representative results than a poorly exposed top-of-the-range system. Worthwhile observations *can* be made with budget instruments in limited exposures, but a very sheltered site may not justify a significant investment in precision instruments, as the site characteristics may limit the accuracy and representativeness of the readings obtained.

- Carefully consider the key decision areas. Should the system be cabled, or wireless? Is it easy to set up and use? How many sensors are offered, and how accurate and reliable will they be? Are all the sensors mounted in one 'integrated' system, or can they be positioned separately for the optimum exposure in each case? Do the records obtained need to conform to WMO CIMO or national meteorological authority standards? Examples and suggestions are given in this chapter.
- Finally – and this should be the last step – match the available budget against the requirements and specifications outlined in previous steps. Consider that a reasonable mid-range or advanced system, when used with care and maintained, should last for 10 or even 20 years, and budget accordingly. There are many 'cheap and cheerful' systems available, but will they outlast their warranty period?

Chapter 3 One-minute summary – *Buying a weather station*

- There are enormous differences in functionality and capability between basic and advanced models. The general rule that 'you get what you pay for' holds true for AWSs as well as most other products, but some systems *are* better than others and it pays to check available products carefully against your requirements to ensure the best fit.
- To simplify selection, this chapter suggests five product and budget categories. Most systems fit within one of these price/performance bands – entry-level systems (single-element, or AWS), budget AWS, mid-range AWS, portable systems, and advanced or professional systems.
- *Entry-level systems.* There are many situations where an entry-level system may perfectly meet the requirements. Provided their limitations in terms of accuracy, capability and lifetime are understood and accepted at the outset, and careful attention is paid to siting and exposure, such systems can represent reasonable value for money for a 'starter' weather monitoring system, or those with limited budgets.
- *Budget AWSs* will meet the needs of many users looking for a system that has tolerable accuracy and covers a reasonably wide range of weather parameters. As with entry-level systems, provided careful attention is paid to siting/exposure and calibration, such 'all-in-one' systems can provide reasonably accurate weather records over a number of years. Some represent good value for money at those price points, while others are best avoided.
- *Mid-range AWSs*, whether pre-configured systems or one built around a combination of core datalogger with a few third-party sensors, will meet the needs of many users looking for a system that has acceptably good accuracy across a wide range of weather parameters. Provided careful attention is paid to siting/ exposure and calibration, such systems can be expected to provide reliable and accurate weather records over a decade or more. A typical mid-range AWS costing three times as much as an entry-level or 'all-in-one' budget system is likely to provide higher-quality records and to outlast its cheaper rival in a similar ratio. Viewed over a typical 10 year period, mid-range systems therefore represent much better value for money.

- *Portable systems* are particularly useful for calibration checks, for fieldwork or for a variety of outdoor users. With a suitable choice of logging interval they can be used for short-term logging or backup system at permanent sites, although battery life and memory limitations mean they are not best suited for permanent installation.
- *Advanced AWSs* tend to be custom-built to a specific requirement, whether for the serious amateur or professional installation, and are capable of almost unlimited expansion. Systems in this price range are accurate, robust and capable of measuring a very wide range of elements, but at a price to match. Provided site and exposure requirements are satisfied and regular calibration checks undertaken, such systems can be relied upon to provide accurate, reliable and high-quality weather measurements over many years, for almost all applications and locations, even in the most remote areas or hostile climates.
- AWS specifications are suggested within four very loose 'user profiles'– Starter, Hobbyist, Amateur and Professional – intended as a pragmatic starting point to what is practical and affordable within various budget and site restraints. As an example, with a limited budget it is probably better to concentrate on air temperature and rainfall observations: wind speed and direction (for instance) are more complex and the site requirements more challenging. These and other elements can probably follow at a later stage as budgets (and perhaps an improved site) allow.

Chapter 4 One-minute summary – *Site and exposure: the basics*

- *Site* refers to 'the area or enclosure where the instruments are exposed', while *exposure* refers to 'the manner in which the sensor or sensor housing is exposed to the weather element it is measuring'.
- Satisfactory site and sensor exposure are fundamental to obtaining representative weather observations. An open well-exposed site is the ideal, of course, but with advance thought and careful positioning of the instruments, good results can often be obtained from all but the most sheltered locations.
- The ideal exposure for one sensor can be the exact opposite for another. For representative wind speed and direction readings, for example, an anemometer mounted on top of a tall mast is ideal, but this would be a very poor exposure for a raingauge owing to wind effects (more on this in Chapter 6, *Measuring precipitation*).
- Based upon World Meteorological Organization (WMO) published guidelines, this chapter outlines preferred site and exposure characteristics for the most common sensor types. No single exposure will provide a perfect fit for the requirements of all sensors, and some compromises may be necessary, particularly for 'all-in-one' personal weather station equipment where all sensors are located in one module. More details on WMO site classifications for individual instrument types follow in subsequent chapters.
- Rooftops or masts may provide much better exposure for some sensors, but carefully consider the accessibility of the site before attempting to install the sensors. If the proposed site cannot be reached safely, fit appropriate safety measures or find another site. **Do not take personal risks, or encourage others to do so, when attempting to install weather station sensors, particularly at height**.

- Brief summaries are included of the main operational weather monitoring networks in the United States and the United Kingdom, whose functions vary from frequent, broad-scale real-time information intended primarily for aviation and forecasting purposes to more spatially dense precipitation-only sites reporting in slower time.

Chapter 5 One-minute summary – *Measuring the temperature of the air*

- Air temperature is one of the most important meteorological quantities, but it is also one most easily influenced by the exposure of the thermometer. Great care needs to be taken in exposing temperature sensors to ensure that, as far as possible, the instrument measures a true and representative value, which is not unduly influenced by the instrument housing, surrounding vegetation or ground cover, the presence of buildings or other objects.
- WMO guidelines specify preference for an open site on level ground, freely exposed to sunshine and wind and not shielded by, or close to, trees, buildings and other obstructions, but recognise within the classification of sites from 1 to 5 that it is rarely possible to find a 'perfect' site. The expected degree of error that may occur in unfavourable circumstances is set out in these class definitions. Certain locations, such as hollows or rooftop sites, are best avoided, as readings obtained in these situations may bear little comparison to observations made elsewhere under standard conditions.
- Some form of shielding for the temperature sensor(s) is essential to provide protection from direct sunshine, infrared radiation from Earth and sky, and from precipitation. The main screen types – louvred (Stevenson screen, Cotton Region Shelter), small plastic radiation screens (typical of AWS systems) and aspirated screens – are covered in some detail, because the thermometer housing (or lack of it) is likely to have the largest impact upon the observed temperature. Almost any form of radiation shelter will provide better results than a bare sensor exposed to sunshine. If the AWS model chosen does not include an effective radiation screen, allow budget to purchase a suitable third-party one and use that instead.
- Traditional louvred screens can accommodate both legacy liquid-in-glass thermometers (where they remain in use) and small electronic sensors, but small AWS radiation shields will hold only the smaller electronic sensors. Aspirated units are preferred by WMO as providing a cost-effective means of measuring 'true air temperature'; they are fast in response and largely free of radiative effects, but they provide a slightly different temperature record from other standard methods. Next-generation climate monitoring networks, such as the US Climate Reference Network USCRN, have adopted multiple-redundancy aspirated methods of measuring air temperature from the outset. Aspirated radiation shields can also provide consistent data between different types and models, important for historical continuity as instruments are changed or updated over time.
- To avoid significant vertical temperature gradients near the Earth's surface, thermometer(s) to measure air temperature should be exposed between 1.25

and 2 m above ground level. In the UK and Ireland, the standard height is 1.25 m; in the United States, between 4 and 6 feet.

- Sites that have long current records of temperature made with legacy instruments in traditional louvred thermometer screens (Stevenson, Cotton Region Shelter) should not substitute an alternative method of measuring temperature (for example, electronic sensors in an aspirated screen) without a substantial overlap period, because doing so risks destroying the continuity and homogeneity of the long record. The overlap period should be a minimum of 12 months, or one-tenth of the existing station record length, whichever is the longer.
- Most air temperature measurements are now made using resistance temperature devices (RTDs), which have superseded traditional liquid-in-glass thermometers. The main types of sensors in use today are the *platinum resistance thermometer* and the *thermistor*. The former is more accurate and more repeatable, but more expensive. Both can be made very small and thus highly responsive.
- Logging intervals of 1 to 5 minutes, with shorter sampling intervals (typically 5 to 15 seconds), are sufficient for most air temperature measurement applications. Running means can be used to smooth out any stray electrical noise together with very short-period temperature fluctuations, which are of little significance in climatological measurements.
- Sheltered sites can introduce significant measurement errors, but with some care given to siting the screen and sensor(s) reasonable air temperature measurements can be made in all but the most restricted locations. Temperature records from suburban sites, even those with limited exposures, can often provide more numerous and perhaps more representative climate records for a town or city than those from more distant sites such as airfields, although the latter may have near-perfect exposures.

Chapter 6 One-minute summary – *Measuring precipitation*

- The term 'precipitation' includes rain, drizzle, snow and snow grains or snow pellets, sleet and hail; minor contributions from dew, frost or fog are also conventionally included in precipitation measurements. Precipitation is highly variable in both space and time, and precipitation measurement networks are usually denser than for other meteorological elements to maximise spatial coverage. There may be as many as 1 million raingauges operating globally, although standards vary from country to country.
- Precipitation measurements are very sensitive to the exposure of the gauge itself – particularly to the wind – and the choice of site is very important to ensure comparable and consistent records. Choose an unsheltered (but not too exposed) spot for the raingauge(s) – loss of catch through wind effects is the greatest single error in precipitation measurements, particularly in snow. A site on short grass or gravel is preferable. Wherever possible, obstructions (particularly upwind obstructions in the direction of the prevailing rain-bearing winds) should be at least twice their height away from the raingauge. Rooftop sites are particularly vulnerable to wind effects and should be avoided. The site should also be secure, but accessible for maintenance (grass cutting, etc.) as required.

- The gauge should be exposed with its rim at the national standard height above ground – in the UK and Ireland, this is 30 cm; in the United States, between 3 and 4 feet (90 to 120 cm). Most countries define a 'standard rim height' as between 50 cm and 150 cm above ground. Take care to set the gauge rim level, and to maintain it accurately so.
- Manual raingauges should have a round, deep funnel to minimise outsplash in heavy rain (shallow funnel gauges should not be used) and should have a capacity sufficient to cope with at least a '1-in-100 year' rainfall event – a minimum of 150 mm in the UK and 500 mm (20 inches) in most parts of the United States. The gauge must be paired with an appropriately graduated and calibrated glass measuring cylinder.
- Most manual raingauges are read once daily, usually at a standard morning observation time, typically between 7 a.m. and 9 a.m. local time. The morning reading should be 'thrown back' to the previous day's date.
- To obtain records of the timing and intensity of rainfall, one or more automated raingauges are often sited alongside the manual storage raingauge. The record from the manual (storage) gauge should be taken as the standard period total and sub-daily records (hourly totals, for instance) taken from the automated gauge adjusted to agree with the daily total taken from the manual gauge, where there is one.
- The preferred resolution of a tipping-bucket raingauge is 0.1 or 0.2 mm; 1 mm capacity devices are too coarse for accurate measurements of small daily amounts. Recording raingauges should be logged at 1 minute or 5 minute resolution (higher frequencies are possible using an event-based logger or modern weighing tipping-bucket gauges). They should be regularly inspected for funnel blockage or any obstruction to the operating mechanism, which will result in the complete loss of useful record if not quickly corrected.
- Snowfall is difficult to measure accurately with most types of raingauge, and without some form of windshield most raingauges will lose 50 per cent or more of the 'true' catch through wind errors introduced by the presence of the gauge, which interferes with the flow of the wind over it, causing a loss of some of the catch.
- Procedures for measuring snow depth and the water equivalent of snowfall are set out.

Chapter 7 One-minute summary – *Measuring atmospheric pressure*

- Atmospheric pressure is the easiest of all of the weather elements to measure, and even basic AWSs, household aneroid barometers or smartphones can provide reasonably accurate readings. It is also the only instrumental weather element that can be observed indoors, making a barometer or barograph – analogue or digital – an ideal instrument for apartment dwellers.
- The units of atmospheric pressure are hectopascals (hPa) – a hectopascal is numerically identical to the more familiar millibar. Inches of mercury (inHg) are still used for some public weather communications within the United States – one inch of mercury is 33.86 hPa.

- Pressure sensors must be located away from places that may experience sudden changes in temperature (direct sunshine, heating appliances or air conditioning outlets) or draughts, which will cause erroneous readings.
- Great accuracy is not required for casual day-to-day observations, as very often the trend of the barometer, whether it is rising or falling, and how rapidly, provides the best single-instrument guide to the weather to be expected over the next 12–24 hours, in temperate latitudes at least.
- Where accurate air pressure records are required, the observed barometer reading needs to be adjusted to a standard level, usually mean sea level (MSL), because air pressure decreases rapidly with altitude. A variety of approaches exist to correct or 'set' a barometer to mean sea level are described in this chapter. The choice of method depends upon accuracy sought (and the accuracy of the sensor) and height above sea level. Downloadable Excel spreadsheets are available to simplify the production of site-specific sea level correction tables where desired.
- The calibration of all barometric pressure sensors, particularly electronic units, should be checked regularly to avoid calibration drift. More details are given in Chapter 15, *Calibration*.
- Because of the twice-daily diurnal cycle of barometric pressure, the hour of observation should always be stated when presenting averages. Automatic weather stations can easily provide 24 hour means, which average out diurnal inequalities in atmospheric pressure.

Chapter 8 One-minute summary – *Measuring humidity*

- 'Humidity' refers to the amount of water vapour in the air, a vital component of the weather machine.
- Various measures are used to quantify the amount of water vapour in the air – relative humidity and dew point being the two most commonly used. Knowledge of any two values can be used to derive other humidity parameters. The amount of water vapour that the air can hold varies significantly with temperature – saturated air at 0 °C holds only a quarter of the amount that saturated air at 20 °C can hold.
- The traditional method of measuring humidity is by using a pair of matched sensors, known individually as dry bulb and wet bulb thermometers and in combination as a dry and wet bulb psychrometer. The wet bulb sensing element is kept permanently moist using a thin close-fitting cotton cap or sleeve. The wet bulb is cooled by evaporation, and the difference in temperature between dry bulb and wet bulb thermometers is a measure of the humidity of the air. Using tables, an online calculator or formulae, the relative humidity (or any other humidity measure) can be quickly and easily determined from simultaneous readings of the two thermometers.
- Electronic humidity sensors provide an alternative method of measuring humidity, which can be used alongside or instead of a traditional psychrometer. Modern sensors are small, economical on power, more reliable at temperatures below freezing and datalogger-friendly, although subject to greater uncertainty and slower response times at high humidities.

- Establishing and maintaining reasonably accurate calibration can be difficult; even the best humidity sensors are no better than ± 2 % RH. Calibration drift is a problem (regular calibration checks are essential) and working lifetimes can be limited. Combined temperature/RH sensors are popular, but can become expensive and inconvenient if the relatively short working lifetime of the humidity component mandates replacement (and recalibration) of the temperature sensor too. The combination of the two sensors also precludes ice-bath calibration checks being made on the temperature sensor (see Chapter 15, *Calibration*).
- Humidity sensors are normally exposed alongside temperature sensors in a thermometer screen (Stevenson screens or similar, AWS radiation screens or aspirated units).
- Logging intervals should be the same as those for temperature observations, although sampling intervals can be reduced (once per minute is ample) as response times for humidity sensors are greater than those for temperature.

Chapter 9 One-minute summary – *Measuring wind speed and direction*

- The wind is highly variable in both speed and direction, and obtaining good measurements of the wind poses particular challenges for instruments, logging equipment and site requirements.
- Wind is a *vector* quantity – it has both direction and speed. Wind *direction* refers to where the wind is coming from. A wind vane needs to be accurately aligned to true north, which is slightly different to magnetic north indicated by a magnetic compass.
- Mean wind *speeds* normally refer to 10 minute periods, gust speeds to 3 seconds. For accurate determination of gust speeds, a high sampling interval (no more than a few seconds) is essential, although the logging interval can be much longer than this.
- Wind direction and speed have traditionally been measured with separate instruments, typically a cup anemometer and a digital wind vane, although modern one-piece ultrasonic anemometers will output both direction and speed in digital form. Sonic anemometers have no moving parts (and thus are more reliable/less vulnerable to mechanical wear) and are much more sensitive: they are gradually replacing traditional instruments.
- WMO CIMO site guidelines for wind observations are set out. An ideal exposure for wind sensor(s) will be in an open location, well away from obstacles, at 10 m above ground level. However, such ideal sites are hard to come by, particularly in urban or suburban areas, and wind records are therefore necessarily more site-specific than most other weather measurements. Some corrections for the variation of mean wind speed with height are possible, and these are described in this chapter. Gust speeds should not be corrected.
- If a position 10 m above ground is not feasible, one as high as possible is preferred, commensurate with both safety and accessibility for installation and maintenance. An elevated exposure will increase the vulnerability of the instruments to extreme weather conditions, particularly snow or ice, lightning and of course high winds. Great care should be taken in installation and cabling to minimise the potential for subsequent weather-related reliability issues.

- The absolute accuracy of wind speed measurements is more likely to be limited by the exposure of the anemometer, rather than the accuracy of the sensor(s) themselves. The accuracy of wind direction measurements also depends critically upon careful alignment during installation.
- Planning permission or zoning approval is not normally required for domestic rooftop-mounted anemometers or wind vanes, and local authority case precedents exist within the UK. Specialist legal advice should be taken if in doubt.
- **Never take risks with personal safety when installing or servicing weather sensors at height.**

Chapter 10 One-minute summary – *Measuring grass and earth temperatures*

- Grass and earth temperatures are the most commonly observed temperature measurements, after screen or air temperature.
- The lowest temperatures on a clear night will be recorded at or close to ground level. Where the surface is covered by short grass, the lowest temperatures are attained just above the tips of the grass blades. The so-called 'grass minimum temperature' is measured using a sensor freely exposed in this position. A 'ground frost' occurs when the grass minimum falls below 0 °C.
- Temperatures are also often measured above concrete or tarmac surfaces, or using sensors buried in road surfaces at roadside AWSs, to provide information on road surface temperatures for road forecasting models.
- Surface temperatures can easily be measured with a suitable trailing-lead electrical sensor (thermistor or platinum resistance thermometer, PRT) connected to a datalogger. Such sensors need to be small (for speed of response), weatherproof and robust as they will be exposed to all extremes of weather.
- WMO guidelines indicate that grass and surface minimum temperatures should relate to the period 'sunset to the morning observation on the following day', although the greater prevalence of unmanned sites is leading more locations to adopt the conventional 'morning to morning' 24 hour period. It is easy enough to program a datalogger to record the minimum temperature over a shorter period, for instance from near sunset to the next morning's observation.
- Earth temperatures are typically measured at depths of 5, 10, 20, 30, 50 and 100 cm below ground level, although few sites will include all depths. The location chosen for such measurements should remain fully exposed to sunshine, wind and rainfall.
- Earth temperatures have traditionally been measured by means of mercury thermometers to specific designs, bent-stem glass thermometers at depths of 20 cm or shallower or specially lagged thermometers hung on chains in steel tubes at greater depths. The use of steel tubes can cause bias in temperature measurements owing to enhanced heat flux along the tube itself. Existing liquid-in-glass earth thermometers in plastic or steel tubes are being progressively replaced by electrical sensors, although access for recalibration or replacement can be problematic unless carefully considered prior to installation. Waterproofed sensor connections are more important considerations for earth temp sensors than fast response times.

- Earth temperatures are normally quoted for a morning observation hour, although hourly values can easily be derived from logged electrical sensors. Hourly values provide useful insights into diurnal temperature variations below the earth's surface.
- Grass temperatures should be sampled and logged at the same interval as used for air or screen temperatures; for earth temperatures, particularly at depth, hourly logging intervals are normally sufficient.

Chapter 11 One-minute summary – *Measuring sunshine and solar radiation*

- Radiation from the Sun consists of a wide range of wavelengths, from extreme ultraviolet to the far infrared, peaking in the visible region. Solar radiation is amongst the most variable of all weather elements and consists of two main components – *direct solar radiation* from the solar disk, and *diffuse solar radiation* from the rest of the sky, the latter as a result of the scattering and reflection of the direct beam in its passage through the atmosphere.
- The most common measurements made are of *sunshine duration*, using a sunshine recorder, and/or *global solar radiation on a horizontal surface*, using a pyranometer. 'Sunshine' is defined in terms of the intensity of a perpendicular beam of visible wavelength solar radiation from the solar disk. The intensity of solar radiation is measured in watts per square metre (W/m^2), and daily totals in megajoules per square metre (MJ/m^2). Daily sunshine durations are measured in hours, to 0.1 h precision, or quoted as a percentage of the maximum possible duration.
- There are numerous models of sunshine recorder. The iconic Campbell–Stokes sunshine recorder has been in use since the late 1870s, although it is steadily being replaced by datalogger-friendly electronic sensors, which give slightly different readings – the Campbell–Stokes unit tending to over-record in broken sunshine. Estimates of sunshine can be derived from pyranometer data, although no method for doing this has yet been shown to provide consistent agreement with dedicated sunshine recorders. Changes in recorder types over time (for instance, the transition from the Campbell–Stokes device to modern electronic sensors using more precisely defined 'sunshine'/'no sunshine' thresholds) mean that today's measurements are not directly comparable with measurements made using different instruments in previous years.
- All solar radiation instruments require an open exposure, one with as clear a horizon as possible: a flat rooftop or a mast are often suitable locations. The WMO CIMO guidelines separately define five site classes for solar radiation and sunshine sensors. The effects of obstructions can be assessed using a solar elevation diagram in conjunction with a site survey, although obstructions within about 3 degrees of the horizon have little effect on the record. The instruments must also be accurately levelled, and most also require some form of azimuth alignment and/or latitude setting.
- Calibrations for solar radiation instruments tend to be based upon field comparisons with reference instruments. WMO organises instrument intercomparisons amongst national meteorological services every 5 years to ensure consistent and transferable measurement standards.

- A high sampling interval is advisable for electronic sensors, as solar radiation is amongst the most variable of all weather elements. The logging interval can be much less frequent than the sampling interval, and hourly totals or means will be sufficient for many applications.
- Sunshine and solar radiation instruments tend to be slightly more variable in their outputs than other meteorological sensors, and even adjacent instruments can be expected to vary somewhat in their readings. For this reason, all but the highest-specification sunshine and solar radiation measurements should be regarded as liable to errors of a few per cent in either direction.

Chapter 12 One-minute summary – *Observing hours and time standards*

- By convention, weather measurements throughout the world are made to a common time standard – Coordinated Universal Time (UTC). For all practical meteorological purposes, UTC is identical to Greenwich Mean Time (GMT).
- For weather measurements to be comparable between different locations, the time(s) at which observations are made, and the period covered by the measurements, should be the same. WMO provides guidance on observation times for the main international synoptic observing networks, while observing practices for other station networks tend to be defined at a country or regional level.
- Many if not most countries around the world have adopted a once-daily morning observation as standard practice, the time typically between 7 a.m. and 9 a.m. Where AWS data are available, it is straightforward to adjust records to conform more closely to the 'nominal' standard morning observation time, even if it is perhaps inconvenient to make manual observations at that hour. Adopting the standard observing time (or close to it) greatly simplifies comparisons of weather observations with other sites – particularly daily rainfall records.
- The start and end times of these recording periods are known as the 'terminal hours' of that measurement. The term 'terminal hour' refers to the time of day at which one 'observation day' ends and the following one commences.
- The once-daily morning observation naturally establishes a standard 24 hour period over which many 'once-daily' values are tabulated, such as daily maximum and minimum air temperatures. However, WMO guidance is that the grass minimum temperature should refer instead to the period from just before sunset to the following morning observation terminal hour. Some other elements, such as sunshine, fall more naturally within the 'civil day' (midnight to midnight local regional time). Professional synoptic reporting sites will use observing and reporting times defined by WMO at regional levels.
- By convention, 24 hour minimum temperatures read or logged at the standard morning hour are entered to the day on which they were read, whereas the 24 hour maximum temperature and total rainfall are entered to the day *prior* to the observation (they are said to be 'thrown back'). Although this occasionally leads to some bizarre anomalies, a 'civil day' (midnight-to-midnight) record period would be difficult to introduce at sites where instruments are read manually (particularly at rainfall-only locations).
- Terminal hours based around 'day maximum' and 'night minimum' temperatures (where the extremes span only 12 hour periods) will give results which are

incompatible with '24 hour' sites to a greater or lesser extent, particularly in temperate latitudes in the winter months, and accordingly should be avoided.

Chapter 13 One-minute summary – *AWS data flows, display and storage*

- Today's AWS products are sophisticated data acquisition and processing systems which enable straightforward collection and processing of almost any type of meteorological or environmental data. Host computer platform and power source are important early considerations: if a mains/utility power supply is used, an uninterruptible power supply with a surge protector is a wise investment.
- Most products follow a broadly similar set of basic processes, which can be briefly summarised as system maintenance, sensor management and sampling, data handling architecture, communications, display (where required), permanent data storage, and access and export facilities. These steps are explained and illustrated using three different approaches to 'system architecture'. Which to choose is very much dependent upon requirement, budget and expected or planned future expansion: all three have their advantages and disadvantages.
- 'Black box' systems usually carry out the required functions and processes within firmware, perhaps via an external communications controller module. A host computer regularly uploads data for post-processing and display (if required) and local storage, using manufacturer or third-party software running on the host, and from here records can be archived, exported, written to websites and so on. This approach is most typical of entry-level and budget systems and has the advantage of simplicity and ease of use.
- The 'black box combined with Internet and cloud storage' system type is an extension of the above, coupled with access to an always-on Internet connection. Sensor interfacing and basic data handling/datalogging processes are managed within a wireless console, and data are regularly uploaded to a remote proprietary data aggregation service. Post-processing and storage are managed 'in the cloud', and data viewed from anywhere using a web browser. Such systems are also straightforward in installation and use, but may incur subscription costs.
- The 'programmable datalogger' approach uses a dedicated datalogger for all sensor interface and data handling processes, and data are uploaded through a host computer and communications network as and when required for onward processing, storage, display and cloud storage or data aggregation services, including web browser access. This approach offers almost unlimited flexibility, including advanced capabilities such as high-resolution event-based logging. Such devices are more expensive than the other types, however, and may require specialist programming.
- Finally, it is important to check and test all sensor / system / upload communication links thoroughly, over a period of at least a few days, before permanent hardware installation or embarking on any long-term data collection. This is particularly important if access to some or all of the sensors is more difficult – wind sensors at height, for instance, or remote or relatively inaccessible sites.

Chapter 14 One-minute summary – *Non-instrumental weather observing*

- Instrumental readings are of course vital in making observations of the weather, but for a complete picture non-instrumental and 'narrative' weather observations are equally important, especially so for the analysis of severe weather events.
- A once-daily 'morning observation' is the best time to read/reset any manual instruments in use, as well as perform visual checks on the continuing operation of all instruments, including the AWS sensors: raingauge funnels are especially likely to become blocked, and the obstruction may not become obvious until the next time rain falls. A manual observation also provides a convenient opportunity to note current weather details such as the amount and types of cloud, the surface visibility, present weather, the occurrence of lying snow and so on. Weather observing need not be restricted to viewing graphical or tabular output on a computer screen!
- With a little practice, maintaining a near 24 hour weather watch becomes second nature, and with some assistance from friends and family, colleagues or neighbours a 365 day, 24 hour coverage of significant weather is not difficult. When combined with the instrumental observations from an AWS and a brief daily descriptive weather diary, a high-quality combined weather record quickly builds up.

Chapter 15 One-minute summary – *Calibration*

- Instrument calibrations are one of the most important, yet sometimes one of the most neglected, areas of weather measurement. Making accurate weather measurements requires accurately calibrated instruments.
- Recording raingauges can be easily and accurately calibrated by passing a known volume of water through the gauge and comparing with the indicated measurement. 'Out of the box' errors for some AWS tipping-bucket raingauges of this type can exceed 20 per cent, so this is a vital test for all new instruments at first installation. Recording raingauges should not be adjusted merely to attempt exact agreement, or near-agreement, with a standard raingauge, because instrumental and exposure differences will always lead to slight variations in the amount of rainfall recorded.
- Two calibration methods are described for temperature sensors. The first is a quick and easy method based on the fixed point of melting ice at 0.0 °C. An extension of the approach can extend the range of calibration points from −5 °C to +40 °C when used with an accurately calibrated reference thermometer. However, this method is not suitable for certain types of sensor, and on some AWS models the temperature elements may not be accessible to allow this test to be undertaken.
- The second temperature calibration method involves careful comparison over a period with a portable reference unit of known calibration. Both sensors (calibrated reference and test) are exposed in identical adjacent surroundings exposures for a period (days to weeks). Plotting differences between the test instrument and the calibrated standard can then be used to prepare a calibration

table or datalogger adjustment algorithm, which is then used to apply the corrections obtained to the sensor readings going forward.

- Calibration checks, and checks for calibration drift or scale magnification errors, on pressure sensors can be made using plotted pressure reports from nearby synoptic station reports over a period of a few days or weeks.
- Make a note in the site metadata of all calibrations applied, and the date. Keep a copy of the calibration table or algorithms used in the metadata file. Retain the calibration test results.
- Calibrations can drift over time, so calibrations should be checked (and adjusted if necessary) regularly – at least once every 6 months for pressure sensors and humidity probes, and every 12 months for temperature sensors.

Chapter 16 One-minute summary – *Metadata – what is it, and why is it important?*

- Metadata is literally 'data about data'. In the context of weather records, it is a description of the site and its surroundings, the instruments in use and any changes over time, information about observational databases and units used, and any other details about the measurements that may be relevant.
- Metadata statements are important because they provide the essential information for any other user of the records to understand more about the location and characteristics of weather records made at any site, thereby enabling more informed use of the data to be made. The 'other user' may be weeks or decades in the future, quite possibly long after all observer recollection of the circumstances of the site and its record have faded.
- A metadata statement is best prepared as a short, structured text document, and retained alongside data files in soft copy or hard copy. A copy or link should also be included on the site weather website, if there is one. Links should also be provided to site photographs, instrument calibration certificates and other related documents.
- Review the metadata statement whenever instrument or site details change, and at least annually. Update as required. Retain previous site descriptions and photographs, which will assist in documenting site, instrument and exposure changes over the years.

Chapter 17 One-minute summary – *Collecting and storing data*

- Making weather measurements, particularly using an AWS, can quickly generate vast amounts of data and these can become unmanageable without some thought being given to how records are to be kept and used.
- Spreadsheets are ideal for archiving weather records and provide more comprehensive analysis tools than the AWS software used to log the sensors. Holding and archiving data in separate sub-hourly, hourly, daily and monthly spreadsheets is easy to do, simplifies record-keeping and makes subsequent analysis much more straightforward.
- Each spreadsheet should include an integral metadata sheet or 'tab' detailing the instruments used, their exposure, units of measurement, record length and any other essential information.

- Months or years of data can be lost in an instant if held in a single file on a single computer disk. An entire lifetime's manuscript record could just as easily be lost forever in a house fire or burglary. Taking simple steps in advance, including putting in place a multiple backup strategy, will greatly improve the chances that records (and instruments) will survive to be available for future users.

Chapter 18 One-minute summary – *Making the data avalanche work for you*

- There are many advantages in storing data in computer-accessible formats such as spreadsheets. Holding and archiving data in hourly, daily and monthly spreadsheets is easy to do, simplifies record-keeping and makes subsequent analysis much more straightforward.
- Spreadsheets are ideal for archiving weather records, and provide more comprehensive analysis and presentation tools than the AWS software more typically employed to read, display and store current and recent sensor output.
- Develop a format and structure that works for you – and stick with it. The files will build rapidly into useful datasets, and even a few months' observations can reveal interesting local weather patterns and peculiarities. Don't forget a 'metadata' sheet giving details of the records in the spreadsheet contents.
- Current local records can often be augmented and compared with historical records from national climate archives. In many countries, online access and downloads are free or available at a nominal charge.
- The examples in this chapter and on www.measuringtheweather.net suggest a few ideas for analysis and how to perform them using Microsoft Excel spreadsheet software. Excel's Pivot Table function is particularly useful for analysing weather records.
- Other specialist graphics plotting software is also relevant for certain types of analysis, such as the preparation of wind roses.
- As with any software, practice builds experience. Experiment with simple graphing and analysis to become familiar with the spreadsheet functions, then experiment with question-based analysis along the lines of the examples given in this chapter. Potential topics and questions are infinite.

Chapter 19 One-minute summary – *Sharing your observations*

- Weather knows no boundaries. The inherent interest and benefit of making weather observations is greatly enhanced by exchanging and comparing observations with others locally, nationally or internationally. There are three main methods of doing so: online or real-time sharing using the Internet, online or offline reporting to informal or voluntary networks, and more formal co-operation with national weather services and other official bodies.
- Sharing real-time weather information from a digital weather station over the Internet via a site-specific website, or submitting the output automatically to one or more data aggregation sites, the largest of which store and display observations from many thousands of locations across the world, can help build a clearer

picture of weather conditions within a town, city or country, help pin down the tracks of showers or thunderstorms, or map an urban heat island.

- With a relatively dense network of reporting locations in populated areas, together with a fast update/refresh rate, highly detailed mesoscale displays of current weather conditions are almost instantly available on the web and on smartphones.

- National forums and publications to assist the exchange of data between those with an interest in 'measuring the weather' are available in several countries: the Climatological Observers Link (COL) in the UK and the Community Collaborative Rain, Hail, and Snow Network (CoCoRaHS) network in the United States are good examples.

- Most national weather services welcome and encourage the contribution of weather observations made by private individuals or organisations, as these provide a richer network of observing points to supplement the wider spacing of professional observing networks. For more than 120 years in the United States, the Cooperative Observer Program has proven itself as a cost-effective method in weather data collection, and currently administers about 8,700 observing sites. The Australian Bureau of Meteorology oversees in excess of 6,000 daily-read rainfall stations across the continent, the majority of which are operated through voluntary co-operating observer arrangements.

- Agreeing to provide observations to a state meteorological service requires minimum standards of site, exposure and instrumentation: the controlling agency may provide the instruments on a free loan basis where the observing site fills a gap in the network. For observers collecting data for a state meteorological agency, they also have the benefit of contributing to their community, and knowing their observations become a part of the nation's permanent weather archive.

- Voluntary observers provide the backbone of most countries observing networks and tend to do so for many years. There are examples within the UK and the United States of a few individuals completing 70 years or more of high-quality weather records. Without doubt, the longer the record, the more interesting it becomes to look back upon notable events.

Metrology and meteorology: An instrument theory primer

Metrology is the science of instruments and their behaviour. *Meteorology* is the science of the atmosphere and its phenomena. The two are intimately related by far more than having all but two letters in common, for meteorology depends upon instrumentation to provide accurate and precise quantitative measurements of the state of the atmosphere at any time or over a period of time. A joint World Meteorological Organization (WMO) and Bureau International des Poids et Mesures (BIPM) initiative in 2011 recognised the need for improved measurements and resulted in the 'Meteomet' initiative, which brought together national meteorological and metrological institutes for mutual benefit [318].

For most users of meteorological instruments, it is certainly not essential to possess a detailed knowledge of the mathematical and physical principles behind the theory and design of any particular sensor, but it can be helpful to understand a few basic concepts and terms as they apply to both the sensors themselves and particularly the output of measurement systems. A knowledge of how sensors react to the elements they measure, what the outputs are and how they are interpreted into useful formats, what uncertainty or limitations there may be on those outputs, and how key sensor characteristics can be compared, all help to create a clearer understanding of the way any measurement system performs and its applicability to the task at hand.

This Appendix provides a very simplified overview of some of the basics of measurement theory relevant to meteorological sensors. There is a wide and extensive technical bibliography on all aspects of meteorological metrology, and the first chapter of the WMO CIMO guide provides a very useful introduction to the field with specific relevance to meteorological instruments in operational contexts [4]. Readers who seek more detailed information beyond the necessarily brief topics outlined here are recommended to consult two useful single-volume works providing an excellent overview of the field at undergraduate or postgraduate module level: Giles Harrison's *Meteorological measurements and instrumentation* (Wiley Advancing Weather and Climate Science series, 2014) [140], and *Meteorological measurement systems* by Fred Brock and Scott J. Richardson (Oxford University Press, 2001) [139]. Some of the following material has been adapted and summarised from these volumes.

Components of a measurement system

Any measurement system, whether it be a simple mercury thermometer or a complex, multi-site instrumented AWS network, consists of some or all of the following components:

- Sensor or transducer
- Analogue output
- Signal processing
- Data transmission
- Data display
- Data storage

The *sensor* (or *transducer*) is the component which reacts to the element being measured. Most generate an *analogue output* which varies in a known manner with changes in the element being measured. This is often a change in physical properties – the expansion of alcohol in a liquid-in-glass thermometer with rising temperature, for example, or the expansion and contraction of an aneroid barometer capsule with changes in atmospheric pressure.

This output is given form by the *signal processing* component. In a liquid-in-glass thermometer, the expansion or contraction of the alcohol in the bulb of the thermometer is amplified by movement within the much smaller capillary tube which forms the stem of the instrument. In the aneroid capsule, tiny expansions or contractions result in changes of electrical capacitance across a circuit, which is then converted into an oscillator frequency output by a second circuit. There may be several sequential signal processing stages, involving for example the eventual conversion of the oscillator frequency into units of atmospheric pressure, or the resistance of a thermistor into units of temperature. An *analogue to digital convertor* may then convert the analogue signal (for example, the voltage change across a resistance bridge) into a digital value. This latter step may take place within an 'on-chip' combined 'smart sensor', or at a later stage when the sensor output is connected to and processed by a datalogger as described in Chapter 13, *AWS data flows, display and storage*.

The raw or processed signals may be *transmitted* elsewhere. This may be a simple cable connecting a sensor to the datalogger, or output from one or more loggers being sent over a communications system, whether a direct cabled connection to a host computer, a Wi-Fi connected console unit with an Internet connection to a remote database, or a complex multi-stage network connection involving radio or satellite links. There may be several transmission links before the information arrives at its final destination.

At the end of this chain there will usually be some way of *displaying* and presenting the information from the sensor. In the alcohol thermometer, this function is carried out by the graduated scale adjacent to the alcohol column in the capillary tube, and the level of the liquid column on that scale is read and noted manually. If the graduated scale is one expressed in terms of standard units, and which has been previously calibrated against known fixed points or reference instruments, then the scale value noted is referred to as the 'temperature' of the instrument and, by implication, the temperature of the medium in which the thermometer bulb is immersed, whether that be air, water or some other substance. In digital systems such as modern AWS models, the display stage may usefully combine outputs from several sensors, or multiple observing locations.

Finally, there is usually a *data storage* step, whereby the processed output from the sensor or combination of sensors is stored in some form. For systems generating digital output directly, this will normally be storage on a computer system. Data storage may be in two or more stages – a temporary (real-time) store in working memory, and a long-term computer archive using data management software such as a database or spreadsheet to facilitate subsequent data retrieval and analysis. In a well-designed archive system, the latter may take place as easily whether the measurement was made seconds or decades earlier.

Examples of meteorological measurement systems

Of course, not all of these steps apply to all types of instrument. The simple example above of the alcohol thermometer is typical of many 'traditional' instruments. In this example there are no communications links, while the first data storage step results from the manual transcription of the observed reading of the thermometer by the observer in the manuscript observation register, which may subsequently be scanned or otherwise digitised into a computer archive.

A modern cup anemometer provides a more typical example of what may be referred to as a 'digital' instrument. Here the sensor converts changes in wind speed into variations in the rotation rate of a vertical shaft (see Chapter 9, *Measuring wind speed and direction*), which are approximately linear with wind speed. One class of anemometer uses a small, low-power light source and photodiode sensor mounted close to the shaft. As the shaft rotates, a suitable optical window alternately transmits or interrupts the light beam, thereby generating a pulsed or frequency output. These pulses are counted by the signal processing system, usually at the datalogger interface, and converted into appropriate units. If it is known from calibration tests that 10 pulses per second correspond to 1 metre of 'wind run', then a count of 223 pulses in a sampling interval of 1 second indicates a wind speed of 22.3 metres per second (m/s) over that sample. Storing every sample would generate an enormous and largely unnecessary volume of wind speed data, and usually a second stage of signal processing at the datalogger averages a number of samples to generate a mean wind speed over any pre-programmed time interval. This value would then be displayed and stored as required. The data acquisition, processing, display and storage routines would then be repeated for each subsequent sampling interval, which may be every ¼ second.

One important facet of instrument performance to consider is that the sensor does not directly 'measure' the element being sampled. Instead, it is variations in some physical property of the sensor itself that are being measured. This property will have a known relationship with the element being sampled, and the output signal will be processed into appropriate units at some later stage in the chain. The relationship may be linear, or non-linear. In *linear systems* the output is directly proportional to changes in the sensed element. An example would be the variation of resistance of a standard 'Pt100' 100 Ω platinum resistance thermometer (PRT), which is accurately made to be 100.0 Ω at 0 °C and 138.5 Ω at 100 °C. Linear interpolation by measurement software will deduce that a PRT showing a resistance of 107.7 Ω indicates a sensor temperature of 20.0 °C. In *non-linear systems* the signal processing routine must include suitable scaling to achieve the appropriate mathematical conversion of the output signal into relevant units as required. Thermistors (sensors whose resistance varies significantly with temperature) are an example of a type of non-linear sensor. Not all sensors generate a continuous analogue output. The sensor output from a tipping-bucket raingauge, for example, is also a pulse. In the same fashion as the pulsed anemometer above, the output is then processed into useful units by the multiplication of sensor counts by the known capacity of the tipping bucket to determine the rainfall amount in a given period of time.

Sensor characteristics

Any particular sensor possesses a number of output characteristics, which Brock and Richardson subdivide into *static* and *dynamic performance*, *calibration drift* and *exposure effects*. Each is briefly considered below.

Static performance

As the term implies, the static performance of a sensor refers to the characteristics of a sensor – sometimes the combined performance of the sensor/signal processing/output components combined, if they are combined into one unit – under steady-state conditions. An example of a static performance metric would be the calibration of a temperature sensor, which could be stated as an offset at a particular temperature ('calibration offset at 20.0 °C is –0.19 degC'). Such calibration would normally be obtained by comparing the sensor output against a known reference under static conditions. For temperature sensors this would typically be obtained using a stirred water bath (or other appropriate liquid) maintained at a particular temperature, although for many meteorological temperature sensors the use of a liquid calibration bath is less than ideal as their working medium is air rather than liquid. At each calibration point, both sensors – the 'reference' and the one being calibrated – must be allowed to attain equilibrium before comparisons are made. Differences between the 'reference' and the sensor under test are then noted. The ambient temperature is then adjusted to the next calibration point, and the process repeated.

Dynamic performance

When the input to a sensor changes, we expect the output to change too – after all, that is the function of a sensor, and of course meteorological sensors would be of little use if they were unable to react quickly to changes in the element which they are measuring. For this reason it is important to quantify how quickly any sensor reacts to a change in environmental conditions: this is termed its *response time* or *time constant*. For routine meteorological observations there is little benefit in using sensors with a very short time constant since the temperature of the air may sometimes fluctuate by a degree or so within a few seconds. Obtaining a representative reading with such a thermometer requires taking the mean of a number of readings, whereas a thermometer with a longer time constant tends to smooth out the rapid fluctuations. Too long a time constant, however, will result in lag errors when rapid changes of temperature occur.

The time constant, τ

The rate of change of many quantities is a 'first order' response, meaning that the rate of change of the measurement is proportional to the difference between the observed value and the ambient value. The time constant concept is perhaps easiest to visualise in terms of a sensor's response change in temperature. For a sensor with heat capacity C in thermal contact with air at temperature T_{air} through an effective thermal resistance R_{th}, the rate of change with time t of the thermometer temperature T is given by:

$$\frac{\mathrm{d}T}{\mathrm{d}t} = \frac{(T_{air} - T)}{R_{th}C}, \tag{A1.1}$$

where $R_{th}C$ is known as the *time constant*, τ [113]. A thermometer will respond to an instantaneous step change in ambient temperature from T_0 to T_1 according to:

$$T(t) = T_0 + (T_1 - T_0)\left[1 - \exp\left(-\frac{t}{\tau}\right)\right]. \tag{A1.2}$$

For $t \gg \tau$ the exponential term will diminish and the sensor's temperature T will approach T_1. When $t = \tau$, the sensor will have registered 63 per cent (τ_{63}) of the incremental change $(T_1 - T_0)$, while after 3τ, it will have registered 95 per cent of the change (τ_{95}). It follows that we can evaluate the time constant τ by observing how quickly a sensor responds to a change in temperature, and by doing so under controlled laboratory conditions we can understand how this varies with ambient conditions, specifically the speed of airflow over the sensor, if its response in air is to be measured.

Step changes are unusual in meteorological air temperature measurements, and instead the effect of the finite sensor time-constant is more obvious in causing the sensor to lag behind the actual air temperature by:

$$T(t) = T_{air}(t) - \tau_{63}\frac{\mathrm{d}T_{air}}{\mathrm{d}t} \tag{A1.3}$$

The responsiveness of many meteorological sensors (and combined systems, such as a temperature sensor within a Stevenson screen) can thus be assessed and compared once their time constants are known, enabling better matching of sensors to the application. Manufacturers often quote time constants in sensor specification literature, although care needs to be taken to distinguish whether performance is quoted to 63 per cent (τ) or 95 per cent levels (3τ), or sometimes 50 per cent, and of course the conditions under which the measurement was carried out (for example, whether a temperature sensor response was measured in liquid or air, and flow/airflow speed across the sensor). For temperature sensors, time constants measured in water (or similar liquids) will be much shorter than those in air, and thus of little relevance for air temperature studies.

Careful laboratory measurements have shown that, for a cylindrical sensor such as a Pt100 platinum resistance thermometer (PRT), the 63 per cent response time τ_{63} varies approximately as

$$\tau_{63} \approx 5.6\frac{d^{3/2}}{v^{1/2}} \tag{A1.4}$$

where d is expressed in millimetres and v is expressed in metres per second [113]. From this, it can be seen that time constants will be reduced by using small sensors (maximum 3 mm diameter) together with increased ventilation rates (at least 3 m/s) over the sensor. Such a combination will result in $\tau_{63} \sim 17$ s, which is within the WMO CIMO recommended response time for air temperature measurements of

20 s ([4], section 2.1.3.3). In contrast, τ_{63} for a 6 mm diameter sensor with an airflow across the sensor of just 0.2 m/s, typical of the interior of a Stevenson screen [112], would be approximately 180 s or ten times as long.

Figure A.1 and **Table A1.1** have been prepared to show the effect of these two differing but realistic values of time constant, such that sensor A has a 63 per cent time constant $\tau = 17$ s and sensor B $\tau = 180$ s. The former is typical of fast-reacting temperature sensors which are in good contact with the surrounding medium (for example, an aspirated air temperature sensor), while the latter is more typical of a temperature sensor exposed within a Stevenson screen. (The screen itself may impose an even larger response time [114, 116]: and where any system consists of two components with differing time constants, the time constant of the combined system will equal that of the slower component.) The example shown is one where the temperature falls 10 degrees Celsius (10 K) instantly; knowing the time constants of the two sensors, we can calculate their temperature every few seconds, out to 20 minutes after the step change. (The actual value of the change in temperature is

Table A1.1 *Temperature response from sensor A ($\tau = 17$ s) and sensor B ($\tau = 180$ s) to a sudden temperature change from 25 °C to 15 °C; data plotted in Figure A1.1*

Time (s)	Time (min)	Sensor A reading (°C)	Sensor B reading(°C)
0		25.0	25.0
10		20.6	24.5
20		18.1	23.9
30	0.5	16.7	23.5
40		16.0	23.0
50		15.5	22.6
60	1.0	15.3	22.2
70		15.2	21.8
80		15.1	21.4
90	1.5	15.1	21.1
120	2.0	15.0	20.1
	3.0	15.0	18.7
	4.0	15.0	17.6
	5.0	15.0	16.9
	6.0	15.0	16.4
	7.0	15.0	16.0
	8.0	15.0	15.7
	9.0	15.0	15.5
	10.0	15.0	15.4
	11.0	15.0	15.3
	12.0	15.0	15.2
	13.0	15.0	15.1
	14.0	15.0	15.1
	15.0	15.0	15.1
	16.0	15.0	15.0
	17.0	15.0	15.0
	18.0	15.0	15.0
	19.0	15.0	15.0
	20.0	15.0	15.0

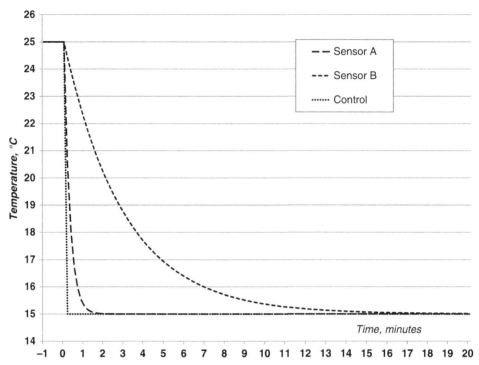

Figure A1.1 Illustrating the response of two sensors A and B, to an instantaneous temperature fall from 25 °C to 15 °C. Sensor A has a time constant τ of 17 s, B 180 s (see also Table A1.1).

immaterial, as the shape of the curve is the same whatever increment is chosen.) The table and graph clearly demonstrate *sensor lag*; sensor B *lags* sensor A by an amount which increases with the difference in their time constants.

Sensor A From **Table A1.1**, we can see that, 50 seconds (3τ) after the 10 K step change from 25 °C to 15 °C, sensor A has responded to 95 per cent of the change (9.5 K). The sensor display will therefore show 15.5 °C at that instant. It attains near-equilibrium (within 0.1 K, a typical sensor accuracy specification) just under 80 seconds after the step change.

Sensor B With a slower time constant, after 50 seconds sensor B has responded to just 2.4 K of the sudden 10 K change, and so will read 22.6 °C (remember that sensor A reads 15.5 °C at this instant). With its 180 s time constant, sensor B does not attain 95 per cent of the step change until 3τ or 540 s (9 minutes) has elapsed. It will not reach near-equilibrium with the new ambient temperature until almost 14 minutes after the drop.

Without being aware beforehand that the time constants (or lag) of the sensors involved differed appreciably, the details of a real-world step change could be interpreted very differently. While instantaneous step changes in temperature of 10 K such as this example are rare in the real atmosphere, changes of similar magnitude occasionally occur within time periods of no more than a few minutes as a result of frontal passages, gust fronts associated with convective storms, turbulent mixing, mountain winds and other phenomena. It is clear that Sensor A would provide a much more realistic record of any such event. It is also evident that a casual examination of the records of the two sensors at anything up to 15 minutes after the step change would give the impression that Sensor B was incorrectly calibrated.

A similar lag issue has been identified in wet bulb sensors, where the wet bulb consists of (typically) a Pt100 PRT covered with a thin cotton sleeve (details in Chapter 8, *Measuring humidity*). The wet bulb sleeve acts to slow response time by a factor of two or three, and rapid changes in temperature will result in considerable lag in the wet bulb reading, which in turn results in temporarily misleading values of dew point and relative humidity derived from the dry and wet bulb psychrometer [112].

Of course, real-world temperature changes rarely take place in such clearly defined steps. A sudden temperature fall might be followed by a second, less rapid one, or a rise, or a period of rapidly varying temperatures, perhaps in a turbulent mixed boundary layer in strong sunshine. We can extend the model to simulate such conditions by assuming realistic 5 K stepwise rises then falls of the 'control' temperature at 5 minute intervals, and again calculating Sensor A and Sensor B temperatures every 10 s, and then calculating 60 s averages (last 6 × 10 s) per WMO CIMO guidelines (**Figure A1.2**). In this sawtooth scenario, it is clear that Sensor A is able to react quickly enough to capture correctly both minimum and maximum temperatures over each period (15.0 °C and 20.0 °C, respectively), even when averaged over 60 s periods. In contrast, Sensor B's greater time constant means that there is insufficient time to react fully to the 5 K change in each 5 minute cycle, and as a result the amplitudes of both maximum and minimum temperatures are

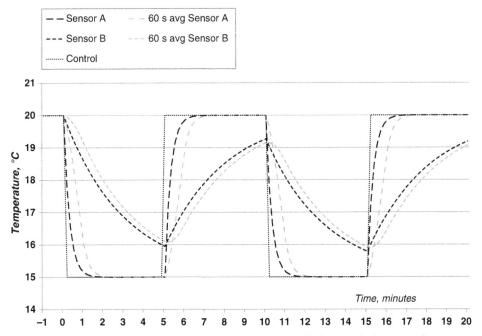

Figure A1.2 Illustrating the response of two sensors A and B in a realistic sawtooth temperature cycle of amplitude 5 K, period 5 minutes. As before, Sensor A has a time constant τ of 17 s, B 180 s. Temperatures are plotted every 10 s (dark lines), together with 60 s means per WMO guidelines (grey lines, same dash pattern).

reduced when compared with the control and Sensor A – during the period following the initial 5 K fall, Sensor B's minimum indicated temperature is 15.8 °C and the maximum 19.3 °C. When averaged over 60 s periods, the amplitude is further reduced, such that the minimum and maximum temperatures become 15.9 °C and 19.2 °C. This modelled scenario is similar to real-world examples comparing records from temperature sensors exposed within a Stevenson screen and an aspirated radiation shield as shown in both **Figures 5.11** and **5.12**, allowing of course for external variations in solar radiation, wind speed and thermal inertia of the screen structure. In each case, the amplitudes of the resulting temperature oscillations are (as expected) smaller in the system with the slower response time. However, these results also underline a key point from Chapter 5, *Measuring the temperature of the air*, which is that the differences in observed temperatures between different models of aspirated radiation shields are smaller than the differences between a Stevenson screen and an aspirated unit (**Table 5.4**, and **Figure 5.10**).

In an ideal measurement system, the sensor response time should be similar to the timescale of the most rapid changes expected. Some elements, particularly wind speed and solar radiation, can experience changes of an order of magnitude within seconds, and the need for fast response is evident. The converse is also true, in that where a rapid response is not required – measuring earth temperatures at all but the shallowest of depths for example – there is no benefit in specifying a more expensive rapid-response sensor.

Hysteresis

Hysteresis is the term given when the response (output) of a sensor varies according to whether the input is increasing or decreasing. It arises in systems or sensors whose response depends not only on the current environment, but also on recent past state. The effects tend to be more obvious on humidity sensors. At high humidity levels (close to saturation), a reduction in ambient humidity may not be immediately reflected in the sensor output.

When a sensor type is known to exhibit hysteresis, it is essential to ensure that any calibration comparisons are undertaken under steady-state conditions, both sensors being allowed to settle to at least 5τ of the slower unit.

A special case – the distance constant

In the special case of cup anemometers, the standard form of the time constant equation results in a decrease of time constant with wind speed – clearly a contradiction in terms and less helpful for defining system characteristics. For this reason, a related parameter known as the response length λ is normally quoted for cup anemometers: λ can be shown to depend upon the mass of the anemometer cups and their cross-sectional area (the lighter the cups and the larger they are, the smaller the distance constant).

The response length can be considered as approximately the passage of a length of airflow (in metres) required for the output of a wind speed sensor to indicate 63 per cent of a step-function change of the input speed. Typical cup anemometer distance constants are between 1 m and 10 m.

In the same way as the more normal time constant parameter for other sensors, the distance constant is a measure of anemometer performance, and 95 per cent of the response can be expected to occur within 3λ. While the lower the distance constant the better the response of the instrument, cup anemometer design is necessarily a compromise between response and robustness. Ensuring the instrument can withstand high wind speeds may introduce constraints on the 'large, lightweight cups' approach.

Calibration drift

Calibration drift is caused by changes in output resulting from physical changes in the sensor itself, one example being the slow settling of components in an aneroid barometer capsule (see Chapter 7, *Measuring atmospheric pressure*). Drift is usually considered separately from static and dynamic characteristics. It can become a problem with all sensors, although it tends to be more rapid (and thus more troublesome) on certain types. It is not always a slow process – the calibration of some RTDs can change quite suddenly for no very apparent reason, for example. It is usually unpredictable, and can generally only be identified by comparing the sensor against a known 'good' reference, either continuously in operational monitoring or in formal calibration tests. The US Climate Reference Network (see Chapter 5, *Measuring the temperature of the air*) employs three aspirated temperature sensors, to provide a '2-against-1' checking method for all measurements. Where one sensor differs significantly from the others on a regular basis, its calibration will be checked and adjusted as necessary. Clearly, 'buddy checking' is not possible with single sensors, and in a comparison between two instruments it will not necessarily be clear which is at fault. However, care should be taken to distinguish calibration drift from dissimilar response times (see above). For this reason, three identical sensors, in identical exposures, are preferable for any measurement, although clearly cost alone means that this ideal is rarely achievable in most meteorological measurement systems.

Exposure effects

Exposure errors can, and do, amplify or dwarf many sensor or system errors. The recommendations on sensor exposure given in the relevant chapters of this book should be adopted wherever possible to minimise exposure errors.

Recommended products

Following publication of the first edition of this book, I was asked many times – and continue to be asked – for product recommendations. 'What is the best tipping-bucket raingauge?', 'What anemometer would you recommend?', and the like. Over the years, I have tested a great many meteorological instruments, and I am happy to share details of those that I have found to be amongst the best in their class. These recommendations are my *independent personal opinion*, and based upon objective, long-term quantitative trials with reference-level instruments in standard exposures: none have been sponsored or otherwise promoted by their manufacturers. In every case the instrument in question has either been purchased in its own right or supplied on loan on a 'no strings' review/evaluation basis. For obvious reasons of geography, the list tends to be slanted towards European suppliers or manufacturers, although in most cases the products detailed are available through distributors worldwide. Product evaluations are normally undertaken over a period of 6 months minimum in order to compare performance in different seasons, considering both winter and summer conditions. Some of the products in the list below have been, and remain, in continuous daily use for a decade or more.

Criteria for inclusion inevitably vary somewhat by product, but include long-term reliability, ease of installation, ease of use, and ease and stability of calibration, and of course accuracy against reference instruments. Product pricing does enter into the equation, but considered over the product's lifetime: a $1000 product that can be expected to last 20 years or more is better value for money than a $500 product that will probably require replacement (and repeat installation costs and resource) within 5 years. Sometimes it may be possible to buy a $1000 instrument second-hand, and if it has been only lightly used (particularly electronic components) this may be an attractive option.

All instruments will eventually fail, become damaged by continuous exposure to the weather, or simply wear out. Anemometers and other instruments sited on tall masts or rooftops, and their associated cabling and connections, are particularly vulnerable to wear and deterioration caused by ultraviolet radiation, static buildup in thundery conditions (and perhaps direct lightning strikes), water ingress caused by rainfall and snowfall, and strong winds amongst other hazards: for such reasons (and a combination of them) they tend to fail somewhat earlier than other sensors. Regular checks and maintenance will extend the life of all external equipment, but even the best instruments require attention from time to time to remain in optimum condition, whether that be a replacement battery or a complete clean/overhaul/ recalibration. Such 'service intervals' should be allowed for (and costed) within an instrument's working lifetime.

This Appendix is arranged by chapter order and element, with recommended instruments following within each category. A brief outline of the instrument/device is given, together with reasons for inclusion and comparative performance statistics where relevant. Manufacturer details and web links are included (more details in **Appendix 3**); these should be consulted for current product specifications, pricing and suchlike, as these can be expected to change over the lifetime of this book.

Portable instruments

Nielsen-Kellerman Kestrel 5500

Portable AWSs are small, light and entirely self-contained, and are therefore particularly useful for fieldwork and field courses, for walkers, for outdoor sports enthusiasts including rowers, glider pilots and the like, together with others who require current on-the-spot wind and weather conditions with the convenience of a handheld unit. They can also come in handy for calibration checks and short-term logging or a temporary backup system at permanent sites. The market leader in the field is the Kestrel range of handheld weather meters, manufactured and sold by Nielsen-Kellerman (www.nkhome.com), and their top-of-the-range model is the **Kestrel 5500** (**Figures 3.8** and **9.9**), described in more detail in Chapter 3, *Buying a weather station*. A full review of this unit appears on www.measuringtheweather.net.

Temperature

Metspec Stevenson-type thermometer screen

If you still have a wooden Stevenson-type screen or Cotton Region Shelter, save yourself the annual chore of sanding down and repairing/repainting by investing in an aluminium and plastic or fibreglass screen. **MetSpec instrument shelters** (www.metspec.net, also **Appendix 3**) have become an established industry standard and are sold to national meteorological services and industrial customers worldwide. Plastic Stevenson-type screens such as those manufactured by Metspec are almost maintenance-free, requiring little more than an annual wash inside and out. The bright, shiny white exterior finish is much more resistant to the elements, including UV radiation and chemical attack, than traditional gloss paint on wooden screens, and retains near-constant radiative properties over decades. They are not cheap – but then neither is a new wooden screen and 20 years of annual painting costs (and the time involved) – but will last for decades. (The author fully expects his two Metspec screens, vintage 2010 and 2011, to outlive him.) Careful side-by-side trials conducted in several locations – by the UK Met Office in particular, who adopted Metspec plastic screens as standard throughout their network in 2006 – have shown that differences between plastic and wooden screens are insignificant [110]. Metspec screens are available in both 'standard' and double width' models (see **Figures 1.6**, **5.1** and **5.2**).

Aspirated radiation shields

The most cost-effective means of measuring 'true air temperature', and also WMO CIMO best practice for measuring air temperatures, is by means of an aspirated PRT sensor in a radiation shield (see Chapter 5, *Measuring the temperature of the air* for more details). **RM Young's model 43502 aspirated radiation shield (Figure 5.6 –** www.youngusa.com, also **Appendix 3**) is compact, relatively inexpensive and very reliable – at the time of writing the author's unit has provided 14 years of almost unbroken operation, once requiring a replacement fan which took a couple of days to source (now I keep a spare handy). It is also easy to clean, an important mainten- ance aspect of aspirated radiation shields which, because of their continuous airflow, can become very grubby – a deterioration which will itself begin to affect record quality if not dealt with promptly. Examples of records made using the device appear in numerous places within this book (**Table 5.3, Figures 5.6** and **5.7**), while this unit is also used in reference climate stations in various countries.

An alternative product, particularly where power supplies are limited, is **Apogee Instruments' aspirated radiation shield**, which requires just 1 W (at 12 v DC) (www .apogeeinstruments.com). A 2 year comparison between the Young and Apogee radiation shields (**Table 5.4**) showed no significant differences between the two. If other aspirated temperature measurement systems are similarly interchangeable, as might be expected, then this greatly simplifies the network manager's concerns about interoperability and 'future proofing' when considering and deploying current and next-generation AWS units and/or reference climate stations.

Platinum Resistance Thermometers (PRTs)

PRTs are widely and relatively cheaply available from a range of manufacturers and resellers, and a brief description of the various industry standard types was given in Chapter 5, *Measuring the temperature of the air*. For optimum response times, the sensor sheath should be no more than 3 mm in diameter, smaller if possible.

Precipitation

The UK standard 'five-inch' storage gauge

Described more fully in Chapter 6, *Measuring precipitation*, **the 'five-inch' manual rain- gauge** (**Figures 6.6** and **6.10**) has been the UK reference instrument since 1874, and continues to be the British Standard today [147]. Although increasingly eclipsed by more sophisticated automated instruments, the instrument is simple, reliable and cost- effective, and with a little care in installation and maintenance it will provide consistent records for decades. The author's gauge has been in daily use since installation in 1975, and I fully expect it to outlive me. It is available from various sources including Fairmount Weather Systems of Cambridge (www.fairmountweather.com). If your site is in a gap in the UK's raingauge network, your local hydrological agency may well provide you with one free of charge: see Chapter 19, *Sharing your observations* for more details.

The CoCoRaHS plastic gauge

Also described more fully in Chapter 6, the 'four-inch' **CoCoRaHS plastic raingauge** (**Figure 6.9**) has over 25,000 users within the CoCoRaHS network. The gauge is made of clear, tough butyrate and the US model has a capacity of 11 inches (275 mm) of precipitation: the internal measuring tube is graduated to 0.01 inch, with a capacity of 1 inch. Precipitation greater than 1 inch overflows into the outer cylinder and is measured by pouring into the measuring tube. Metric versions of the CoCoRaHS gauge are available from European resellers. A comparison of the metric model at two climatically different sites within the UK against standard 'five-inch' raingauges over a 12 month period [319] demonstrated excellent agreement (within 2 per cent) between the CoCoRaHS gauge and the reference, despite the former being only about one-fifth of the price of a standard copper gauge. Full details of the trials and a more detailed analysis can be found on www.measuringtheweather.net. Some regional hydrology authorities within the UK now use this gauge, or very similar alternatives, to increase station density cost-effectively within their networks.

Lambrecht meteo rain[e] digital raingauge

The **Lambrecht meteo rain[e] digital raingauge** is described in more detail in Chapter 6, *Measuring precipitation*, and illustrated in **Figure 6.8**: manufactured by Lambrecht in Germany (www.lambrecht.net), it is available through distributors worldwide. This highly sensitive instrument (resolution 0.01 mm) is equally adept at measuring hours of barely perceptible drizzle or torrential thunderstorm rainfall. Its digital output can be harnessed to provide any combination of real-time period totals and intensity outputs, from second-by-second rainfall intensity data to hourly, daily and monthly totals, all to a high and consistent degree of accuracy. It is ideally suited for use within an existing gauge network, and examples of output from the gauge are given in **Figures 6.19** and **6.20**. Field calibration is also quick and easy. Although a relatively new product, Lambrecht's rain[e] gauge is already and deservedly becoming widely adopted within hydrological and meteorological networks in several European countries. The author's unit has been in place and recording faultlessly for more than 8 years at the time of writing; over a recent 12 month period, the total catch was within 1.1 per cent of the standard UK 'five-inch' reference checkgauge (**Figure A2.1(a)**), with only a handful of daily totals differing by as much as ±5 per cent from the standard gauge.

EML's Kalyx and ARG314 tipping bucket raingauges

Environmental Measurements Ltd (EML: www.emltd.net) manufacture a wide range of hydrological monitoring equipment, including a number of raingauges. EML's compact TBR, their popular 0.2 mm/tip **Kalyx raingauge** (**Figure 6.17**), follows the dimensions of the traditional 'five-inch' Snowdon raingauge by using a five inch (127 mm) diameter funnel. This gauge proved easy to install, well designed, very reasonably priced and has given many months of good service. One step up is EML's **ARG314 TBR**, with a 200 cm^2 funnel and choice of 0.1 or 0.2 mm resolution, itself widely deployed within UK hydrological and meteorological agencies (**Figure 6.17**). Examples of output are given in **Table 6.4** and **Figure 6.13**. Either gauge is ideally suited for siting alongside a 'five-inch' checkgauge, and both offer the added benefit of EML's unique and proven aerodynamic profile to improve catch efficiency by reducing wind eddies around the gauge structure. Over a 12 month comparison period, the performance of a loan ARG314 unit was exemplary, recording within 1 per cent of the accumulated total precipitation from a nearby 'five-inch' UK reference gauge (743 mm against 751 mm: **Figure A2.1(b)**). Over the same period, the smaller Kalyx gauge accumulated 715 mm, within WMO's ±5 per cent accuracy guideline (**Figure A2.1(c)**).

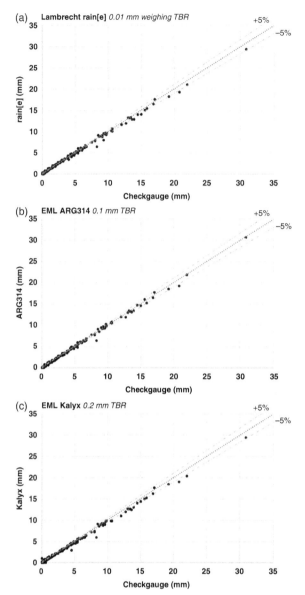

Figure A2.1 Scatterplot comparing daily rainfall totals over a 12 month period from three automated gauges against the reference daily fall from a UK standard 'five-inch' manual gauge. The 5 per cent accuracy limits of the WMO CIMO standard are indicated on either side of the 1:1 correspondence line. Figure (a) Lambrecht meteo rain[e] weighing TBR, (b) EML ARG314 0.1 mm TBR, and (c) EML Kalyx 0.2 mm TBR. Source notes: [320]

Wind speed and direction

Gill Instruments Windsonic anemometer

Gill Instruments manufacture and supply a range of 2D and 3D sonic anemometers. Amongst these is the **Gill Windsonic 2D anemometer**, described in more detail in Chapter 9, *Measuring wind speed and direction* (**Figure 9.7**). The sensor itself consists of a tough corrosion-free polycarbonate body (an anodised aluminium body is also available), and having no moving parts it is ideally suited to harsh environmental conditions or exposure in 'awkward to reach' locations. In use it has proved both reliable and exquisitely sensitive, continuing to record wind speed and direction at airflow velocities well below the stopping speed of conventional cup anemometers (see **Figures 9.2** and **9.10**), even in unconventional exposures such as measuring airflow within a Stevenson screen [112]. Without any exaggeration, this is one of the most impressive environmental sensors available today. It is expensive, although the purchase price compares favourably with that of a separate cup anemometer and wind vane set of equivalent quality, and with a mean time between failures around 15 years, it should comfortably outlast other instruments. My only suggestion to the manufacturer was to fit bird spikes to the 'roof' of the unit, because this elevated platform is otherwise attractive as a resting post for birds, and the accumulation of bird droppings that results can foul the instrument's sensors. Gill Instruments contact details and website appear in **Appendix 3**.

Sunshine

Kipp & Zonen CSD3 sunshine recorder

The **Kipp & Zonen CSD3 sunshine recorder** is described more fully in Chapter 11, *Measuring sunshine and solar radiation*, and is illustrated in **Figure 11.10**. The UK Met Office adopted this instrument in place of the traditional Campbell–Stokes sunshine recorder in 2003, and it has since been adopted as a 'standard instrument' in other countries too. Without the requirement for daily access necessary to change the recording cards on the Campbell–Stokes device, it is easier to install 'at height' in order to achieve a clear horizon. Once installed, it operates reliably – the author's unit has worked faultlessly for more than a decade at the time of writing. Although expensive in purchase price, the only consumables required are desiccant modules which need to be renewed every few months (safe access to do this should be allowed for when siting the instrument). In use, it is an impressive and reliable instrument, very fast in operation and capable of responding to short bursts of sunshine lasting only a second or two. The only word of caution, however, relates to the threshold trigger level of the device, which does drift over time. Manufacturer recalibration is expensive. If there was a way in which Kipp & Zonen could offer rapid turnaround and less expensive single-point calibration checks in-country or offer a swap-out scheme for a pre-calibrated sensor every 5 years or so, this would indeed be close to the 'perfect' automatic sensor.

Datalogger

Campbell Scientific dataloggers

Campbell Scientific are a leading designer and manufacturer of data loggers, data acquisition systems, and measurement and control products used worldwide, many of which are ideally suited to meteorological applications. The author has used Campbell Scientific dataloggers and software in both professional and personal applications for over 20 years, and without hesitation has found them utterly reliable. It will be obvious from this book that an AWS built around a reliable, expandable and programmable datalogger is capable of measuring almost any environmental element for which a suitable sensor exists, and Campbell Scientific dataloggers prove this statement. Almost all of the logged data examples in this book originated from one of two Campbell Scientific CR1000 dataloggers which have remained in constant use (literally every second of every day) for more than a decade at the time of writing, updating a system built around the previous model CR10X logger in 2012 which itself was in use for over 11 years (and remains in place as a test system for new sensors). The CR1000 has since been updated to the CR1000X, and no doubt this will in turn be updated during the lifetime of this book, together with the impressive and highly capable Loggernet software. The author has yet to see Campbell's UK technical support team stumped for an answer to any question! The products are expensive, but in common with other products in this Appendix, they can be confidently expected to both outperform and outlast less expensive alternatives, and ultimately show a better return on investment over time.

Further product reviews

The author's review of the Davis Instruments Vantage Pro2 and the Vantage Vue AWS can be found on www.measuringtheweather.net. Unfortunately, most Internet 'product reviews' of meteorological instruments or AWSs lack sufficient detail or long-term comparisons against reference instruments. The following published materials are more useful in those respects:

[124] Lacombe, M., et al., 2011: Instruments and observing methods report no. 106: WMO field intercomparison of thermometer screens/shields and humidity measuring instruments, Ghardaïa, Algeria, November 2008 – October 2009. *Instruments and Observing Methods Report No. 106*, WMO/TD-No. 1579, Geneva, Switzerland.
[233] Jenkins, G., 2014: A comparison between two types of widely used weather stations. *Weather*, **69**(4): pp. 105–10 doi:https://doi.org/10.1002/wea.2158.
[321] Bell, S., D. Cornford, and L. Bastin, 2013: The state of automated amateur weather observations. *Weather*, **68**(2): pp. 36–41 doi:https://doi.org/10.1002/wea.1980.
[322] Bell, S., D. Cornford, and L. Bastin, 2015: How good are citizen weather stations? Addressing a biased opinion. *Weather*, **70**(3): pp. 75–84 doi:https://doi.org/10.1002/wea.2316.

Others will be added to www.measuringtheweather.net over time – check back for updates.

Useful sources

Contact details for meteorological societies and manufacturers and suppliers of meteorological instruments. All details are correct at the time of going to press.

Meteorological societies – by region and country

There is a comprehensive list of meteorological societies worldwide on the website of the International Forum of Meteorological Societies (IFMS) at ifms.org; at the time of writing, this listed 43 members. A selection of a few of the largest societies worldwide is given in the table below. Priority has been given to those societies with membership open to individuals, rather than those open only to National Meteorological Services or other corporate bodies. As well as physical meetings, many societies now offer virtual meetings by video link, particularly since the COVID pandemic in 2020/21.

NORTH AMERICA

UNITED STATES OF AMERICA	**American Meteorological Society** 45 Beacon Street Boston, MA 02108–3693, USA www.ametsoc.org	The American Meteorological Society (AMS) is a global community committed to advancing weather, water, and climate science and service. Founded 1919.
CANADA	**Canadian Meteorological and Oceanographic Society** P.O. Box 3211 Station D Ottawa, ON, K1P 6H7, Canada www.cmos.ca	The Society exists for the advancement of meteorology and oceanography in Canada. Founded in 1939, originally as the Canadian Branch of the Royal Meteorological Society.

ASIA

CHINA	**Chinese Meteorological Society (CMS)** cms1924.org	The CMS aims to promote scientific advancement in meteorology and to facilitate development of weather and climate observations. Established 1924.
	Hong Kong Meteorological Society c/o Hong Kong Observatory 134A Nathan Road Kowloon, Hong Kong www.meteorology.org.hk	Established in 1988, Hong Kong Meteorological Society aims to develop and disseminate knowledge of meteorology and related oceanic, hydrologic and geographic sciences and promote and advance the professional application of meteorology.

(cont.)

ASIA

INDIA **Indian Meteorological Society**
Room No. 605, VI Floor, Satellite
Meteorological Building, Mausam Bhavan
Complex, Lodi Road, New Delhi-110 003
www.indianmetsoc.com

The Indian Meteorological Society is a non-profit scientific organisation with the purposes of advancement of meteorological and allied sciences in all their aspects and dissemination of knowledge of such sciences both amongst the scientific workers and amongst the public. Established 1956.

JAPAN **Meteorological Society of Japan**
c/o Japan Meteorological Agency
1–3–4, Ote-machi, Chiyoda-ku,
Tokyo 100–0004 JAPAN
www.metsoc.jp

Originally founded as the Meteorological Society of Tokyo in 1882, today the society has more than 3000 members. The purpose of the society is to activate meteorological research and promote the progress and development of such research in cooperation with related academic societies both in Japan and around the world.

AUSTRALASIA

AUSTRALIA **Australian Meteorological and Oceanographic Society**
6/416 Gore Street
FITZROY VIC 3065
www.amos.org.au

The Australian Meteorological and Oceanographic Society (AMOS) is an independent society representing the atmospheric and oceanographic sciences in Australia and has over 500 members with regional centres in ACT, Adelaide, Darwin, Melbourne, NSW, Perth, Queensland and Tasmania.

NEW ZEALAND **Meteorological Society of New Zealand**
673 The Coastal Highway
RD 1
Upper Moutere 7173
www.metsoc.org.nz

The Meteorological Society of New Zealand is an independent group of weather enthusiasts with the stated aim to encourage an interest in the atmosphere, weather and climate, particularly as related to the New Zealand region. Anyone with an interest in the atmosphere, weather and climate can join.

EUROPE

EUROPEAN METEOROLOGICAL SOCIETY **European Meteorological Society**
c/o Insitut für Meteorologie, FU Berlin
C-H-Becker-Weg 6–10
12165 Berlin
ems-sec@emetsoc.org
www.emetsoc.org

The EMS is an umbrella organisation for the various national or regional meteorological societies in Europe. At the time of writing, the Society has 38 Member Societies: contact details for all EMS members are given on the EMS website.

BELGIUM **Flemish Association for meteorology Vlaamse Vereniging voor Weerkunde (VVW)**
www.weerkunde.be/

VVW is a membership organisation which provides a monthly magazine (HALO), meetings, presentations, access to mailing lists and contacts with like-minded people interested in meteorology and climate.

(*cont.*)

EUROPE

FRANCE	**Météo et Climat** 73, avenue de Paris 94165 Saint-Mandé CEDEX, France meteoetclimat.fr	Météo et Climat is a community concerned with preserving and developing knowledge in the fields of atmospheric and climate sciences, and in particular climate change. Individual members as well as organisations and companies are welcome as partners or associate members.
GERMANY	**Deutsche Meteorologische Gesellschaft** c/o Institut für Meteorologie Freie Universität Berlin C-H-Becker-Weg 6–10 12165 Berlin, Germany dmg-ev.de	DMG was established in 1883. It is a forum for communication and exchange, and acting in the interest of its members. It organises topical conferences, runs a DACH-wide triennial conference, and publishes *Meteorologische Zeitschrift*.
	Ring europäischer Hobbymeteorologen e.V. (Ring of European hobby meteorologists – ReH eV) www.reh-ev.org	The 'Ring Europäische Hobbymeteorologe eV' (ReH for short) was founded in 1985 by Hans-Martin Goede. It publishes a monthly newsletter for hobby meteorologists and for everyone interested in weather and meteorology, and has members from both Europe and North America.
ICELAND	**Veðurfræðifélagið** Veðurstofu Íslands Bustadavegur 9 150 Reykjavik www.vedur.org	The Icelandic Meteorological Society is open to all with a genuine interest for weather and related disciplines. The purpose of the society is to improve and deepen the knowledge of meteorology and related fields in Iceland, and three national afternoon meetings are held each year.
IRELAND	**Irish Meteorological Society** c/o Met Éireann Glasnevin Hill Dublin 9 D09 Y921 Ireland www.irishmetsociety.org	The Irish Meteorological Society was founded in 1981. Its main aims are the promotion of an interest in meteorology and the dissemination of meteorological knowledge, pure and applied. The Society includes members not only from Ireland but from all over the world who are interested in weather and weather-related topics.

(cont.)

EUROPE		
NETHERLANDS	**Nederlandse Vereniging voor Beroeps Meteorologen** Postbus 464 6700 AL Wageningen, the Netherlands www.nvbm.nl	The Dutch Association for the Promotion of Meteorology (NVBM) was founded in 1991. Members are or have been meteorological professionals in university, weather room, research and development, industry, management, climate research.
	Vereniging voor Weerkunde en Klimatologie *(Society for Weather and Climatology)* www.vwkweb.nl	VWK was founded in 1974 and has around 600 members. Publishes a monthly magazine *Weerspiegel* ('*Weather mirror*'), and holds regular meetings, both nationally and regionally.
UNITED KINGDOM	**Royal Meteorological Society** 104 Oxford Road READING RG1 7LJ, UK www.rmets.org	Anyone who is interested or involved in meteorology or associated sciences can join the Royal Meteorological Society, which was founded in 1850. Worldwide membership is made up of professionals and academics, students and teachers, enthusiasts and observers, and there is a category to suit everyone.
	Climatological Observers Link (COL) 16 Wootton Way MAIDENHEAD Berkshire SL6 4QU, UK www.colweather.org Facebook page *Weather Group – UK & Ireland*	The Climatological Observers Link is an organisation of amateur meteorologists, founded in 1970. Its membership is mostly drawn from within the British Isles, although membership is open to anyone. COL publishes a monthly weather summary of British weather and an online weather forum, and organises events and conferences for those interested in practical weather observing.
	Tornado and Storm Research Organisation (TORRO) www.torro.org.uk	Founded in 1974, TORRO is a privately supported research body specialising in severe convective weather in Britain and Ireland, supported by several hundred voluntary observers, investigators and other contributors. Two conferences are held every year in Oxford.

Suppliers of meteorological instruments

This listing includes most of the instruments and sensors which are referenced in this book, but it is by no means a complete list of all manufacturers and suppliers worldwide. The website of the Association of Hydro-Meteorological Equipment Industry (HMEI) provides a more complete and current industry contact list – see hmei.org *for more information. Suppliers and manufacturers are listed alphabetically by brand name. In most cases, products are sold and supported internationally, but check with the supplier before placing an order. Inclusion in this does not constitute endorsement by the author.*

   Contact details shown are correct at the time of going to press.

AEM	**AEM** 12410 Milestone Center Drive, Suite 300, Germantown, MD 20876, USA https://aem.eco/	AEM is a global company, serving customers from offices across North America and Europe. Brands include Davis Instruments, Earth Networks, FTS Forest Technology Systems Ltd, High Sierra Electronics, Lambrecht meteo GmbH, OneRain, and Vieux & Associates.
APOGEE INSTRUMENTS	**Apogee Instruments, Inc.** 721 West 1800 North Logan, Utah 84321 USA www.apogeeinstruments.com	Apogee Instruments was founded in 1996 by Dr Bruce Bugbee, an eager scientist and avid inventor who began creating and manufacturing his own research-quality instruments. Since then, Apogee Instruments has grown to become a respected leader in the manufacture of innovative, durable, and accurate environmental instruments renowned for cost-effective measurement technology. Distributors worldwide.
CAMPBELL SCIENTIFIC	**Campbell Scientific, Inc.** 815 West 1800 North Logan, Utah 84321–1784 USA www.campbellsci.com **Campbell Scientific Ltd** Campbell Park, 80 Hathern Road Shepshed, Loughborough LE12 9GX, UK www.campbellsci.co.uk	Campbell Scientific are a leading designer and manufacturer of data loggers, data acquisition systems, and measurement and control products used worldwide in a variety of applications related to weather, water, energy, gas flux and turbulence, specialising in rugged, low-power systems for long-term, stand-alone monitoring and control. Founded in Logan, Utah, USA in 1974, they have offices in Australia, Brazil, Canada, China, Costa Rica, France, Germany, India, South Africa, Spain, Thailand and UK.
DAVIS INSTRUMENTS	**Davis Instruments Corporation** 3465 Diablo Ave. Hayward, California 94545 USA www.davisnet.com	Davis Instruments, based in Hayward, California, have been manufacturing weather stations since 1989, and today they have tens of thousands of users around the world, from far northern Alaska to Antarctica. In 2019 they became part of the Advanced Environmental Monitoring (AEM) group – see AEM entry.

(cont.)

	Prodata Weather Systems	John Dann's Prodata Weather Systems is the leading supplier of Davis Instruments equipment within the United Kingdom.
	Prodata Weather Systems Unit 12, Espace North Building 181 Wisbech Road Littleport, Ely, Cambridgeshire CB6 1AE, UK sales@weatherstations.co.uk www.weatherstations.co.uk	
ELITECH	**Elitech Technology, Inc. (USA)** 2528 Qume Dr Milpitas, CA, 95035, USA www.elitechlog.com	Founded in 1996, Elitech's main product lines are portable temperature and humidity dataloggers for cold chain supply and refrigeration industries.
EML LTD	**Environmental Measurements Ltd** 7 Jupiter Court Orion Business Park North Shields NE29 7SE, UK www.emltd.net	Environmental Measurements Ltd (EML) designs and develops instrumentation for meteorological and environmental monitoring, including aerodynamic raingauges, wind speed and wind direction sensors, temperature humidity probes, radiation sensor shields, barometric pressure sensors, surface wetness probes, data loggers and automatic weather stations. EML is also a systems integrator, and supplies products worldwide.
EPPLEY	**The Eppley Laboratory, Inc.** 12 Sheffield Avenue, PO Box 419 Newport, Rhode Island 02840, USA www.eppleylab.com	Eppley, founded in 1917, are specialists in solar radiation measurement, manufacturing radiometers, pyranometers, pyrheliometers and pyrgeometers.
EKO	**EKO Instruments B.V.** 1–21–8 Hatagaya Shibuya-ku 151–0072 Tokyo, Japan www.eko-instruments.com	Established in Tokyo in 1927, EKO manufactured sensors for the Japanese meteorological and environmental market, and today manufactures and sells a range of professional-quality solar radiation sensors worldwide.
FAIRMOUNT WEATHER SYSTEMS	**Fairmount Weather Systems** Unit 4, Whitecroft Road Meldreth, Hertfordshire SG8 6NE, UK www.fairmountweather.com	Fairmount Weather Systems has been providing Meteorological instruments to National Hydro-Meteorological Institutes since 1986. They are both the largest and leading manufacturer of traditional British Standard meteorological equipment in the UK, together with 'mercury replacement' systems under the Intellisense® brand.

(cont.)

GEMINI DATALOGGERS (TINYTAG)	**Gemini Data Loggers (UK) Ltd** Scientific House Terminus Road Chichester West Sussex PO19 8UJ, UK www.geminidataloggers.com	Gemini Data Loggers (established in 1984) develop and sell the Tinytag range of dataloggers worldwide, offering loggers for a wide range of measurements in addition to those for weather applications.
GEONOR	**Geonor AS** Grini Næringspark 10, 1361 Østerås, Norway www.geonor.no	Geonor AS manufactures and supplies all-weather precipitation gauges for accurate measurement of snow and rain. The company also manufactures and markets equipment and instruments for geotechnical and civil engineering applications.
GILL INSTRUMENTS	**Gill Instruments** Saltmarsh Park 67 Gosport Street Lymington, Hampshire SO41 9EG, UK gillinstruments.com	Gill Instruments has over 35 years experience in ultrasonic air flow measurement, including anemometers for measuring wind speed and direction. Gill currently manufacture and supply the most extensive range of ultrasonic anemometers on the market, together with a comprehensive range of weather stations from compact varieties to reference grade stations.
INSTROMET WEATHER SYSTEMS LTD	**Instromet Weather Systems Ltd** Unit 10B, Lyngate Industrial Estate North Walsham Norfolk NR28 0AJ, UK www.instromet.co.uk	Instromet manufacture and supply a range of weather monitoring equipment; they are best known for their sunshine recorder (see Chapter 11 for details).
KIPP & ZONEN	**Kipp & Zonen B.V.** Delftechpark 36 2628 XH Delft The Netherlands www.kippzonen.com	Kipp & Zonen (a product brand within OTT HydroMet) provides class-leading instruments for measuring solar radiation and atmospheric properties in meteorology, climatology, hydrology, industry, renewable energy, agriculture and public health.
LAMBRECHT METEO	**Lambrecht meteo GmbH** Friedländer Weg 65–67 37085 Göttingen GERMANY www.lambrecht.net	Lambrecht meteo, an AEM brand, develops and manufactures world-class meteorological sensors and measurement solutions for wind, precipitation, pressure, temperature, and humidity serving various meteorological and environmental end-markets.
MET ONE INSTRUMENTS	**Met One Instruments, Inc.** 1600 NW Washington Blvd Grants Pass, OR 97526, USA www.metone.com	Met One Instruments, Inc. provide instruments and systems for both meteorological and air quality applications.

(*cont.*)

METSPEC	**Metspec** Unit 1, Thistleton Block Market Overton Industrial Estate Ironstone Lane Market Overton, Oakham LE15 7TP, UK www.metspec.net	MetSpec Instrument Shelters are an established industry standard and are sold to national meteorological services and industrial customers worldwide. Their innovative design features a durable white powder–coated frame and robust outer louvers which are resistant to UV radiation and chemical attack.
NIELSEN- KELLERMAN	**Nielsen-Kellerman** 21 Creek Circle Boothwyn, PA 19061, USA www.nkhome.com *UK distributor* www.r-p-r.co.uk	Nielsen-Kellerman Company designs, manufactures and distributes rugged, waterproof environmental and sports performance instruments for active lifestyles and technical applications, including Kestrel Pocket Weather Meters.
NOVALYNX	**NovaLynx Corporation** 431 Crown Point Cir Ste 120 Grass Valley, CA 95945–9531, USA www.novalynx.com	Weather monitoring instruments and systems, including US-standard Cotton Region Shelters and eight-inch raingauges.
ONSET	**Onset Computer Corporation** 470 MacArthur Blvd Bourne, MA 02532, USA www.onsetcomp.com	Onset is one of the world's leading suppliers of data loggers, used around the world in a broad range of applications including weather and climate monitoring.
R M YOUNG	**R. M. Young Company** 2801 Aero Park Drive Traverse City, Michigan 49686 USA www.youngusa.com	Founded in 1964, the R. M. Young Company specialises in the development and manufacture of professional meteorological instruments through a network of international resellers.
RS ONLINE	**RS Components Ltd** Birchington Road Corby Northants NN17 9RS, UK uk.rs-online.com/web/	Founded in 1937, RS Components are a global omni-channel provider of product and service solutions for designers, builders and maintainers of industrial equipment and operations. RS offer more than 700,000 stocked and three million unstocked high-quality industrial and electronic products, sourced from over 2,500 suppliers. RS are an excellent source for many sensors and components relevant to meteorological measurement, particularly PRTs and thermistors.
SETRA	**Setra Systems, Inc.** 159 Swanson Rd Boxborough, MA 01719, USA www.setra.com	Setra Systems has designed and manufactured premium pressure sensors for over 50 years: its product range includes several sensors ideal for meteorological applications.

(*cont.*)

TEMPCON	**Tempcon Instrumentation Ltd** Unit 19 Ford Lane Business Park Ford Lane, Ford Nr Arundel, West Sussex BN18 0UZ www.tempcon.co.uk	Tempcon Instrumentation Ltd (established 1980) specialises in instrumentation and sensors (probes) for measuring, controlling and logging temperature, humidity, pressure, water level, water temperature, air quality/CO_2, voltage, current, energy (kWh), wind speed/direction, and others. Manufacturer of PRTs.
UK WEATHERSHOP	**Weather Shop** a division of **Tempcon Instrumentation Ltd** www.weathershop.co.uk	A bricks-and-mortar family-owned business located in Ford, near Arundel, West Sussex, offering a wide range of weather monitoring instruments and systems, including all the consumer brands (Accur8, Bresser, Davis Instruments, Technoline, etc.).
VAISALA	**Vaisala Oyj** Vanha Nurmijärventie 21 01670 Vantaa Helsinki, Finland www.vaisala.com	Vaisala is a leading global supplier of environmental and industrial measurement systems. Vaisala Oyj was founded in Helsinki, Finland in 1936, and today has offices in 15 countries. The company's worldwide customer base includes many national meteorological and hydrological institutes, aviation and road organisations, defence forces and wind parks.
VECTOR INSTRUMENTS	**Windspeed Limited (Vector Instruments)** 115 Marsh Road, RHYL Denbighshire, LL18 2AB, UK www.windspeed.co.uk	Vector Instruments manufacture and sell wind sensors (anemometers and windvanes) widely used in professional AWS systems.
WEATHER-YOUR-WAY	**WeatherYourWay** 2966 Gateway Avenue Hartford, WI 53027, USA www.weatheryourway.com	WeatherYourWay are a friendly and helpful US-based supplier of a wide range of consumer meteorological instruments, run by a qualified meteorologist (an ex-NWS forecaster). They are the official supplier to the CoCoRaHS network (see Chapter 19 for details) and supply (amongst many other products) the low-cost plastic raingauges used in the CoCoRaHS network.

APPENDIX 4

Mercury-based legacy thermometers and barometers

Two of the most significant and far-reaching changes in meteorological instrumentation have taken place within the past two decades: the first being the widespread adoption of automation, using digital instrumentation and logging methods, and the second the progressive withdrawal of traditional mercury-based instruments. The latter change has come about as a result of statutory restrictions on the manufacture and supply of mercury-based instruments, primarily traditional thermometers and barometers in the meteorological context, under UN Environment Programme regulations following the Minamata Convention which recognised mercury and its compounds as long-lasting environmental toxins. The Convention was signed by 128 countries in 2013 and came into force in 2020 (www.mercuryconvention.org/en) [5], and has resulted in the almost complete cessation of manufacture and supply of meteorological instruments containing mercury. WMO advice remains to eliminate the use of mercury-based instruments as soon as possible, and in many countries this has largely been achieved. There are a few legitimate exceptions to complete withdrawal, however, one of which is at long-period sites where an overlap period between 'traditional' and 'electronic/digital' instrumentation is beneficial to assess any risks to record homogeneity brought about by the change in instrumentation. For whatever reason, it is likely that some 'legacy' instruments will continue to remain in use for the immediate future, although it should be appreciated that repair or replacements in kind will no longer be possible for instruments that contain mercury should any breakages occur.

This brief Appendix covers a few important topics relevant to users of liquid-in-glass thermometers and barometers, which might otherwise become increasingly difficult to track down in the literature. One recently introduced alternative product to mercury thermometers is briefly outlined.

Liquid-in-glass (LiG) thermometers

'Traditional' thermometers use the thermal expansion of a liquid in a narrow-bore glass tube to provide a measure of temperature by means of a graduated and calibrated scale. For over 300 years, mercury was most commonly used, although alcohol is normally used for minimum thermometers, or for those used in cold climates, because it has a lower freezing point (–115 °C / –175 °F) than mercury (–38 °C / –36 °F). Long use and familiarity produced reliable, accurate and reasonably robust instruments, which could be read quickly and accurately by eye to 0.1 degC. Not all 'legacy' thermometers are mercury-based – as stated, traditional minimum thermometers use alcohol as the working liquid, and of course are not affected by mercury legislation – while thermometers utilising coloured alcohol have now been developed as alternatives to other types of mercury-based thermometers.

In construction, there are two main patterns of meteorological thermometer – the *sheathed* and *solid stem*. Sheathed thermometers are encased in an outer glass sheath, hence the name: the thermometer scale is normally engraved on the thermometer stem, protected from weathering. Solid stem thermometers usually have the graduations marked on the thermometer stem or on a separate plastic, metal or wooden scale attached to the thermometer. Because the scale is exposed to the elements, it can be subject to wear, expansion and contraction in varying temperature and humidity, and fading over time. 'Attached' scales inevitably move slightly over time, rendering the thermometer calibration less certain: 'sheathed' thermometers are both more robust and easier to handle.

Types of LiG thermometer

Three main types of thermometer remained in widespread use until the first decade of the twenty-first century, namely the 'dry bulb', the maximum thermometer and the minimum thermometer. The 'dry bulb' thermometer was used to register the current temperature, and the 'wet bulb' variant – identical in form to the dry bulb except that its bulb was kept permanently moist by a wick or sleeve fed by capillary action from a water reservoir – was used to provide a measure of the water vapour content of the air (see Chapter 8, *Measuring humidity*). Both are easily replicated with electrical PRT sensors, or conventional LiG thermometry using coloured alcohol may be used instead.

Once reset, the mercury maximum thermometer (**Figure A4.1**) also indicates the current temperature as the temperature rises, but by virtue of a small constriction near the bulb the column of mercury will remain at its highest point once the temperature begins to fall. The minimum thermometer (**Figure A4.2**) indicates the lowest temperature reached as the alcohol meniscus carries a small glass index

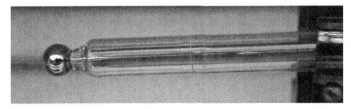

Figure A4.1 Sheathed mercury maximum thermometer, showing the constriction in the stem. (Photograph by the author)

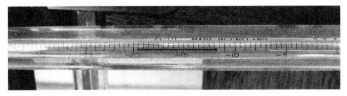

Figure A4.2 Sheathed alcohol minimum thermometer index: the minimum temperature is reading –11.8 °C, and the current temperature is –6.8 °C. (Photograph by the author)

in the thermometer stem down with it, leaving it in place as the temperature rises once more.

Because of their size, conventional mercury or alcohol-based thermometry require exposure within a suitable thermometer screen, such as a Stevenson screen or Cotton Region Shelter (Chapter 5, *Measuring the temperature of the air*). Normally both dry and wet bulb thermometers are hung vertically within the screen or shelter, while both maximum and minimum thermometers are exposed almost horizontally.

Resetting maximum and minimum thermometers

A mercury maximum thermometer is reset by firmly grasping the end of the thermometer furthest from the bulb and shaking the mercury column back down past the constriction towards the bulb in the manner of a clinical thermometer, taking care to avoid any obstacles in doing so to avoid breaking the thermometer. The minimum thermometer is reset by gently tilting the thermometer bulb-end upwards until the index rests once more on the end of the alcohol meniscus or 'bubble'. Note that both screen thermometers are most vulnerable to breakage while being reset.

For use in meteorological or climatological applications, the calibration of liquid-in-glass thermometry should be checked at least every 5 years. The expected uncertainty of liquid-in-glass thermometers over the normal range of temperatures at the observing location should remain within ± 0.2 degC.

'Debubbling' alcohol-based thermometers

Alcohol-based minimum thermometers can suffer from a break-up of the alcohol column within the thermometer stem, a condition known as 'bubbling'. This can happen spontaneously (for some reason, certain thermometers are more prone to bubbling than others), or from careless handling: a common error in inexperienced or relief observers is to attempt to reset the *minimum* thermometer by shaking it down in the same fashion as the *maximum* thermometer. Thermometers that are left exposed to sunshine during the day, such as grass or concrete minimum thermometers, can also develop 'bubbling' as a result of evaporation and distillation of the alcohol within the thermometer stem caused by solar heating: subsequent condensation back into the alcohol column can result in the breaking-up of the spirit column. Whatever the reason, while it lasts bubbling will result in an error in the observed minimum temperature – too high *or* too low, depending on the relative size and position of the 'bubbles' and the index. To minimise the effects of solar radiation on thermometers left out all day, a short black shield is normally placed on the end of the thermometer furthest from the bulb. In sunshine this becomes warmer than the glass body of the thermometer, reducing the risk of alcohol condensation in the expansion reservoir at the top of the thermometer.

Bubbling can be rectified by *very gently* heating the thermometer in warm water (not hot), or over a gentle heat source, to drive the alcohol column close to the upper reservoir, at which point the 'bubbles' will safely fold back into the upper chamber. This procedure has to be undertaken with great care, for heating too rapidly, or continuing to heat for too long, risks fracturing the thermometer. As soon as all the bubbles have cleared, remove the heat source and allow the thermometer to cool *slowly* back to ambient room temperatures. Too rapid cooling will simply reintroduce bubbles in the thermometer stem. If bubbles re-appear during cooling, allow to cool and then repeat the process. If a particular thermometer bubbles frequently for no obvious reason, it should be replaced.

Alternatives to 'traditional' liquid-in-glass maximum and minimum thermometers

Logging screen- or radiation-shield based platinum resistance thermometers (PRTs), as described more fully in Chapter 5, will of course provide daily maximum and minimum temperatures to continue conventional climatological records, but for some the option of a programmable datalogger may be too complex, too expensive or simply too dissimilar from existing manual observation procedures. One new product which may be a better fit in such circumstances is Metspec's new Mercure system, illustrated in **Figure A4.3**. This product, available only in late-model prototype form at the time of going to press, will particularly appeal to both individuals and regional or national meteorological services who seek a cost-effective but familiar replacement for existing liquid-in-glass maximum and minimum thermometers within a similar 'manual morning observation' routine.

The Mercure product consists of two PRTs, configured as a dry and wet bulb psychrometer, connected to and logged by a control unit which houses two large easy-to-read front-mounted alphanumeric display panels. The unit itself is mounted within a Stevenson screen, where it replaces the four traditional liquid in glass thermometers (dry bulb, wet bulb, maximum and minimum). It is powered by a small long-life battery, itself charged by means of a small solar panel mounted on the screen roof. A brief button push on the front panel displays the current air temperature and relative humidity; second and subsequent button pushes display, in turn, the currently logged maximum and minimum temperatures since the unit was last reset, followed by the maximum and minimum temperature for the previous

Figure A4.3 Metspec Mercure unit housed within Stevenson screen, a cost-effective replacement to conventional liquid-in-glass thermometry. Photograph by the author.

24 hour period. The unit is configured to reset maximum and minimum temperatures automatically at 0900 UTC (default, but configurable by the user). The various options, including the automatic reset and the 'previous 24 hours' recall facility, makes it easy to maintain a standard daily 0900–0900 UTC terminal hour observational routine even where it may not always be possible to make manual observations at or close to 0900 UTC.

The Mercure unit configuration also stores one minute mean temperatures to an internal logger, and these can be quickly and easily downloaded at any convenient interval, perhaps weekly or monthly, for more detailed examination of particular occasions or to maintain a high-resolution temperature record, as required. The unit is supplied pre-calibrated, and initial field tests confirm that accuracy and compatibility within ±0.1 degrees Celsius of existing thermometry is expected.

At the time of going to press the unit was still undergoing final acceptance trials, but is expected to be made available for sale during 2024. Metspec's contact details are given in **Appendix 3**.

Mercury barometers

Mercury has been used as the working liquid in barometers since Torricelli in 1644, but today there are many viable electronic sensors with as good or better accuracy than traditional instruments, and almost all operational measurements of atmospheric pressure are now made with such sensors (see Chapter 7, *Measuring atmospheric pressure*). It is unlikely that mercury barometers will disappear entirely, however, for two reasons – firstly, they are very much more valuable items than individual thermometers, and many are collector's pieces in their own right: many older instruments also qualify as 'antique instruments', which are exempt from some aspects of the mercury legislation.

One item of feedback from readers of the first edition of this book was that it lacked any explanation of how to read the vernier scale on a barometer (named after the French mathematician Pierre Vernier, 1580–1637). Almost every barometer manufactured within the last 200 years uses a vernier scale (**Figure A4.4**), and knowing how to adjust and read the scale quickly and accurately is essential. Perhaps it is simply a generation of school students brought up on digital displays rather than vernier scales, but in my experience of teaching university students surprisingly few seem to be familiar with the latter. In fact, using a vernier scale is very straightforward and with a little practice becomes second nature.

Figure A4.4 shows the vernier scale on two mercury barometers, the one on the left being a Kew pattern barometer with a millibar (hPa) scale, the one on the left a Fortin barometer with an inches of mercury (inHg) scale: in both cases the reference point (the straight base of the movable vernier scale) has already been adjusted to form a tangent to the *top* of the meniscus of the mercury column in the tube (both vernier scales have an optical parallax compensation mechanism to minimise parallax errors). Reading the vernier scale essentially involves firstly reading off an 'incremental unit' value adjacent to the reference point on the movable vernier scale (the straight base of the scale on both barometers pictured here), and then reading off the 'sub-unit' from the point where the main scale and the vernier scale line graduations coincide to form an unbroken straight line; the two values are then added.

(a)   (b)

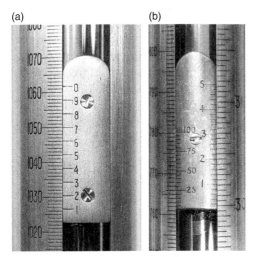

Figure A4.4 Reading the vernier scale on a mercury barometer; see text for explanation. Figure A4.4a is a Kew barometer, scale in hPa; A4.4b a Fortin barometer, scale in inches of mercury (inHg). (Photographs by the author)

In **Figure A4.4a**, the 'main unit' adjacent to the base of the vernier is 1022 hPa, while the point of coincidence of the graduation lines on the main scale and the movable scale occurs at scale point 7; each line on the movable scale marks 0.1 hPa, thus the 'as read' reading of the barometer is 1022 + (7 × 0.1) = 1022.7 hPa.

Things are slightly more complicated with an inch scale, where as well as the major incremental unit the tenths and hundredths of an inch also have to be read off. In **Figure A4.4b**, the reference point at the base of the vernier scale lies just under the 30 (inch) graduation; the longer graduation lines between the main 'inch' labels are 0.1 inch markers, and the shorter lines 0.05 inches. Therefore the starting point is 29.95 inches. The value to one-thousandth of an inch is then indicated by the point of coincidence between the graduation lines on the main scale and the moveable vernier scale. Each of the longer lines on the moveable vernier represents 0.01 inHg, and the smaller lines 0.002 inHg. In this case the coincidence of the lines occurs at exactly '2', and therefore the 'as read' barometer reading is 29.95 + (2 × 0.01) or 0.02 inches exactly, thus 29.970 inHg. If instead the next smaller line above the 2 graduation was exactly coincident with the main scale, the reading would have been 29.972 inHg; if the one below, the reading would have been 29.968 inHg. If the coincidence point was midway between the 2 and the next smaller graduation above it, the reading would instead have been 29.971 inHg.

A short video clip makes the point much easier than 400 words, and for examples the reader is referred to the 'vernier' entry on Wikipedia and to numerous 'reading a vernier' clips on YouTube and the like. The principle of reading remains the same, whether reading a barometer or a micrometer or any other instrument, and once grasped becomes automatic.

Correcting mercury barometer readings to mean sea level (MSL)

Accurate corrections of barometric pressure to MSL are required for many purposes, particularly aviation briefings and climatological averages, where precision and accuracy to 0.1 hPa are essential. As set out in Chapter 7, *Measuring atmospheric pressure*, WMO provide a general reduction formula suitable for sites up to about 750 m above MSL in the WMO CIMO guide [4], section 3.7. However, different methods are required for mercury barometers because both the temperature of the mercury and the barometer scale itself must also be corrected for in precise work. More details regarding the construction of MSL correction tables for mercury barometers can be found in reference [218], section 12.7, pages 428–449. This publication has been out of print for many years, but a PDF is available in the Met Office Archive, the link to which is given in the reference.

Useful functions

Vector mean winds

The 'vector mean wind' is a useful way to combine wind speed and direction records to come up with a *resultant wind flow* from a series of varying wind velocities over time. The calculation resolves individual samples of wind velocity into east-west and north-south components, which can then be averaged numerically in the normal manner. The averaged value of the two components are then converted back into the resultant (think 'average') wind direction and speed.

This method of calculation is necessary because the use of polar co-ordinates (compass bearings) means they cannot be simply averaged numerically – the 'mean' of a north-westerly wind (315°) and a north-easterly wind (045°) is clearly not a southerly wind, as would be indicated by the numerical average of the two wind directions ((315 + 45)/2 = 180). The calculations can be performed over a minute, a day, a year or for any other time period.

The details of the method are given below (from reference [323]) together with a listing of an Excel macro (also downloadable from www.measuringtheweather .net).

Advanced loggers include a vector mean wind option to summarise sampled wind speeds and directions in logged output. Where AWS software output is given as compass points rather than as degrees, a simple Excel macro (also downloadable from www.measuringtheweather.net) can be used within an Excel spreadsheet to convert compass bearings into equivalent degrees (so that, for example, all southerly winds will be converted to 180 degrees, south-south-westerly to 202.5°, and so on). The vector mean wind calculation can then be run as above.

Vector mean wind theory

Given a sequence of N observations of direction θ_i and velocity u_i, the mean east-west, V_e, and north-south, V_n, components of the wind are

$$V_e = -\frac{1}{N}\sum u_i \sin(\theta_i) \qquad\qquad \text{A5.1}$$

$$V_n = -\frac{1}{N}\sum u_i \cos(\theta_i) \qquad\qquad \text{A5.2}$$

The resultant mean wind speed and direction are:

$$\overline{U}_{RV} = \left(V_e^2 + V_n^2\right)^{1/2} \qquad\qquad \text{A5.3}$$

$$\overline{\theta}_{RV} = \arctan(V_e/V_n) + FLOW \qquad\qquad \text{A5.4}$$

$$\text{where } FLOW \quad \begin{aligned} &= +180; && \text{for } \arctan(V_e/V_n) < 180 \\ &= -180; && \text{for } \arctan(V_e/V_n) > 180 \end{aligned} \qquad \text{A5.5}$$

Equation A5.4 assumes the angle returned by the arctan function is in degrees.

Calculation using Excel

The listing in **Table A5.1** will return the two components, east and north, from two cells containing the scalar mean wind speed (in chosen units) and the wind direction in degrees (0° to 360°). This code is downloadable from www.measuringtheweather.net.

Table A5.1 *Excel code to calculate a vector mean wind. Calculate the north and east components (cells A4 and A5) from every observation of scalar mean wind speed (cells A1 and A2). Average these components over the period required, then evaluate cells A6 to A14 from the period averages to derive the vector mean wind.*

Cell	Content or action	Excel code
	***Evaluate for the range of observations required* ...**	
A1	Scalar mean wind speed, in chosen units	
	From AWS	
A2	Wind direction, in degrees	
	From AWS	
A3	*Blank cell*	
A4	N component	=IF(A1>0,A1*COS(A2*(PI()/180)),A3)
	Check first that wind is not calm, then evaluate cosine of wind direction	
A5	E component	=IF(A1>0,A1*SIN(A2*(PI()/180)),A3)
	Check first that wind is not calm, then evaluate sine of wind direction	
	...* then evaluate for the period chosen, using the average of the N and E components as derived above	
A6	Sin/Cos ratio	=A4/A5
A7	Take modulus of sin/cos ratio	=SQRT(A6*A6)
A8	Take arctan of modulus	=ATAN(A7)*180/PI()
A9	Quadrant 1	=IF(AND(A4>0,A5>0),90-A8,0)
	A value will fall in this sector if arctan modulus is between 0° and 90°	
A10	Quadrant 2	=IF(AND(A4<0,A5>0),90+A8,0)
	A value will fall in this sector if arctan modulus is between 90° and 180°	
A11	Quadrant 3	=IF(AND(A4<0,A5<0),270-A8,0)
	A value will fall in this sector if arctan modulus is between 180° and 270°	

Table A5.1 (*cont.*)

Cell	Content or action	Excel code
A12	Quadrant 4 *A value will fall in this sector if arctan modulus is between 270° and 360°*	=IF(AND(A4>0,A5<0),270+A8,0)
A13	Vector mean wind angle (degrees)	=MAX(A9:A13)
A14	Vector mean wind speed (original units)	=SQRT((A4*A4)+(A5*A5))

Conversion of compass point wind directions to degrees

Some brands of AWS output wind direction only as compass points rather than as degrees of azimuth. The latter is required for a vector mean wind calculation. A small Excel script downloadable from www.measuringtheweather.net will convert a selection of cells from compass points to degrees. (Note this will permanently change the cell values; if you wish to retain the compass points, copy the column first and apply the macro to the copy.)

Unit conversions

The following tables provide conversions for units of temperature, precipitation, barometric pressure and wind speed.

Temperature

Table A6.1 *Temperature conversions*

°C	°F	°F	°C
−40	−40	−40	−40.0
−35	−31	−30	−34.4
−30	−22	−20	−28.9
−25	−13	−10	−23.3
−20	−4	0	−17.8
−15	5	5	−15.0
−10	14	10	−12.2
−5	23	15	−9.4
0	32	20	−6.7
5	41	25	−3.9
10	50	30	−1.1
15	59	35	1.7
20	68	40	4.4
25	77	45	7.2
30	86	50	10.0
35	95	55	12.8
40	104	60	15.6
45	113	65	18.3
50	122	70	21.1
55	131	75	23.9
		80	26.7
		85	29.4
		90	32.2
		95	35.0
		100	37.8
		105	40.6
		110	43.3
		115	46.1
		120	48.9
		130	54.4

Table A6.2 *Precipitation conversions*

mm	inches	inches	mm
0.1	0.004	0.01	0.25
0.2	0.01	0.02	0.5
0.5	0.02	0.03	0.8
1	0.04	0.04	1.0
2	0.08	0.05	1.3
3	0.12	0.1	2.5
4	0.16	0.2	5.1
5	0.20	0.5	12.7
10	0.39	1	25.4
20	0.79	2	50.8
30	1.18	5	127
40	1.57	10	254
50	1.97	20	508
100	3.94	50	1 270
200	7.87	100	2 540
500	19.69	200	5 080
1 000	39.37	500	12 700
2 000	78.74		
5 000	196.85		
10 000	393.70		

Table A6.3 *Pressure conversions (at 0 °C)*

hPa	inches of mercury (inHg)	inches of mercury (inHg)	hPa
950	28.05	28.00	948.2
960	28.35	28.25	956.7
970	28.64	28.50	965.1
975	28.79	28.75	973.6
980	28.94	29.00	982.1
985	29.09	29.10	985.4
990	29.23	29.20	988.8
995	29.38	29.30	992.2
1 000	29.53	29.40	995.6
1 005	29.68	29.50	999.0
1 010	29.83	29.60	1002.4
1 015	29.97	29.70	1005.8
1 020	30.12	29.80	1009.1
1 025	30.27	29.90	1012.5
1 030	30.42	30.00	1015.9
1 035	30.56	30.10	1019.3
1 040	30.71	30.20	1022.7
1 045	30.86	30.30	1026.1
1 050	31.01	30.40	1029.5
1 055	31.15	30.50	1032.8
		30.60	1036.2
		30.70	1039.6
		30.80	1043.0
		30.90	1046.4
		31.00	1049.8
		31.25	1058.2

Table A6.4 *Conversions between various units of wind speed*

	knots (kn)	metres per second (m/s)	miles per hour (mph)	kilometres per hour (km/h)
Convert from	*Multiply by*			
knots (kn)	1	0.5144	1.152	1.853
metres per second (m/s)	1.943	1	2.237	3.600
miles per hour (mph)	0.868	0.447	1	1.609
kilometres per hour (km/h)	0.540	0.278	0.621	1

References and Further Reading

All citations, website links, DOIs and URLs in the following reference listing were checked and updated where necessary and were believed correct as at late 2023. However, WMO library references in particular appear to change frequently: if the link is found to be broken, entering the citation details into Google search is normally more reliable than the search function on the WMO website.

1. Matthews, T., et al., 2020: Going to extremes: Installing the world's highest weather stations on Mount Everest. *Bulletin of the American Meteorological Society*, **101**: pp. E1870–E1890. doi:10.1175/bams-d-19-0198.1

2. Wilkinson, Freddie, 2022: Next-gen weather station installed near Everest's summit. *National Geographic*, 25 May 2022 www.nationalgeographic.com/environment/article/perpetual-planet-next-gen-weather-station-installed-near-everests-summit

3. Photographs of the six AWS sites and the archive of (lightly) quality-controlled data from the Mt Everest AWSs are available at www.nationalgeographic.org/projects/perpetual-planet/everest/weather-data/. The most recent data can be viewed at the low-bandwidth page: https://everest-pwa.nationalgeographic.org. Note, however, that data transmission lapses and instrument faults may mean that near-real-time data are occasionally unavailable.

4. World Meteorological Organization (WMO), 2021: *WMO No. 8 – Guide to Meteorological Instruments and Methods of Observation (CIMO guide). 2021 edition – Volume I: Measurement of Meteorological Variables*. Geneva: World Meteorological Organization. 581 pp. WMO permalink: https://library.wmo.int/idurl/4/68695. English: also available in French, Spanish and Russian.

5. Burt, Stephen, 2017: Mercury sunset. *Meteorological Technology International*, **2017**(4): pp. 16–20

6. Middleton, W. E. K., 1969: *Invention of the meteorological instruments*. Baltimore, MD: The Johns Hopkins Press. 362 pp.

7. Lawrence, E. N., 1973: Merle's weather. *Weather*, **28**(3): pp. 127–30. doi:10.1002/j.1477-8696.1973.tb02248.x

8. Meaden, G. T., 1973: Merle's weather diary and its motivation. *Weather*, **28**(5): pp. 210–211. doi:10.1002/j.1477-8696.1973.tb02267.x

9. Fiebrich, Christopher A., 2009: History of surface weather observations in the United States. *Earth-Science Reviews*, **93**(3): pp. 77–84. doi:10.1016/j.earscirev.2009.01.001

10. Tinniswood, Adrian, 2001: *His invention so fertile: A life of Christopher Wren*. London: Jonathan Cape. 463 pp.

11. Bennett, J. A., 2002: *The mathematical science of Christopher Wren*. Cambridge: Cambridge University Press. 148 pp.

12. Tinniswood, Adrian, 2019: *The Royal Society and the invention of modern science*. London: Head of Zeus. 208 pp.

13. Gribbin, John, 2006: *The fellowship: The story of the Royal Society and a scientific revolution*. London: Allen Lane. 336 pp.

14. Mills, A., 2009: Dr Hooke's 'Weather-Clock' and its self-emptying bucket. *Bulletin of the Scientific Instrument Society*, **102**: pp. 29–30: see also Mihailescu, I., 2021: Graphical details: the secret life of Christopher Wren's drawing of the weather clock. *Notes and Records of the Royal Society*, **77**: pp. 355–378

15. Folland, C. K. and B. G. Wales-Smith, 1977: Richard Towneley and 300 years of regular rainfall measurement. *Weather*, **32**(12): pp. 438–445. doi:10.1002/j.1477-8696.1977.tb04501.x

16. Biswas, A. K., 1967: The automatic rain-gauge of Sir Christopher Wren, F.R.S. *Notes and Records of the Royal Society of London*, **22**(1): pp. 94–104. doi:10.1098/rsnr.1967.0009

17. Inwood, Stephen, 2002: *The man who knew too much: The strange and inventive life of Robert Hooke 1635–1703*. London: Macmillan. 497 pp.

18. Bennett, J. A., 1980: Robert Hooke as mechanic and natural philosopher. *Notes and Records of the Royal Society of London*, **35**(1): pp. 33–48. doi:10.1098/rsnr.1980.0003

19. Austin, Jillian F. and Anita McConnell, 1980: James Six F.R.S.: Two hundred years of the Six's self-registering thermometer. *Notes and Records of the Royal Society of London*, **35**(1): pp. 49–65

20. Hamblyn, Richard, 2001: *The invention of clouds: How an amateur meteorologist forged the language of the skies*. New York: Farrar, Straus and Giroux. 403 pp.

21. Blench, B. J. R., 1963: Luke Howard and his contribution to meteorology. *Weather*, **18**(3): pp. 83–92. doi:10.1002/j.1477-8696.1963.tb01977.x

22. Details of the Alexander Cumming barometer in the Royal Collection, including photographs and a full description, can be found online at www.royalcollection.org.uk/eGallery. King George III paid the enormous sum of £1,178 for this instrument in 1763, to which he added a payment of £150 and an annual retainer to Cumming of £37 10s for maintaining the barograph. The king kept the Cumming barograph with the Pinchbeck astronomical clock in the Passage Room of his apartments at Buckingham House; today it is displayed in the Throne Room at Buckingham Palace. Luke Howard's instrument is in the Science Museum in London.

23. Middleton, W. E. K., 1966: *A history of the thermometer and its use in meteorology*. Baltimore, MD: Johns Hopkins Press. 249 pp. Historical details regarding aspirated thermometers are given on pp. 234–238. Middleton rightly credits the pioneering work of Aitken and Assman in the 1880s, who were more than a century ahead of their time in this area.

24. Bathurst, Bella, 1999: *The Lighthouse Stevensons*. London: HarperCollins. 284 pp.

25. Morrison-Low, A. D., 2010: *Northern Lights: The age of Scottish lighthouses*. Edinburgh: NMSE Publishing Ltd. 262 pp.

26. Stevenson, Thomas, 1864: New description of box for holding thermometers. *Journal of the Scottish Meteorological Society*, **1**: p. 122.

27. Burt, Stephen, 2013: An unsung hero in meteorology: Charles Higman Griffith (1830–1896). *Weather*, **68**(5): pp. 135–138. doi:10.1002/wea.2059

28. Council of the Royal Meteorological Society, 1884: Report of the Council for the year 1883: Appendix 1, Report of the thermometer screen committee. *Quarterly Journal of the Royal Meteorological Society*, **10**(50): pp. 92–94. doi:10.1002/qj.4970105003

29. Multhauf, R. P., 1961: The introduction of self-registering meteorological instruments. *Washington DC: United States National Museum Bulletin*, **228**: pp. 95–116. Multhauf made the interesting speculation that the technology of the 1880s (mechanical sensors, levers to magnify small movements and a clock mechanism to drive a paper chart) was largely available in the 1660s, and that Hooke could probably have built a 'modern' clock-driven single-element recorder, such as a thermograph, had he instead progressed his ideas along those lines.

30. Strangeways, I. C. and S. W. Smith, 1985: Development and use of automatic weather stations. *Weather*, **40**(9): pp. 277–285 . doi:10.1002/j.1477-8696.1985.tb06900.x

31. Camuffo, Dario and Chiara Bertolin, 2012: The earliest temperature observations in the world: the Medici Network (1654–1670). *Climatic Change*, **111**(2): pp. 335–363. doi:10.1007/s10584-011-0142-5

32. Camuffo, D., C. Bertolin, P. D. Jones, R. Cornes and E. Garnier, 2010: The earliest daily barometric pressure readings in Italy: Pisa AD 1657–1658 and Modena AD 1694, and the weather over Europe. *The Holocene*, **20**(3): pp. 337–349

33. Cornes, Richard C., et al., 2012: A daily series of mean sea-level pressure for Paris, 1670–2007. *International Journal of Climatology*, **32**(8): pp. 1135–1150 doi:10.1002/joc.2349.

34. Cornes, Richard C., et al., 2012: A daily series of mean sea-level pressure for London, 1692–2007. *International Journal of Climatology*, **32**(5): pp. 641–656. doi:10.1002/joc.2301: also Cornes, Richard C., et al., 2023: The London, Paris and De Bilt sub-daily pressure series. *Geoscience Data Journal.*, online version doi: 10.1002/gdj3.226

35. Manley, Gordon, 1953: The mean temperature of central England, 1698–1952. *Quarterly Journal of the Royal Meteorological Society*, **79**(342): pp. 558–567 doi:10.1002/qj.49707934222

36. Manley, Gordon, 1974: Central England temperatures: Monthly means 1659 to 1973. *Quarterly Journal of the Royal Meteorological Society*, **100**(425): pp. 389–405 doi:10.1002/qj.49710042511

37. The Central England Temperature (CET) database and the England and Wales Precipitation (EWP) series (and others) are available from the Hadley Centre website: metoffice.gov.uk/hadobs/index.html

38. Parker, D. E., T. P. Legg, and C. K. Folland, 1992: A new daily central England temperature series, 1772–1991. *International Journal of Climatology*, **12**(4): pp. 317–342. doi:10.1002/joc.3370120402

39. For more information on the Centennial Observing Stations and an updated list (nominations are sought and reviewed every two years), see https://wmo.int/centennial-observing-stations

40. Bergström, Hans and Anders Moberg, 2002: Daily air temperature and pressure series for Uppsala (1722–1998). *Climatic Change*, **53**(1): pp. 213–252 doi:10.1023/a:1014983229213. Further information is available on the University of Uppsala's website at www.uu.se/en/about-uu/celsius-300

41. Burt, Stephen, 2023: The Celsius Symposium: 300 years of weather statistics. *Weather*, **78**(1): pp. 30–31 doi:10.1002/wea.4325

42. Cocheo, Claudio and Dario Camuffo, 2002: Corrections of systematic errors and data homogenisation in the daily temperature Padova series (1725–1998). *Climatic Change*, **53**(1): pp. 77–100

43. Moberg, Anders and Hans Bergström, 1997: Homogenization of Swedish temperature data. Part III: the long temperature records from Uppsala and Stockholm. *International Journal of Climatology*, **17**(7): pp. 667-699 doi: 10.1002/(SICI)1097-0088(19970615)17:7<667::AID-JOC115>3.0.CO;2-J. Also: Moberg, Anders, et al., 2002: Daily Air Temperature and Pressure Series for Stockholm (1756–1998). *Climatic Change*, **53**(1): pp. 171–212 doi: 10.1023/A:1014966724670.

44. Böhm, Reinhard, et al., 2010: Early instrumental warm-bias: a solution for long central European temperature series 1760–2007. *Climatic Change*, **101**(1–2): pp. 41–67. doi:10.1007/s10584-009-9649-4

45. Maugeri, Maurizio, Letizia Buffoni, and Franca Chlistovsky, 2002: Daily Milan temperature and pressure series (1763–1998): History of the observations and data and metadata recovery. *Climatic Change*, **53**(1): pp. 101–117. doi:10.1023/A:1014970825579

46. Maugeri, M., et al., 2002: Daily Milan temperature and pressure series (1763–1998): Completing and homogenising the data. *Climatic Change*, **53**: pp. 119–149

47. Czech Hydrometeorological Institute. *Meteorological observations at the Prague Clementinum*. Available from: http://portal.chmi.cz/historicka-data/pocasi/praha-klementinum?l=en

48. Prague Klementinum details – on the Czech Hydrometeorological Institute site at www.chmi.cz/historicka-data/pocasi/praha-klementinum?l=en

49. Winkler, P., 2009: Revision and necessary correction of the long-term temperature series of Hohenpeissenberg, 1781–2006. *Theoretical and Applied Climatology*, **98**(3): pp. 259–268 doi: 10.1007/s00704-009-0108-y.

50. Deutscher Wetterdienst. *Meteorological Observatory Hohenpeissenberg*. Available from: www.dwd.de/EN/aboutus/locations/observatories/mohp/mohp.html

51. Pappert, Duncan, et al., 2021: Unlocking weather observations from the Societas Meteorologica Palatina (1781–1792). *Climate of the Past*, **17**(6): pp. 2361–2379 doi:10.5194/cp-17-2361-2021

52. Bennett, J. A., 1990: *Church, state and astronomy in Ireland: 200 years of Armagh Observatory*. Armagh: The Armagh Observatory. 277 pp.

53. Butler, C. J., et al., 2005: Air temperatures at Armagh Observatory, Northern Ireland, from 1796 to 2002. *International Journal of Climatology*, **25**(8): pp. 1055–1079. doi:10.1002/joc.1148

54. Butler, C. J. and A. M. García-Suárez, 2012: Relative humidity at Armagh Observatory, 1838–2008. *International Journal of Climatology*, **32**(5): pp. 657–668. doi:10.1002/joc.2302

55. Butler, John and Michael Hoskin, 1987: The archives of Armagh Observatory. *Journal for the History of Astronomy*, **18**: pp. 295–307. (List of meteorological records on page 9 of the paper.)

56. Guest, Ivor, 1991: *Dr John Radcliffe and his Trust*. London: The Radcliffe Trust. 595 pp. PDF copy available from The Radcliffe Trust website

57. Burt, Stephen and Timothy Burt, 2019: *Oxford weather and climate since 1767*. Oxford: Oxford University Press. 544 pp.

58. Burt, Stephen, 2021: Two hundred years of thunderstorms in Oxford. *Weather*, **76**: pp. 212–222. doi:10.1002/wea.3884

59. Wallace, J. G., 1997: *Meteorological observations at the Radcliffe Observatory, Oxford: 1815–1995*. Oxford: University of Oxford – School of Geography. 77 pp.

60. Manley, Gordon, 1941: The Durham meteorological record, 1847–1940. *Quarterly Journal of the Royal Meteorological Society*, **67**(292): pp. 363–380. doi:10.1002/qj.49706729209

61. Burt, Stephen and Timothy Burt, 2022: *Durham weather and climate since 1841*. Oxford: Oxford University Press. 580 pp.

62. Details of the Minneapolis records can be found at www.climatestations.com/minneapolis-2

63. Slonosky, Victoria, 2018: *Climate in the age of empire: weather observers in colonial Canada*. American Meteorological Society, Boston, MA. 325 pp.

64. Slonosky, Victoria C., 2015: Daily minimum and maximum temperature in the St Lawrence Valley, Quebec: two centuries of climatic observations from Canada. *International Journal of Climatology*, **35**(7): pp. 1662–1681. doi:10.1002/joc.4085

65. Details of the New York Central Park historical records can be found at www.weather.gov/okx/CentralParkHistorical

66. The Blue Hill Meteorological Observatory website contains much interesting information about the Observatory and its climate record since 1885: see https://bluehill.org/climate-weather.

67. Wang, Pao K. and Zhang De'er, 1988: An introduction to some historical governmental weather records of China. *Bulletin of the American Meteorological Society*, **69**(7): pp. 753–758. doi:10.1175/1520-0477(1988)069<0753:aitshg>2.0.co;2

68. Brönnimann, Stefan, et al., 2019: Unlocking Pre-1850 instrumental meteorological records: A global inventory. *Bulletin of the American Meteorological Society*, **100**(12): pp. ES389–ES413. doi:10.1175/bams-d-19-0040.1

69. Ren, Yuyu, Guoyu Ren, Rob Allan, Jiao Li et al., 2022: The 1757–62 temperature observed in Beijing. *Bulletin of the American Meteorological Society*, **103**(11): pp. E2470–E2483. doi:10.1175/bams-d-21-0245.1

70. Ashcroft, Linden, 2018: *Meteorology at Sydney Observatory*, in *The story of Sydney Observatory*. Ultimo, New South Wales, Australia: Museum of Applied Arts and Sciences. 61 pp.

71. Ashcroft, Linden, et al., 2021: The world's longest known parallel temperature dataset: A comparison between daily Glaisher and Stevenson screen temperature data at Adelaide, Australia, 1887–1947. *International Journal of Climatology*. doi:10.1002/joc.7385

72. Gergis, Joëlle, et al., 2021: A historical climate dataset for southwestern Australia, 1830–1875. *International Journal of Climatology*, **41**(10): pp. 4898–4919. doi:10.1002/joc.7105

73. Diamond, Howard J., et al., 2013: US Climate Reference Network after one decade of operations: Status and assessment. *Bulletin of the American Meteorological Society*, **94**(4): pp. 485–498. doi:10.1175/bams-d-12-00170.1

74. Hubbard, K. G., X. Lin, and C. B. Baker, 2005: On the USCRN temperature system. *Journal of Atmospheric and Oceanic Technology*, **22**: pp. 1095–1100

75. Hubbard, K. G., et al., 2004: Air temperature comparison between the MMTS and the USCRN temperature systems. *Journal of Atmospheric and Oceanic Technology*, **21**: pp. 1590–1597

76. Body, D. and F. Kuik, 2021: *Generic Automatic Weather Station (AWS) tender specifications*. Geneva: World Meteorological Organization (WMO). Available from https://library.wmo.int/records/item/57830-generic-automatic-weather-station-aws-tender-specifications.

77. Vuerich, E., et al., 2009: *WMO field intercomparison of rainfall intensity gauges*. Geneva, Switzerland: World Meteorological Organization: Instruments and observing methods report no. 99, WMO/TD-No. 1504. Available from https://library.wmo.int/records/item/50453-wmo-field-intercomparison-of-rainfall-intensity-gauges. 290 pp.

78. Coney, Jonathan, et al., 2022: How useful are crowdsourced air temperature observations? An assessment of Netatmo stations and quality control schemes over the United Kingdom. *Meteorological Applications*, **29**(3): doi:10.1002/met.2075

79. Burt, Stephen. 2013: *User review of the Davis Vantage Vue automatic weather station*. Available from: www.measuringtheweather.net

80. Burt, Stephen. 2009: *User review of the Davis Vantage Pro2 automatic weather station*. Available from: www.measuringtheweather.net

81. Sun, B., et al., 2005: A comparative study of ASOS and USCRN temperature measurements. *Journal of Atmospheric and Oceanic Technology*, **22**(6): pp. 679–686. doi:10.1175/jtech1752.1

82. More information, and access to current USCRN observations, is available at www.ncei.noaa.gov/access/crn/.

83. Leeper, Ronald D., Jared Rennie, and Michael A. Palecki, 2015: Observational perspectives from U.S. Climate Reference Network (USCRN) and Cooperative Observer Program (COOP) network: Temperature and precipitation comparison. *Journal of Atmospheric and Oceanic Technology*, **32**(4): pp. 703–721. doi:10.1175/jtech-d-14-00172.1

84. Trewin, B. 2018: *Station catalogue – Australian Climate Observations Reference Network – Surface Air Temperature (ACORN-SAT)*. Available from www.bom.gov.au/metadata/catalogue/19115/ANZCW0503900725

85. Ren, Yuyu and Guoyu Ren, 2011: A remote-sensing method of selecting reference stations for evaluating urbanization effect on surface air temperature trends. *Journal of Climate*, **24**(13): pp. 3179–3189. doi:10.1175/2010jcli3658.1

86. Strangeways, Ian, 2015: A global climate reference network. *Weather*, **70**(4): pp. 124–129 doi:10.1002/wea.2460

87. Thorne, P. W., et al., 2018: Towards a global land surface climate fiducial reference measurements network. *International Journal of Climatology*, **38**(6): pp. 2760–2774. doi:10.1002/joc.5458

88. Doesken, N., 2005: *The National Weather Service MMTS (Maximum-Minimum Temperature System) – 20 years after*, in *American Meteorological Society conference papers* – available online at ams.confex.com/ams/pdfpapers/91613.pdf

89. Quayle, Robert G., et al., 1991: Effects of recent thermometer changes in the cooperative station network. *Bulletin of the American Meteorological Society*, **72**(11): pp. 1718–1723. doi:10.1175/1520-0477(1991)072<1718:EORTCI>2.0.CO;2

90. The data for Table 4.2 came from NCEI's HOMR database, March 2023 (with grateful thanks to Shelley McNeill, NOAA). Lists of current stations can be extracted from the website www.ncei.noaa.gov/access/homr. Of the 4,472 'temperature' sites listed in Table 4.2, 89% are MMTS sites suggesting that the remainder (11%, 495 sites) continue to use Cotton Region Shelters (CRS). Unsurprisingly, this represents a continuing decline from 31% in February 2009 (data from that date from the website weather.gov/coop) when there were 4,300 listed sites within the continental United States, of which 1,337 were CRS sites (namely 31%).

91. Meyer, Steven J. and Kenneth G. Hubbard, 1992: Nonfederal automated weather stations and networks in the United States and Canada: A preliminary survey. *Bulletin of the American Meteorological Society*, **73**(4): pp. 449–457. doi:10.1175/1520-0477(1992)073<0449:nawsan>2.0.co;2

92. Mahmood, Rezaul, et al., 2019: A technical overview of the Kentucky Mesonet. *Journal of Atmospheric and Oceanic Technology*, **36**(9): pp. 1753–1771. doi:10.1175/jtech-d-18-0198.1

93. Mahmood, Rezaul, et al., 2020: The Total Solar Eclipse of 2017: Meteorological observations from a statewide mesonet and atmospheric profiling systems. *Bulletin of the American Meteorological Society*, **101**(6): pp. E720–E737. doi:10.1175/BAMS-D-19-0051.1

94. Details of the Met Office land surface observing network, and a map of the locations, is available at www.metoffice.gov.uk/weather/guides/observations/uk-observations-network

95. Details of the Met Office synoptic and climate station network, including map of the locations and a table of their location, is available at www.metoffice.gov.uk/research/climate/maps-and-data/uk-synoptic-and-climate-stations

96. A list of, and links to, the world's national weather services is available on the WMO website: https://community.wmo.int/members

97. Preston-Thomas, H., 1990: International Practical Temperature Scale of 1990. *Metrologia*, **27**: pp. 3–10

98. The basic requirements were set out over a century ago: see, for example, Gaster, F. (1882) Report on experiments made at Strathfield Turgiss in 1869 with stands or screens of various patterns, devised and employed for the exposing of thermometers, in order to determine the temperature of the air. *Quarterly Weather Report for 1879*, Meteorological Office, London: also Köppen, W. (1913) Uniform thermometer set-up for meteorological stations for the determination of air temperature and humidity. *Meteorol. Zeitschr.*, **30**, pp. 474–88 and 514–23: English translation in *Monthly Weather Review*, August 1915.

99. Foken, Thomas, ed. 2022. *Springer handbook of atmospheric measurements*. doi:10.1007/978-3-030-52171-4. Springer International Publishing, 1748 pp.

100. Park, Jae-Woo, et al., 2022: Air temperature dependencies on the structure of thermometer screens in summer at Daejeon, South Korea. *Meteorological Applications*, **29**(3): doi: 10.1002/met.2064.

101. Barnett, A., D. B. Hatton, and D. W. Jones, 1998: *Recent changes in thermometer [screen] design and their impact*. World Meteorological Organization, Geneva: Instruments and Observing Methods Report no. 66

102. Strangeways, Ian, 2010: *Measuring global temperatures: Their analysis and interpretation*. Cambridge: Cambridge University Press.

103. Oke, T. R., et al., 2017: *Urban Climates*, in *Urban Climates*: Cambridge: Cambridge University Press. 525 pp.

104. Chandler, T. J., 1965: *The climate of London*. London: Hutchinson. 292 pp.

105. Sparks, W. R. 1972: *The effect of thermometer screen design on the observed temperature*. World Meteorological Organization, Geneva, Publication No. 315. Available online at https://library.wmo.int/doc_num.php?explnum_id=8131

106. International Organization for Standardization (ISO), 2007: *ISO 17714 Meteorology – Air temperature measurements – Test methods for comparing the performance of thermometer shields/screens and defining important characteristics*. Geneva: International Organization for Standardization (ISO)

107. Gaster, Frederick, 1882: Report on experiments made at Strathfield Turgiss in 1869 with stands or screens of various patterns, devised and employed for the exposing of thermometers, in order to determine the temperature of the air. *Meteorological Office Quarterly Weather Report, Addendum for 1879 [dated May 1880, published 1882]*.

108. Bilham, E. G., 1937: A screen for sheathed thermometers. *Quarterly Journal of the Royal Meteorological Society*, **63**(271): pp. 309–322. doi:10.1002/qj.49706327104

109. Specifications and drawings to construct Cotton Region shelters are available at several places on the web – try https://drive.google.com/drive/folders/1vHg5oUZMQ5iqmM2iC73PK70dMQiZ4qaQ?usp=sharing. Cotton Region Shelters come in 'medium' and 'large' sizes, the main difference being that the large size usually had a wooden post on which the maximum and minimum thermometers were mounted, with a baffle within the shelter to avoid the readings being affected by the aspirator fan used to ventilate the dry and wet bulb thermometers (source: Grant Goodge, ex NOAA, personal communication March 2023). In the US, NovaLynx sell self-assembly Cotton Region shelter kits (Small Instrument shelter, model no. 380–601 and Large, model no. 380–605) – see https://novalynx.com/brochures/380-shelters.pdf. A copy of the 1960 Met Office/HMSO publication *Instructions for making thermometer screens of the Stevenson type* (Met O No. 670, 19 pages) is available online from the National Meteorological Library: www.metoffice.gov.uk/research/library-and-archive, item Miscellaneous Met Office Publications METDLA/3/4. Another option, with less complicated carpentry involved, is described in McConnell, D. (1988) Making a simple thermometer screen. *Weather*, **43**, pp. 198–203; other do-it-yourself plans can be found on the web.

110. Perry, M. C., M. J. Prior, and D. E. Parker, 2007: An assessment of the suitability of a plastic thermometer screen for climatic data collection. *International Journal of Climatology*, **27**(2): pp. 267–276. doi:10.1002/joc.1381

111. Hubbard, K.G., X. Lin, and E.A. Walter-Shea, 2001: The effectiveness of the ASOS, MMTS, Gill, and CRS air temperature radiation shields. *Journal of Atmospheric and Oceanic Technology*, **18**: pp. 851–864

112. Burt, Stephen, 2022: Measurements of natural airflow within a Stevenson screen and its influence on air temperature and humidity records. *Geoscientific Instrumentation, Methods and Data Systems*, **11**(2): pp. 263–277. doi:10.5194/gi-11-263-2022

113. Burt, Stephen and Michael de Podesta, 2020: Response times of meteorological air temperature sensors. *Quarterly Journal of the Royal Meteorological Society*, **146**: pp. 2789–2800. doi:10.1002/qj.3817

114. Harrison, R. G., 2010: Natural ventilation effects on temperatures within Stevenson screens. *Quarterly Journal of the Royal Meteorological Society*, **136**(646): pp. 253–259. doi:10.1002/qj.537

115. Lopardo, G., et al., 2014: Comparative analysis of the influence of solar radiation screen ageing on temperature measurements by means of weather stations. *International Journal of Climatology*, **34**(4): pp. 1297–1310. doi:10.1002/joc.3765

116. Harrison, R. G., 2011: Lag-time effects on a naturally ventilated large thermometer screen. *Quarterly Journal of the Royal Meteorological Society*, **137**(655): pp. 402–408. doi:10.1002/qj.745

117. Richardson, Scott J., et al., 1999: Minimizing errors associated with multiplate radiation shields. *Journal of Atmospheric and Oceanic Technology*, **16**(11): pp. 1862–1872. doi:10.1175/1520-0426(1999)016<1862:meawmr>2.0.co;2

118. Differences in air temperature between identical sensors housed in a Stevenson screen and an aspirated radiation shelter (RM Young model 43502). Daylight only, 10 year period (49 853 hourly observations) at Stratfield Mortimer Observatory in southern England (51°N, 1°W), minimum class size 5 entries

119. Aitken, John, 1884: Thermometer screens. *Proceedings of the Royal Society of Edinburgh*, **12**: pp. 661–696

120. Aitken, John, 1913: The Stevenson screen. *Symons's Meteorological Magazine*, **48**: pp. 232–233

121. Miller, Samuel H., 1877: On the aspiration of the dry and wet bulb thermometers. *Quarterly Journal of the Royal Meteorological Society*, **3**(19): pp. 150–158. doi:10.1002/qj.4970031906.

122. Assmann, Richard, 1887: Das Aspirationspsychrometer, ein neuer Apparat zur Ermittlung der wahren Temperatur und Feuchtigkeit der Luft. *Wetter*, **4**: pp. 245–286

123. Daily maximum, minimum and mean daily temperatures over the period midnight to midnight UTC from calibrated platinum resistance sensors in adjacent RM Young and Apogee aspirated radiation shields were logged over the two year period November 2016 to December 2018 (765 daily observations) at Stratfield Mortimer Observatory in southern England (51°N, 1°W)

124. Lacombe, M., D. Bousri, M. Leroy and M. Mezred, 2011: Instruments and observing methods report no. 106: WMO field intercomparison of thermometer screens/shields and humidity measuring instruments, Ghardaïa, Algeria, November 2008 – October 2009. *Instruments and Observing Methods Report No. 106*, WMO/TD-No. 1579. Geneva: World Meteorological Organization (WMO) https://library.wmo.int/records/item/50488-wmo-field-intercomparison-of-thermometer-screens-shields-and-humidity-measuring-instruments.

125. Apogee Instruments aspirated radiation shield: www.apogeeinstruments.com/aspirated-radiation-shield

126. One minute mean temperatures in a Stevenson screen and nearby aspirated screen observed at Stratfield Mortimer Observatory in southern England (51°N, 1°W), 0700 to 1000 UTC on 3 September 2022

127. Details of the Hong Kong King's Park meteorological site are available from www.hko.gov.hk/en/wxinfo/aws/100_Upper_Air/kings-park-past-and-present.html

128. Hatton, D. B., 2002: Results of an intercomparison of wooden and plastic thermometer screens. In *Papers presented at the WMO Technical Conference on Meteorological Instruments and Methods of Observation (TECO-2002), Bratislava, Slovak Republic,*

23–25 September 2002. WMO, Geneva, Instruments and Observing Methods Report No. 75 (WMO/TD – No 1123)

129. Lin, X., K. G. Hubbard, and G. E. Meyer, 2001: Airflow characteristics of commonly used temperature radiation shields. *J. Atmospheric and Oceanic Technology*, **18**: pp. 329–339

130. Assessment of the Campbell Scientific 'Met 21' passive thermometer screen. Unpublished report, available at www.measuringtheweather.net

131. Warne, Jane, 1998: *A preliminary investigation of temperature screen [sic] design and their impacts on temperature measurements*. Bureau of Meteorology, Australia: Instrument Test Report No. 649

132. Buisan, Samuel T., Cesar Azorin-Molina, and Yolanda Jimenez, 2015: Impact of two different sized Stevenson screens on air temperature measurements. *International Journal of Climatology*, **35**(14): pp. 4408–4416. doi:10.1002/joc.4287

133. Brandsma, T. and J. P. van der Meulen, 2008: Thermometer screen intercomparison in De Bilt (the Netherlands) – Part II: description and modeling of mean temperature differences and extremes. *International Journal of Climatology*, **28**(3): pp. 389–400. doi:10.1002/joc.1524

134. Aoshima, Tadayoshi, et al., 2010: *RIC-Tsukuba (Japan) Intercomparison of Thermometer Screens/Shields in 2009*, in WMO TECO 2010 conference paper, available at https://community.wmo.int/en/teco-2010-programme

135. Nordli, P. Ø., et al., 1997: The effect of radiation screens on Nordic time series of mean temperature. *International Journal of Climatology*, **17**(15): pp. 1667–1681. doi: 10.1002/(SICI)1097-0088(199712)17:15<1667::AID-JOC221>3.0.CO;2-D

136. Andersson, T. and I. Mattison, 1991: *A field test of thermometer screens*. SMHI Report No. RMK 62, Norrkoping, Sweden

137. Hourly averages by month of Stevenson screen warming vs aspirated screen, over 10 years 2013–22, Stratfield Mortimer Observatory in southern England (51°N, 1°W): 87,608 observations, from possible 87,672 (data availability 99.93%)

138. One minute mean temperatures in a Stevenson screen and nearby aspirated screen observed at Stratfield Mortimer Observatory, southern England (51°N, 1°W). Figure 5.10 dataset 1130-1430 UTC on 16 December 2022: Figure 5.11 1200-1500 UTC on 13 August 2022: Figure 5.12 0300-0600 UTC on 28 November 2022.

139. Brock, Fred V. and Scott J. Richardson, 2001: *Meteorological Measurement Systems*. Oxford: Oxford University Press. 290 pp. doi:10.1093/oso/9780195134513.001.0001.

140. Harrison, R. Giles, 2014: *Meteorological measurements and instrumentation*. Wiley: Advancing weather and climate science series. doi:10.1002/9781118745793

141. de Podesta, Michael, Stephanie Bell, and Robin Underwood, 2018: Air temperature sensors: Dependence of radiative errors on sensor diameter in precision metrology and meteorology. *Metrologia*, **55**: pp. 229–244

142. Trewin, Blair, 2022: A climatology of short-period temperature variations at Australian observation sites. *Journal of Southern Hemisphere Earth Systems Science*. doi:10.1071/es21027

143. Ayers, G. P. and J. O. Warne, 2020: Response time of temperature measurements at automatic weather stations in Australia. *Journal of Southern Hemisphere Earth Systems Science*, **70**(1): pp. 160–165. doi:10.1071/es19032

144. Miller, Ronald J. and Bair, Andrea, 2022: *NOAA State Climate Extremes Committee Memorandum: New value for Washington maximum temperature record at Hanford, WA*. Available from www.ncei.noaa.gov/monitoring-content/extremes/scec/reports/20220210-Washington-Maximum-Temperature.pdf

145. International Organization for Standardization (ISO), 2021: *BS ISO 23350 2021 Hydrometry – Catching-type liquid precipitation measuring gauges*. Geneva: International Organization for Standardization (ISO)

146. Lanza, L. G., et al., 2021: Calibration of non-catching precipitation measurement instruments: A review. *Meteorological Applications*, **28**(3): doi:10.1002/met.2002

147. British Standards Institution, 2024: *BS 7843:2024 Acquisition and management of meteorological precipitation data from a gauge network*. British Standards Institution

148. Sevruk, B. and S. Klemm, 1989: *Catalogue of national standard precipitation gauges*. Vol. Instruments and Observing Methods, Report no. 39: WMO/TD-No. 313. Geneva: World Meteorological Organization

149. Goble, Peter E., et al., 2020: Who received the most rain today?: An analysis of daily precipitation extremes in the contiguous United States using COCORAHS and COOP reports. *Bulletin of the American Meteorological Society*, **101**(6): pp. E710–E719. doi:10.1175/bams-d-18-0310.1

150. Doesken, N. and Henry Reges, 2010 The value of the citizen weather observer. *Weatherwise*, **63**: pp. 30–37. doi:10.1080/00431672.2010.519607

151. The CoCoRaHS website at www.cocorahs.org includes maps of precipitation totals across the United States and Canada updated daily, as well as information on how to join, observing methods, how to buy and site the CoCoRaHS raingauge, and much more

152. Reges, Henry and Nolan Doesken, 2020: A day in the life of the CoCoRaHS network. *Weatherwise*, **73**(4): pp. 32–39. doi:10.1080/00431672.2020.1762416

153. Reges, Henry W., et al., 2016: CoCoRaHS: The evolution and accomplishments of a volunteer rain gauge network. *Bulletin of the American Meteorological Society*, **97**(10): pp. 1831–1846. doi:10.1175/bams-d-14-00213.1

154. The excellent Australian Government Bureau of Meteorology climate website is updated daily at www.bom.gov.au/climate

155. Table 6.1 sources: Information on raingauge site numbers were sought from the various national meteorological services, updated to May 2023. **Australia**: Current rainfall station numbers (6652) were taken from the stations list at www.bom.gov.au/climate/cdo/about/sitedata.shtml selecting for all sites still open at the date accessed (6 March 2023). The same list was used to count all sites still open whose records commenced in 1920 or earlier to count those with > 100 years record (total 1927) although this does include some with breaks in their record. Australian-run sites in Antarctica were excluded from these counts. **Switzerland**: MeteoSwiss site at www.meteoswiss.admin.ch/services-and-publications/applications/measurement-values-and-measuring-networks.html#param=messnetz-partner&lang=en&table=true, selecting aviation, climate, precipitation and partner stations and de-duplicating sites that appear in more than one category. **Republic of Ireland**: Met Eireann rainfall site numbers at end 2022 (enquiries team email 10 March 2023). **USA** NOAA – NOAA/NWS network: see also Table 4.2 and references therein. USA CoCoRaHS numbers are sites in US GCHN daily dataset from Table 4.2. **UK**: information provided by Melyssa Wright, Climate and Pollen Networks Manager, Met Office. **Germany** (Deutscher Wetterdienst, DWD): personal communication from Stegfan Gilge and Carola Grundmann 22 March 2023: see also websites 'Bodenmessnetzkarte des DWD – automatische Niederschlagsstationen (Nst A)' which shows the automatic precipitation sites, and the 'Bodenmessnetzkarte des DWD – konventionelle Niedrschlagsstationen (Nst k)', the conventional ones. Together with the main and aviation sites and the German army ones 'Hauptamtliches Messntz, GeoInfoDBW und Aerologie' and automatic weather stations 'automatische Wetterstationen [WST III]' there are 1868 precipitation sites in Germany. **The Netherlands**: 336 raingauges plus 34 synoptic AWS – thanks to Gerard van der Schrier (KNMI). **India** (India Meteorological Department): information provided by Sreejith Op. The total includes 209 climate stations (Synoptic Observatory) managed by India

Meteorological Department and 338 part-time observatories. No reply was received to my enquiries to **France** (MeteoFrance), and the numbers are those supplied for the first edition of this book in 2012.

156. Personal communication from Christopher C. Burt, e-mail dated 26 August 2011

157. I am indebted to Michael Kendon from the UK Met Office National Climate Information Centre for this figure, and the detailed rationale that went in to defining it (personal communication, 21 March 2023)

158. Burt, Stephen, et al., 2016: Cumbrian floods, 5/6 December 2015. *Weather*, **71**(2): pp. 36–37. doi:10.1002/wea.2704

159. Groisman, Pavel Ya and David R. Legates, 1994: The accuracy of United States precipitation data. *Bulletin of the American Meteorological Society*, **75**(2): pp. 215–227. doi:10.1175/1520-0477(1994)075<0215:taousp>2.0.co;2. Xref 3135a

160. Pan, Xicai, et al., 2016: Bias corrections of precipitation measurements across experimental sites in different ecoclimatic regions of western Canada. *The Cryosphere*, **10**(5): pp. 2347–2360. doi:10.5194/tc-10-2347-2016

161. Essery, Charles I. and David N. Wilcock, 1991: The variation in rainfall catch from standard UK Meteorological Office raingauges: a twelve year case study. *Hydrological Sciences Journal*, **36**(1): pp. 23–34. doi:10.1080/02626669109492482

162. Green, M.J., 1970: Some factors affecting the catch of raingauges. *Meteorological Magazine*, **99**: pp. 10–20

163. Groisman, Pavel Ya, Eugene l. Peck, and Robert G. Quayle, 1999: Intercomparison of recording and standard nonrecording U.S. gauges [raingauges]. *Journal of Atmospheric and Oceanic Technology*, **16**: pp. 602–609

164. Kurtyka, J. C., 1953: *Precipitation measurement study. Report of investigation No 20.* Illinois, USA: State Water Survey. 178. Available from https://citeseerx.ist.psu.edu/viewdoc/download?doi=10.1.1.461.3364&rep=rep1&type=pdf

165. Heberden, William, 1769: XLVII. Of the different quantities of rain, which appear to fall, at different heights, over the same spot of ground. *Philosophical Transactions*, **59**: pp. 359–362. doi:10.1098/rstl.1769.0047

166. Muchan, Katie and Harry Dixon, 2019: Insights into rainfall undercatch for differing raingauge rim heights. *Hydrology Research*, **50**(6): pp. 1564–1576. doi:10.2166/nh.2019.024

167. Daly, Christopher, et al., 2007: Observer bias in daily precipitation measurements at United States Cooperative Network Stations. *Bulletin of the American Meteorological Society*, **88**(6): pp. 899–912. doi:10.1175/bams-88-6-899

168. BBC News, 18 January 2024: Anglesey man has measured the rainfall every day since 1948. www.bbc.co.uk/news/uk-wales-68005382

169. More information on the US SRG can be found at www.weather.gov/iwx/coop_8inch.

170. Liu, X. C., T. C. Gao, and L. Liu, 2013: A comparison of rainfall measurements from multiple instruments. *Atmospheric Measurement Techniques*, **6**(7): pp. 1585–1595. doi:10.5194/amt-6-1585-2013

171. Peck, E. L., 1993: Biases in precipitation measurements: an American experience, in *American Meteorological Society: Eighth symposium on meteorological observations and instrumentation*. Anaheim, CA, 17–22 January 1993

172. Strangeways, Ian 2007: *Precipitation: Theory, measurement and distribution*. Cambridge: Cambridge University Press. 290 pp.

173. Lanza, Luca G. and Arianna Cauteruccio, 2022: *Chapter 1 – Accuracy assessment and intercomparison of precipitation measurement instruments*, in *Precipitation Science*, S. Michaelides, Editor: Elsevier. pp. 3–35

174. Acreman, Mike, 1989: Extreme rainfall in Calderdale, 19 May 1989. *Weather*, **44**(11): pp. 438–446. doi:10.1002/j.1477-8696.1989.tb04980.x.Xref1032

175. Collinge, V. K., et al., 1990: Radar observations of the Halifax storm, 19 May 1989. *Weather*, **45**(10): pp. 354–365. doi:10.1002/j.1477-8696.1990.tb05554.x

176. Burt, Stephen, 2005: Cloudburst upon Hendraburnick Down: The Boscastle storm of 16 August 2004. *Weather*, **60**(8): pp. 219–227. doi:10.1256/wea.26.05

177. Holley, Dan, James Dent, and Colin Clark, 2021: Thunderstorms and extreme rainfall in south Norfolk, 16 August 2020: meteorological analysis. *Weather*. doi:10.1002/wea.4038

178. Dent, James, Colin Clark, and Dan Holley, 2022: The Brettenham, East Anglia storm of 25 July 2021: hydrological response and implications for PMP. *Weather*. doi: 10.1002/wea.4206

179. Holley, Dan, James Dent, and Colin Clark, 2022: The Brettenham storm of 25 July 2021. *Weather*. doi:10.1002/wea.4145

180. Burt, Christopher, 2007: *Extreme weather: A guide and record book*. Second ed.: W W Norton & Co, New York. 303 pp.

181. Burt, Christopher. 2019: *A summary of U.S. State historical precipitation extremes*. Available from www.wunderground.com/cat6/Summary-US-State-Historical-Precipitation-Extremes

182. Quetelard, Hubert, et al., 2009: Extreme Weather: World-Record rainfalls during tropical cyclone *Gamede*. *Bulletin of the American Meteorological Society*, **90**(5): pp. 603–608. doi:10.1175/2008bams2660.1

183. Costello, T. A. and H. J. Williams, 1991: Short duration rainfall intensity measured using calibrated time-of-tip data from a tipping bucket raingage [raingauge]. *Agricultural and Forest Meteorology*, **57**: pp. 147–155. Xref 3134

184. Table 6.4 and Figure 6.13: rainfall record from EML ARG314 0.1 mm resolution TBR within Stratfield Mortimer Observatory in southern England (51°N, 1°W), 31 October 2022

185. EML's field calibration kit for raingauges is described at www.emltd.net/rfvk.html

186. Sušin, Nejc and Peter Peer, 2018: Open-source tool for interactive digitisation of pluviograph strip charts. *Weather*, **73**(7): pp. 222–226. doi:10.1002/wea.3001

187. Lambrecht rain[e] raingauge: details at www.lambrecht.net/en/products/precipitation/weighing-precipitation-sensor-rain-e

188. Burt, Stephen, 2016: Field test: Lambrecht rain[e] high-precision precipitation sensor. *Envirotech-Online*. Available from www.envirotech-online.com/article/air-monitoring/6/lambrecht/field-test-lambrecht-raine-high-precision-precipitation-sensor/2045

189. Figure 6.19: Rainfall record from Stratfield Mortimer Observatory in southern England (51°N, 1°W) on 23 October 2022

190. Figure 6.20: Rainfall record from Stratfield Mortimer Observatory in southern England (51°N, 1°W) on 10 January 2023

191. EML SW120R precipitation detector www.emltd.net/sw120r.html

192. Pollock, M. D., et al., 2018: Quantifying and mitigating wind-induced undercatch in rainfall measurements. *Water Resources Research*, **54**(6): pp. 3863–3875. doi:10.1029/2017wr022421

193. Strangeways, Ian, 2004: Improving precipitation measurement. *International Journal of Climatology*, **24**(11): pp. 1443–1460. doi:10.1002/joc.1075

194. Cauteruccio, Arianna and Luca G. Lanza, 2020: Parameterization of the collection efficiency of a cylindrical catching-type rain gauge based on rainfall intensity. *Water*, **12**(12): pp. 3431. doi:10.3390/w12123431

195. British Standards Institution, 2010: *BS EN 13798:2010 Hydrometry – Specification for a reference raingauge pit.*

196. Hudleston, F., 1933: A summary of seven years' experiments with raingauge shields in exposed positions, 1926–32 at Hutton John, Penrith. *British Rainfall 1933*: pp. 274–293.

197. Goodison, B. E., P. Y. T. Louie, and D. Yang, 1998: *WMO solid precipitation measurement intercomparison – final report.* Geneva: World Meteorological Organization (WMO), TD No. 872

198. Goodison, B. E., B. Sevruk, and S. Klemm, 1989: *WMO solid precipitation measurement intercomparison: Objectives, methodology and analysis.* International Association of Hydrological Sciences, 1989: Atmospheric deposition. Proceedings, Baltimore Symposium (May 1989). IAHS Publication No. 179, Wallingford, Oxon, UK

199. Kochendorfer, John, et al., 2022: How well are we measuring snow post-SPICE? *Bulletin of the American Meteorological Society*, **103**(2): pp. E370–E388. doi:10.1175/bams-d-20-0228.1

200. Nitu, R., et al., 2021: *WMO Solid Precipitation Intercomparison Experiment (SPICE) (2012–2015).* Geneva: World Meteorological Organization. 451 pp. Available from https://library.wmo.int/viewer/56317?medianame=iom_131_en_1_#page=1&viewer=picture&o=bookmark&n=0&q=.

201. Smith, Craig D., et al., 2020: Evaluation of the WMO Solid Precipitation Intercomparison Experiment (SPICE) transfer functions for adjusting the wind bias in solid precipitation measurements. *Hydrology and Earth System Sciences*, **24**(8): pp. 4025–4043. doi:10.5194/hess-24-4025-2020

202. World Meteorological Organization (WMO), 1992: *Snow cover measurements and areal assessment of precipitation and soil moisture (WMO-No. 749)*, ed. B. Sevruk. Geneva: World Meteorological Organization (WMO). 312 pp. Available from https://library.wmo.int/records/item/33781-snow-cover-measurements-and-areal-assessment-of-precipitation-and-soil-moisture.

203. Doesken, Nolan and Arthur Judson, 1996: *The Snow booklet: A guide to the science, climatology and measurement of snow in the United States.* Colorado Climate Center, Colorado State University

204. Whipple, G. M., 1881: The relative frequency of given heights of the barometer readings at the Kew Observatory, during the ten years, 1870 to 1879. *Quarterly Journal of the Royal Meteorological Society*, **7**(37): pp. 52–59. doi:10.1002/qj.4970073710

205. Smith, John, 1688: *A compleat Discourse of the Nature, Use and Right Managing of that wonderful instrument the Baroscope. or Quicksilver Weather-Glass.* London, printed for Joseph Watts, at the Angel in St Paul's Churchyard. Reference from Whipple, 1881 q.v.

206. Middleton, W. E. K., 1964: *The history of the barometer.* Baltimore, MD: Johns Hopkins Press. 489 pp.

207. Lamb, H. H., 1986: Ancient units used by the pioneers of meteorological measurements. *Weather*, **41**(7): pp. 230–233. doi:10.1002/j.1477-8696.1986.tb03842.x

208. Cornes, Richard, 2010: *Early meteorological data from London and Paris.* PhD thesis – University of East Anglia, 233 pp.

209. Harrison, R. Giles, 2020: Make your own met measurements: build a digital barometer for about £10. *Weather*, **76**: pp. 45–47. doi:10.1002/wea.3857

210. Amores, Angel, et al., 2022: Numerical simulation of atmospheric Lamb Waves generated by the 2022 Hunga-Tonga volcanic eruption. *Geophysical Research Letters*, **49**(6): doi:10.1029/2022gl098240

211. Burt, Stephen, 2022: Multiple airwaves crossing Britain and Ireland following the eruption of Hunga Tonga–Hunga Ha'apai on 15 January 2022. *Weather*, **77**: pp. 76–81

212. Li, Chunyan, 2022: Global shockwaves of the Hunga Tonga-Hunga Ha'apai volcano eruption measured at ground stations. *iScience*, **25**(11): pp. 105356. doi:10.1016/j.isci.2022.105356

213. Matoza, Robin S., et al., 2022: Atmospheric waves and global seismoacoustic observations of the January 2022 Hunga eruption, Tonga. *Science*. doi: 10.1126/science.abo7063. (Published online, 12 May 2022)

214. Terry, James P., et al., 2022: Tonga volcanic eruption and tsunami, January 2022: globally the most significant opportunity to observe an explosive and tsunamigenic submarine eruption since AD 1883 Krakatau. *Geoscience Letters*, **9**(1): doi:10.1186/s40562-022-00232-z

215. Harrison, Giles, 2022: Pressure anomalies from the January 2022 Hunga Tonga-Hunga Ha'apai eruption. *Weather*, **77**(3): pp. 87–90. doi:10.1002/wea.4170

216. Smart, David, 2022: The first hour of the paroxysmal phase of the 2022 Hunga Tonga-Hunga Ha'apai volcanic eruption as seen by a geostationary meteorological satellite. *Weather*, **77**(3): pp. 81–82. doi:10.1002/wea.4173

217. Meteorological Office, 1980: *Handbook of Meteorological Instruments: Volume 1, Measurement of atmospheric pressure*, in *Handbook of Meteorological Instruments*: London: Her Majesty's Stationery Office

218. Meteorological Office, 1956: *Handbook of Meteorological Instruments: Part 1, Instruments for surface observations*. Fifth impression 1969 ed. London: Her Majesty's Stationery Office. PDF copy available in Met Office Archive at https://digital.nmla .metoffice.gov.uk/SO_3c42ff4e-ac2a-4f5c-9d8c-1d7716f3c211

219. A variety of online humidity calculators are available – for example, www.omnicalculator .com/physics/absolute-humidity. There are also numerous smartphone apps providing similar functionality available for both Android and Apple devices

220. Harrison, R. G. and C. R. Wood, 2012: Ventilation effects on humidity measurements in thermometer screens. *Quarterly Journal of the Royal Meteorological Society*, **138**(665): pp. 1114–1120. doi:10.1002/qj.985

221. McIlveen, Robin, 2010: *Fundamentals of weather and climate*. Second ed, Oxford: Oxford University Press. 632 pp., specifically Chapter 5.4; also Bolton, D., 1980: The computation of equivalent potential temperature. *Mon. Weather Rev.*, **108**: pp. 1046–1053

222. Farahani, Hamid, Rahman Wagiran, and Mohd Hamidon, 2014: Humidity sensors principle, mechanism, and fabrication technologies: a comprehensive review. *Sensors*, **14**(5): pp. 7881–7939. doi:10.3390/s140507881

223. Burt, Stephen, 2011: Exceptionally low relative humidity in northern England, 2–3 March 2011. *Weather*, **66**(7): pp. 197–199. doi:10.1002/wea.833

224. Malyon, Charles, 2015: *Comparative performance of surface relative humidity measurements using commercially–available sensors within an automatic weather screen*. MSc dissertation, University of Reading Department of Meteorology

225. One minute mean relative humidity records, comparing a digital sensor and RH derived from dry and wet bulb psychrometer readings within a Stevenson screen: Stratfield Mortimer Observatory in southern England (51°N, 1°W), 17 August 2022

226. Burt, Stephen and Ed Hawkins, 2019: Near-zero humidities on Ben Nevis, Scotland, revealed by pioneering nineteenth century observers and modern volunteers. *International Journal of Climatology*, **39**: pp. 4451–4466. doi:10.1002/joc.6084

227. One minute mean dry and wet bulb temperatures, and relative humidity measured using a digital sensor and derived from dry and wet bulb psychrometer readings, in a Stevenson screen observed at Stratfield Mortimer Observatory, southern England (51°N, 1°W), 9–10 December 2022

228. The US Heat Index is given (in °F only) at www.nsis.org/weather/heatindex.html (accessed 29 November 2022)

229. The Canadian humidex index can be found at www.csgnetwork.com/canhumidexcalc.html, which gives an online humidity calculator for temperatures in °C and relative humidity as % (accessed 29 November 2022)

230. Burt, Stephen, 2022: The gust that never was: a meteorological instrumentation mystery. *Weather*, **77**: pp. 123–126

231. Strangeways, Ian, 2003: *Measuring the natural environment*. Second Edition ed. Cambridge: Cambridge University Press. 534 pp.

232. Meteorological Office, 1981: *Handbook of Meteorological Instruments: Volume 4, Measurement of surface wind*, in *Handbook of Meteorological Instruments*: London: Her Majesty's Stationery Office

233. Jenkins, Geoff, 2014: A comparison between two types of widely used weather stations. *Weather*, **69**(4): pp. 105–110. doi:10.1002/wea.2158

234. Product specifications sourced from www.gill.co.uk/products/anemometer/windsonic.htm.

235. Beljaars, A. C. M., 1987: The influence on sampling and filtering on measured wind gusts. *Journal of Oceanic and Atmospheric Technology*, **4**: pp. 613–626. doi: 10.1175/1520-0426 (1987)004<0613:TIOSAF>2.0.CO;2

236. Beljaars, A. C. M., 1987: *Instruments and Observing Methods Report No. 31: The measurement of gustiness at routine wind stations: a review*. Instruments and Observing Methods Report No. 31. Geneva: World Meteorological Organization (WMO). Available from https://library.wmo.int/index.php?lvl=notice_display&id=15514

237. Curran, J. C., et al., 1977: Cairngorm summit automatic weather station. *Weather*, **32**(2): pp. 61–63. doi:10.1002/j.1477-8696.1977.tb04513.x

238. Barton, J. S., 1984: Observing mountain weather using an automatic station. *Weather*, **39**(5): pp. 140–145. doi:10.1002/j.1477-8696.1984.tb07480.x

239. Barton, J. S., 2020: The Cairngorm automatic weather station. *Weather*, **75**(4): pp. 129–130. doi:10.1002/wea.3514

240. The Cairngorm AWS is online at cairngormweather.eps.hw.ac.uk and the site contains some details of the instruments together with a multi-year data archive. The summit instruments and webcam were set up by funded research projects, which finished a number of years ago. Heriot-Watt University continues to support these facilities as much as possible, but regret that there may well be extended periods when they are unavailable.

241. Putnam, William Lowell, 1991: *The worst weather on Earth: A history of the Mount Washington Observatory*. New York: American Alpine Club. 265 pp.

242. Smith, Alan A., 1982: The Mount Washington Observatory—50 Years Old. *Bulletin of the American Meteorological Society*, **63**(9): pp. 986–995. doi: 10.1175/1520-0477(1982) 063<0986:TMWOYO>2.0.CO;2

243. Kelsey, Eric, et al., 2014: Blown Away: Interns experience science, research, and life on top of Mount Washington. *Bulletin of the American Meteorological Society*, **96**(9): pp. 1533–1543. doi:10.1175/BAMS-D-13-00195.1

244. Philippoff, K. 2023: *Brutal cold on Mount Washington: A weather story*. Available from www.mountwashington.org/experience-the-weather/observer-comments.aspx?id=59058

245. The Mount Washington Observatory website is at www.mountwashington.org/weather

246. Rappaport, Edward N., 1994: Hurricane Andrew [August 1992]. *Weather*, **49**(2): pp. 51–61. doi:10.1002/j.1477-8696.1994.tb05974.x

247. Details of the Barrow Island world wind speed extreme are available online at wmo.asu.edu/content/world-maximum-surface-wind-gust, along with other WMO 'extremes' investigation summaries

248. Meteorological Office, 1982: *Observer's Handbook*. HMSO

249. Burt, Stephen, 2007: A comparison of traditional and modern methods of measuring earth temperatures. *Weather*, **62**(12): pp. 331–336. doi:10.1002/wea.165

250. McIlveen, Robin, 2010: *Fundamentals of weather and climate*. Second ed. Oxford: Oxford University Press. 632 pp.

251. Vignola, Frank, Joseph Michalsky, and Thomas Stoffel, 2012: *Solar and Infrared radiation measurements*. Boca Raton, Florida: Taylor & Francis. 394 pp.

252. Figure 11.1 shows the American Society for Testing and Materials (ASTM) G-173–03 Reference Spectra ('Standard Tables for Reference Solar Spectral Irradiance at Air Mass 1.5: Direct Normal and Hemispherical for a 37 Tilted Surface'), also known as ISO 9845–1, 2022 which is available at www.iso.org/obp/ui/#iso:std:iso:9845:-1:ed-2:v1:en. The two plots were prepared using the SMARTS model version 2.9.2 (SMARTS: Simple Model of the Radiative Transfer of Sunshine). More information about the SMARTS model is available online at www.nrel.gov/rredc/smarts; full details of the ASTM G-173–03 Reference Spectra standard and complete references can be found at http://rredc.nrel.gov/solar/spectra/am1.5 and linked references therein

253. Gueymard, Christian A., 2018: A reevaluation of the solar constant based on a 42-year total solar irradiance time series and a reconciliation of spaceborne observations. *Solar Energy*, **168**: pp. 2–9. doi:10.1016/j.solener.2018.04.001

254. Wild, Martin, et al., 2019: The cloud-free global energy balance and inferred cloud radiative effects: an assessment based on direct observations and climate models. *Climate Dynamics*, **52**(7–8): pp. 4787–4812. doi:10.1007/s00382-018-4413-y

255. More information on the Australian Bureau of Meteorology's use of satellite measurements for surface solar radiation estimates is given online at www.bom.gov.au/climate/austmaps/solar-radiation-glossary.shtml#globalexposure

256. Good, Elizabeth, 2010: Estimating daily sunshine duration over the UK from geostationary satellite data. *Weather*, **65**(12): pp. 324–328. doi:10.1002/wea.619

257. Matuszko, Dorota, et al., 2020: Sunshine duration in Poland from ground- and satellite-based data. *International Journal of Climatology*, **40**(9): pp. 4259–4271. doi:10.1002/joc.6460

258. For more detail the reader is referred to World Meteorological Organization (1984) *Radiation and Sunshine Duration Measurements: Comparison of Pyranometers and Electronic Sunshine Duration Recorders of RA VI* (Budapest, Hungary, 1984). WMO/TD No. 146: also WMO (1989) *Automatic Sunshine Duration Measurement Comparison, Hamburg (Germany)*, IOM 42 (results not formally published but available from WMO). See also the reports of the WMO International Pyrheliometer Intercomparisons, conducted by the World Radiation Centre at Davos, Switzerland (www.pmodwrc.ch/en/ipc-xiii) and carried out at five-yearly intervals, also distributed by WMO. At the time of writing, the most recent intercomparison event took place in September-October 2021

259. Foster, N. B. and L. W. Foskett, 1953: A photoelectric sunshine recorder. *Bulletin of the American Meteorological Society*: pp. 212–215

260. Michalsky, J. J., 1992: Comparison of a National Weather Service Foster sunshine recorder and the World Meteorological Organization standard for sunshine duration. *Solar Energy*, **48**(2): pp. 133–141. doi:10.1016/0038-092X(92)90041-8

261. See, for example, Olivieri, Jean (1998) *Sunshine measurements using a pyranometer* and Fiore, J. V. et al. (1998) *A comparison between a proposed ASOS sunshine sensor and a pyrheliometer*: both papers in Instruments and Observing Methods, Report No. 70: Papers presented at the WMO Technical Conference on meteorological and environmental instruments and methods of observation (TECO-98) held in Casablanca, Morocco, 13–15 May 1998: also Dyson, Paul (2005) *Investigation of the accuracy of the*

Delta-T devices BF3 sunshine sensor. Australian Bureau of Meteorology, Instrument Test Report No. 700

262. Massen, Francis, 2016: *Sunshine duration from pyranometer readings.* doi:10.13140/RG.2.1.2255.7042

263. Wang, Yuchang, et al., 2021: A review on sunshine recorders: Evolution of operation principle and construction. *Measurement*, **186**: pp. 110138. doi:10.1016/j.measurement.2021.110138

264. Slob, W. H. and W. A. A. Monna, 1991: *Bepaling van directe en diffuse straling en van zonneschijndurr uit 10-minuutwaarden van globale straling [Determination of direct and diffuse radiation and of sunshine duration from 10-minute values of global radiation].* De Bilt, the Netherlands: Koninklijk Nederlands Meteorologisch Instituut (KNMI)

265. Groen, G., et al., 2011: *Implementation of the Hinssen-Knap algorithm for the calculation of sunshine duration.* De Bilt: KNMI. Technical report TR-319

266. Stokes, George Gabriel, 1880: Description of the card supporter for sunshine recorders adopted at the Meteorological Office. *Quarterly Journal of the Royal Meteorological Society*, **7**: pp. 83–94

267. Stanhill, G., 2003: Through a glass brightly: Some new light on the Campbell–Stokes sunshine recorder. *Weather*, **58**(1): pp. 3–11. doi:10.1256/wea.278.01

268. Sanchez-Lorenzo, A., et al., 2013: New insights into the history of the Campbell–Stokes sunshine recorder. *Weather*, **68**(12): pp. 327–331. Doi:10.1002/wea.2130

269. Wood, Curtis R. and R. Giles Harrison, 2011: Scorch marks from the sky. *Weather*, **66**(2): pp. 39–41. Doi:10.1002/wea.657

270. Painter, H. E., 1981: The performance of a Campbell–Stokes sunshine recorder compared with a simultaneous record of the normal incidence irradiance. *Meteorological Magazine*, **110**: pp. 102–109.

271. Baumgartner, D. J., et al., 2017: A comparison of long-term parallel measurements of sunshine duration obtained with a Campbell–Stokes sunshine recorder and two automated sunshine sensors. *Theoretical and Applied Climatology.* doi:10.1007/s00704-017-2159-9

272. Kerr, Andrew and Richard Tabony, 2004: Comparison of sunshine recorded by Campbell–Stokes and automatic sensors. *Weather*, **59**(4): pp. 90–95. Doi:10.1256/wea.99.03

273. Legg, Tim, 2014: Comparison of daily sunshine duration recorded by Campbell–Stokes and Kipp and Zonen sensors. *Weather*, **69**(10): pp. 264–267. Doi: 10.1002/wea.2288

274. Sanchez-Romero, A., et al., 2015: Using digital image processing to characterize the Campbell–Stokes sunshine recorder and to derive high-temporal resolution direct solar irradiance. *Atmospheric Measurement Techniques*, **8**(1): pp. 183–194. Doi:10.5194/amt-8-183-2015

275. Shearn, P. D., 1999: *Automatic sunshine sensor trial report.* Bracknell, Met Office: Observations, Logistics and Automation Branch (unpublished internal report: copy available from National Meteorological Library, Exeter, UK)

276. Matuszko, Dorota, 2015: A comparison of sunshine duration records from the Campbell-Stokes sunshine recorder and CSD3 sunshine duration sensor. *Theoretical and Applied Climatology*, **119**(3–4): pp. 401–406. Doi:10.1007/s00704-014-1125-z

277. Koninklijk Nederlands Meteorologisch Instituut – KNMI, 2002: *Klimaatatlas van Nederland 1971–2000.* De Bilt, the Netherlands: Koninklijk Nederlands Meteorologisch Instituut – KNMI. Zonneschijn, pp. 68–70

278. Steurer, P. M. and T. R. Karl, 1991: *Historical sunshine and cloud data in the United States.* Office of Scientific and Technical Information, Oak Ridge National Laboratory, Oak Ridge, TN 37831. 158 pp.

279. For details, see Baker, D. C. and D. A. Haines, 1969: *Solar radiation and sunshine duration relationships in the north-central region and Alaska.* North Central Regional

Research Publication 195, Technical Bulletin 262, Agricultural Experiment Station, University of Minnesota, Minneapolis (as quoted in Steurer and Karl, 1991, Oak Ridge National Laboratory): 0.12 cal cm^{-2} min^{-1} is about 87 W/m^2. Note that the threshold for the Marvin sunshine recorder was determined as approximately 255 W/m^2 (0.37 cal cm^{-2} min^{-1}) in Brooks, C. F. and Brooks, E. S. (1947) *Sunshine recorders: A comparative study of the burning-glass and thermometric systems. J. Meteorol*, **4**, pp. 105–115

280. Stanhill, Gerald and Shabtai Cohen, 2005: Solar radiation changes in the United States during the twentieth century: Evidence from sunshine duration measurements. *Journal of Climate*, **18**(10): pp. 1503–1512. Doi:10.1175/jcli3354.1

281. Quinlan, F. T., 1985: *A history of sunshine data in the United States*, in *Handbook of applied meteorology*, D. D. Houghton, Editor., John Wiley and Sons, New York. 1199–1201

282. Cerveny, Randall S. and Robert C. Balling, 1990: Inhomogeneities in the long-term United States' sunshine record. *Journal of Climate*, **3**(9): pp. 1045–1048

283. International Organization for Standardization (ISO), 2023: *ISO 9847:2023 Solar energy – Calibration of field pyranometers by comparison to a reference pyranometer*

284. World Meteorological Organization (WMO), 2019: *Manual on Codes. International Codes, Volume I.1: Annex II to the WMO Technical Regulations, Part A – Alphanumeric Codes*. Geneva, Switzerland: WMO. 480. Available from https://library.wmo.int/doc_num.php?explnum_id=10235

285. This information was collated from the Climatological Observers Link *Bulletin* for November 2022. Station terminal hours form one part of the station grading code for COL stations. Details are available at www.colweather.org.uk

286. Table 13.3 contains a short extract from the datalogger record of gust speeds at the Stratfield Mortimer Observatory in southern England (51°N, 1°W) during windstorm *Eunice* on 18 February 2022. Entries shown are the highest 3 s mean wind speeds, near-duplicate table entries within 3 s of those shown have been omitted for clarity

287. Bilham, E. G., 1948: Washington codes. *Meteorological Magazine*, **77**: pp. 217–20. This is the first instance of the word 'okta' in the index of the *Meteorological Magazine*. The attribution of the term to Colonel Ernest Gold CB, DSO, OBE, FRS (1882–1976) is well known but apparently poorly documented: Gold was the Deputy Director of the Meteorological Office upon his retirement in 1947.

288. Hamblyn, Richard, 2021: *The Cloud Book: how to understand the skies*. Second ed. Exeter: David & Charles. 176 pp.

289. Pretor-Pinney, Gavin, 2006: *The Cloudspotter's guide*. London: Hodder & Stoughton / Sceptre. 320 pp.

290. Dunlop, Storm, 2008: *The Oxford dictionary of weather*. Second ed.: Oxford University Press

291. Met Office, 1991: *Meteorological glossary*. Sixth ed. London: Her Majesty's Stationery Office (HMSO)

292. Meaden, G. T., 1976: Practical hail gauges for climatological stations. *Journal of Meteorology, UK*, **1**: pp. 313–319

293. International Electrotechnical Commission, 2022: *IEC 60751:2022: Industrial platinum resistance thermometers and platinum temperature sensors*. Geneva: International Electrotechnical Commission. 22 pp.

294. McArthur, L. J. B, 2005: *World Climate Research Programme – Baseline Surface Radiation Network (BSRN): Operations manual, version 2.1*. Geneva: World Meteorological Organization. 176. Available from https://epic.awi.de/id/eprint/30644/

295. In the UK, the Climatological Observers Link (www.colweather.org.uk) archives all monthly and annual summaries submitted by members, together with numerous long runs of monthly station data added since. These datasets have been used to create and publish 10 and 30 year averages for all sites with sufficiently long records. Membership

also includes free access to up to 1 GB for members' weather and climate records on a commercial cloud storage facility.

296. Figure 18.2. Microsoft Excel scatterplot showing the relationship between daily sunshine duration and daily global solar radiation totals during 10 years in June. Records from Stratfield Mortimer Observatory in southern England (51°N, 1°W), period 2013–22

297. Hourly sunshine records from Stratfield Mortimer Observatory in southern England (51°N, 1°W), 20 years period 2003–22. Instromet sunshine recorder 2003–2011 inclusive, Kipp & Zonen CSD3 unit 2012 onwards. Open exposure at 10 m above ground. Data availability 100%, minor gaps filled with estimates from alternate instruments

298. Hourly rainfall and wind direction records from Stratfield Mortimer Observatory in southern England (51°N, 1°W), 20 year period 2003–22. Wind direction and speed from instruments at 10 m above ground level, rainfall from a 0.2 mm tipping-bucket gauge located within the nearby instrument enclosure

299. Frequency of days with snow or sleet observed to fall, and thunder heard, at Stratfield Mortimer Observatory in southern England (51°N, 1°W), by week number over the 30 years standard averaging period 1991–2020

300. Hourly rainfall and 10 m wind speed data at Stratfield Mortimer Observatory (51°N, 1°W) over the 25 year period 1997 to 2022 inclusive. Wind speeds in m/s

301. Tufte, Edward R., 1983: *The visual display of quantitative information.* Cheshire, CT: Graphics Press. 197 pp.

302. Sunshine data for 2022 at Stratfield Mortimer Observatory in southern England (51°N, 1°W), 5 minute data in seconds plotted using an Excel pivot table

303. Tufte, Edward R., 1990: *Envisioning information.* Cheshire, CT: Graphics Press. 126 pp.

304. Tufte, Edward R., 1997: *Visual explanations.* Cheshire, CT: Graphics Press. 157 pp.

305. Lakes Environmental Software: www.weblakes.com. WRPLOT view is free to use, although registration is required

306. Met Office. 2019: *Met Office MIDAS Open: UK Land Surface Stations Data (1853–current).* Centre for Environmental Data Analysis (CEDA). Available from http://catalogue.ceda.ac.uk/uuid/dbd451271eb04662beade68da43546e1

307. Hawkins, Ed, et al., 2022: Rainfall Rescue: Millions of historical monthly rainfall observations taken in the UK and Ireland rescued by citizen scientists. *Geoscience Data Journal,* **10**(2): pp. 246–261. doi:10.1002/gdj3.157

308. Burt, Stephen and Roger Brugge, 2011: *Climatological averages for 1981–2010 and 2001–2010 for stations appearing in the monthly bulletin of the Climatological Observers Link.* Maidenhead, UK: Climatological Observers Link, 434 pp.

309. Climatological Observers Link (COL), 2021: *Climatological averages for 1991–2020 and 2011–2020 for stations appearing in the monthly bulletin of the Climatological Observers Link (COL).* 71 pp. Available from www.colweather.org.uk

310. Kirk, Peter J., Matthew R. Clark, and Ellie Creed, 2021: Weather Observations Website. *Weather,* **76**(2): pp. 47–49 doi: 10.1002/wea.3856

311. For information on WOW inputs into a storm on 31 October 2021, see https://weatherobservationswebsite.blogspot.com/2022/02/using-wow-observations-to-capture-high.html

312. Burt, Stephen, 2020: The Climatological Observers Link (COL) at 50. *Weather,* **75**(5): pp. 137–144

313. https://media.bom.gov.au/social/blog/10/a-short-history-of-the-bureau-of-meteorology

314. Details of the US voluntary COOP observers awards scheme (updated November 2019) can be found at www.nws.noaa.gov/directives/sym/pd01013014curr.pdf

315. Carter, H. E., 1948: Seventy-five years of rainfall recording. *Meteorological Magazine*: pp. 234–235

316. More details on the US COOP programme can be found at www.weather.gov/coop/overview

317. Ogden, R.J., 1978: Co-operating observers and the climatological network. *Meteorological Magazine*, **107**: pp. 209–218

318. Merlone, A., G. Lopardo, F. Sanna et al., 2015: The MeteoMet project – metrology for meteorology: challenges and results. *Meteorological Applications*, **22**(S1): pp. 820–829. doi: 10.1002/met.1528

319. Burt, Stephen, 2014: Trials of the low-cost, high-accuracy CoCoRaHS raingauge. *Climatological Observers Link Bulletin*, **532**: pp. 29–30

320. Daily rainfall totals over the 12 months ending April 2023 at the Stratfield Mortimer Observatory in central southern England (51°N, 1°W). Minor breaks in record for maintenance purposes, and for periods of snowfall, have been estimated to complete the records

321. Bell, Simon, Dan Cornford, and Lucy Bastin, 2013: The state of automated amateur weather observations. *Weather*, **68**(2): pp. 36–41. doi:10.1002/wea.1980

322. Bell, Simon, Dan Cornford, and Lucy Bastin, 2015: How good are citizen weather stations? Addressing a biased opinion. *Weather*, **70**(3): pp. 75–84. doi:10.1002/wea.2316

323. Brooks, C. E. P. and N. Carruthers, 1953: *Handbook of statistical methods in meteorology*. London: HMSO. Vector mean winds, pp. 178–191. More details on the method can be obtained from Webmet.com: www.webmet.com/met_monitoring/62.html

Index

Printed in the United States
by Baker & Taylor Publisher Services